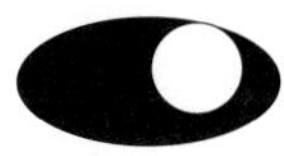

Current Topics in Membranes, Volume 43

Membrane Protein–Cytoskeleton Interactions

Current Topics in Membranes, Volume 43

Series Editors

Arnost Kleinzeller
Department of Physiology
University of Pennsylvania
School of Medicine
Philadelphia, Pennsylvania

Douglas M. Fambrough
Department of Biology
The Johns Hopkins University
Baltimore, Maryland

Dale J. Benos
Department of Physiology and Biophysics
University of Alabama
Birmingham, Alabama

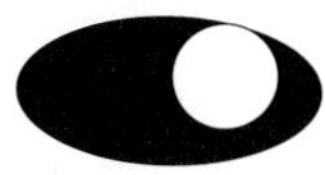

Current Topics in Membranes, Volume 43

Membrane Protein-Cytoskeleton Interactions

Edited by

W. James Nelson

Department of Molecular and Cellular Physiology
Stanford University School of Medicine
Stanford, California

ACADEMIC PRESS

San Diego London Boston New York Sydney Tokyo Toronto

This book is printed on acid-free paper. ♾

Academic Press, Inc.
525 B Street, Suite 1900, San Diego, California 92101-4495, USA
http://www.apnet.com

Academic Press Limited
24-28 Oval Road, London NW1 7DX, UK
http://www.hbuk.co.uk/ap/

International Standard Serial Number: 1063-5823

International Standard Book Number: 0-12-153343-3

PRINTED IN THE UNITED STATES OF AMERICA
96 97 98 99 00 01 EB 9 8 7 6 5 4 3 2 1

Contents

CHAPTER 8 Molecular and Genetic Dissection of the Membrane Skeleton in *Drosophila*
Ronald R. Dubreuil

CHAPTER 9 Membrane-Cytoskeleton Interactions with Cadherin Cell Adhesion Proteins: Roles of Catenins as Linker Proteins
Margaret J. Wheelock, Karen A. Knudsen, and Keith R. Johnson

CHAPTER 10 The Desmosome: A Component System for Adhesion and Intermediate Filament Attachment
Andrew P. Kowalczyk and Kathleen J. Green

Contributors

Numbers in parentheses indicate the pages on which the authors' contributions begin.

James Melvin Anderson (211), Departments of Internal Medicine and Cell Biology, Yale University School of Medicine, New Haven, Connecticut 06520

Robert Bacallao (397), Department of Medicine, Indiana University School of Medicine, Indianapolis, Indiana 46202

Kenneth A. Beck (15), Department of Molecular and Cellular Physiology, Beckman Center for Molecular and Genetic Medicine, Stanford University School of Medicine, Stanford, California 94305

Vann Bennett (129), Departments of Cell Biology and Biochemistry, Howard Hughes Medical Institute, Duke University Medical Center, Durham, North Carolina 27710

Dale J. Benos (345), Department of Physiology and Biophysics, University of Alabama at Birmingham, Birmingham, Alabama 35294

Jennifer D. Black (313), Department of Molecular Immunology, Roswell Park Cancer Institute, Buffalo, New York 14263

Lilly Y. W. Bourguignon (293), Department of Cell Biology and Anatomy, School of Medicine, University of Miami, Miami, Florida 33101

Alexandra R. Brecher (211), Department of Cell Biology, Yale University School of Medicine, New Haven, Connecticut 06520

David R. Burgess (53), Department of Biological Sciences, University of Pittsburgh, Pittsburgh, Pennsylvania 15260

James T. Campanelli (237), Department of Molecular and Cellular Physiology, Howard Hughes Medical Institute, Stanford University, Stanford, California 94305

Horacio F. Cantiello (373), Renal Unit, Massachusetts General Hospital East, Charlestown, Massachusetts 02129; and Department of Medicine, Harvard Medical School, Boston, Massachusetts

Prasad Devarajan (97), Department of Pediatrics, Yale University School of Medicine, New Haven, Connecticut 06510

R. Brian Doctor (397), Department of Cell Biology, Division of Physiology and Cellular Biophysics, Duke University Medical Center, Durham, North Carolina 27710

Ronald R. Dubreuil (147), Committee on Cell Physiology and Department of Pharmacological and Physiological Sciences, University of Chicago, Chicago, Illinois 60637

Michael Edidin (1), Department of Biology, The Johns Hopkins University, Baltimore, Maryland 21218

Martin Eigenthaler (265), Department of Vascular Biology, The Scripps Research Institute, La Jolla, California 92037

Alan S. Fanning (211), Department of Internal Medicine, Yale University School of Medicine, New Haven, Connecticut 06520

Karl R. Fath (53), Department of Biological Sciences, University of Pittsburgh, Pittsburgh, Pennsylvania 15260

John G. Forte (73), Department of Molecular and Cell Biology, University of California, Berkeley, California 94720

Gregory G. Gayer (237), Department of Molecular and Cellular Physiology, Howard Hughes Medical Institute, Stanford University, Stanford, California 94305

Steven R. Gill (27), Department of Biology, The Johns Hopkins University, Baltimore, Maryland 21218

Kathleen J. Green (187), Departments of Pathology, Dermatology and the R. H. Lurie Cancer Center, Northwestern University Medical School, Chicago, Illinois 60611

Keith R. Johnson (169), Department of Biology, University of Toledo, Toledo, Ohio 43606

Karen A. Knudsen (169), Department of Cell Biology, Lankenau Medical Research Center, Wynnewood, Pennsylvania 19096

Andrew P. Kowalczyk (187), Departments of Pathology, Dermatology, and the R. H. Lurie Cancer Center, Northwestern University Medical School, Chicago, Illinois 60611

Stephen Lambert (129), Departments of Cell Biology and Biochemistry, Howard Hughes Medical Institute, Duke University Medical Center, Durham, North Carolina 27710

Lynne A. Lapierre (211), Department of Internal Medicine, Yale University School of Medicine, New Haven, Connecticut 06520

Lazaro J. Mandel (397), Department of Cell Biology, Division of Physiology and Cellular Biophysics, Duke University Medical Center, Durham, North Carolina 27710

Jon S. Morrow (97), Department of Pathology, Yale University School of Medicine, New Haven, Connecticut 06510

W. James Nelson (15), Department of Molecular and Cellular Physiology, Beckman Center for Molecular and Genetic Medicine, Stanford University School of Medicine, Stanford, California 94305

Adriana G. Prat (373), Renal Unit, Massachusetts General Hospital East, Charleston, Massachusetts 02129; and Department of Medicine, Harvard Medical School, Boston, Massachusetts

Elizabeth A. Repasky (313), Department of Molecular Immunology, Roswell Park Cancer Institute, Buffalo, New York 14263

Richard H. Scheller (237), Department of Molecular and Cellular Physiology, Howard Hughes Medical Institute, Stanford University, Stanford, California 94305

Trina A. Schroer (27), Department of Biology, The Johns Hopkins University, Baltimore, Maryland 21218

Sanford J. Shattil (265), Departments of Vascular Biology and Molecular and Experimental Medicine, The Scripps Research Institute, La Jolla, California 92037

Peter R. Smith (345), Department of Physiology, Medical College of Pennsylvania and Hahnemann University, Philadelphia, Pennsylvania 19129

Christina M. Van Itallie (211), Department of Internal Medicine, Yale University School of Medicine, New Haven, Connecticut 06520

Margaret J. Wheelock (169), Department of Biology, University of Toledo, Toledo, Ohio 43606

Xuebiao Yao (73), Department of Molecular and Cell Biology, University of California, Berkeley, California 94720

Preface

The plasma membrane acts as both a boundary and a site of exchange between the outside and the inside of the cell. Correct protein insertion and organization in the plasma membrane are critical for these functions. How are membrane proteins targeted to the plasma membrane, and how, after delivery, are the distribution and function of membrane proteins regulated? The cytoskeleton plays essential roles in delineating membrane protein distributions, defining structural and functional domains in the plasma membrane, and regulating membrane protein and, ultimately, cell function.

This volume of *Current Topics in Membranes* brings together leaders in several overlapping research areas who have focused on the function of the cytoskeleton in regulating membrane protein targeting, distribution, and function. The volume provides unique insights into many topical problems in cell biology related to vesicle trafficking to the plasma membrane, regulation of membrane protein distributions, formation of membrane domains in polarized cells, regulation of ion and solute transport, cell–cell and cell–ECM adhesion complexes, and signal transduction.

The volume is introduced by two short essays that focus the topics of the volume on membrane protein organization in lipid bilayers (Edidin) and the role of the membrane cytoskeleton in regulating protein sorting and retention in membranes (Beck and Nelson).

The second section examines the role of the cytoskeleton in regulating intracellular vesicle trafficking to the plasma membrane: Schroer and Gill discuss the interactions between mircotubules and intracellular membranes and the roles of microtubule-based motors and accessory proteins in vesicle trafficking to the plasma membrane; Fath and Burgess describe the identification of motor proteins on transport vesicles derived from the Golgi complex which are targeted to specific membrane domains in polarized epithelial cells; and Yao and Forte discuss the role of membrane–cytoskeleton interactions in regulated exocytosis and apical insertion of vesicles in epithelial cells.

The third section focuses on the role of the cytoskeleton in regulating protein distributions and formation of structural and functional domains in the plasma membrane of polarized cells: Devarajan and Morrow discuss the role of the spectrin cytoskeleton in regulating the organization of polarized epithelial cell membranes; Lambert and Bennett examine the roles of axonal ankyrins and ankyrin-binding proteins in the formation of lateral membrane domains and transcellular connections at the node of Ranvier; and Dubreuil describes a molecular and genetic dissection of the membrane skeleton in polarized cells in *Drosophila.*

The fourth section examines the interactions between membrane proteins and the cytoskeleton associated with cell–cell and cell–extracellular matrix adhesion complexes: Wheelock, Knudsen, and Johnson examine the role of catenins in regulating membrane–cytoskeleton interactions with cadherin cell adhesion proteins; Kowalczyk and Green describe the organization of the desmosome as a component system for adhesion and intermediate filament attachment; Fanning, Lapierre, Brecher, Van Itallie, and Anderson examine protein interactions in the tight junction and the role of MAGUK proteins in regulating tight junction organization and function; Gayer, Campanelli, and Scheller discuss the regulation of membrane protein organization at the neuromuscular junction; and Eigenthaler and Shattil describe the platelet cytoskeleton and its role in integrin signaling.

The fifth section examines the roles of the membrane–cytoskeleton in cellular structure and function: Bourguignon describes the interactions between the membrane–cytoskeleton and CD44 during lymphocyte signal transduction and cell adhesion; Repasky and Black further examine the dynamic properties of the lymphocyte membrane cytoskeleton in relation to lymphocyte activation status, signal transduction, and protein kinase C; Smith and Benos discuss the role of the membrane cytoskeleton in regulating epithelial ion channel activity; Cantiello and Prat describe the role of actin filament organization on ion channel activity and cell volume regulation; and Doctor, Bacallao, and Mandel examine the role of the cytoskeleton in membrane alterations in ischemic and anoxic renal epithelia.

W. J. Nelson

Previous Volumes in Series

Current Topics in Membranes and Transport

Volume 19 Structure, Mechanism, and Function of the Na/K Pump* (1983)
Edited by Joseph F. Hoffman and Bliss Forbush III

Volume 20 Molecular Approaches to Epithelial Transport* (1984)
Edited by James B. Wade and Simon A. Lewis

Volume 21 Ion Channels: Molecular and Physiological Aspects (1984)
Edited by Wilfred D. Stein

Volume 22 The Squid Axon (1984)
Edited by Peter F. Baker

Volume 23 Genes and Membranes: Transport Proteins and Receptors* (1985)
Edited by Edward A. Adelberg and Carolyn W. Slayman

Volume 24 Membrane Protein Biosynthesis and Turnover (1985)
Edited by Philip A. Knauf and John S. Cook

Volume 25 Regulation of Calcium Transport across Muscle Membranes (1985)
Edited by Adil E. Shamoo

Volume 26 Na^{+}–H^{+} Exchange, Intracellular pH, and Cell Function* (1986)
Edited by Peter S. Aronson and Walter F. Boron

Volume 27 The Role of Membranes in Cell Growth and Differentiation (1986)
Edited by Lazaro J. Mandel and Dale J. Benos

Volume 28 Potassium Transport: Physiology and Pathophysiology* (1987)
Edited by Gerhard Giebisch

Volume 29 Membrane Structure and Function (1987)
Edited by Richard D. Klausner, Christoph Kempf, and Jos van Renswoude

Volume 30 Cell Volume Control: Fundamental and Comparative Aspects in Animal Cells (1987)
Edited by R. Gilles, Arnost Kleinzeller, and L. Bolis

Volume 31 Molecular Neurobiology: Endocrine Approaches (1987)
Edited by Jerome F. Strauss, III and Donald W. Pfaff

Volume 32 Membrane Fusion in Fertilization, Cellular Transport, and Viral Infection (1988)
Edited by Nejat Düzgünes and Felix Bronner

Volume 33 Molecular Biology of Ionic Channels* (1988)
Edited by William S. Agnew, Toni Claudio, and Frederick J. Sigworth

Volume 34 Cellular and Molecular Biology of Sodium Transport* (1989)
Edited by Stanley G. Schultz

Volume 35 Mechanisms of Leukocyte Activation (1990)
Edited by Sergio Grinstein and Ori D. Rotstein

Volume 36 Protein–Membrane Interactions* (1990)
Edited by Toni Claudio

Volume 37 Channels and Noise in Epithelial Tissues (1990)
Edited by Sandy I. Helman and Willy Van Driessche

Current Topics in Membranes

Volume 38 Ordering the Membrane Cytoskeleton Tri-Layer* (1991)
Edited by Mark S. Mooseker and Jon S. Morrow

Volume 39 Developmental Biology of Membrane Transport Systems (1991)
Edited by Dale J. Benos

Volume 40 Cell Lipids (1994)
Edited by Dick Hoekstra

Volume 41 Cell Biology and Membrane Transport Processes* (1994)
Edited by Michael Caplan

Volume 42 Chloride Channels (1994)
Edited by William B. Guggino

** Part of the series from the Yale Department of Cellular and Molecular Physiology*

CHAPTER 1

Getting There Is Only Half the Fun

Michael Edidin
Department of Biology, The Johns Hopkins University, Baltimore, Maryland 21218

Plasma membranes then and now

Then: *Motto:* Drunkard: "Will I ever, ever get home again?"
Polya (1921): "You can't miss; just keep going and stay out of 3D!" (Adam and Delbruck, 1968)

Now: "You can't go home again." (Wolfe, 1942)

I. INTRODUCTION

Cell membranes are elaborations of a fluid phospholipid bilayer. Lipid composition, inclusion of small molecules, and application and insertion of proteins all contribute to the baroque architecture of fully differentiated and functioning membranes, but do not change their character; they are 2-dimensional fluids. Given this character, we expect that integral membrane proteins can diffuse freely, random walking to sample all areas of a membrane and so randomizing over the entire membrane surface.

After 25 years of probing protein dynamics in cell surface membranes, it appears that the fluid properties of the phospholipid bilayer have some unexpected consequences for the dynamics of these proteins. Lateral diffusion of membrane proteins is readily measured on scales of hundreds of square nanometers to many square micrometers. However, unhindered square-micrometer-scale lateral diffusion seems to be the exception, not the rule, for membrane proteins. The lateral diffusion of individual protein molecules is impeded by other proteins. Furthermore, instead of randomizing molecules in the plane of the membrane, lateral diffusion brings them to associate with one another. Reduction of dimensionality, from 3 dimensions to 2, and the high surface concentration of membrane proteins enhances their encounters; these encounters often keep proteins from wandering freely through a membrane. Lateral diffusion of proteins on a smaller scale, hundreds of square Angstroms—a scale that is important for coupling reactions—is less well characterized than longer range lateral diffusion, but it too seems to be limited by molecular interactions.

II. A BASELINE FOR LATERAL DIFFUSION OF MEMBRANE PROTEINS

The lateral dynamics of proteins in membranes are best developed from a simple case, an artificial fluid phospholipid bilayer containing a low concentration of membrane proteins, say 1 protein per 5000 lipid molecules. We assume that the bulk of each protein is embedded in the bilayer and that the proteins do not interact with one another. In this case, the observed lateral diffusion coefficient for the protein will be determined mainly by the lipid viscosity. A typical diffusion coeficient (D) for multispan membrane proteins in synthetic phospholipid bilayers is $\sim 1 \times 10^{-8}$ $cm^2 \cdot s^{-1}$, about 100 times smaller than the diffusion coefficient of a protein in water (Vaz *et al.,* 1982). This can be used to infer that the viscosity of the phospholipid bilayer is ~100 times greater than water, 100 cp versus 1 cp. A viscosity of 100–500 cp is measured for 3-dimensional solutions of oils, so the estimate of lipid viscosity seems reasonable.

The method for measuring D usually is some form of photobleaching and recovery (for reviews, see Jovin and Vaz, 1988; Wolf, 1989). An area of the membrane is defined by a focused light beam, and the concentration of protein in the area is measured either in terms of light absorption or, usually, in terms of fluorescence from a small fluorophore labeling the protein of interest. Lateral mobility is measured by bleaching a fraction of the label (using a brief pulse of intense light) and then following the changes in the area concentration of unbleached molecules with time after bleaching. Recovery of fluorescence or increase in absorbance in the bleached area may be due largely to flow (especially in cell membranes) or to lateral diffusion. The dominant transport process can be seen in the shape of the recovery curve. If it is diffusional, D can be calculated from the extent and the half-time for recovery (R) of fluorescence. R is commonly termed the "mobile fraction" of labeled molecules. An R value of $<100\%$ is attributed to anchoring or trapping of some of the labeled proteins. In the artificial system that we are considering here (fluorescence/absorbance return to their prebleach levels), R is $\sim 100\%$ as long as the area bleached is only a small fraction (1–10%) of the total bilayer. This is consistent with the structure of the bilayer that we have assumed—an isotropic, continuous fluid. No protein molecules are trapped in pockets or domains. Hence all are mobile and can contribute to the recovery of fluorescence after photobleaching. The scale of the photobleaching recovery measurement is square micrometers, so it is comparable to the scale for measurement of the macroviscosity of solutions. Since the synthetic bilayer is made of a single lipid species, D measured in this way is also likely to be an accurate estimate of the lateral diffusion of protein molecules in smaller areas (square nanometers).

Once any of the conditions set for our artificial membrane system are changed, both D and R change. Lowering the system temperature increases the viscosity of fluid lipid and so reduces D, without affecting R. Lowering the temperature to or below the phase transition temperature of the phospholipids (the temperature at which they shift from liquid crystalline (fluid) to gel state) upsets our assumption of a completely fluid bilayer. Both D and R decrease (Vaz *et al.,* 1981). The decrease in D is due at least in part to the greatly increased viscosity of the gel phase formed as the system is cooled. The simplest explanation of the decrease in R is that a fraction of the protein is frozen in place, trapped in gel lipids. However, both D and R of proteins still in the fluid lipid regions of the cooled bilayer are affected by the number of solid obstacles (gel lipid domains) that they encounter. The percolation of a protein through a mixture of fluid and solid regions is governed by the proportions of the two lipid phases in the bilayer and by the size and shape of the solid obstacles. If gel domains are abundant, proteins may be trapped in isolated lakes of fluid lipid unconnected to other fluid domains and hence unable to contribute to recovery after photobleaching. (See Thompson *et al.,* 1995, for a discussion of this point in a slightly different context, aggregation of membrane proteins.)

Increasing the protein concentration in a fluid phospholipid bilayer also decreases D, again because diffusing proteins meet impenetrable obstacles (other proteins), so the path of their random walk over the surface is increased. R is not affected by protein concentration unless, as has proved to be the case in some model systems, the protein molecules aggregate at high surface concentrations. In that case, increasing protein concentration results in decreases in R as well as D. The large obstacles presented by protein aggregates isolate diffusing proteins in membrane subregions and prevent their percolation from one membrane region to another (Saxton, 1992).

III. LATERAL DIFFUSION IN BIOLOGICAL MEMBRANES

The first quantitative data on lateral diffusion of membrane proteins were obtained in a biological membrane system that closely approximates the ideal synthetic membrane bilayer described above. Two laboratories used the distinctive light-induced changes in the absorption spectrum of vertebrate rhodopsin to monitor the diffusion of this proteins in the disc membranes of rod outer segments (Liebman and Entine, 1974; Poo and Cone, 1974). The bulk of the rhodopsin molecule is buried in the membrane bilayer, and the lipids of the disc membrane are rich in unsaturated fatty acids. They are fluid and are not likely to contain any gel lipid. The biological membrane departs from our model only in that the concentration of rhodopsin is high; the protein–lipid ratio is ~1:1. D measured in the biological membrane was $0.3–0.5 \times 10^{-8}$ $cm^2 \cdot s^{-1}$, not far from the values measured for proteins in synthetic lipid systems. All of the rhodopsin was mobile; R approached 100%. The same D and R were measured about a decade later using fluorescence photobleaching and recovery (Wey *et al.,* 1981).

The gratifying agreement between results for lateral diffusion of rhodopsin and results for diffusion in the system described earlier proved to be exceptional. Most all other measurements of protein lateral diffusion in cell membranes report D values that are 10- to 100-fold smaller than those measured for rhodopsin and R values that are often significantly less than 100% (for review, see Edidin, 1994). Some of the lower values reported for D and R are artifacts of the labels commonly used for photobleaching recovery experiments: fluorescent antibodies and antibody fragments (Bierer *et al.,* 1987). However, most reflect the structure of biological membranes. These seem to be filled with obstacles, barriers, and traps that effectively prevent unhindered lateral diffusion of proteins on a scale of square micrometers.

IV. CONSTRAINTS TO THE LATERAL DIFFUSION COEFFICIENT (D)

We must account for a 10- to 100-fold difference between D of proteins measured in pure lipid bilayers and D measured in cell surface membranes.

Logically, D must be affected by molecular interactions occurring in one or more of three regions of the greater membrane—the bilayer itself, the cytoplasm below the membrane bilayer, and the external volume just above the bilayer. We have already noted that in a model membrane the viscosity of the bilayer limits D. In native membranes this viscosity sets an upper limit to D.

The first locus shown to hinder lateral diffusion was the spectrin membrane skeleton of erythrocytes. A number of experiments comparing lateral diffusion of band 3 (a multispan membrane protein) in normal and mutant erythrocytes, and also in ghosts with chemically disrupted spectrin, showed that D increased 10- to 50-fold when the spectrin mesh was incomplete or damaged; R also increased (Sheetz *et al.,* 1980; Tsuji and Ohnishi, 1986; Golan, 1989). The cyto- or membrane skeleton also appears to anchor proteins, reducing R, in several differentiated cell types (Pollerberg *et al.,* 1986; Madreperla *et al.,* 1989; Venkatakrishnan *et al.,* 1991), as well as in activated cells (Liu *et al.,* 1995; see Edidin, 1994, for a review of other work on R and activated cells).

Many studies of constraints to lateral diffusion suggest that the cyto- or membrane skeleton is not an important constraint to D. These studies follow the behavior of single-span membrane proteins with large exodomains and cytoplasmic domains ranging from ~30 to over 500 amino acids in length. Truncation of the cytoplasmic domains proves to have little or no effect on D (reviewed in Edidin, 1994; but see Sheets *et al.,* 1995). The most extreme case, truncating the epidermal growth factor receptor's cytoplasmic domain from 523 residues to none, does not alter D (1.2×10^{-10} cm$^2 \cdot$ s^{-1} for both wild-type and completely truncated forms). R increases slightly with truncation, from ~60% to ~80% (Livneh *et al.,* 1986). The only truncation of cytoplasmic domains that changed D by a factor of ~10 (from 1×10^{-10} to 8×10^{-10} cm$^2 \cdot$ s^{-1}) was of a protein with two cytoplasmic domains that interact with signaling molecules of the cytoplasm (Wade *et al.,* 1989). Increased D and truncation correlated with decreased signaling.

Even transferring a protein's exodomain from its transmembrane and cytoplasmic domains to a lipid anchor does not change D. The converse, transferring the exodomain of a lipid-anchored protein to a peptide sequence of transmembrane and cytoplasmic amino acids, also has no effect (Zhang *et al.,* 1991,1993). However, truncating exodomains or mutating proteins to remove sites of N-glycosylation increases D to near the limit set by lipid viscosity (Wier and Edidin, 1989). It appears that D for proteins with bulky exodomains is most constrained by interactions of these domains with neighboring proteins. The cell surface is crowded, with perhaps half its area or more occupied by proteins, and so the random walk of a diffusing molecule is impeded by its neighbors. A given molecule cannot move from one area (say 100 Å^2) to another unless the receiving area is unoccupied by protein.

Of course, taken together, the data indicate that different constraints to lateral diffusion are most effective in different cells or for different membrane proteins. Inded, in one instance we were able to show that, in Chinese hamster ovary cells, the glycosylation state of major histocompatibility complex (MHC) class I molecules has little effect on D while cytoskeleton does have some effect on lateral mobility (R). When the same MHC-I molecules are expressed in L-cells, glycosylation state has large effects on D and there seems to be no significant interaction with cytoskeleton (Barbour and Edidin, 1992; Wier and Edidin, 1989).

V. MOLECULAR CLUSTERING AS A CONSEQUENCE OF LATERAL DIFFUSION

Molecular crowding of the membrane surface not only constrains random walks, it also leads to capture of diffusing molecules by neighbors in the membrane. We can make a rough estimate of the equivalant surface concentrations of all membrane integral proteins. Assuming that a typical protein explores a sphere of 100 Å diameter and that the surface concentration of protein is 10^4 to 10^5 molecules/μm^2, we find that the surface concentration is equivalant to that in a 100-Å slice of a 10- to 100-m*M* solution. This and the restrictions on orientation of proteins imposed in 2 dimensions implies that proteins with very low affinities for one another will associate with one another (Grasberger *et al.,* 1986). Indeed, it may be that lateral diffusion is constrained not only by the surface concentration of proteins, but by the balance of repulsive and attractive forces between them. We have found (S. Barbour and M. Edidin, unpublished data) that D decreases significantly after intact cell surfaces are treated with neuraminidase, removing negative charge from glycoproteins and so decreasing charge repulsion between them. D returns to control levels if surfaces are resialylated by the action of exosialytransferase in the presence of added sugar donor.

Dimers and higher oligomers often form despite charge repulsion. The most striking example of this is glycophorin, which exists in a concentration-dependent equilibrium between a monomer and a dimer (Adair and Engelman, 1992). The monomers associate through their transmembrane regions, despite the fact that the exodomain of this protein bears many negatively charged sialic acids. Concentration-dependent heterodimers and hetero-oligomers have also been detected. In one notable case the 2-dimensional equilibrium constant for the association has been calculated (Poo *et al.,* 1995). The K_{2d} for the specific association of a complement receptor, CR3, and the Fc receptor FcγRIIB was estimated as 2×10^{-12} mol/dm^2. In the presence of a soluble inhibitor, the K_{2d} is reduced to 1×10^{-11} mol/dm^2.

Although their associations have not been well characterized or quantitiated, it is evident that a number of other membrane integral proteins also form heterologous associations. The insulin receptor and class I MHC molecules associate in a reversible manner. Cross-linking and immobilizing one of the pair reduces, but does not abolish, lateral diffusion of the other (Edidin and Reiland, 1990). This tactic also shows a reversible association between class I MHC molecules and one of the chains of the interleukin-2 (IL-2) receptor complex (Edidin *et al.,* 1988). The argument for reversibility is again made from the fact that cross-linking MHC-I molecules with antibody reduces D of the IL-2 receptor chain. It is not clear if another example, immobilization of one chain of nerve growth factor receptor by the other, also represents an equilibrium association between the two chains (Wolf *et al.,* 1995). However, since only partial immobilization results, this is likely. Most other evidence for the presence of molecular clusters in cell surface membranes comes from static measurements, either coprecipitation of heterologous chains or detection of molecular proximity using fluorescence energy transfer.

Once clustered, membrane proteins may internalize, thus ending their diffusion over the cell surface. If they persist at the surface, molecular clusters may in turn impede the lateral diffusion of other membrane proteins. Saxton (1992) has shown that ragged clusters of molecules form greater impediments to lateral diffusion of a tracer than do the same number of molecules when dispersed. The surface concentration of clusters will also affect the measured diffusion coefficient, so that D depends on the distance scale over which it is measured (Saxton, 1994). At high cluster density, above the percolation threshold, molecules may be isolated so that they are unable to contribute to observed recovery of fluorescence in a photobleaching recovery experiment, even though they are free to diffuse over short distances. These effects may account for the observation that even artificial proteins, stearoylated dextrans integrated into the cell surface, show an appreciable immobile fraction (~30%) (Wolf *et al.,* 1980). Immobility is almost total if the dextrans are aggregated in the plane of the membrane by antibody. This suggests that the average distance between clusters or other barriers to lateral mobility is too small to allow passage of the average cluster of stearoyl dextrans.

Lateral diffusion does not only bring molecules together to cluster; it is also required to disperse molecular clusters. Membrane proteins are delivered to the surface by fusion of transport vesicles with plasma membrane. Fusion inserts a patch of molecules, which disperse from the fusion site with time. For transmembrane proteins, the only model we have for this is the dispersion of viral proteins after fusion of enveloped viruses with the cell (Lowy *et al.,* 1995). Lateral diffusion of viral proteins away from the region of envelope fusion is too slow to be unhindered. It remains to be seen if the obstacles to lateral diffusion of these proteins are in the fusion pore, or if they reflect the state of the cell plasma membrane near the fusion site.

One specialized form of membrane proteins, lipid-anchored proteins, is delivered to the surface in a cluster. This cluster may also include transmembrane proteins involved in targeting the lipid-anchored proteins to a particular surface of morphologically polarized cells. The complex, detected in terms of changes in mobile fraction (R), disperses with time at the cell surface. The scale for dispersion is tens of minutes, implying dispersion by hindered lateral diffusion (Hannan *et al.,* 1993).

VI. BARRIERS TO LATERAL DIFFUSION, MEMBRANE DOMAINS

The results summarized in earlier sections of this chapter speak to molecular anchoring, molecular crowding, and cluster formation in cell surface membranes. The dynamics of these processes are detected in lateral diffusion measurements. As noted, these constraints to lateral diffusibility arise mainly from the probability of a given protein encountering other membrane proteins (which themselves may be immobile, tethered to the cytoskeleton). A series of studies of lateral diffusion and lateral mobility detected another impediment to lateral diffusion, confinement of membrane proteins to domains with apparent diameters ranging from ~200 nm to >1 μm.

The first study to suggest membrane domains was that already cited on lateral diffusion of band 3 in erythrocytes (Sheetz *et al.,* 1980). Since D and R were much greater in spectrin-defective erythrocytes (spherocytes) than in normal erythrocytes, it was suggested that the spectrin meshes caged membrane proteins into separate domains. Low D and R, measured on a scale much larger than that of a single spectrin mesh, implied that diffusing molecules were trapped in meshes whose sides could open to allow further lateral diffusion. The idea of domains of membrane proteins in cells other than erythrocytes was developed from the observation that the observed mobile fraction (R) depended upon the size of the area observed and bleached in a photobleaching recovery experiment (Yechiel and Edidin, 1987). The larger the area bleached, the smaller the value of R. It was noted early in this chapter that R may be less than 100% if the area of the bleached spot is large relative to the total area of the membrane. Since even the largest area bleached was much less than the total area of the cells examined (not more than ~10% at most), the reduction in R with increased D could only mean that the membrane continuum was divided into smaller regions (domains) whose diameters were less than that of the region observed and bleached (1–2 μm).

VII. PARTICLE TRACKING, PARTICLE DRAGGING, AND MEMBRANE DOMAINS

Photobleaching recovery measurements report on the behavior of populations of molecules. Recovery of fluorescence after bleaching involves the random walks of thousands of molecules.This limits its effectiveness in further defining membrane domains. However, other methods allow us to follow the lateral diffusion of single molecules or of small clusters of molecules. These single particle tracking (SPT) methods label one or a few membrane proteins with a small gold bead or fluorescent particle. The behavior of the particle, and hence of the membrane proteins to which it is attached, can be followed by interference contrast or fluorescence microscopy. The earliest development of SPT followed single molecules, low-density lipoprotein (LDL) receptors labeled with fluorescent LDL, a monovlaent ligand (Barak and Webb, 1981; Ghosh and Webb, 1994). However, most SPT labels are 50-nm gold beads coated with antibodies to the membrane proteins of interest. These coated are multivalent; they may bear from 30 to 100 molecules of immunoglobulin G (see De Roe *et al.*, 1987; Edidin *et al.*, 1994), a fraction of which can bind to the surface, creating a small cluster of proteins. Since lateral diffusion is only weakly sensitive to molecular weight of the diffusing species ($D \propto 1/\ln MW$), the bead-labeled aggregates appear to diffuse much as single particles. D values estimated by photobleaching recovery and from random walks of bead-labeled molecules are usually in good agreement (see, for example, Edidin *et al.*, 1991).

Lateral diffusion coefficients describe the area visited by the diffusing object per unit time; their units (cgs) are square centimeters per second. Tracking beads gives us a plot of area visited versus time; this is linear for unconfined lateral diffusion. If the molecules labeled by a bead are laterally confined, the plot of area versus time will level off. Although real data are noisy, statistical methods have been developed for analyzing these plots to detect confinement of molecules to domains. Depending upon the molecule examined, membrane domains have diameters of between 200 and 1100 nm. Molecules appear to dwell in a domain for periods in the range of 3–30 sec before jumping to another domain; apparently the molecular barriers creating a domain of this sort are dynamic (reviewed in Sheets *et al.,* 1995). A recent model for this behavior makes predictions that could allow experimental determination of the times between barrier openings (Saxton, 1995).

The size of membrane domains and the nature of the barriers that form them have also been characterized by dragging bead-labeled molecules along the membrane surface using a laser optical trap (a laser tweezers). In early experiments, the technique reported barrier-free paths (BFP) of ~1 μm for bead-labeled MHC-I molecules observed at room temperature. Much longer BFP were observed at 36°C, consistent with the dynamic nature of the barriers. Lipid-anchored forms of MHC-I molecules had far longer BFP than conventional transmembrane MHC, indicating that the barriers were either in the membrane bilayer itself or in the cell cytoplasm. This was consistent with fluorescence photobleaching and recovery measurements, suggesting that lipid-anchored proteins were not confined to domains (Edidin and Stroynowski, 1991). Later experiments on MHC-I molecules with truncated cytoplasmic domains showed that the barriers are in the cytoplasm, not in the bilayer (Edidin *et al.,* 1994).

BFP and SPT data are statistical measures of the confinement of diffusing particles. They increase with increased temperature and decrease with increased valence of the labeling particle, or with decreased trapping forces. Hence estimates of domain size depend upon the measurement conditions. Membrane domains, although most easily conceived of as ponds or lakes with fixed perimeters, are better thought of as transient, chaotic features of lateral membrane organization. They may well be fractals, with features of barriers and gaps repeated on several size scales. Indeed, photobleaching recovery measurements have been analyzed in terms of fractal time tails (Brust-Mascher *et al.,* 1992) and, as mentioned, lateral diffusion in the presence of aggregates can be analyzed in these terms (Saxton, 1994). The multiple scales of domains imply that lateral diffusion and lateral confinement measured on the scale of micrometers may not report accurately on smaller scales that are important in diffusional coupling of reactions (e.g., the coupling of hormone receptors to adenylate cyclase). However, recent work correlates protein mobility, measured by large-scale photobleaching recovery, and agonist-stimulated adenylate cyclase activity, coupled by lateral diffusion on a nanometer scale (Zakharova *et al.,* 1995).

VIII. CONCLUSION

The dynamics of biological membranes are more complex than was first imagined when the fluid properties of these membranes were first characterized. A protein molecule diffusing in the membrane is rarely free to explore the entire area in which it resides. As it diffuses, it can meeet many fates, with different probabilities. There is a small probability that it can go home again. There are higher probabilities that it will be trapped, aggregated, or internalized long before it reaches that goal. It is likely that the balance between these probabilities is an important determinant of membrane function.

References

Adair, B. D., and Engelman, D. M. (1994). Glycophorin A helical transmembrane domains dimerize in phospholipid bilayers: A resonance energy transfer study. *Biochemistry* **33,** 5539–5544.

Adam, G., and Delbruck, M. (1968). Reduction of dimensionality in biological diffusion processes. *In* "Structural Chemistry and Molecular Biology" (A. Rich and N. Davidson, eds.), pp. 198–215. W. H. Freeman, San Francisco.

Barak, L. S., and Webb, W. W. (1981). Fluorescent low density lipoprotein for observation of dynamics of individual receptor complexes on cultured human fibroblasts. *J. Cell Biol.* **90,** 595–604.

Barbour, S., and Edidin, M. (1992). Cell-specific constraints to the lateral diffusion of a membrane glycoprotein. *J. Cell. Physiol.* **150,** 526–533.

Bierer, B. E., Herrmann, S. H., Brown, C. S., Burakoff, S. J., and Golan, D. E. (1987). Lateral mobility of class I histocompatibility antigens in B lymphoblastoid cell membranes: Modulation by cross-linking and effect of cell density. *J. Cell Biol.* **105,** 1147–1152.

Brust-Mascher, I., Feder, T. J., Slattery, J. B., Baird, B., and Webb, W. W. (1992). Constrained diffusion or immobile fraction on the cell surface: A new interpretation. *Mol. Cell. Biol.* **3,** 306a.

De Roe, C., Courtoy, P. J., and Baudhuin, P. (1987). A model of protein-colloidal gold interactions. *J. Histochem. Cytochem.* **35,** 1191–1198.

Edidin, M. (1994). Fluorescence photobleaching and recovery, FPR, in the analysis of membrane structure and dynamics. *In* "Mobility and Proximity in Biological Membranes" (S. Damjanovich, M. Edidin, J. Szollosi, and L. Tron, eds.), pp. 109–135. CRC Press, Boca Raton, FL.

Edidin, M., and Reiland, J. J. (1990). Dynamic measurements of associations between class I MHC antigens and insulin receptors. *Mol. Immunol.* **27,** 1313–1317.

Edidin, M., and Stroynowski, I. (1991). Differences between the lateral organization of conventional and inositol phospholipid-anchored membrane proteins: A further definition of micrometer scale membrane domains. *J. Cell Biol.* **112,** 1143–1150.

Edidin, M., Aszalos, A., Damjanovich, S., and Waldmann, T. (1988). Lateral diffusion measurements give evidence for association of the Tac peptide of the IL-2 receptor with the T27 peptide in the plasma membrane of HUT-102-B2 T cells. *J. Immun.* **141,** 1206–1210.

Edidin, M., Kuo, S. C., and Sheetz, M. (1991). Lateral movements of membrane glycoproteins restricted by dynamic cytoplasmic barriers. *Science* **254,** 1379–1382.

Edidin, M., Zuniga, M. C., and Sheetz, M. P. (1994). Truncation mutants define and locate cytoplasmic barriers to lateral mobility of membrane glycoproteins. *Proc. Natl. Acad. Sci. U.S.A.* **91,** 3378–3382.

Ghosh, R. N., and Webb, W. W. (1994). Automated detection and tracking of individual and clustered cell surface low density lipoprotein molecules. *Biophys. J.* **66,** 1301–1318.

Golan, D. (1989). Red blood cell membrane protein and lipid diffusion. *In* "Red Blood Cell Membranes" (P. Agre and J. C. Parker, eds.), pp. 367–400. Marcel Dekker, New York.

Grasberger, B., Minton, A. P., DeLisi, C., and Metzger, H. (1986). Interaction between proteins localized in membranes. *Proc. Natl. Acad. Sci. U.S.A.* **83,** 6258–6262.

Hannan, L. A., Lisanti, M. P., Rodriguez-Boulan, E., and Edidin, M. (1993). Correctly sorted molecules of a GPI-anchored protein are clustered and immobile when they arrive at the apical surface of MDCK cells. *J. Cell Biol.* **120,** 353–358.

Jovin, T. M., and Vaz, W. L.C. (1988). Rotational and translational diffusion in membranes measured by fluorescence and phosphorescence methods. *Methods Enzymol.* **172,** 471–513.

Liebman, P. A., and Entine, G. (1974). Lateral diffusion of pigment in photoreceptor disk membranes. *Science* **185,** 457–459.

Liu, S-q. J., Hahn, W. C., Bierer, B. E., and Golan, D. E. (1995). Intracellular mediators regulate CD2 lateral diffusion and cytoplasmic Ca^{2+} mobilization upon CD2-mediated T cell activation. *Biophys. J.* **68,** 459–470.

Livneh, E., Benveniste, M., Prywes, R., Felder, S., Kam, Z., and Schlessinger, J. (1986). Large deletions in the cytoplasmic kinase domain of the EGF-receptor do not affect its lateral mobility. *J. Cell Biol.* **103,** 327–331.

Lowy, R. J., Sarkar, D. P., Whitnall, M. H., and Blumenthal, R. (1995). Differences in dispersion of influenza virus lipids and proteins during fusion. *Exp. Cell Res.* **216,** 411–421.

Madreperla, S. A., Edidin, M., and Adler, R. (1989). Na^+,K^+-adenosine triphosphatase polarity in retinal photoreceptors: A role for cytoskeletal attachments. *J. Cell Biol.* **109,** 1483–1493.

Pollerberg, G. E., Schachner, M., and Avoust, J. (1986). Differentiation state-dependent surface mobilities of two forms of the neural cell adhesion molecule. *Nature* (*London*) **324,** 462–465

Poo, H., Krauss, J. C., Mayo-Bond, L., Todd, R. F. III, and Petty, H. R. (1995). Interaction of Fc gamma receptor type IIIB with complement receptor type 3 in fibroblast transfectants: Evidence from lateral diffusion and resonance energy transfer studies. *J. Mol. Biol.* **247,** 597–603.

Poo, M-M., and Cone, R. A. (1974). Lateral diffusion of rhodopsin in the photoreceptor membrane. *Nature* (*London*) **247,** 438–441.

Saxton, M. J. (1992). Lateral diffusion and aggregation. A Monte Carlo study. *Biophys. J.* **61,** 119–128.

Saxton, M. J. (1994). Anomalous diffusion due to obstacles: A Monte Carlo study. *Biophys. J.* **66,** 394–401.

Saxton, M. J. (1995). Single-particle tracking: Effect of corrals. *Biophys. J.* **69,** 389–398.

Sheets, E. D., Simson, R., and Jacobson, K. (1995). New insights into membrane dynamics from the analysis of cell surface interactions by physical methods. *Curr. Topics Cell Biol.* **7,** 707–714.

Sheetz, M. P., Schindler, M., and Koppel, D. (1980). Lateral mobility of integral membrane proteins is increased in spherocytic erthrythrocytes. *Nature* (*London*) **285,** 510–512

Thompson, T. E., Sankaram, M. B., Biltonen, R. L., Marsh, D., and Vaz, W. L.C. (1995). Effects of domain structure on in-plane reactions and interactions. *Mol. Membr. Biol.* **12,** 157–62.

Tsuji, A., amd Ohnishi, S. (1986). Restriction of the lateral motion of band 3 in the erythrocyte membrane by the cytoskeletal network: Dependence on spectrin association state. *Biochemistry* **25,** 6133–6139.

Vaz, W. L. C., Kapitza, H. G., Stumpel, J., Sackmann, E., and Jovin, T. M. (1981). Translational mobility of glycophorin in bilayer membranes of dimyristoylphosphatidylcholine. *Biochemistry* **20,** 1392–1396.

Vaz, W. L. C., Criado, M., Madeira, V. M. C., Schoellmann, G., and Jovin, T. M. (1982). Size dependence of the translational diffusion of large integral membrane proteins in liquid-crystalline phase lipid bilayers. *Biochemistry* **21,** 5608–5612.

Venkatakrishnan, G., McKinnon, C. A., Pilpal, C. G., Wolf, D. E., and Ross, A. H. (1991). Nerve growth factor receptors are preaggregated and immobile on responsive cells. *Biochemistry* **30,** 2748–2753.

Wade, W. F., Freed, J. H., and Edidin, M. (1989). Translational diffusion of class II major histocompatibility complex molecules is constrained by their cytoplasmic domains. *J. Cell. Biol.* **109,** 3325–3331.

Wey, C.-L., Edidin, M. A., and Cone, R. A. (1981). Lateral diffusion of rhodopsin in photoreceptor cells measured by fluorescence photobleaching and recovery (FPR). *Biophys. J.* **33,** 225–232.

Wier, M., and Edidin, M. (1989). Constraint of the translational diffusion of a membrane glycoprotein by its external domains. *Science* **242,** 412–414.

Wolf, D. E. (1989). Designing, building and using a fluorescence recovery after photobleaching instrument. *Methods Cell Biol.* **30,** 271–306.

Wolf, D. E., Henkart, P., and Webb, W. W. (1980). Diffusion, patching and capping of stearoylated dextrans on 3T3 cell plasma membranes. *Biochemistry* **19,** 3893–3904.

Wolf, D. E., McKinnon, C. A., Daou, M. C., Stephens, R. M., Kaplan, D. R., and Ross, A. H. (1995). Interaction with TrkA immobilizes gp75 in the high affinity nerve growth factor receptor complex. *J. Biol. Chem.* **270,** 2133–2138.

Wolfe, T. (1942). "You Can't Go Home Again." Sundial Press, New York.

Yechiel, E., and Edidin, M. (1987). Micrometer scale domains in fibroblast plasma membranes. *J. Cell Biol.* **105,** 755–760.

Zakharova, O. M., Rosenkranz, A. A., and Sobolev, A. S. (1995). Modification of fluid lipid and mobile protein fractions of reticulocyte plasma membrane affects agonist-stimulated adenylate cyclase: Application of percolation theory. *Biochim. Biophys. Acta* **1236,** 177–184.

Zhang, F., Crise, B., Su, B., Hou, Y., Rose, J. K., Bothwell, A., and Jacobson, K. (1991). Lateral diffusion of membrane-spanning and glycosylphosphatidylinositol-linked proteins: Toward establishing rules governing the lateral mobility of membrane proteins. *J. Cell Biol.* **115,** 75–84.

Zhang, F., Lee, G. M., and Jacobson, K. (1993). Protein lateral mobility as a reflection of membrane microstructure. *Bioessays* **15,** 579–588.

CHAPTER 2

Once There, Making the Descision To Stay or Leave

Kenneth A. Beck and W. James Nelson
Department of Molecular and Cellular Physiology, Beckman Center for Molecular and Genetic Medicine, Stanford University School of Medicine, Stanford, California 94305

I. INTRODUCTION

Protein sorting is a complex process occurring at all stages of the secretory cycle. Newly synthesized membrane proteins are transported through a series of common compartments, the endoplasmic reticulum and Golgi cisternae, regardless of their ultimate destination. In the *trans*-Golgi network (TGN), proteins are packaged into transport vesicles that are targeted to different membrane domains in polarized epithelial cells. Since proteins enter the TGN through a common pathway but leave in discrete vesicle packages, it is expected that protein-sorting mechanisms in the TGN drive the segregation of different classes of membrane proteins. Upon arrival at the plasma membrane, proteins have different fates: Some proteins are internalized and then, in the endosome, are either sorted for degradation in lysosomes or recycled back to the plasma membrane; other resident proteins are sequestered from internalization pathways and retained in the plasma membrane.

The identity and mechanisms of action of the protein machineries that drive these sorting events in the TGN, plasma membrane, and endosomes are poorly understood. However, several known protein complexes that form membrane coat structures can be viewed as candidate membrane protein-sorting machines. These include the clathrin–AP complex (Keen, 1990; Robinson, 1994), the coatamer (COP) complex (Kreis and Pepperkok, 1994; Rothman and Warren, 1994), caveolae (Anderson, 1993), p200 (Narula and Stow, 1995), and spectrin (Beck and Nelson, 1996).

The clathrin–AP complex provides a well-described membrane coat structure involved in protein sorting (Keen, 1990; Robinson, 1994). Clathrin lattices are thought to be important for sorting membrane proteins of both the plasma membrane and the TGN. Clathrin–AP complexes assemble into a 2-dimensional, oligomeric lattice that associates with the cytoplasmic surface of membranes to form a clathrin-coated pit. Specific classes of membrane proteins are collected within the clathrin lattice through direct binding to the clathrin–AP complex. Other membrane proteins are excluded from the clathrin lattice. Invagination of the clathrin lattice forms a coated transport vesicle with a protein composition different from that of the membrane from which it was derived.

In this chapter, we examine the structural and functional properties of the spectrin-based membrane skeleton to demonstrate its potential as a protein-sorting machine.

II. DOES THE SPECTRIN-BASED MEMBRANE SKELETON HAVE PROPERTIES OF A PROTEIN-SORTING MACHINE?

Proteins of the spectrin-based membrane skeleton form a 2-dimensional, oligomeric lattice that binds to specific classes of integral membrane proteins (Bennett and Gilligan, 1993; Fig. 1). The major structural component of the membrane skeleton is spectrin, a flexible, rod-shaped molecule composed of homologous, but nonidentical, α- and β-subunits. Spectrin heterodimers self-associate end-to-end to form heterotetramers (length 200 nm), which, in erythrocytes, are cross-linked by a complex of short actin oligomers, protein 4.1, and adducin. High-resolution electron micrographs of the spectrin membrane skeleton in erythrocytes reveal a densely packed lattice that extends over the entire cytoplasmic surface of the plasma membrane.

The distribution of the membrane skeleton in polarized neurons and epithelia is restricted to functionally and structurally distinct plasma mem-

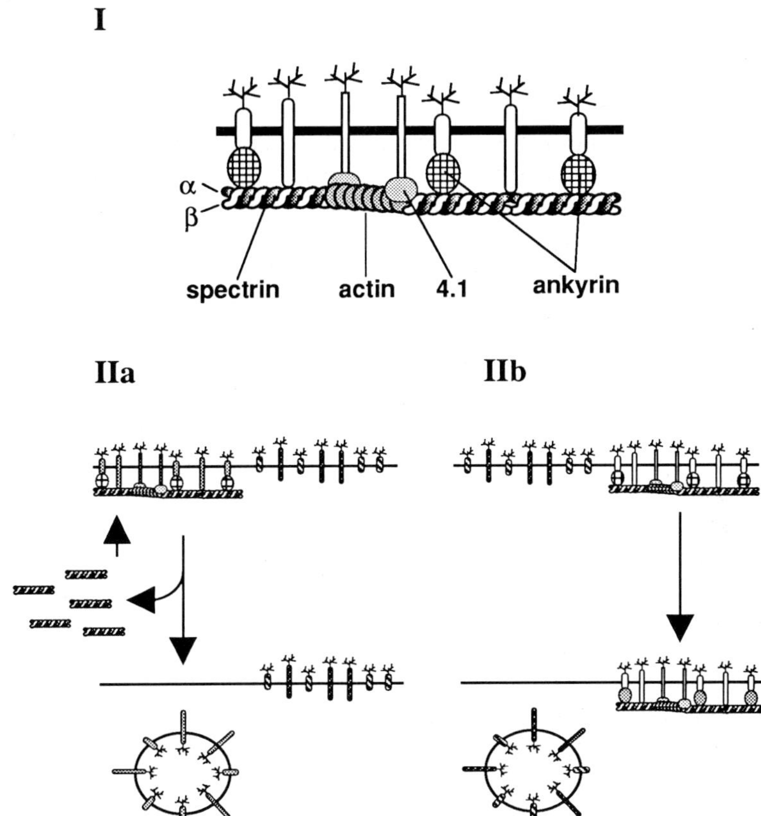

FIGURE 1 Protein sorting events mediated by the membrane skeleton. The spectrin membrane skeleton (I) exists as an extensive 2-dimensional lattice that binds to specific classes of membrane proteins. The spectrin lattice is composed of spectrin heterotetramers cross-linked by short actin filaments. Membrane protein interactions are mediated by spectrin, ankyrin, and protein 4.1. A 2-dimensional oligomeric matrix, such as the spectrin-based membrane skeleton, can facilitate a membrane protein-sorting event by collecting specific classes of membrane proteins and restricting their distribution to within a discrete membrane domain. Membrane protein trafficking (IIa) requires that transport vesicles be formed from the domain defined by the oligomeric matrix, whereas membrane protein retention (IIb) is achieved when transport vesicles are derived from membranes outside of the matrix.

brane domains, although much less is known about the 3-dimensional structure of the spectrin-based membrane skeleton in nonerythroid cells. In Purkinje cells of the cerebellum, isoforms of spectrin and ankyrin homologous to the erythroid proteins localize to the cell body and dendritic processes, whereas isoforms distinct from the erythroid proteins localize to the axonal plasma membrane (Lazarides and Nelson, 1983). Within myelinated axons, the distribution of membrane skeleton proteins is restricted to nodes of Ranvier (Kordeli *et al.,* 1990). In polarized epithelial cells (e.g., Madin–Darby canine kidney (MDCK) cells), the distribution of a nonerythroid isoform of spectrin is restricted to the basolateral plasma membrane domain (Nelson and Veshnock, 1986). An isoform of erythroid β-spectrin has been identified in association with the Golgi complex (Beck *et al.,* 1994). These restricted distributions of the membrane skeleton imply that, if component proteins have preferential affinities for specific membrane proteins, then the membrane skeleton may restrict membrane proteins to specialized membrane domains.

The spectrin-based membrane skeleton is tightly coupled to the membrane through direct binding to a diverse subset of membrane proteins. Some of these interactions occur through the membrane skeleton protein ankyrin. Ankyrin is a bifunctional molecule with separate binding sites for membrane proteins and β-spectrin (Bennett, 1992). Many of these ion transport proteins and cell adhesion molecules, like the membrane skeleton, exhibit restricted cell surface distributions in polarized cells. Biophysical studies have shown that proteins bound to the membrane skeleton have reduced lateral diffusion in the plane of the lipid bilayer (Sheetz *et al.,* 1980; Dahl *et al.,* 1994), suggesting that the membrane skeleton can physically restrict the distribution of membrane proteins within a membrane compartment (see Fig. 1). Thus, the co-restriction of both the membrane skeleton and associated membrane proteins to discrete domains of the plasma membrane in polarized cells suggests a causal relationship in the formation of functionally and structurally distinct membrane domains.

III. FORMATION OF MEMBRANE DOMAINS AND PROTEIN-SORTING AT THE LEVEL OF THE PLASMA MEMBRANE

In polarized epithelial cells (e.g., MDCK cells), ankyrin and fodrin become restricted to the sites of cell–cell contact, and their detergent insolubility and metabolic half-life increase, indicating assembly into a cytoskeletal complex (Nelson and Veshnock, 1986,1987b). Significantly, assembly of ankyrin and fodrin at cell–cell contacts in "nonpolarized" fibroblasts (McNeill *et al.,* 1990) and retinal pigmented epithelial (RPE) cells (Marrs *et al.,* 1995) is dependent upon expression of E-cadherin. Deletion of the cytoplasmic domain of E-cadherin in transfected fibroblasts inhibits Na,K-ATPase accumulation at sites of cell–cell contact, indicating that linkage to the cytoskeleton is important in protein recruitment to the adhesion site (McNeill *et al.,* 1990).

Na,K-ATPase and E-cadherin are directly linked to different membrane–cytoskeleton complexes. Na,K-ATPase binds directly to ankyrin, which in turn binds to fodrin; *in vitro* studies with purified proteins demonstrated the specificity and high-affinity binding of these proteins (Nelson and Veshnock, 1987a; Morrow *et al.,* 1989; Davis and Bennett, 1990; Devarajan *et al.,* 1994), and cell fractionation studies demonstrated that these complexes exist in MDCK cells (Nelson and Hammerton, 1989) and other polarized epithelial cells (Gundersen *et al.,* 1991; Marrs *et al.,* 1993,1995). In addition to ankyrin and fodrin, the membrane–cytoskeleton complex comprises several well-characterized proteins (adducin, protein 4.1, and actin) that localize to cell–cell contacts in MDCK cells (Kaiser *et al.,* 1989). Posttranslational modifications of these proteins may play an important role in regulating membrane–cytoskeleton assembly in response to E-cadherin–mediated cell–cell adhesion. A complex of Na,K-ATPase, ankyrin, and fodrin has been detected in association with E-cadherin following fractionation of MDCK cell extracts (Nelson *et al.,* 1990), but the protein interactions involved in the linkage between the Na,K-ATPase complex and E-cadherin are not known. In MDCK cells, E-cadherin directly associates with two cytoplasmic proteins, termed α- and β-catenin (Kemler and Ozawa, 1989). Our results show that α-catenin binds to the E-cadherin–β-catenin complex at the cell surface coincident with the titration of the complex into a detergent-insoluble fraction (Hinck *et al.,* 1994). These findings indicate that α-catenin is a candidate protein for linking the cadherin–catenin complex to the Na,K-ATPase–ankyrin–fodrin complex, perhaps by binding directly to fodrin (Lombardo, *et al.,* 1993). Defining the interactions between the cadherin–catenin and membrane–cytoskeleton (Na,K-ATPase) complexes will be important in determining mechanisms involved in the recruitment of the membrane–cytoskeleton complex, and perhaps the cellular machinery for vesicle docking, to active sites of cadherin-mediated cell adhesion.

Selective incorporation of specific membrane proteins (e.g., Na,K-ATPase) into the membrane–cytoskeleton may restrict protein diffusion away from those sites, resulting in their localized retention and accumulation (Nelson, 1992). This is supported by several observations. The distributions of other proteins that bind to the membrane–cytoskeleton are restricted to specific plasma membrane domains (voltage-sensitive Na^+ channel in the node of Ranvier (Kordeli *et al.,* 1990), anion exchanger on the basolateral membrane of kidney epithelia (Drenckhahn *et al.,* 1985)). Loss of membrane–cytoskeleton protein interactions in erythrocytes (Sheetz *et al.,* 1980) and RBL cells (Dahl *et al.,* 1994) increases the diffusion of proteins in the membrane. Although deletion of α-spectrin in *Drosophila* does not affect the polarized distribution of Na,K-ATPase in embryonic epithelial cells (Lee *et al.,* 1993), β-spectrin and ankyrin remain co-localized with Na,K-ATPase (R. Dubreil, personal communication), indicating that maternal α-spectrin is present in the complex (Lee *et al.,* 1993). Finally, accumulation of Na,K-ATPase at E-cadherin adhesion sites in MDCK (Nelson and Veshnock, 1986) and RPE (Marrs *et al.,* 1995) cells correlates with >20-fold longer half-lives of cadherin, Na,K-ATPase and membrane–cytoskeletal proteins at sites of cell–cell contact than at the noncontacting (apical) membrane.

IV. FORMATION OF MEMBRANE DOMAINS AND PROTEIN-SORTING AT THE LEVEL OF THE GOLGI COMPLEX

The Golgi complex is composed of sequentially arranged compartments of unique membrane protein compositions (Pfeffer and Rothman, 1987). This organization is maintained in the face of continuous transport of newly synthesized membrane proteins through each of these compartments. In the TGN, multiple sorting events occur in which classes of membrane proteins are packaged into specific transport vesicles that are delivered to different organelles (Griffiths and Simons, 1986). Thus, there is a requirement for mechanisms in the Golgi complex that facilitate a number of different membrane protein-sorting processes (see Fig. 1).

A homologue of the erythroid isoform of β-spectrin localizes to the Golgi apparatus of nonerythroid cells (Beck *et al.,* 1994), indicating that this family of structural proteins plays a broad role in membrane organization throughout the cell. β-Spectrin is likely to play a fundamental role in Golgi complex structure and function. When the Golgi complex was fragmented without a loss of function (using nocodazole), β-spectrin association with Golgi membranes was ultimately maintained. However, when both steady-state Golgi structure and function were abolished (mitotic cells and cells treated with Brefeldin A (BFA)), β-spectrin dissociated from Golgi membranes and appeared to be diffusely distributed in the cytoplasm (Beck *et al.,* 1994). These results indicate that β-spectrin may form a cytoskeletal meshwork, analogous to the one found on the erythrocyte plasma membrane, that associates with Golgi membranes and serves to maintain the structural integrity of the Golgi complex. Dissociation of this meshwork, revealed by the accumulation of spectrin in the cytoplasm, would be expected to coincide with the loss of morphological integrity of the Golgi complex and subsequent fragmentation, as observed in our experiments.

It is also possible that β-spectrin may play a role in direct sorting of proteins into the vesicular transport pathway (see Fig. 1). After very short incubations in BFA (2 min), both β-spectrin and β-COP are rapidly lost from Golgi membranes (Beck *et al.,* 1994). β-COP is a constituent of the coatamer protein complex that mediates vesicular transport between the endoplasmic reticulum and Golgi complex, and within the Golgi stack (Kreis and Pepperkok, 1994). Although the molecular mechanism of coatamer function has not been completely defined, it is likely that it functions in a way analogous to clathrin-coated pits of the plasma membrane and the TGN. Coat proteins, therefore, appear to direct the sorting of membrane proteins through their ability to perform two key functions: selective membrane protein binding and protein oligomerization. Both of these functions are characteristic of the spectrin-based membrane skeleton (see Fig. 1).

V. THE MEMBRANE SKELETON AS A MEMBRANE PROTEIN SORTING MACHINE

We have defined four functional criteria anticipated for a membrane protein-sorting machine based on the structural and functional properties of a model sorting complex, the clathrin-coated pit: restricted distribution, 2-dimensional oligomerization, specific membrane protein binding, and a dynamic assembly state. The hallmark of these properties is the ability of a protein complex to collect classes of membrane proteins and restrict their distribution within a discrete membrane domain. The known properties of the spectrin membrane skeleton satisfy these criteria:

The spectrin membrane skeleton localizes to the plasma membrane and Golgi complex, compartments with an extensive requirement for membrane protein sorting.

Membrane skeleton proteins assemble into a 2-dimensional lattice.

The membrane skeleton lattice is restricted in its distribution to specialized membrane domains in polarized cells.

The membrane skeleton has a diversity of binding sites for membrane proteins. The extent of this repertoire is amplified by multiple isoforms of membrane skeleton proteins. The distributions of these isoforms are different, implying unique specificities for different classes of membrane proteins.

Membrane proteins that bind to the membrane skeleton exhibit restricted lateral diffusion in the plane of the lipid bilayer.

The membrane skeleton localizes to the plasma membrane and Golgi complex. Both of these membrane compartments are partitioned into structural and functional subdomains that require restricted membrane protein distributions, and both participate in extensive exchange of membrane and proteins with other organelles, requiring the segregation of resident and transient membrane proteins.

The membrane skeleton, like clathrin–AP complexes, forms a 2-dimensional oligomeric lattice with the capability to bind a selected cohort of membrane proteins, an essential step in the sorting of membrane proteins (see Fig. 1). At the plasma membrane of polarized epithelial cells, the membrane skeleton regulates the distribution of one membrane skeleton-binding protein, Na,K-ATPase, by restricting its access to the endocytic pathway, thus preserving its retention within this domain (Hammerton *et al.,* 1991). This process could also preserve resident protein populations in Golgi compartments by blocking their access to the vesicular transport machinery (see Fig. 1).

That the Golgi membrane skeleton has a dynamic assembly state is shown by treatment of cells with BFA, which results in rapid dissociation of spectrin from Golgi membranes (Beck *et al.,* 1994). BFA blocks association of other coat protein molecules (γ-adaptin, β-COP, and p200) with Golgi membranes by disrupting their normal cycle of assembly and dissociation (Orci *et al.,* 1991; Narula *et al.,* 1992; Wong and Brodsky, 1992). The effect of BFA on Golgi spectrin dissociation occurs with kinetics similar to that of other coat proteins (Beck *et al.,* 1994), suggesting that Golgi spectrin cycles on a time scale that could accompany a vesicular transport event. This leaves open the possibility that the function(s) of the membrane skeleton are not limited to retention of resident proteins (see Fig. 1).

Transport of membrane proteins from one cellular compartment to another requires the coupling of membrane protein sorting with transport vesicle formation. The extent to which the membrane skeleton can drive a membrane vesiculation process, however, is not immediately evident. As noted earlier, the membrane skeleton facilitates the retention of classes of plasma membrane proteins by serving to restrict membrane protein access to the vesicular transport machinery, a processes that is expected to be less dependent on lattice dissociation (see Fig. 1). It is noteworthy that erythrocytes derived from patients with hereditary spherocytosis, a membrane skeleton deficiency, are characterized by severe membrane destabilization resulting in membrane blebbing and shedding of membrane fragments (Palek, 1987). Therefore, in the context of the Golgi complex and the plasma membrane of nonerythroid cells, the membrane skeleton could serve to first collect classes of membrane proteins within a defined domain. Subsequent dissociation of the membrane skeleton may destabilize this membrane domain, thereby facilitating formation of membrane vesicles or tubules enriched in a discrete subset of membrane proteins.

The TGN represents a site in which a variety of classes of membrane proteins are sorted, including future resident proteins of the plasma membrane. Since the membrane skeleton binds plasma membrane proteins, it may serve to collect newly synthesized plasma membrane proteins as they are transported through the TGN. Thereby, in the TGN, membrane skeleton and clathrin–AP complexes may sequester different populations of membrane proteins for targeting to different compartments: the plasma membrane and endosome–lysosome, respectively (see Fig. 1). Furthermore, since the membrane skeleton binds a specific class of membrane proteins that is restricted in its distribution in epithelial cells, formation of these complexes in the TGN may serve to sort membrane proteins for delivery to distinct cell surface domains. Disassembly of the membrane skeleton and subsequent membrane destabilization would allow formation of transport vesicles enriched in this class of membrane proteins. In this way the membrane skeleton could serve as a membrane protein-sorting machine that directs the segregation of membrane proteins with polarized cell surface distributions (see Fig. 1).

Acknowledgments

K. A. Beck was supported by PHS grant CA09302 from the National Cancer Institute and GM15809 from the National Institutes of Health. Work in the Nelson lab was supported by grants from the National Institutes of Health and the March of Dimes Foundation.

References

Anderson, R. (1993). Caveolae: Where incoming and outgoing messengers meet. *Proc. Natl. Acad. Sci. U.S.A.* **90,** 10909–10913.

Beck, K. A., and Nelson, W. J. (1996). The spectrin-based membrane skeleton as a membrane protein sorting machine. *Am. J. Physiol.* (*Cell*). **270,** C1263–1270.

Beck, K. A., Malhotra, V., and Nelson, W. J. (1994). Golgi spectrin: Identification of an erythroid beta-spectrin homolog associated with the Golgi complex. *J. Cell Biol.* **127,** 707–723.

Bennett, V. (1992). Ankyrins: Adaptors between diverse plasma membrane proteins and the cytoplasm. *J. Biol. Chem.* **267,** 8703–8706.

Bennett, V., and Gilligan, D. (1993). The spectrin-based membrane skeleton and micron-scale organization of the plasma membrane. *Annu. Rev. Cell Biol.* **9,** 27–66.

Dahl, S. C., Greib, R. W., Fox, M. T., Edidin, M., and Branton, D. (1994). Rapid capping in alpha-spectrin deficient MEL cells from mice afflicted with hereditary hemolytic anemia. *J. Cell Biol.* **125,** 1057–1066.

Davis, J., and Bennett, V. (1990). The anion exchanger and Na^+,K^+-ATPase interact with distinct sites on ankyrin in *in vitro* assays. *J. Biol. Chem.* **265,** 17252–17256.

Devarajan, P., Scaramuzzino, D. A., and Morrow, J. S. (1994). Ankyrin binds to two distinct cytoplasmic domains of Na, K-ATPase α subunit. *Proc. Natl. Acad. Sci. U.S.A.* **91,** 2965–2969.

Drenckhahn, D., Schulter, K., Allen, D., and Bennett, V. (1985). Colocalization of band 3 with ankyrin and spectrin at the basal membrane of intercalated cells in the rat kidney. *Science* **230,** 1287–1289.

Griffiths, G., and Simons, K. (1986). The *trans* Golgi network: Sorting at the exit site of the Golgi complex. *Science* **234,** 438–443.

Gundersen, D., Orlowski, J., and Rodriguez-Boulan, E. (1991). Apical polarity of Na,K-ATPase in retinal pigment epithelium is linked to a reversal of the ankyrin-fodrin submembrane cytoskeleton. *J. Cell Biol.* **112,** 863–872.

Hammerton, R. W., Krzeminski, K. A., Mays, R. W., Wollner, D. A., and Nelson, W. J. (1991). Mechanism for regulating cell surface distribution of Na/K-ATPase in polarized epithelial cells. *Science* **254,** 847–850.

Hinck, L., Nathke, I. S., Papkoff, J., and Nelson, W. J. (1994). Dynamics of cadherin/catenin complex formation: Novel protein interactions and pathways of complex formation. *J. Cell Biol.* **125,** 1327–1340.

Kaiser, H. W., O'Keefe, E., and Bennett, V. (1989). Adducin: Ca^{++}-dependent association with sites of cell-cell contact. *J. Cell Biol.* **109,** 557–569.

Keen, J. H. (1990). Clathrin and associated assembly and disassembly proteins. *Annu. Rev. Biochem.* **59,** 415–438.

Kemler, R., and Ozawa, M. (1989). Uvomorulin-catenin complex: Cytoplasmic anchorage of a Ca^{2+}-dependent cell adhesion molecule. *Bioessays* **11,** 88–91.

Kordeli, E., Davis, J., Trapp, B., and Bennett, V. (1990). An isoform of ankyrin is localized at nodes of Ranvier in myelinated axons of central and peripheral nerves. *J. Cell Biol.* **110,** 1341–1352.

Kreis, T., and Pepperkok, R. (1994). Coat proteins in intracellular membrane transport. *Curr. Opin. Cell Biol.* **6,** 533–537.

Lazarides, E., and Nelson, W. (1983). Erythrocyte and brain forms of spectrin in cerebellum: Distinct membrane-cytoskeletal domains in neurons. *Science* **220,** 1295–1296.

Lee, J. K., Coyne, R. S., Dubreuil, R. R., Goldstein, L. S. B., and Branton, D. (1993). Cell shape and interaction defects in α-spectrin mutants of *Drosophila melanogaster. J. Cell Biol.* **123,** 1797–1809.

Lombardo, C. R., Rimm, D. L., Kennedy, S. P., Forget, B. G., and Morrow, J. S. (1993). Ankyrin independent membrane sites for non-erythroid spectrin. *Mol. Biol. Cell* **4,** 57a.

Marrs, J. A., Napolitano, E. W., Murphy-Erdosh, C., Mays, R. W., Reichardt, L. F., and Nelson, W. J. (1993). Distinguishing roles of the membrane-cytoskeleton and cadherin mediated cell-cell adhesion in generating different Na^+,K^+-ATPase distributions in polarized epithelia. *J. Cell Biol.* **123,** 149–164.

Marrs, J., Andersson-Fisone, C., Jeong, M., Cohen-Gould, L., Zurzolo, C., Nabi, I., Rodriguez-Boulan, E., and Nelson, W. (1995). Plasticity in epithelial cell phenotype: Modulation by expression of different cadherin cell adhesion molecules. *J. Cell Biol.* **129,** 507–519.

McNeill, H., Ozawa, M., Kemler, R., and Nelson, W. J. (1990). Novel function of the cell adhesion molecule uvomorulin as an inducer of cell surface polarity. *Cell* **62,** 309–316.

Morrow, J. S., Cianci, C. D., Ardito, T., Mann, A. S., and Kashgarian, M. (1989). Ankyrin links fodrin to the alpha subunit of Na^+,K^+-ATPase in Madin-Darby canine kidney cells and in intact renal tubule cells. *J. Cell Biol.* **108,** 455–465.

Narula, N., and Stow, J. (1995). Distinct coated vesicles labeled for p200 bud from trans-Golgi network membranes. *Proc. Natl. Acad. Sci. U.S.A.* **92,** 2874–2878.

Narula, N., McMorrow, I., Plopper, G., Doherty, J., Matlin, K., Burke, B., and Stow, J. (1992). Identification of a 200-kD, Brefeldin-sensitive protein on Golgi membranes. *J. Cell Biol.* **117,** 27–38.

Nelson, W. J. (1992). Regulation of cell surface polarity from bacteria to mammals. *Science* **258,** 948–955.

Nelson, W. J., and Hammerton, R. W. (1989). A membrane-cytoskeletal complex containing Na^+,K^+-ATPase, ankyrin, and fodrin in Madin-Darby canine kidney (MDCK) cells: Implications for the biogenesis of epithelial cell polarity. *J. Cell Biol.* **108,** 893–902.

Nelson, W. J., and Veshnock, P. J. (1986). Dynamics of membrane-skeleton (fodrin) organization during development of polarity in Madin-Darby canine kidney epithelial cells. *J Cell Biol.* **103,** 1751–1766.

Nelson, W. J., and Veshnock, P. J. (1987a). Ankyrin binding to Na+,K+-ATPase and implications for the organization of membrane domains in polarized cells. *Nature* (*London*) **328,** 533–536.

Nelson, W. J., and Veshnock, P. J. (1987b). Modulation of fodrin (membrane skeleton) stability by cell-cell contact in Madin-Darby canine kidney epithelial cells. *J. Cell Biol.* **104,** 1527–1537.

Nelson, W. J., Shore, E. M., Wang, A. Z., and Hammerton, R. W. (1990). Identification of a membrane-cytoskeletal complex containing the cell adhesion molecule uvomorulin (E-cadherin), ankyrin, and fodrin in Madin-Darby canine kidney epithelial cells. *J Cell Biol* **110,** 349–357.

Orci, L., Tagaya, M., Amherdt, M., Perrelet, A., Donaldson, J., Lippincott-Schwartz, J., Klausner, R., and Rothman, J. (1991). Brefeldin A, a drug that blocks secretion, prevents the assembly of non-clathrin-coated buds on Golgi cisternae. *Cell* **64,** 1183–1195.

Palek, J. (1987). Hereditary elliptocytosis, spherocytosis and related disorders: Consequences of a deficiency or a mutation of membrane skeletal proteins. *Blood Rev.* **1,** 147–168.

Pfeffer, S. R., and Rothman, J. E. (1987). Biosynthetic protein transport and sorting by the endoplasmic reticulum and Golgi. *Annu. Rev. Biochem.* **56,** 829–852.

Robinson, M. (1994). The role of clathrin, adaptors and dynamin in endocytosis. *Curr. Opin. Cell Biol.* **6,** 538–544.

Rothman, J. E., and Warren, G. (1994). Implications of the SNARE hypothesis for intracellular membrane topology and dynamics. *Curr. Biol.* **4,** 220–233.

Sheetz, M. P., Schindler, M., and Koppel, D. (1980). Lateral mobility of integral membrane proteins is increased in spherocytic erythrocytes. *Nature* (*London*) **285,** 510–512.

Wong, D. H., and Brodsky, F. M. (1992). 100-kD proteins of Golgi- and trans-Golgi network-associated coated vesicles have related but distinct membrane binding properties. *J. Cell Biol.* **117,** 1171–1179.

CHAPTER 3

Interactions between Microtubules and Intracellular Membranes: The Roles of Microtubule-Based Motors and Accessory Proteins

Trina A. Schroer and Steven R. Gill
Department of Biology, The Johns Hopkins University, Baltimore, Maryland 21218

I. INTRODUCTION

It is well established that the microtubule cytoskeleton both supports and directs the dynamic movements of nearly all subcellular membranes.

Two classes of microtubule motors, the kinesins and cytoplasmic dyneins, are known to provide the driving force behind a wide variety of these motile events. Yet the full complement of cytosolic and membrane-associated molecules that participate in binding, motility, and release of membranes from microtubules remains largely undefined. In this chapter, we discuss recent developments in this area, defining the motors and accessory components thought to be involved in endomembrane motility, and speculate on the structure and composition of the membrane–microtubule interface for stable and motile membranes.

II. THE MICROTUBULE SUBSTRATE

A. Polarity and Organization

On the basis of their subcellular localization and intrinsic structural polarity, microtubules are the cytoskeletal filament best suited to serve as a substrate for rapid transport of intracellular components over long distances. The process of membrane translocation is taken to extreme in neurons, where it is referred to as fast axonal transport, a process upon which the distal regions of the neuron are absolutely dependent for survival. In most cells, cytoplasmic microtubules are found in one of two basic arrangements that derive from the location and precise arrangement of their nucleating material. Microtubule minus (i.e., slow-growing) ends usually remain associated with this nucleating material, and their plus (i.e., rapidly growing and more dynamic) ends project into distal regions of the cell. In the arrangement seen in amebae and a wide variety of cultured cells (most notably fibroblasts), microtubules radiate from a single, perinuclear focus known as the microtubule-organizing center (MTOC). The microtubule plus ends are thus distributed throughout the cell periphery while the minus ends cluster near the nucleus, providing a clearly oriented framework for movement of cytoplasmic components (Heidemann and McIntosh, 1980). In the axons of nerve cells, large numbers of parallel microtubules then project toward the synapse (Euteneuer and McIntosh, 1981; Heidemann *et al.,* 1981). Microtubules in polarized epithelial cells take on a slightly different arrangement. In simple epithelia, the microtubule-nucleating material is not clustered into a single focus but is instead delocalized under the apical membrane (the side facing the external environment; Tucker *et al.,* 1992; Meads and Schroer, 1995; Vogl *et al.,* 1995), so that the microtubules project toward the basal surface (the side facing the body's internal environment), forming a cylinder of filaments running parallel to the lateral walls of the cell (Bacallao *et al.,* 1989). Unorganized "mats" of microtubules lacking a discernible focus are also present in the apical and basal cytoplasm. Epithelial cells with more elaborate morphologies (e.g., retinal pigmented epithelium, Sertoli cells, inner pillar cells of the cochlea) exhibit the same overall arrangement but contain large numbers of microtubules of uniform polarity within their elongated neck regions (Troutt and Burnside, 1988; Tucker *et al.,* 1993; Vogl *et al.,* 1995). Despite the variation in subcellular location and orientation of microtubules in these different cell types, microtubules in all are likely to participate in the movement and anchorage of various internal membranes.

Our knowledge of microtubule orientation allows us to make predictions about the motors that participate in different transport processes. For example, translocation from the center of a fibroblast outward should utilize a plus end–directed motor, while movement from the periphery toward the cell center is predicted to require a minus end–directed motor. Likewise, anterograde (toward the synapse) fast axonal transport is plus end–directed motility while retrograde (back to the cell body) fast axonal transport is minus end–directed. In polarized epithelia, transcellular transport (transcytosis) from the apical to the basal surface is expected to use a plus end–directed motor and transcytosis toward the cell apex a minus end–directed motor; nondirectional movement of vesicles in apical or basal cytoplasm might involve either type of motor. Such predictions have guided research efforts for many years by suggesting useful experimental systems and defining specific testable hypotheses.

B. Dynamics and Stability

Thinking about microtubules as "roadways" for subcellular movements may lend the mistaken impression that these filaments are long-lived, stable structures. While a subset of cytoplasmic microtubules do appear to be stable, the majority (at least in cultured cells) are remarkably dynamic (Schulze and Kirschner, 1986). Direct observation of individual microtubules in living cells reveals that the filaments rarely remain of fixed length but instead are constantly growing and shrinking (Cassimeris *et al.,* 1988; Sammak and Borisy, 1988). Growth occurs at a fairly steady rate for a period of minutes; when growth ceases, the microtubule depolymerizes rapidly and completely. The lifetime of a dynamic microtubule in a fibroblast is, on average, between 10 and 20 min. Microtubules in other cell types, such as epithelia, are less dynamic and consequently longer lived (Shelden and Wadsworth, 1993). That microtubules in cultured cells can have widely varying half-lives, from minutes to hours, suggests that microtubules in postmitotic somatic cells may be even longer lived.

The fact that most cytoplasmic microtubules are transient, ephemeral structures means that specific mechanisms must exist to maintain them in a nondynamic state. The tubulin subunits in stable microtubules are subject to posttranslational, covalent modifications, including acetylation and C-terminal detyrosination (reviewed in Bulinski and Gundersen, 1991). These modifications correlate with stability but are not believed to be its root cause (Maruta *et al.,* 1986; Khawaja *et al.,* 1988; Webster *et al.,* 1990). Rather, microtubules destined to become long lived are somehow targeted for modification; once modified, they may bind proteins that further increase their stability. Potential stabilizing factors include the microtubule-associated proteins (MAPs), which profoundly inhibit filament depolymerization *in vitro* (Pryer *et al.,* 1992). Stable microtubules are usually short and bent or curved and they rarely extend into the cell periphery, in marked contrast to the extended, radial distribution of the dynamic microtubule population (Schulze and Kirschner, 1987). This unusual morphology may be caused by centripetal cytoplasmic flux (Terasaki and Reese, 1994; Mikhailov and Gundersen, 1995) occurring over the filament lifetime.

It is not known whether stable and dynamic microtubules show differences in terms of their associations with membranous organelles. Stabilizing proteins such as MAP2 and tau may impede motor access to the microtubule surface, inhibiting motility (Lopez and Sheetz, 1993; Hagiwara *et al.,* 1994) but favoring binding of membranes to stable microtubules. This may contribute to the localization of the Golgi apparatus and lysosomes near the MTOC (see Section IV,B). Microtubule dynamics need not necessarily interfere with organelle motility but may, in fact, account for some types of intracellular organelle movements. Membranes in *Xenopus* extracts possess specialized subdomains that remain associated with the plus end of a dynamic microtubule, its excursions causing the membranes to be drawn into long tubules (Waterman-Storer *et al.,* 1995a). This may contribute to the motility *in vivo* of tubular membrane compartments such as the endoplasmic reticulum (ER), ER–Golgi intermediate compartment, *trans*-Golgi network, and endosomes.

III. MICROTUBULE-BASED MOTORS

A. Introduction

The last 10 years have seen the discovery of dozens of microtubule-based motors. For our purposes, they can be divided into two classes, those that drive movement toward the plus ends of microtubules (toward the periphery of a fibroblast, a neuronal synapse, or the basal membrane of an epithelial cell) and those that drive movement in the opposite direction (toward the center of a fibroblast, from the synapse to the cell body, or toward the apical surface of an epithelial cell). The first cytosolic microtubule-based motor discovered, kinesin, is plus end directed, as are most members of the rapidly growing kinesin superfamily (for reviews, see Goldstein, 1993; Hoyt, 1994). While some plus end–directed kinesin-related proteins (e.g., the BimC family; Enos and Morris, 1990; Hagan and Yanagida, 1992; Hoyt *et al.,* 1992; Sawin *et al.,* 1992; Heck *et al.,* 1993) are involved in mitotic spindle dynamics, others power organelle movements (Hall and Hedgecock, 1991; Otsuka *et al.,* 1991; Kondo *et al.,* 1994; Nangaku *et al.,* 1994; Noda *et al.,* 1995; Okada *et al.,* 1995; Yamazaki *et al.,* 1995). The ubiquitously distributed cytoplasmic dynein is a soluble relative of axonemal dynein (Lye *et al.,* 1987; Paschal *et al.,* 1987), the enzyme responsible for ciliary and flagellar motility. Both cytosolic and axonemal dyneins drive movement toward the minus ends of microtubules (Sale and Satir, 1977; Paschal and Vallee, 1987). An unusual dynein variant isolated from the slime amoeba, *Reticulomyxa,* has been reported to move bidirectionally on microtubules (Euteneuer *et al.,* 1988). Another class of minus end–directed motor includes the kinesin-related proteins ncd (McDonald *et al.,* 1990; Walker *et al.,* 1990) and Kar3p (Endow *et al.,* 1994; Middleton and Carbon, 1994); while these proteins are believed to participate only in meiosis and mitosis, distantly related cousins may drive membrane motility. To make matters more complicated, all microtubule motors characterized to date contain multiple subunits, leaving us with a dizzying array of polypeptides to consider.

Previous studies of endomembrane movement on microtubules have focused on the roles of the archetypal motors, kinesin and cytoplasmic dynein, and it has only lately become clear that other kinesin- (and perhaps dynein-) related motors participate in intracellular membrane dynamics. Because less is known of the functions of these other motors, the remainder of this chapter focuses on kinesin and cytoplasmic dynein and their involvement in microtubule-based membrane motility.

B. Structure

1. Kinesin

Conventional kinesin is a heterotetramer containing two heavy chains of $M_r \sim 120,000$ and two light chains of $M_r \sim 64,000$ (reviewed in Walker and Sheetz, 1993). Several light-chain splice variants have been characterized (Cyr *et al.,* 1991) but their functional significance remains unknown. The heavy chain contains the so-called motor domain (the site for microtubule and ATP binding) within its amino-terminal ~ 350 amino acids. This part of the molecule comprises the globular heads seen in the electron microscope (Amos, 1987; Hirokawa *et al.,* 1989; Scholey *et al.,* 1989). The middle third of the heavy chain folds into a long α-helix that dimerizes by coiled-coil interactions into a stalk of ~55 nm in length. Near the middle of the heavy chain sequence is a non-α-helical hinge that allows the kinesin molecule to bend back on itself (Hackney *et al.,* 1992). The c-terminal one-third of the heavy chain is predicted to be another globular domain that binds kinesin light chains (Gauger and Goldstein, 1993) to form the amorphous tail domain. Kinesin structure is shown schematically in Fig. 1.

2. Cytoplasmic Dynein

At a M_r of more than 1 million, cytoplasmic dynein (Fig. 1) is a significantly larger and more complex molecule than kinesin. In composition and ultrastructure, it is similar to two-headed outer arm dyneins in flagella and cilia (Vallee *et al.,* 1988). The molecule comprises two M_r ~520,000 heavy chains, three or four M_r ~74,000 intermediate chains, several M_r ~50,000–60,000 light intermediate chains, and at least four M_r 8,000–22,000 light chains (Schroer lab, unpublished data). The heavy, intermediate, and light chains are similar in amino acid sequence to their axonemal counterparts (Holzbaur and Vallee, 1994; King and Patel-King, 1995). Electron microscopic analysis shows the molecule to contain a pair of large globular heads with smaller projecting domains (Amos, 1989) that are believed to be the sites for microtubule and ATP binding and hydrolysis (Goodenough and Heuser, 1982). These are connected via a filamentous stalk domain to a third subdomain, the base, thought to contain the intermediate and associated light chains (Schroer lab, unpublished data; Mitchell and Rosenbaum, 1986; King and Witman, 1989). Based on the model of axonemal outer arm dynein (King and Witman, 1990; King *et al.,* 1991), the base has been proposed to be the site for binding membranes and other cargoes. In keeping with this hypothesis, the cytoplasmic dynein intermediate chains were recently found to interact specifically with dynactin, an accessory factor thought to function as the anchor to cargo (see Section III,C,1,b). The light intermediate chains are tightly associated with the heavy chains (Gill *et al.,* 1994), but their submolecular location has not been determined.

C. Regulation

Video microscopic analysis of living cells reveals that subcellular membranes are in almost constant movement, undergoing both inward and outward excursions on microtubules the frequency of which are subject to

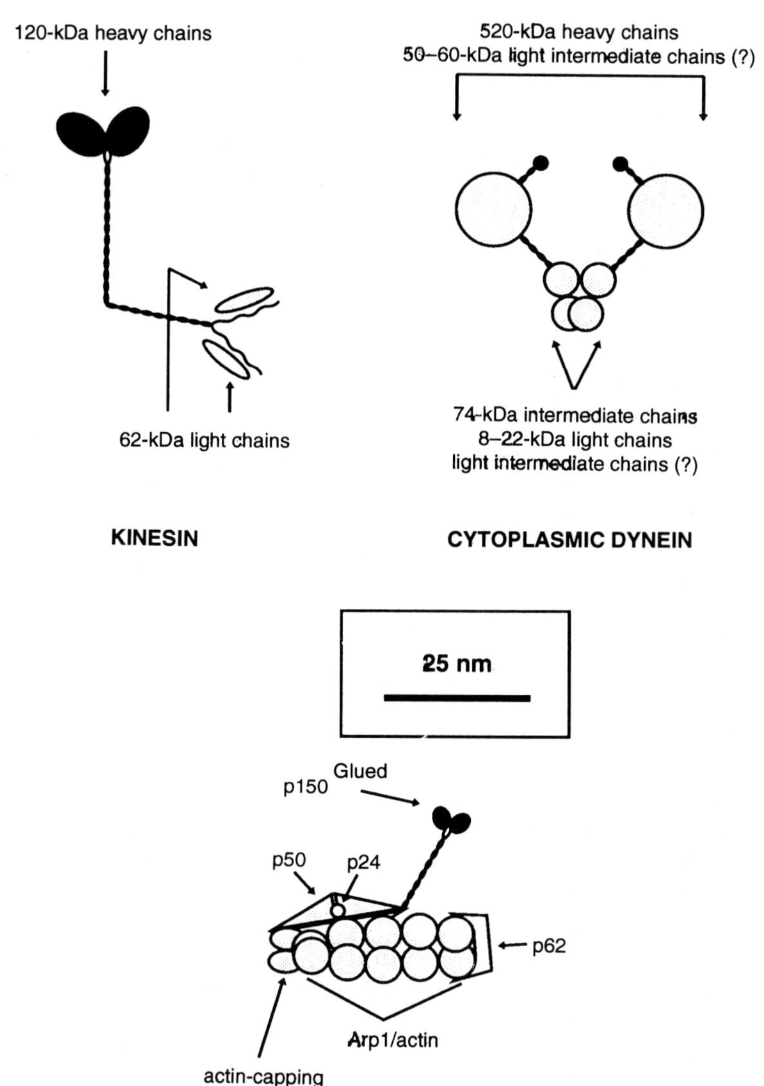

FIGURE 1 Schematic representations of the morphologies of kinesin, cytoplasmic dynein, and dynactin (drawn approximately to size; bar = 25 nm). The overall structures and organization of subunits within each molecule are indicated. Microtubule-binding domains (black) are at the top and cargo binding domains are at the bottom.

physiological regulation (Hamm-Alvarez *et al.,* 1993). This suggests that plus end- and minus end-directed transport activities are carefully controlled. Although the mechanisms for regulation of cytosolic microtubule motor activities are only beginning to be understood, both motor-associated accessory proteins and phosphorylation of motors and their accessory proteins appear to play important roles.

1. Phosphorylation

Protein phosphorylation is a rapid, readily reversible regulatory mechanism that governs the activity of many enzymes and structural proteins. Studies of the effects of phosphorylation on the activities of kinesin and cytoplasmic dynein have taken a number of different tacks. One body of work has focused on treating cells with drugs that perturb phosphorylation balance, most commonly okadaic acid, a cell-permeant inhibitor of phosphatase activity. Effects on microtubule-based motility are then correlated with the implied changes in phosphorylation state. A second approach is to isolate motors from cells under different physiological conditions (e.g., interphase vs mitosis, rapidly growing vs quiescent, etc.) and determine the phosphorylation state of the different subunits. Both approaches have provided information as to which kinesin and dynein subunits are phosphorylated, and we are slowly beginning to gain an understanding of how such modifications affect vesicle motility.

a. Kinesin. Studies in cultured neurons have revealed that the heavy and light chains of kinesin as well as the putative kinesin membrane receptor, kinectin (see Section III,D), are phosphorylated *in vivo* (Hollenbeck, 1993). The sites for heavy and light chain phosphorylation appear to be clustered in what may be regulatory subdomains within each molecule. Phosphorylation has been reported to have differing effects on membrane binding. One study reported a reduction in synaptic vesicle binding (Sato-Yoshitake *et al.,* 1992), but analysis of the effects of phosphorylation of kinesin heavy chain selectively showed this species to be preferentially associated with membranes (Lee and Hollenbeck, 1995). Either way, it seems clear that phosphorylation may provide a mechanism for regulating kinesin association with cargo. It is not yet known whether phosphorylation affects kinesin motor activity directly, although an effect on kinesin ATPase has been observed (Matthies *et al.,* 1993).

b. Cytoplasmic Dynein. The large number of subunits in the cytoplasmic dynein molecule make analysis of phosphorylation challenging, but several groups have made progress on this question. Cell-free extracts of *Xenopus* eggs are an especially good *in vitro* system for analysis of how phosphorylation affects different cell-biological phenomena, as the transitions between interphase and mitosis are absolutely dependent on a set of specific phosphorylation events. *In vitro* vesicle motility studies indicate that cytoplasmic dynein activity predominates in this system, and dynein-based vesicle motility appears to be regulated at two levels by phosphorylation. Membrane movements are dramatically inhibited when cells enter mitosis; at this time, the M_r ~60,000 cytoplasmic dynein light intermediate chain shows increased phosphorylation and dynein dissociates from membranes (Niclas *et al.,* 1996). At the end of mitosis, membranes resume their normal interphase level of activity, presumably by rebinding dynein. Treatment of interphase extracts with okadaic acid (or other phosphatase inhibitors) causes further stimulation of membrane motility that does not correlate with an increase in membrane-associated dynein but is instead thought to result from enhancement of the activity of the membrane-bound pool (Allan, 1995). Here, the target of phosphorylation has not been defined.

Cytoplasmic dynein phosphorylation has also been investigated in systems in which organelle motility is subject to other types of physiological regulation. Serum starvation, which has an inhibitory effect on intracellular vesicle transport (Hamm-Alvarez *et al.,* 1993), is correlated with increased levels of dynein heavy chain phosphorylation (Lin *et al.,* 1994) and the redistribution of dynein from lysosomes to cytosol (Lin and Collins, 1993). Under these circumstances, lysosome motility is not altered, suggesting that heavy chain phosphorylation may alter membrane binding but not activity. Further evidence that heavy chain phosphorylation correlates with reduced cytoplasmic dynein activity comes from neuronal systems. In axons, cytoplasmic dynein must be actively transported from the cell body to the nerve terminal to provide a source of motor for the return journey. It is assumed that anterogradely transported dynein moves in an inactive state, and this pool shows increased levels of heavy chain phosphorylation as compared to dynein in brain (Dillman and Pfister, 1994).

2. Accessory Proteins

a. Kinesin-Associated Proteins. The use of *in vitro* assays in which vesicle motility is reconstituted from isolated components has made it possible to isolate cytosolic factors that influence the activity of kinesin and cytoplasmic dynein. In an assay for microtubule-dependent movement of secretory granules from cytotoxic T lymphocytes, the membranes were found to move via kinesin (Burkhardt *et al.,* 1993). Motility was affected by three polypeptides of M_r 150,000, 79,000, and 73,000 that copurify with kinesin (McIlvain *et al.,* 1994). Treatment of cells with the phosphatase inhibitor okadaic acid caused these *k*inesin-*a*ssociated *p*roteins (KAPs) to become hyperphosphorylated, resulting in a twofold stimulation of the frequency of kinesin-driven movements. KAP phosphorylation specifically affected motor activity and did not appear to alter kinesin–membrane binding.

b. Dynactin. This M_r 1,200,000 multiprotein assembly comprising 10 distinct subunits (reviewed in Schroer, 1996; Schroer *et al.*, 1996) was first identified as an activator of cytoplasmic dynein-based vesicle motility *in vitro* (Gill *et al.*, 1991; Schroer and Sheetz, 1991). Dynactin is required for a basal level of movement of chick embryo fibroblast vesicles on microtubules; with dynein alone, vesicles bind microtubules but do not move. Other cytosolic factors stimulate motility further, increasing both the frequency and velocity of transport. Mutational analysis of the role of different dynactin subunits confirms that the protein is required for cytoplasmic dynein-based motility *in vivo* (Clark and Meyer, 1994; Muhua *et al.*, 1994; Plamann *et al.*, 1994; McGrail *et al.*, 1995).

Molecular shadowing and antibody decoration experiments (Schafer *et al.*, 1994) reveal that dynactin is composed of two structural domains, a 37-nm filament that resembles a short actin polymer and a laterally associated sidearm (Fig. 1). The actin-like filament contains one monomer of conventional actin and nine monomers of the *a*ctin-*r*elated *p*rotein Arp1 (also known as centractin; see Clark and Meyer, 1992). The barbed (plus) end of the Arp1–actin filament contains a molecule of actin-capping protein, an α/β heterodimer of M_r ~69,000. Actin-capping protein blocks barbed-end polymerization of actin filaments but can nucleate actin polymerization *in vitro*. The dynactin subunit p62 is associated with the opposite end of the filament, suggesting that this protein may be a pointed (minus) end–binding factor. p62 appears to be a novel protein, as its amino acid sequence bears no homology to any previously identified protein. The location of the lone actin monomer within the Arp1 filament is unknown. It may copolymerize randomly with Arp1, or it may serve to nucleate Arp1 polymerization by forming a complex with p62 or capping protein. Analysis of the polymerization properties of purified Arp1 with and without these other proteins should shed some light on this question.

The laterally associated dynactin sidearm is composed of two subdomains, a distal rodlike portion that terminates in a pair of small globular heads and a flexible shoulder that lies on the Arp1 filament. On the basis of dissociation (Schroer lab, unpublished data) and antibody decoration (Schafer *et al.*, 1994) studies, the two subdomains together are thought to contain two molecules of p150^{Glued}, four molecules of p50, and the one molecule of p24 present in dynactin. In further support of a direct interaction between p150^{Glued} and p50 within the dynactin sidearm, overexpression of exogenous p50 in cultured cells causes both p50 and p150^{Glued} to dissociate from the Arp1 filament (Echeverri *et al.*, 1996). The two p150^{Glued} molecules are assumed to dimerize via coiled-coil interactions between the two long α-helices present in each monomer. This dimer assembles onto the Arp1 filament with its N-termini extending outward; the N-terminal coiled coil is thought to comprise the rod of the sidearm and the extreme N-terminal domains the globular heads. The shoulder is likely composed of the remainder of the p150^{Glued} dimer and the p50 and p24 subunits.

The sidearm, specifically its p150^{Glued} subunit, is thought to be the site at which dynactin associates with dynein and microtubules. The N-terminus of p150^{Glued} contains a conserved microtubule-binding site also present in the endosome–microtubule *c*ytoplasmic *l*inker *p*rotein CLIP-170 (see Section V,B; Pierre *et al.*, 1992). This region of p150^{Glued} binds microtubules both *in vitro* and when overexpressed in cultured cells (Waterman-Storer *et al.*, 1995b). Like CLIP-170–microtubule binding (Rickard and Kreis, 1991), the p150^{Glued}–microtubule interaction is likely to be regulated by phosphorylation. Using affinity chromatography and blot overlay techniques, p150^{Glued} has also been shown to bind to dynein intermediate chain (Karki and Holzbaur, 1995; Vaughan and Vallee, 1995). The interacting domains have been mapped to the N-terminal 200 amino acids of dynein intermediate chain (Vaughan and Vallee, 1995) and the middle region (approximately amino acids 500–800) of p150^{Glued} (E. Holzbaur, personal communication), suggesting that the base of the dynein molecule may be bound to the shoulder of the dynactin sidearm. This leaves the Arp1 filament as the dynactin component most likely to bind cargo (reviewed in Schroer, 1996; Schroer *et al.*, 1996). It is not yet known which polypeptide subunit of the Arp1 filament interacts with cargo or what cargo components are involved.

D. *Mechanisms for Membrane Binding* (for review, see Vallee and Sheetz, 1996)

Kinesin is thought to bind membranes by associating with the transmembrane protein kinectin (Toyoshima *et al.*, 1992; Kumar *et al.*, 1995; Yu *et al.*, 1995; reviewed in Burkhardt, 1996). The c-terminus of the heavy chain (Skoufias *et al.*, 1994) and the light chains (Yu *et al.*, 1992) both appear to contribute to binding, indicating that the membrane-binding site lies within the feathery tail of the molecule. The kinesin–membrane interaction may be stabilized by other factors, particularly in axons (Schnapp *et al.*, 1992), where cell survival depends on the maintenance of direction preference of anterogradely transported membranes. That a subpopulation of axonal membranes carry tightly associated kinesin is consistent with this hypothesis. The cytoplasmic dynein–membrane interaction is equally complicated. Isolated dynein can bind artificial phospholipid vesicles (Lacey and Haimo, 1994; Ferro and Collins, 1995), suggesting that this motor may have the capacity to bind directly to the membrane bilayer. However, dynactin, a peripherally associated protein, is required for dynein-based motility both *in vitro* and *in vivo*. As mentioned earlier, dynactin is believed to function by binding dynein to the membrane surface; given the size and complexity of the dynein molecule, it is entirely possible that both dynactin and lipid binding contribute to its association with membranes.

IV. MICROTUBULES IN MEMBRANE LOCALIZATION AND DYNAMICS

A. Introduction

For many years, our view of the cell was dominated by the static electron and light microscopic images obtained from fixed and processed cells. Different membrane compartments were seen to occupy distinct subcellular locations that, in many cases, were the same in a wide variety of cell types. As the functions of these various compartments were defined, it became apparent that material passed from one compartment to another, and significant effort has since been aimed at understanding the dynamic behavior of organellar membrane proteins and lipids and the soluble molecules contained within their lumina. It is now clear that the membranes of the central membrane system are in constant communication and that material can move between compartments through tubules as well as discrete vesicles. Microtubules are intimately involved in endomembrane dynamics, serving as the substrate for efficient and targeted movement of vesicles and tubules as well as the means for maintaining and anchoring different organelles in their characteristic locations (for review, see Cole and Lippincott-Schwartz, 1995).

Although many forms of endomembrane motility depend on microtubules, the contribution of actomyosin-based transport to membrane dynamics must not be ignored. Axonally transported membranes can switch between filament systems (Kuznetsov *et al.*, 1992), and membranes in the apical cytoplasm of epthelial cells have been proposed to do the same (discussed by Fath and Burgess in Chapter 4, this volume; Fath *et al.*, 1994). That a myosin-related protein drives particle movements in *Drosophila* embryos (Mermall *et al.*, 1994) further supports the idea that actomyosin-dependent intracellular transport is a ubiquitous process. It would seem that membranes that are destined for movement have the capacity to do so on both microtubules and actin filaments.

B. The Secretory Membrane System

The starting point for newly synthesized components of the central membrane system is the ER, an interconnected network of membrane tubules and cisternae distributed throughout cytoplasm. The ER spreads from the nuclear envelope to the periphery and is excluded only from lamellipodial regions (historically referred to as ectoplasm). Its lacy, reticular structure depends on intact cytoplasmic microtubules (Terasaki *et al.,* 1986). Direct visualization of ER in living cells and of ER-like networks reconstituted *in vitro* has revealed this organelle to be remarkably dynamic, its steady-state distribution maintained by the movement, fusion, and branching of membrane tubules (Dabora and Sheetz, 1988; Lee and Chen, 1988). Both plus end– and minus end–directed motor activities contribute to ER membrane dynamics (Vale and Hotani, 1988; Hollenbeck, 1989; Henson *et al.,* 1992; Feiguin *et al.,* 1994; Allan, 1995).

Membrane exiting the ER moves into the ER–Golgi intermediate compartment (ERGIC), another network of membrane tubules and cisternae (Schweizer *et al.,* 1990; Saraste and Kuismanen, 1992) running from the cell periphery to the *cis* region of the centrally located Golgi apparatus. Its subcellular distribution depends on intact microtubules (Saraste and Svensson, 1991). Movement of material through the ERGIC requires microtubules and, because it represents a type of centripetal movement, would be predicted to utilize a minus end–directed motor. Recycling of membrane from the *cis*-Golgi region to the ER is also microtubule dependent (Lippincott-Schwartz *et al.,* 1990) but occurs in the opposite direction to forward flow. As predicted, Golgi–ER recycling has been shown to utilize kinesin (Feiguin *et al.,* 1994; Lippincott-Schwartz *et al.,* 1995), which may explain the unexpected association of this motor with Golgi membranes (Marks *et al.,* 1994; Schmitz *et al.,* 1994).

In many cell types, the Golgi apparatus is organized around one side of the nucleus, closely associated with the MTOC. Careful inspection reveals the Golgi apparatus to be composed of a set of individual stacks of membrane cisternae, each of which is interconnected by a network of tubules and vesicles (Rambourg and Clermont, 1990; Ladinsky *et al.,* 1994). In the absence of microtubules, these fully functional, individual Golgi stacks disperse throughout the cell (Thyberg and Moskalewski, 1985; Turner and Tartakoff, 1989). When microtubules repolymerize, the Golgi membranes can be seen to move back toward the MTOC (Ho *et al.,* 1989), suggesting that the maintenance of the Golgi apparatus at this site depends on minus end–directed, microtubule-based transport. In support of this hypothesis, recruitment of exogenously added Golgi membranes to the juxtanuclear region of permeabilized cells was found to depend on cytoplasmic dynein (Corthésy-Theulaz *et al.,* 1992).

Because Golgi function does not seem to depend on its subcellular location, the question of why the Golgi apparatus normally resides near the cell center is a mystery. This region of the cell also contains stable microtubules that appear to be intimately associated with Golgi membranes (Skoufias *et al.,* 1990; Letourneau and Wire, 1995) as well as the ERGIC (Mizuno and Singer, 1994). Microtubule-stabilizing and other proteins bound to these microtubules may anchor the Golgi apparatus near the center of the cell.

The next stop on the pathway is the *trans*-Golgi network (TGN), another interconnected network of membrane tubules and cisternae adjacent to the Golgi apparatus that is also assumed to rely on cytoplasmic dynein activity to maintain its juxtanuclear location. The TGN is an extremely dynamic structure whose morphology is, in part, the result of frequent extension of tubules that appear to correspond to transport vesicle precursors (Allan and Vale, 1994; Ladinsky *et al.,* 1994). As the extension of tubules from the TGN and earlier components of the Golgi apparatus are most likely driven by a plus end–directed motor such as kinesin (Lippincott-Schwartz *et al.,* 1995), it would appear that the TGN also has the capacity to bind and move via both motors.

After leaving the TGN, membranes have a choice of destinations. Lysosomal proteins and components of the endosomal system traffic to endosomes, where they are sorted and redistributed. Secretory and membrane proteins move in vesicles either to a storage location or directly to the cell surface. Although most secreted and plasma membrane proteins will eventually reach the surface even in the absence of microtubules, export from the Golgi apparatus is made considerably more efficient by microtubule-based transport. Such movement would be predicted to utilize kinesin or another plus end–directed motor. Secretory granules isolated from cytotoxic T lymphocytes (Burkhardt *et al.,* 1993) and chromaffin cells (Urrutia *et al.,* 1991) interact with kinesin *in vitro,* as they are assumed to do *in vivo.* Vesicles exiting the TGN in the polarized Madin–Darby canine kidney epithelial cell line require the activities of both kinesin and cytoplasmic dynein to arrive at the apical surface (Lafont *et al.,* 1994); a similar export mechanism is likely to operate in other epithelia (see Fath and Burgess, Chapter 4, this volume; Fath *et al.,* 1994). An analogous, although not identical, type of centrifugal organelle movement, pigment granule dispersion in chromatophores, also appears to utilize kinesin (Rodionov *et al.,* 1991).

While many of the events described above have been shown to involve conventional kinesin, other plus end–directed motors must be kept in mind. Different types of anterogradely transported organelles appear to utilize different kinesin-related proteins. The proteins unc104 in *Caenorhabditis elegans* (Hall and Hedgecock, 1991) and the closely related KIF1A in mouse (Okada *et al.,* 1995) drive anterograde transport of synaptic vesicles and their precursors exclusively. The related KIF1B, also from mouse, is thought to be specific for mitochondrial motility (Nangaku *et al.,* 1994), while murine KIF2 (Noda *et al.,* 1995) and KIF3 (Yamazaki *et al.,* 1995) drive movement of yet other populations of axonal membranes. These (Pesavento *et al.,* 1994) and other kinesin-related proteins (Sekine *et al.,* 1994) are good candidates for plus end–directed membrane motors in all cells.

C. The Endocytic Membrane System

Lysosomes are found in the central region of many cells, a subcellular localization that depends on microtubules (Matteoni and Kreis, 1987). Like the Golgi apparatus, lysosome positioning likely involves both minus end–directed motility and stable binding to microtubules. The tubular lysosomes that extend to the periphery of macrophages also move on microtubules (Swanson *et al.,* 1987), but in this case via kinesin (Hollenbeck and Swanson, 1990). In other cells, this process can be mimicked by acidifying the extracellular medium (Heuser, 1989); under these circumstances lysosome movement to the periphery is microtubule (Heuser, 1989) and kinesin (Feiguin *et al.,* 1994) dependent. Lysosomes thus appear to have the capacity for translocation by kinesin and cytoplasmic dynein.

Early light microscopic analysis of cells undergoing endocytosis revealed that endocytic markers move from peripheral regions of the cell to the cell center (Freed and Lebowitz, 1970; Willingham and Pastan, 1978; Herman and Albertini, 1984). These two morphologically defined compartments are now known to correspond to early and late endosomes; early endosomes localize in the periphery while late endosomes reside near lysosomes at the cell center (reviewed in Gruenberg and Maxfield, 1995). This microtubule-dependent, centripetal movement utilizes a vesicular intermediate known as the endosomal carrier vesicle (Gruenberg *et al.,* 1989) that is transported by cytoplasmic dynein (Goltz *et al.,* 1992; Aniento *et al.,* 1993; Oda *et al.,* 1995). In polarized epithelia that contain two distinct populations of early endosomes, one derived from the basal surface and one from the apical surface (Bomsel *et al.,* 1989), movement to the common late endosome (Parton *et al.,* 1989) requires both kinesin and cytoplasmic dynein (Bomsel *et al.,* 1990). As endocytic sorting is a critical intermediate step in transcellular movement (i.e., transcytosis), this most likely explains the dependence of transcytosis on microtubules (Breitfeld *et al.,* 1990; Hunziker *et al.,* 1990). At present, the role of microtubules earlier in endocytosis remains unclear (Gruenberg *et al.,* 1989; but see Jin and Snider, 1993; Thatte *et al.,* 1994).

Because the endosome network is remarkably dynamic (Hopkins *et al.,* 1990), these organelles may not remain stably bound to microtubules for long periods of time. Rather, peripheral early endosomes and endosomal carrier vesicles have been proposed to be loaded onto microtubules transiently via CLIP-170 before they acquire the capacity for minus end–directed motility (see Section V,B; Pierre *et al.,* 1992). The behavior of phagosomes, a specialized endocytic organelle in macrophages and other phagocytic cells, has many features in common with endosome traffic (Desjardins *et al.,* 1994). *In vitro* motility studies with isolated phagosomes indicate that they exhibit stable binding to microtubules (Blocker *et al.,* 1996) as well as bidirectional movement on microtubules driven by cytoplasmic dynein and an undefined kinesin-related protein (Griffiths lab, unpublished data).

V. THE MEMBRANE–MICROTUBULE INTERFACE

A. Structural Features

Long before the molecules mediating the interactions of membranes with microtubules were identified, the morphological features of this junction were studied by electron microscopy, commonly in nerve cell axons. Here, membranous organelles such as mitochondria, lysosomes (multivesicular bodies), smooth ER, and synaptic vesicle precursors are seen to connect to microtubules via fine, filamentous bridges (Tsukita and Ishikawa, 1980; Hirokawa, 1982; Schnapp and Reese, 1982). The number of connections appears to be dependent on the size of the organelle, with elongated structures such as smooth ER and mitochondria displaying more linkages than small, spherical membranes. Small vesicles often have an appendage protruding from one end (Schnapp and Reese, 1982), suggesting that these membranes are deformed as they are pulled through axoplasm. More extensive analysis of vesicle–microtubule cross-bridges indicated they were between 15 and 30 nm in length and 5–6 nm in diameter (Miller and Lasek, 1985; Langford *et al.,* 1987).

Possible functional homologues of the protrusions seen on axonally transported vesicles are the swellings at the tips of membrane tubules moving on microtubules *in vitro* (Allan and Vale, 1994). These globular structures are proposed to be stable membrane subdomain structures enriched in microtubule motors. They are heterogeneous in shape; some are the same diameter as the tubule itself (~100 nm) and blunt, some are round and between ~150 and 500 nm in diameter, and others are complex, containing what appear to be multiple spherical subdomains. The complex, multilobed structures were only seen on tubules formed from crude Golgi membranes, suggesting that they might correspond to the precursors of vesicles that bud from tubules of the TGN (Ladinsky *et al.,* 1994). These domains may be analogous to the tip–attachment complexes that allow membrane tubules to remain bound to the plus ends of dynamic microtubules (Waterman-Storer *et al.,* 1995a).

B. Molecular Components

As alluded to earlier, membranes can associate with microtubules either stably or in a dynamic fashion, depending on the binding properties of the microtubule cross-link. Dynamic interactions are assumed to involve microtubule motors (or tip–attachment complexes) while stable interactions utilize microtubule-binding proteins. The CLIPs are a class of protein isolated by virtue of their ability to bind both microtubules and membranes. CLIP-170 (see also Section III,C,2,b) is a candidate for a protein that loads endocytic membranes onto microtubules in the cell periphery so that subsequent recruitment of minus end–directed motor activity will yield centripetal movement (Pierre *et al.,* 1992). The CLIP-170–microtubule interaction is regulated by phosphorylation, with only the dephosphorylated form being capable of microtubule binding (Rickard and Kreis, 1991). Other CLIPs may perform this function for other endomembranes such as phagosomes. Mitochondria bind MAP2 at specific sites on their surface (Linden *et al.,* 1989), providing another mechanism for membrane-microtubule interactions.

C. The Switch between Stable and Motile Binding

Once it has become stably bound to a microtubule, how does an organelle become motile? Phosphorylation is known to reduce the affinities of both CLIP-170 and MAP2 (Lopez and Sheetz, 1995) for microtubules, providing a handy mechanism for releasing associated membranes. Nucleotide hydrolysis by heterotrimeric and/or small GTP-binding proteins (Pfeffer, 1992) may also regulate motility (Bloom *et al.,* 1993), perhaps by converting a docked vesicle to a motile one. In either case, motor activities must be available nearby, either already bound to the docked membrane or in its immediate vicinity. There are many cases in which membranes are thought to carry inactive motors (e.g., high cytoplasmic dynein concentrations on lysosomes (Lin *et al.,* 1994); kinesin on the ERGIC in the cell periphery

(Lippincott-Schwartz *et al.,* 1995)), control of whose activity may depend on these regulatory mechanisms.

VI. SUMMARY

The dynamics and motility of intracellular membranes are complicated processes that utilize at least two classes of cytoskeletal filament, each of which supports movements driven by a number of cytosolic motor proteins. Cells have the capacity to regulate their endomembrane dynamics in many ways—for example, as they pass through the cell cycle or during differentiation into specific somatic tissues. Many of the molecules involved in these processes have been defined, but countless more await identification and functional analysis. The next decade is sure to be as fruitful and illuminating as the last in terms of our understanding of the molecular mechanisms that underlie intracellular membrane motility.

References

Allan, V. (1995). Protein phosphatase 1 regulates the cytoplasmic dynein-driven formation of endoplasmic reticulum networks in vitro. *J. Cell Biol.* **128,** 879–891.

Allan, V., and Vale, R. (1994). Movement of membrane tubules along microtubules in vitro: Evidence for specialised sites of motor attachment. *J. Cell Sci.* **107,** 1885–1897.

Amos, L. A. (1987). Kinesin from pig brain studied by electron microscopy. *J. Cell Sci.* **87,** 105–111.

Amos, L. A. (1989). Brain dynein crossbridges microtubules into bundles. *J. Cell Sci.* **93,** 19–28.

Aniento, F., Emans, N., Griffiths, G., and Gruenberg, J. (1993). Cytoplasmic dynein-dependent vesicular transport from early to late endosomes. *J. Cell Biol.* **123,** 1373–1387.

Bacallao, R., Antony, C., Dotti, C., Karsenti, E., Stelzer, E. H. K., and Simons, K. (1989). The subcellular organization of Madin-Darby canine kidney cells during the formation of a polarized epithelium. *J. Cell Biol.* **109,** 2817–2832.

Blocker, A., Severin, F. F., Habermann, A., Hyman, A. A., Griffiths, G., and Burkhardt, J. K. (1996). MAP-dependent binding of phagosomes to microtubules. *J. Biol. Chem.* **271,** 1–9.

Bloom, G. S., Richards, B. W., Leopold, P. L., Ritchey, D. M., and Brady, S. T. (1993). GTP-gamma-S inhibits organelle transport along axonal microtubules. *J. Cell Biol.* **120,** 467–476.

Bomsel, M., Prydz, K., Parton, R. G., Gruenberg, J., and Simons, K. (1989). Endocytosis in filter-grown Madin-Darby canine kidney cells. *J. Cell Biol.* **109,** 3243–3258.

Bomsel, M., Parton, R., Kuznetsov, S. A., Schroer, T. A., and Gruenberg, J. (1990). Microtubule- and motor-dependent fusion in vitro between apical and basolateral endocytic vesicles from MDCK cells. *Cell* **62,** 719–731.

Breitfeld, P. P., McKinnon, W. C., and Mostov, K. E. (1990). Effect of nocodazole on vesicular traffic to the apical and basolateral surfaces of polarized MDCK cells. *J. Cell Biol.* **111,** 2365–2373.

Bulinski, J. C., and Gundersen, G. G. (1991). Stabilization of post-translational modification of microtubules during cellular morphogenesis. *Bioessays* **13,** 285–293.

Burkhardt, J. K. (1996). In search of membrane receptors for microtubule-based motors: Is kinectin a kinesin receptor? *Trends Cell Biol.* **6,** 127–131.

Burkhardt, J. K., McIlvain, J. M. Jr., Sheetz, M. P., and Argon, Y. (1993). Lytic granules from cytotoxic T cells exhibit kinesin-dependent motility on microtubules in vitro. *J. Cell Sci.* **104,** 151–162.

Cassimeris, L., Pryer, N. K., and Salmon, E. D. (1988). Real-time observations of microtubule dynamic instability in living cells. *J. Cell Biol.* **107,** 2223–2231.

Clark, S. W., and Meyer, D. I. (1992). Centractin is an actin homologue associated with the centrosome. *Nature* (*London*) **359,** 246–250.

Clark, S. W., and Meyer, D. I. (1994). ACT3: A putative centractin homolog in S. cerevisiae is required for proper orientation of the mitotic spindle. *J. Cell Biol.* **127,** 129–138.

Cole, N. B., and Lippincott-Schwartz, J. (1995). Organization of organelles and membrane traffic by microtubules. *Curr. Opin. Cell Biol.* **7,** 55–64.

Corthésy-Theulaz, I., Pauloin, A., and Pfeffer, S. R. (1992). Cytoplasmic dynein participates in centrosomal localization of the Golgi complex. *J. Cell Biol.* **118,** 1333–1345.

Cyr, J. L., Pfister, K. K., Bloom, G. S., Slaughter, C. A., and Brady, S. T. (1991). Molecular genetics of kinesin light chains: Generation of isoforms by alternative splicing. *Proc. Natl. Acad. Sci. U.S.A.* **88,** 10114–10118.

Dabora, S. L., and Sheetz, M. P. (1988). The microtubule-dependent formation of a tubulovesicluar network with charcteristics of the ER from cultured cell extracts. *Cell* **54,** 27–35.

Desjardins, M., Huber, L. A., Parton, R. G., and Griffiths, G. (1994). Biogenesis of phagolysosomes proceeds through a sequential series of interactions with the endocytic apparatus. *J. Cell Biol.* **124,** 677–688.

Dillman, J. F., and Pfister, K. K. (1994). Differential phosphorylation in vivo of cytoplasmic dynein associated with anterogradely moving organelles. *J. Cell Biol.* **127,** 1671–1681.

Echeverri, C. J., Paschal, B. M., Vaughan, K. T., and Vallee, R. B. (1996). Molecular characterization of 50kD subunit of dynactin reveals function for the complex in chromosome alignment and spindle organization during mitosis. *J. Cell Biol.* **132,** 617–633.

Endow, S. A., Kang, S. J., Satterwhite, L. L., Rose, M. D., Skeen, V. P., and Salmon, E. D. (1994). Yeast Kar3 is a minus-end microtubule motor protein that destabilizes microtubules preferentially at the minus ends. *EMBO J.* **13,** 2708–2713.

Enos, A. P., and Morris, N. R. (1990). Mutation of a gene that encodes a kinesin-like protein blocks nuclear division in *A. nidulans. Cell* **60,** 1019–1027.

Euteneuer, U., and McIntosh, J. R. (1981). Polarity of some motility-related microtubules. *Proc. Natl. Acad. Sci. U.S.A.* **78,** 372–376.

Euteneuer, U., Koonce, M. P., Pfister, K. K., and Schliwa, M. (1988). An ATPase with properties expected for the organelle motor of the giant amoeba, *Reticulomyxa. Nature* (*London*) **332,** 176–178.

Fath, K. R., Trimbur, G. M., and Burgess, D. R. (1994). Molecular motors are differentially distributed on Golgi membranes from polarized epithelial cells. *J. Cell Biol.* **126,** 661–675.

Feiguin, F., Ferreira, A., Kosik, K. S., and Caceres, A. (1994). Kinesin-mediated organelle translocation revealed by specific cellular manipulations. *J. Cell Biol.* **127,** 1021–1039.

Ferro, K. L., and Collins, C. A. (1995). Microtubule-independent phospholipid stimulation of cytoplasmic dynein ATPase activity. *J. Biol. Chem.* **270,** 4492–4496.

Freed, J. J., and Lebowitz, M. M. (1970). The association of a class of saltatory movements with microtubules in cultured cells. *J. Cell Biol.* **45,** 334–354.

Gauger, A. K., and Goldstein, L. S. (1993). The *Drosophila* kinesin light chain. Primary structure and interaction with kinesin heavy chain. *J. Biol. Chem.* **268,** 13657–13666.

Gill, S. R., Schroer, T. A., Szilak, I., Steuer, E. R., Sheetz, M. P., and Cleveland, D. W. (1991). Dynactin, a conserved, ubiquitously expressed component of an activator of vesicle motility mediated by cytoplasmic dynein. *J. Cell Biol.* **115,** 1639–1650.

Gill, S. R., Cleveland, D. W., and Schroer, T. A. (1994). Characterization of DLC-A and DLC-B, two families of cytoplasmic dynein light chain subunits. *Mol. Biol. Cell* **5,** 645–654.

Goldstein, L. S. (1993). With apologies to Scheherazade: Tails of 1001 kinesin motors. *Annu. Rev. Genet.* **27,** 319–351.

Goltz, J. S., Wolkoff, A. W., Novikoff, P. M., Stockert, R. J., and Satir, P. (1992). A role for microtubules in sorting endocytic vesicles in rat hepatocytes. *Proc. Natl. Acad. Sci. U.S.A.* **89,** 7026–7030.

Goodenough, U. W., and Heuser, J. E. (1982). Substructure of the outer dynein arm. *J. Cell Biol.* **95,** 798–815.

Gruenberg, J., and Maxfield, F. R. (1995). Membrane transport in the endocytic pathway. *Curr. Opin. Cell Biol.* **7,** 552–563.

Gruenberg, J., Griffiths, G., and Howell, K. E. (1989). Characterization of the early endosome and putative endocytic carrier vesicles in vivo and with an assay of vesicle fusion in vitro. *J. Cell Biol.* **108,** 1301–1316.

Hackney, D. D., Levitt, J. D., and Suhan, J. (1992). Kinesin undergoes a 9S to 6S conformational transition. *J. Biol. Chem.* **267,** 8696–8701.

Hagan, I., and Yanagida, M. (1992). Kinesin-related cut7 protein associates with mitotic and meiotic spindles in fission yeast. *Nature* (*London*) **356,** 74–76.

Hagiwara, H., Yorifuji, H., Satoyoshitake, R., and Hirokawa, N. (1994). Competition between motor molecules (kinesin and cytoplasmic dynein) and fibrous microtubule-associated proteins in binding to microtubules. *J. Biol. Chem.* **269,** 3581–3589.

Hall, D. H., and Hedgecock, E. M. (1991). Kinesin-related gene unc-104 is required for axonal transport of synaptic vesicles in *C. elegans. Cell* **65,** 837–847.

Hamm-Alvarez, S. F., Kim, P. Y., and Sheetz, M. P. (1993). Regulation of vesicle transport in CV-1 cells and extracts. *J. Cell Sci.* **106,** 955–66.

Heck, M. M. S., Pereira, A., Pesavento, P., Yannoni, Y., Spradling, A. C., and Goldstein, L. S. B. (1993). The kinesin-like protein klp61F is essential for mitosis in *Drosophila. J. Cell Biol.* **123,** 665–679.

Heidemann, S. R., and McIntosh, J. R. (1980). Visualization of the structural polarity of microtubules. *Nature* (*London*) **286,** 517–519.

Heidemann, S. R., Landers, J. M., and Hamborg, M. A. (1981). Polarity orientation of axonal microtubules. *J. Cell Biol.* **91,** 661–665.

Henson, J. H., Nesbitt, D., Wright, B. D., and Scholey, J. M. (1992). Immunolocalization of kinesin in sea urchin coelomocytes: Association of kinesin with intracellular organelles. *J. Cell Sci.* **103,** 309–320.

Herman, B., and Albertini, D. F. (1984). A time-lapse video image intensification analysis of cytoplasmic organelle movements during endosome translocation. *J. Cell Biol.* **98,** 565–576.

Heuser, J. (1989). Changes in lysosome shape and distribution correlated with changes in cytoplasmic pH. *J. Cell Biol.* **108,** 855–864.

Hirokawa, N. (1982). Cross-linker system between neurofilaments, microtubules, and membranous organelles in frog axons revealed by the quick-freeze, deep-etching method. *J. Cell Biol.* **94,** 129–142.

Hirokawa, N., Pfister, K. K., Yorifuji, H., Wagner, M. C., Brady, S. T., and Bloom, G. S. (1989). Submolecular domains of bovine brain kinesin indentified by electron microscopy and monoclonal antibody decoration. *Cell* **56,** 867–878.

Ho, W. C., Allan, V. J., van Meer, G., Berger, E. G., and Kreis, T. E. (1989). Reclustering of scattered Golgi elements occurs along microtubules. *Eur. J. Cell Biol.* **48,** 250–263.

Hollenbeck, P. J. (1989). The distribution, abundance and subcellular localization of kinesin. *J. Cell Biol.* **108,** 2335–2342.

Hollenbeck, P. J. (1993). Phosphorylation of neuronal kinesin heavy and light chains in vivo. *J. Neurochem.* **60,** 2265–2275.

Hollenbeck, P. J., and Swanson, J. A. (1990). Radial extension of macrophage tubular lysosomes supported by kinesin. *Nature* (*London*) **346,** 864–866.

Holzbaur, E. L., and Vallee, R. B. (1994). Dyneins: Molecular structure and cellular function. *Annu. Rev. Cell Biol.* **10,** 339–372.

Hopkins, C. R., Gibson, A., Shipman, M., and Miller, K. (1990). Movement of internalized ligand-receptor complexes along a continuous endosomal reticulum. *Nature* (*London*) **346,** 335–339.

Hoyt, M. A. (1994). Cellular roles of kinesin and related proteins. *Curr. Opin. Cell Biol.* **6,** 63–68.

Hoyt, M. A., He, L., Loo, K. K., and Saunders, W. S. (1992). Two *Saccharomyces cerevisiae* kinesin-related gene products required for mitotic spindle assembly: *J. Cell Biol.* **118,** 109–120.

Hunziker, W., Male, P., and Mellman, I. (1990). Differential microtubule requirements for transcytosis in MDCK cells. *EMBO J.* **9,** 3515–3525.

Jin, M., and Snider, M. D. (1993). Role of microtubules in transferrin receptor transport from the cell surface to endosomes and the Golgi complex. *J. Biol Chem.* **268,** 18390–18397.

Karki, S., and Holzbaur, E. L. F. (1995). Affinity chromatography demonstrates a direct binding between cytoplasmic dynein and the dynactin complex. *J. Biol. Chem.* **270,** 28806–28811.

Khawaja, S., Gundersen, G. G., and Bulinski, J. C. (1988). Enhanced stability of microtubules enriched in detyrosinated tubulin is not a direct function of detyrosination level. *J. Cell Biol.* **106,** 141–149.

King, S. M., and Patel-King, R. S. (1995). The M(R) = 8,000 and 11,000 outer arm dynein light chains from *Chlamydomonas* flagella have cytoplasmic homologues. *J. Biol. Chem.* **270,** 11445–11452.

King, S. M., and Witman, G. B. (1989). Molecular structure of *Chlamydomonas* outer arm dynein. *In* "Cell Movement" (F. D. Warner, P. Satir, and I. R. Gibbons, eds.). New York, Alan R. Liss.

King, S. M., and Witman, G. B. (1990). Localization of an intermediate chain of outer arm dynein by immunoelectron microscopy. *J. Biol. Chem.* **265,** 19807–19811.

King, S. M., Wilkerson, C. G., and Witman, G. B. (1991). The Mr 78,000 intermediate chain of *Chlamydomonas* outer arm dynein interacts with alpha-tubulin in situ. *J. Biol. Chem.* **266,** 8401–8407.

Kondo, S., Sato-Yoshitake, R., Noda, Y., Aizawa, H., Nakata, T., Matsuura, Y., and Hirokawa, N. (1994). KIF3A is a new microtubule-based anterograde motor in the nerve axon. *J. Cell Biol.* **125,** 1095–1107.

Kumar, J., Yu, H., and Sheetz, M. P. (1995). Kinectin, an essential anchor for kinesin-driven vesicle motility. *Science* **267,** 1834–1837.

Kuznetsov, S. A., Langford, G. M., and Weiss, D. G. (1992). Actin-dependent organelle movement in squid axoplasm. *Nature* (*London*) **356,** 722–725.

Lacey, M. L., and Haimo, L. T. (1994). Cytoplasmic dynein binds to phospholipid vesicles. *Cell Motil. Cytoskeleton* **28,** 205–212.

Ladinsky, M. S., Kremer, J. R., Furcinitti, P. S., McIntosh, J. R., and Howell, K. E. (1994). HVEM tomography of the trans-Golgi network: Structural insights and identification of a lace-like vesicle coat. *J. Cell Biol.* **127,** 29–38.

Lafont, F., Burkhardt, J. K., and Simons, K. (1994). Involvement of microtubule motors in basolateral and apical transport in kidney cells. *Nature* (*London*) **372,** 801–803.

Langford, G. M., Allen, R. D., and Weiss, D. G. (1987). Substructure of sidearms on squid axoplasmic vesicles and microtubules visualized by negative contrast electron microscopy. *Cell Motil. Cytoskeleton* **7,** 20–30.

Lee, C., and Chen, L. B. (1988). Dynamic behavior of endoplasmic reticulum in living cells. *Cell* **54,** 37–46.

Lee, K. D., and Hollenbeck, P. J. (1995). Phosphorylation of kinesin in vivo correlates with organelle association and neurite outgrowth. *J. Biol. Chem.* **270,** 5600–5605.

Letourneau, P. C., and Wire, J. P. (1995). Three-dimensional organization of stable microtubules and the Golgi apparatus in the somata of developing chick sensory neurons. *J. Neurocytol.* **24,** 207–223.

Lin, S. X. H., and Collins, C. A. (1993). Regulation of the intracellular distribution of cytoplasmic dynein by serum factors and calcium. *J. Cell Sci.* **105,** 579–588.

Lin, S. X., Ferro, K. L., and Collins, C. A. (1994). Cytoplasmic dynein undergoes intracellular redistribution concomitant with phosphorylation of the heavy chain in response to serum starvation and okadaic acid. *J. Cell Biol.* **127,** 1009–1019.

Linden, M., Nelson, B. D., Loncar, D., and Leterrier, J. F. (1989). Studies on the interaction between mitochondria and the cytoskeleton. *J. Bioenerg. Biomembr.* **21,** 507–518.

Lippincott-Schwartz, J., Donaldson, J. G., Schweizer, A., Berger, E. G., Hauri, H.-P., Yuan, L. C., and Klausner, R. D. (1990). Microtubule-dependent retrograde transport of proteins into the ER in the presence of Brefeldin A suggests an ER recycling pathway. *Cell* **60,** 821–836.

Lippincott-Schwartz, J., Cole, N. B., Marotta, A., Conrad, P. A., and Bloom, G. S. (1995). Kinesin is the motor for microtubule-mediated Golgi-to-ER membrane traffic. *J. Cell Biol.* **128,** 293–306.

Lopez, L. A., and Sheetz, M. P. (1993). Steric inhibition of cytoplasmic dynein and kinesin motility by MAP2. *Cell Motil. Cytoskeleton* **24,** 1–16.

Lopez, L. A., and Sheetz, M. P. (1995). A microtubule-associated protein (MAP2) kinase restores microtubule motility in embryonic brain. *J. Biol. Chem.* **270,** 12511–12517.

Lye, R. J., Porter, M. E., Scholey, J. M., and McIntosh, J. R. (1987). Identification of a microtubule-based cytoplasmic motor in the nematode *C. elegans. Cell* **51,** 309–318.

Marks, D. L., Larkin, J. M., and McNiven, M. A. (1994). Association of kinesin with the Golgi apparatus in rat hepatocytes. *J. Cell Sci.* **107,** 2417–2426.

Maruta, H., Greer, K., and Rosenbaum, J. L. (1986). The acetylation of alpha-tubulin and its relationship to the assembly and disassembly of microtubules. *J. Cell Biol.* **103,** 571–579.

Matteoni, R., and Kreis, T. E. (1987). Translocation and clustering of endosomes and lysosomes depends on microtubules. *J. Cell Biol.* **105,** 1253–1265.

Matthies, H. J., Miller, R. J., and Palfrey, H. C. (1993). Calmodulin binding to and cAMP-dependent phosphorylation of kinesin light chains modulate kinesin ATPase activity. *J. Biol. Chem.* **268,** 11176–11187.

McDonald, H. B., Stewart, R. J., and Goldstein, L. S. (1990). The kinesin-like ncd protein of Drosophila is a minus end-directed microtubule motor. *Cell* **63,** 1159–1165.

McGrail, M., Gepner, J., Silvanovich, A., Ludmann, S., Serr, M., and Hays, T. S. (1995). Regulation of cytoplasmic dynein function in vivo by the Drosophila Glued complex. *J. Cell Biol.* **131,** 411–425.

McIlvain, J. M. Jr., Burkhardt, J. K., Hamm-Alvarez, S., Argon, Y., and Sheetz, M. P. (1994). Regulation of kinesin activity by phosphorylation of kinesin-associated proteins. *J. Biol. Chem.* **269,** 19176–19182.

Meads, T., and Schroer, T. A. (1995). Polarity and nucleation of microtubules in polarized epithelial cells. *Cell Motil. Cytoskeleton* **32,** 273–288.

Mermall, V., McNally, J. G., and Miller, K. G. (1994). Transport of cytoplasmic particles catalysed by an unconventional myosin in living Drosophila embryos. *Nature* (*London*) **369,** 560–562.

Middleton, K., and Carbon, J. (1994). KAR3-encoded kinesin is a minus-end-directed motor that functions with centromere binding proteins (CBF3) on an in vitro yeast kinetochore. *Proc. Natl. Acad. Sci. U.S.A.* **91,** 7212–7216.

Mikhailov, A. V., and Gundersen, G. G. (1995). Centripetal transport of microtubules in motile cells. *Cell Motil. Cytoskeleton* **32,** 173–186.

Miller, R. H., and Lasek, R. J. (1985). Cross-bridges mediate anterograde and retrograde vesicle transport along microtubules in squid axoplasm. *J. Cell Biol.* **101,** 2181–2193.

Mitchell, D. R., and Rosenbaum, J. L. (1986). Protein-protein interactions in the 18S ATPase of *Chlamydomonas* outer dynein arms. *Cell Motil. Cytoskeleton.* **6,** 510–520.

Mizuno, M. and Singer, S. J. (1994). A possible role for stable microtubules in intracellular transport from the endoplasmic reticulum to the Golgi apparatus. *J. Cell Sci.* **107,** 1321–1331.

Muhua, L., Karpova, T. S., and Cooper, J. A. (1994). A yeast actin-related protein homologous to that found in vertebrate dynactin complex is important for spindle orientation and nuclear migration. *Cell* **78,** 669–679.

Nangaku, M., Sato-Yoshitake, R., Okada, Y., Noda, Y., Takemura, R., Yamazaki, H., and Hirokawa, N. (1994). KIF1B, a novel microtubule plus end-directed monomeric motor protein for transport of mitochondria. *Cell* **79,** 1208–1220.

Niclas, J., Allan, V. J., and Vale, R. D. (1996). Cell cycle regulation of dynein association with membranes modulates microtubule-based organelle transport. *J. Cell Biol.* **133,** 585–593.

Noda, Y., Sato-Yoshitake, R., Kondo, S., Nangaku, M., and Hirokawa, N. (1995). KIF2 is a new microtubule-based anterograde motor that transports membranous organelles distinct from those carried by kinesin heavy chain or KIF3A/B. *J. Cell Biol.* **129,** 157–167.

Oda, H., Stockert, R. J., Collins, C., Wang, H., Novikoff, P. M., Satir, P., and Wolkoff, A. W. (1995). Interaction of the microtubule cytoskeleton with endocytic vesicles and cytoplasmic dynein in cultured rat hepatocytes. *J. Biol. Chem.* **270,** 15242–15249.

Okada, Y., Yamazaki, H., Sekine, Y., Aizawa, H., and Hirokawa, N. (1995). The neuron-specific kinesin superfamily protein KIF1a is a unique monomeric motor for anterograde axonal transport of synaptic vesicle precursors. *Cell* **81,** 769–780.

Otsuka, A. J., Jeyaprakash, A., Garcia-Anoveros, J., Tang, L. Z., Fisk, G., Hartshorne, T., Franco, R., and Born, T. (1991). The C. elegans unc-104 gene encodes a putative kinesin heavy chain-like protein. *Neuron* **6,** 113–122.

Parton, R. G., Prydz, K., Bomsel, M., Simons, K., and Griffiths, G. (1989). Meeting of the apical and basolateral endocytic pathways of the Madin-Darby canine kidney cell in late endosomes. *J. Cell Biol.* **109,** 3259–3272.

Paschal, B. M., and Vallee, R. B. (1987). Retrograde transport by the microtubule-associated protein MAP 1C. *Nature* (*London*) **330,** 181–183.

Paschal, B. M., Shpetner, H. S., and Vallee, R. B. (1987). MAP 1C is a microtubule-activated ATPase which translocates microtubules in vitro and has dynein-like properties. *J. Cell. Biol.* **105,** 1273–1282.

Pesavento, P. A., Stewart, R. J., and Goldstein, L. S. (1994). Characterization of the KLP68D kinesin-like protein in *Drosophila*: Possible roles in axonal transport. *J. Cell Biol.* **127,** 1041–1048.

Pfeffer, S. (1992). GTP-binding proteins in intracellular transport. *Trends Cell Biol.* **2,** 41–46.

Pierre, P., Scheel, J., Rickard, J., and Kreis, T. E. (1992). CLIP-170 links endocytic vesicles to microtubules. *Cell* **70,** 887–900.

Plamann, M., Minke, P. F., Tinsley, J. H., and Bruno, K. S. (1994). Cytoplasmic dynein and actin-related protein Arp1 are required for normal nuclear distribution in filamentous fungi. *J. Cell Biol.* **127,** 139–149.

Pryer, N. K., Walker, R. A., Skeen, V. P., Bourns, B. D., Soboeiro, M. F., and Salmon, E. D. (1992). Brain microtubule-associated proteins modulate microtubule dynamic instability in vitro: Real-time observations using video microscopy. *J. Cell Sci.* **103,** 965–976.

Rambourg, A., and Clermont, Y. (1990). Three-dimensional electron microscopy: Structure of the Golgi apparatus. *Eur. J. Cell Biol.* **51,** 189–200.

Rickard, J. E., and Kreis, T. E. (1991). Binding of pp170 to microtubules is regulated by phosphorylation. *J. Biol. Chem.* **266,** 17597–17605.

Rodionov, V. I., Gyoeva, F. K., and Gelfand, V. I. (1991). Kinesin is responsible for centrifugal movement of pigment granules in melanophores. *Proc. Natl. Acad. Sci. U.S.A.* **88,** 4956–4960.

Sale, W. S., and Satir, P. (1977). Direction of active sliding of microtubules in *Tetrahymena* cilia. *Proc. Natl. Acad. Sci. U.S.A.* **74,** 2045–2049.

Sammak, P. J., and Borisy, G. G. (1988). Direct observation of microtubule dynamics in living cells. *Nature (London)* **332,** 724–726.

Saraste, J., and Kuismanen, E. (1992). Pathways of protein sorting and membrane traffic between the rough endoplasmic reticulum and the Golgi complex. *Semin. Cell Biol.* **3,** 343–355.

Saraste, J., and Svensson, K. (1991). Distribution of the intermediate elements operating in ER to Golgi transport. *J. Cell Sci.* **100,** 415–430.

Sato-Yoshitake, R., Yorifuji, H., Inagaki, M., and Hirokawa, N. (1992). The phosphorylation of kinesin regulates its binding to synaptic vesicles. *J. Biol. Chem.* **267,** 23930–23936.

Sawin, K. E., LeGuellec, K., Philippe, M., and Mitchison, T. J. (1992). Mitotic spindle organization by a plus-end-directed microtubule motor. *Nature (London)* **359,** 540–543.

Schafer, D. A., Gill, S. R., Cooper, J. A., Heuser, J. E., and Schroer, T. A. (1994). Ultrastructural analysis of the dynactin complex: An actin-related protein is a component of a filament that resembles f-actin. *J. Cell Biol.* **126,** 403–412.

Schmitz, F., Wallis, K. T., Rho, M., Drenckhahn, D., and Murphy, D. B. (1994). Intracellular distribution of kinesin in chromaffin cells. *Eur. J. Cell Biol.* **63,** 77–83.

Schnapp, B. J., and Reese, T. S. (1982). Cytoplasmic structure in rapid-frozen axons. *J. Cell Biol.* **94,** 667–679.

Schnapp, B. J., Reese, T. S., and Bechtold, R. (1992). Kinesin is bound with high affinity to squid axon organelles that move to the plus-end of microtubules. *J. Cell Biol.* **119,** 389–399.

Scholey, J. M., Heuser, J., Yang, J. T., and Goldstein, L. S. B. (1989). Identification of globular mechanochemical heads of kinesin. *Nature (London)* **338,** 355–357.

Schroer, T. A. (1996). Structure and function of dynactin. *Semin. Cell Biol.* In press.

Schroer, T. A., and Sheetz, M. P. (1991). Two activators of microtubule-based vesicle transport. *J. Cell Biol.* **115,** 1309–1318.

Schroer, T. A., Bingham, J. B., and Gill, S. R. (1996). Actin-related protein 1 and cytoplasmic dynein-based motility–whats the connection? *Trends Cell Biol.* **6,** 212–215.

Schulze, E., and Kirschner, M. (1986). Microtubule dynamics in interphase cells. *J. Cell Biol.* **102,** 1020–1031.

Schulze, E., and Kirschner, M. (1987). Dynamic and stable populations of microtubules in cells. *J. Cell Biol.* **104,** 277–288.

Schweizer, A., Fransen, J. A., Matter, K., Kreis, T. E., Ginsel, L., and Hauri, H. P. (1990). Identification of an intermediate compartment involved in protein transport from endoplasmic reticulum to Golgi apparatus. *Eur. J. Cell Biol.* **53,** 185–196.

Sekine, Y., Okada, Y., Noda, Y., Kondo, S., Aizawa, H., Takemura, R., and Hirokawa, N. (1994). A novel microtubule-based motor protein (KIF4) for organelle transports, whose expression is regulated developmentally. *J. Cell Biol.* **127,** 187–201.

Shelden, E., and Wadsworth, P. (1993). Observation and quantification of individual microtubule behavior in vivo: Microtubule dynamics are cell-type specific. *J. Cell Biol.* **120,** 935–945.

Skoufias, D. A., Burgess, T. L., and Wilson, L. (1990). Spatial and temporal colocalization of the Golgi apparatus and microtubules rich in detyrosinated tubulin. *J. Cell Biol.* **111,** 1929–1937.

Skoufias, D. A., Cole, D. G., Wedaman, K. P., and Scholey, J. M. (1994). The carboxyl-terminal domain of kinesin heavy chain is important for membrane binding. *J. Biol. Chem.* **269,** 1477–1485.

Swanson, J., Bushnell, A., and Silverstein, S. C. (1987). Tubular lysosome morphology and distribution within macrophages depend on the integrity of cytoplasmic microtubules. *Proc. Natl. Acad. Sci. U.S.A.* **84,** 1921–1925.

Terasaki, M., and Reese, T. S. (1994). Interactions among endoplasmic reticulum, microtubules, and retrograde movements of the cell surface. *Cell Motil. Cytoskeleton* **29,** 291–300.

Terasaki, M., Chen, L. B., and Fujiwara, K. (1986). Microtubules and the endoplasmic reticulum are highly interdependent structures. *J. Cell Biol.* **103,** 1557–1568.

Thatte, H. S., Bridges, K. R., and Golan, D. E. (1994). Microtubule inhibitors differentially affect translational movement, cell surface expression, and endocytosis of transferrin receptors in K562 cells. *J. Cell Physiol.* **160,** 345–357.

Thyberg, J., and Moskalewski, S. (1985). Microtubules and the organization of the Golgi complex. *Exp. Cell Res.* **159,** 1–16.

Toyoshima, I., Yu, H., Steuer, E. R., and Sheetz, M. P. (1992). Kinectin, a major kinesin-binding protein on ER. *J. Cell Biol.* **118,** 1121–1131.

Troutt, L. L., and Burnside, B. (1988). The unusual microtubule polarity in teleost retinal pigment epithelial cells. *J. Cell Biol.* **107,** 1461–1464.

Tsukita, S., and Ishikawa, H. (1980). The movement of membranous organelles in axons. Electron microscopic identification of anterogradely and retrogradely transported organelles. *J. Cell Biol.* **84,** 513–530.

Tucker, J. B., Paton, C. C., Richardson, G. P., Mogensen, M. M., and Russell, I. J. (1992). A cell surface-associated centrosomal layer of microtubule-organizing material in the inner pillar cell of the mouse cochlea. *J. Cell Sci.* **102,** 215–226.

Tucker, J. B., Paton, C. C., Henderson, C. G., and Mogensen, M. M. (1993). Microtubule rearrangement and bending during assembly of large curved microtubule bundles in mouse cochlear epithelial cells. *Cell Motil. Cytoskeleton* **25,** 49–58.

Turner, J. R., and Tartakoff, A. M. (1989). The response of the Golgi complex to microtubule alterations: The roles of metabolic energy and membrane traffic in Golgi complex organization. *J. Cell Biol.* **109,** 2081–2088.

Urrutia, R., McNiven, M. A., Albanesi, J. P., Murphy, D. B., and Kachar, B. (1991). Purified kinesin promotes vesicle motility and induces active sliding between microtubules in vitro. *Proc. Natl. Acad. Sci. U.S.A.* **88,** 6701–6705.

Vale, R. D., and Hotani, H. (1988). Formation of membrane networks in vitro by kinesin-driven microtubule movement. *J. Cell Biol.* **107,** 2233–2241.

Vallee, R. B., and Sheetz, M. P. (1996). Targeting of motor proteins. *Science* **271,** 1539–1544.

Vallee, R. B., Wall, J. S., Paschal, B. M., and Shpetner, H. S. (1988). Microtubule-associated protein 1C from brain is a two-headed cytosolic dynein. *Nature (London)* **332,** 561–563.

Vaughan, K. T., and Vallee, R. B. (1995). Cytoplasmic dynein binds dynactin through a direct interaction between the intermediate chains and p150^{Glued}. *J. Cell Biol.* **131,** 1507–1516.

Vogl, A. W., Weis, M., and Pfeiffer, D. C. (1995). The perinuclear centriole-containing centrosome is not the major microtubule organizing center in Sertoli cells. *Eur. J. Cell Biol.* **66,** 165–179.

Walker, R. A., and Sheetz, M. P. (1993). Cytoplasmic microtubule-based motors. *Annu. Rev. Biochem.* **62,** 429–451.

Walker, R. A., Salmon, E. D., and Endow, S. A. (1990). The *Drosophila* claret segregation protein is a minus-end directed motor molecule. *Nature* (*London*) **347,** 780–782.

Waterman-Storer, C. M., Gregory, J., Parsons, S. F., and Salmon, E. D. (1995a). Membrane/microtubule tip attachment complexes (TACs) allow the assembly dynamics of plus ends to push and pull membranes into tubulovesicular networks in interphase Xenopus egg extracts. *J. Cell Biol.* **130,** 1161–1169.

Waterman-Storer, C. M., Karki, S., and Holzbaur, E. L. (1995b). The p150Glued component of the dynactin complex binds to both microtubules and the actin-related protein centractin (Arp-1). *Proc. Natl. Acad. Sci. U.S.A.* **92,** 1634–1638.

Webster, D. R., Wehland, J., Weber, K., and Borisy, G. G. (1990). Detyrosination of alpha tubulin does not stabilize microtubules in vivo. *J. Cell Biol.* **111,** 113–122.

Willingham, M. C., and Pastan, I. (1978). The visualization of fluorescent proteins in living cells by video intensification microscopy (VIM). *Cell* **14,** 501–507.

Yamazaki, H., Nakata, T., Okada, Y., and Hirokawa, N. (1995). Kif3a/B—a heterodimeric kinesin superfamily protein that works as a microtubule plus end-directed motor for membrane organelle transport. *J. Cell Biol.* **130,** 1387–1399.

Yu, H., Toyoshima, I., Steuer, E. R., and Sheetz, M. P. (1992). Kinesin and cytoplasmic dynein binding to brain microsomes. *J. Biol. Chem.* **267,** 20457–20464.

Yu, H., Nicchitta, C. V., Kumar, J., Becker, M., Toyoshima, I., and Sheetz, M. P. (1995). Characterization of kinectin, a kinesin-binding protein: Primary sequence and N-terminal topogenic signal analysis. *Mol. Biol. Cell* **6,** 171–183.

CHAPTER 4

Role of the Cytoskeleton and Molecular Motors in Transport between the Golgi Complex and Plasma Membrane

Karl R. Fath and David R. Burgess
Department of Biological Sciences, University of Pittsburgh, Pittsburgh, Pennsylvania 15260

I. INTRODUCTION

The synthesis and degradation pathways in eukaryotic cells require the constant movement of materials between various membrane compartments and organelles. Many times these materials must move relatively great distances through the cytoplasm. The highly cross-linked cytoplasm obviates a major role of diffusion in moving membranes and other cellular organelles over significant distances in the cytoplasm (see review by Luby-Phelps, 1994). Thus a facilitated transport is required to deliver membranes in a timely fashion from their sites of synthesis to their places of utilization and to target components to the proper destination. It has become apparent in the past 5–10 years that cytoskeletal elements, microtubules (MTs), and actin filaments, initially thought of as only a structural framework, also serve as tracks or substrates on which cellular cargo is translocated. MTs apparently serve as cellular "airways," for they are substrates for membranes moving long distances in the interior of the cell (e.g., from the neuronal cell body to the synapse). MTs form regular and long tracks whose organization is relatively easily to discern. Actin filaments, by contrast, do not generally form long tracks and are thought to function primarily in short movements near the plasma membrane (PM) in an area called the cell cortex. In the transportation analogy, actin filaments may be thought of as highways or routes for local delivery.

It has been proposed that the delivery of newly synthesized material from the Golgi complex to the PM, much like mailed parcels, must use both the long-distance and local routes because both tracks and motors translocating cargo along the tracks are distinct, but work in harmony. In other words, these new materials perhaps travel along MTs using one class of mechanochemical motors from the Golgi complex to the cell cortex. Once in the cortex, they release from MTs and bind to and move along actin filaments via a distinct class of motor molecule to the PM. In this chapter we first discuss the role of MTs and the MT-based motors kinesin and cytoplasmic dynein in Golgi-to-PM movements. Then we discuss the roles of actin filaments and unconventional myosins (Mooseker and Cheney, 1995) in these movements. Finally, we discuss possible interactions of these two systems.

II. MT-BASED MOLECULAR MOTORS

MTs are polar structures, with the plus end preferred for assembly over the minus end. This polarity is important not only to the assembly and dynamics of MTs, but also in their role as polarized tracks for the translocation of membranes. Cytoplasmic organelles with associated molecular motors move specifically toward one end or the other of the MT depending upon the family of molecular motors. Members of the kinesin superfamily are plus end–directed MT-based motors (reviewed by Walker and Sheetz, 1993) that are elongate molecules ~80 nm in length and composed of two heavy chains (~120 kDa) and two light chains (~65 kDa). Kinesin was first discovered in the squid giant axon but has since been found in such organisms as yeast, *Drosophila*, nematodes, and humans. There are a large number of kinesin genes and multiple kinesins in a single organism. Although kinesins generally transport their cargo toward the plus ends of

MTs, there are at least two MT minus end–directed kinesins (ncd from *D. melanogaster* and KAR3 from *Saccharomyces cerevisiae*) that are thought to be involved in mitotic–meiotic movements (reviewed by Hoyt, 1994).

The other class of MT-based motors are members of the cytoplasmic dynein superfamily (reviewed by Holzbaur and Vallee, 1994; Walker and Sheetz, 1993). Cytoplasmic dyneins, currently thought to be a single gene product, are ~45-nm-long molecules composed of two heavy chains (~500 kDa), several intermediate chains (~70 kDa), and a variable number of light chains (40–60 kDa). Like kinesin, dynein appears to be found in nearly all organisms and cell types. There are many reports that cytoplasmic dynein is involved in organellar movement toward the MT minus end (Euteneur *et al.,* 1988; Hirokawa *et al.,* 1990; Lacey and Haimo, 1992; Lin and Collins, 1992; Paschal *et al.,* 1987; Schnapp and Reese, 1989; Schroer *et al.,* 1989). The activity of cytoplasmic dynein may be controlled by the multisubunit dynactin complex that can bind to dynein intermediate chains (Karki and Holzbaur, 1995; Vaughan and Vallee, 1995) and may mediate dynein interactions with membranes. (For a more in-depth discussion of the dynactin complex, see Chapter 3, this volume.)

III. ORIENTATION OF THE MT CYTOSKELETON RELATIVE TO MEMBRANE COMPARTMENTS

Because of the importance of polarity in MT-based motility, in order to understand the types of motors involved in specific membrane movements, one must first understand the organization of the MT cytoskeleton relative to the membrane compartments. It is a widely accepted model that kinesin is the motor for Golgi-to-PM movements because MTs are polarized with their minus ends at the nucleus, where the Golgi complex resides. While this simplified model is likely appropriate for fibroblasts, wherein the MT-organizing center (MTOC) is adjacent to the nucleus and Golgi complex, it is incorrect for polarized epithelial cells (Fig. 1). However, in all cell types the MT minus ends are associated with the MTOC; it is the MTOC and not the Golgi complex that organizes the assembly and polarity of MTs.

MTs in polarized epithelia are arranged in parallel bundles with their minus ends and MTOCs in the apical cytoplasm and their plus ends toward the basal cytoplasm and Golgi complex (Fig. 1) (Achler *et al.,* 1989; Bacallao *et al.,* 1989; Drenckhahn and Dermietzel, 1988; Gilbert *et al.,* 1991; Sandoz *et al.,* 1985). There is also a collection of MTs of mixed polarity running transversely in the apical cytoplasm (Bacallao *et al.,* 1989). In neurons, the centrioles and Golgi complexes of these highly polarized cells reside near the nucleus on the side of the nucleus where the axon originates (Fig. 1).

The MTs nucleate from the centrioles but then detach as they are transported into the axon and dendrites. The MTs in the cell body and axon are oriented with their plus ends away from the nucleus, whereas the MTs in dendrites are of mixed polarity (Sharp *et al.,* 1995). In interphase fibroblast-like cells, centrioles lie near the nucleus and the MTs radiate out with their plus ends toward the cell periphery (Fig. 1).

IV. ROLE OF MTS AND MT-BASED MOLECULAR MOTORS IN GOLGI-TO-PM TRANSPORT: THREE CASE STUDIES

A. *Polarized Epithelia*

Of all cell types, the role of MTs in Golgi-to-PM transport is best understood in polarized epithelia. The cell polarity, which is reflected in the formation of an epithelial sheet, has assisted in the dissection of membrane synthesis and distribution pathways. The PM of polarized epithelial cells is divided into two spatial and functional domains that are separated by cell–cell junctions (Fig. 1). The protein and lipid components of the PM and the organization of the underlying cytoskeleton are distinct for these domains. Newly synthesized materials are transported to the apical domain by two pathways; both the identity of the protein and the cell type influence the pathway utilized. In the Madin–Darby canine kidney– (MDCK-) derived cell line, proteins are sorted in the *trans*-Golgi network (Simons and Wandinger-Ness, 1990) and then transported in carrier vesicles directly from the *trans*-Golgi network to the apical or basolateral PM. In hepatocytes, both apical- and basolateral-destined membrane constituents are inserted initially in the basolateral membrane. After insertion, apically targeted proteins are collected and transcytosed as endosomes apically (Bartles and Hubbard, 1988). The intestinal cell line Caco-2 (Gilbert *et al.,* 1991) and intestinal cells *in vivo* (Achler *et al.,* 1989; Trahair *et al.,* 1989) use both a direct and an indirect pathway for transport of materials apically. We are beginning to understand some of the sorting signals that may target a protein into basolaterally or apically targeted carrier vesicles; however, discussion of these signals is beyond the scope of this chapter, and the reader is referred to two excellent reviews on this subject (Mays *et al.,* 1994; Mellman *et al.,* 1995).

Many laboratories have taken a pharmacological approach to gain an understanding of roles of MTs in membrane delivery in polarized epithelial cells. These studies use the drugs nocodazole or colchicine to depolymerize MTs and then follow the time course of appearance of newly synthesized proteins on the apical and basolateral surfaces. These studies generally

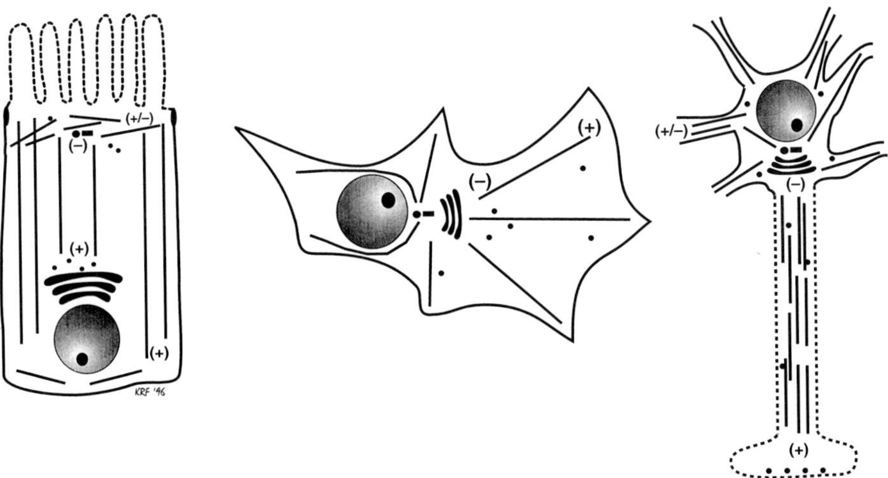

FIGURE 1 Organization of the MT and PM domains in polarized epithelial cells, fibroblasts, and neurons. In polarized epithelia there are parallel bundles of MTs along the apical–basolateral axis with the MT minus ends nearest the apical PM (dashed membrane) and centrioles and the MT plus ends nearest the Golgi and basolateral PM (solid membrane). There is also an apical dense mat of transverse MTs of mixed polarities. The MTs of fibroblasts are oriented with their minus ends at the MTOC near the nucleus, and their plus ends radiating toward the cell periphery. The MTs in axons are organized in an overlapping linear array with their plus-ends distal. The MTs in dendrites are of mixed polarities. Studies of the sorting of viral glycoproteins suggest that the axon PM (dashed membrane) corresponds to the polarized epithelial cell apical PM and the somatodendritic PM (solid membrane) corresponds to the polarized epithelial cell basolateral PM (Dotti *et al.*, 1991).

agree that MTs play a crucial role in facilitating the transport of apically targeted membrane protein transport vesicles to the apical surface by both the direct (Achler *et al.*, 1989; Bennett *et al.*, 1984; Breitfeld *et al.*, 1990; Eilers *et al.*, 1989; Gilbert *et al.*, 1991; Hugon *et al.*, 1987; Hunziker *et al.*, 1990; Lafont *et al.*, 1994; Matter *et al.*, 1990; Parczyk *et al.*, 1989; Rindler *et al.*, 1987) and the indirect (Achler *et al.*, 1989; Breitfeld *et al.*, 1990; Hunziker *et al.*, 1990) pathways. The delivery of basolaterally targeted materials to the basolateral membrane was originally considered to be MT and actin independent (Breitfeld *et al.*, 1990; Hunziker *et al.*, 1990; Parczyk *et al.*, 1989; Rindler *et al.*, 1987; Salas *et al.*, 1986). When polarized epithelia are treated with MT-disruptive drugs, there is a delay in delivery of materials to the apical plasma membrane, although 50–80% of the materials eventually reach the apical domain. The remaining materials are missorted to the basolateral domain. These MT disruption studies assume that all MTs are removed by drug treatment, although it has been shown that it is difficult to completely depolymerize MTs in polarized epithelia (Eilers *et al.*, 1989). In more recent studies, both permeabilization and nocodazole treatment were used to completely disrupt MTs in MDCK cells (Lafont *et al.*, 1994). In these studies, it was found that intact MTs are required for transport to both the apical and basolateral domains. These results are consistent with the finding that isolated basolateral and apical carrier vesicles from MDCK cells can bind MTs *in vitro* (van der Sluijs *et al.*, 1990).

Because the plus ends of MTs in polarized epithelia are nearest the Golgi complex in the apical region of the cell, just subjacent to the brush border terminal web, a MT-based system for translocating membranes from the Golgi complex for the direct pathway of membrane protein delivery or for translocating endosomes from the basolateral surface in the transcytotic pathway to the apical cortex would probably require a minus end–directed motor, such as dynein (Fig. 2). Consistent with this polarity, we have found that apically targeted Golgi-derived vesicles that were isolated from intestinal epithelial cells possess dynein on their surface (Fath *et al.*, 1994). In the aforementioned permeabilized MDCK cell studies (Lafont *et al.*, 1994), polarized transport was initiated by the addition of exogenous cytosol and ATP. Transport to the apical surface was inhibited when either kinesin or dynein was immunodepleted from the cytosol, while transport to the basolateral domain was inhibited by kinesin removal (Lafont *et al.*, 1994). These workers propose that basolaterally targeted vesicles use kinesin to reach the PM along both the longitudinal and transverse MTs. The apically targeted vesicles may use dynein to move apically on the longitudinal MTs and then both kinesin and dynein to traverse the apical transverse network of MTs (Fig. 2).

B. Neurons

The neuron, the most highly polarized cell type in the body, has become a rich source of knowledge of membrane trafficking. Work on the fast axonal transport of membranous organelles in the earlier years by metabolic labeling and most recently by video-enhanced microscopy have added greatly to our understanding of membranous movements. Research by Allen, Lasek, Sheetz, and others in the squid giant axon led to the initial identification of kinesin as an anterograde transport motor. The polarity of MTs in axons is consistent with evidence suggesting that the movement of membranes anterogradely from the cell body toward the synapse uses the plus end–directed motor kinesin (Vale *et al.*, 1985), while membranes returning from the synapse via retrograde transport likely use dynein (Paschal *et al.*, 1987; Schnapp and Reese, 1989; Schroer *et al.*, 1989). The associa-

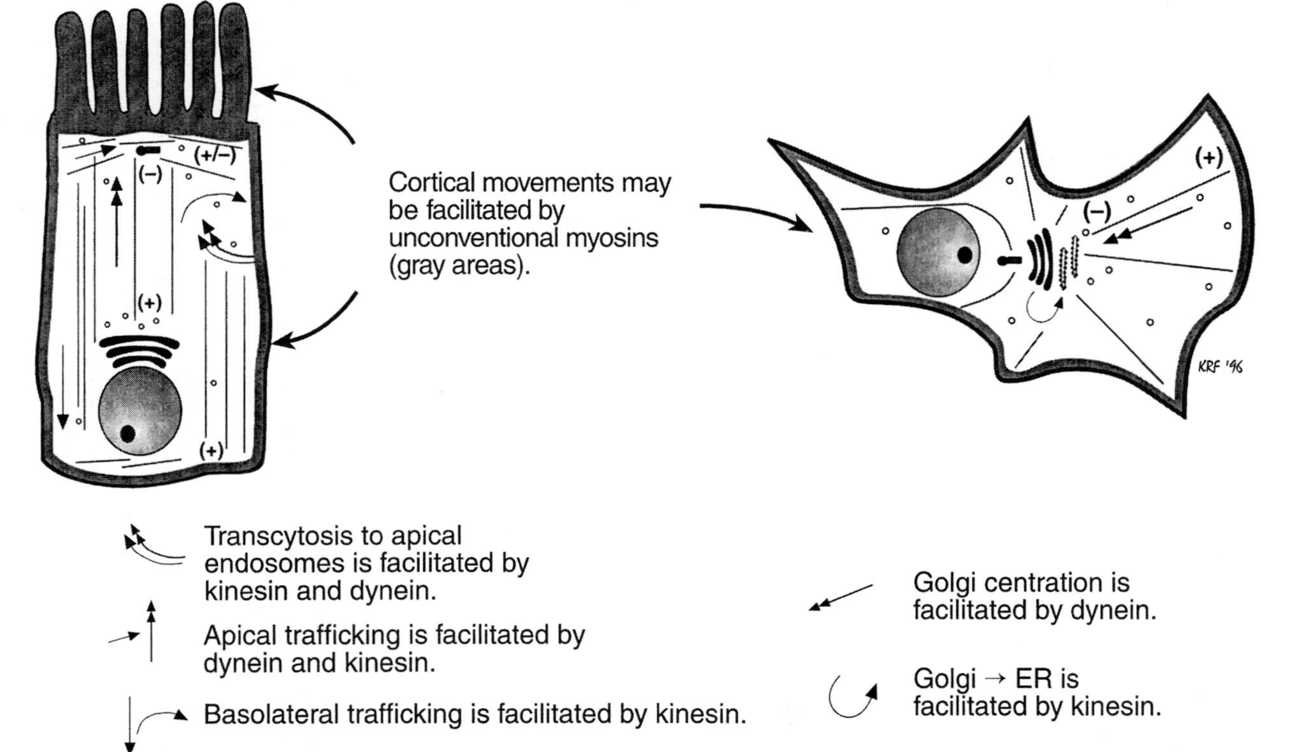

FIGURE 2 Roles of molecular motors in membrane movements. Movement of membranes in the deep regions of the cytoplasm of polarized epithelia and fibroblasts likely occurs on MTs using kinesins (single-headed arrows) and cytoplasmic dynein (double-headed arrows). In the actin-rich, MT-poor cell cortex (gray shading) actin-based motors may transport vesicles near the PM.

tion of these motors with moving organelles was elegantly demonstrated by the nerve ligation studies of Hirokawa *et al.* In these studies, the movement of membranes was blocked by a localized constriction in the axons of a mouse peripheral nerve. Immunolocalization analysis of the membranes that accumulated on either side of the constriction showed that both kinesin and dynein were associated with anterogradely moving organelles (Hirokawa *et al.,* 1990), while dynein and significantly less kinesin were associated with retrogradely moving organelles (Hirokawa *et al.,* 1991). It is thought that the anterogradely moving dynein is inactive, passive cargo that becomes activated as a motor in the axonal terminal by phosphorylation of the heavy chains (Dillman and Pfister, 1994).

Although movements of membranes in dendrites is thought to be driven by a fast axonal transport (Grafstein and Forman, 1980), it is unclear how directional movement can occur in the presence of MTs of both polarities. It has been proposed that perhaps the MTs of opposite polarities are physically separated into channels and only one group is used for motility, or that the differential distribution of MT-associated proteins confers motility onto only one polarity of MTs (Sheetz and Martenson, 1991).

C. *Fibroblasts*

Much of what we know about the structural dynamics of the endoplasmic reticulum (ER) and Golgi apparatus has come from studies in fibroblasts. The Golgi complex in fibroblasts is actively concentrated near the minus ends of MTs at the MTOC (Ho *et al.,* 1989; Thyberg and Moskalewski, 1985) by the action of dynein (Fig. 2) (Corthésy-Theulaz *et al.,* 1992). The intimate association of the MTOC and Golgi complex is indicated by the fact that the MTOC and Golgi complex in migrating fibroblasts are localized to the side of the nucleus that is facing the leading lamella (Kupfer *et al.,* 1983). This positioning of the Golgi complex is thought to be important in the MT-dependent insertion of newly synthesized material that is directed to the growing edge of fibroblasts (Rogalski *et al.,* 1984). Because of the MT orientation in fibroblasts, the transport of materials from the ER to the Golgi complex would require a dynein-like motor while movements post-Golgi (and in the Golgi-to-ER salvage pathway) require kinesin (Lippincott-Schwartz and Cole, 1995). The terms "retrograde transport" and "anterograde transport" have recently crept into the literature referring to MT-dependent Golgi → ER and ER → Golgi → cell surface movement in fibroblasts.

Unfortunately, this simplification is likely to be confusing because the MT polarity, as defined by the location of the MTOC, relative to Golgi complex location is different for fibroblasts than for epithelial cells. To avoid such confusion, it might be more appropriate to reserve the use of these terms to the axon, where they define functional aspects of neuronal cell biology.

V. ROLE OF ACTIN FILAMENTS AND UNCONVENTIONAL MYOSINS IN MEMBRANE MOVEMENTS

Actin filaments, like MTs, are polar structures as to end-specific assembly kinetics and transport properties. Nearly all eukaryotic cells contain a dense meshwork of actin filaments just beneath the PM that excludes organelles from the area immediately subjacent to the PM. This meshwork, constituting the cell cortex, is thought to be integral in such roles as PM support, integral membrane protein immobilization, receptor-mediated endocytosis, phagocytosis, receptor clustering, and cytokinesis. In addition to this meshwork, the cortical actin in many cells is also collected into bundles that extend the PM into surface specializations such as filopodia and microvilli (Fath and Burgess, 1995). The plus ends of actin filaments in these bundles and in the meshwork are associated with the PM from which they may nucleate. Although it is not commonly appreciated, there are a significant number of actin filaments in the cytoplasm away from the PM. Because these filaments are not as densely packed or as organized as the cortical filaments, they are difficult to identify by electron microscopy or immunofluorescence. For example, approximately 92% of the total actin in the squid giant axon is found in the interior of the axon, and 60% of this is polymerized (Morris and Lasek, 1984).

Although both plus end– and minus end–directed MT-based motors have been identified, to date only plus end–directed actin-based motors have been found. These motors are all members of the myosin superfamily, which at present contains at least 11 members (Mooseker and Cheney, 1995). Although myosins are a diverse class of enzymes, all myosins are mechanoenzymes that bind to actin in an ATP-dependent manner, possess an Mg-ATPase activity that is activated by actin under physiological conditions, and can use the energy liberated from ATP hydrolysis to move toward the membrane-associated plus end of actin filaments. Myosins share a structurally conserved "motor end" or head end of the molecule that has an ATP-binding and an actin-binding domain. Rayment *et al.*'s (1993a,b) solution for the 3-dimensional structure of the myosin head and how it may interact with actin provides a structural basis for how ATP hydrolysis may lead to a conformational change necessary for force production. The myosins are most divergent in the sequence in the carboxyl tail region that is thought to be specialized to bind to other myosin tails, to bind to actin filaments, or to bind to membranes. It is the tail that may determine the intracellular localization of these motor proteins and position them to perform specialized tasks (Mooseker, 1993). Thus it may be through the regulation of the binding of myosin to certain membranes that their transport is determined. What do myosins bind to in the membrane? Although the tails of some forms of myosin-I can bind directly to acidic phospholipids (Adams and Pollard, 1989), it is thought that some protein(s) most likely control this binding. For example, the contractile vacuole of *Acanthamoeba* selectively binds only the myosin-IC isoform, although other isoforms are present in the cell (Baines *et al.,* 1995; Doberstein *et al.,* 1993). Xu *et al.* (1995) have identified several proteins in *Acanthamoeba* that bind to the SH3 domain in the tail of myosin-IC. One of these proteins, called Acan125 (from its 125-kDa molecular mass), apparently can form complexes with myosin-I in the cytoplasm, for it co-immunoprecipitates with myosin-I from *Acanthamoeba* cytosol. The immunofluorescent co-localization of Acan125 and myosin-I on cellular organelles (likely amebastomes) further supports an *in vivo* association. Because the biochemical and immunofluorescence results suggest that there is abundant soluble Acan125 (and myosin-IC), it is unlikely that Acan125 is a simple membrane linker. Since the SH3 domains of other proteins interact at the membrane, it is tempting to propose that proteins such as Acan125 may regulate the binding of myosin-I to membranes in response to cell signals. Unfortunately, the SH3 domain is lacking from all forms of metazoan myosin-I (Mooseker and Cheney, 1995), including intestinal brush border myosin-I, which does bind to membranes.

Because myosins are actin-activated ATPases that are concentrated around membranes and can move isolated membranes relative to actin filaments in *in vitro* motility assays (Adams and Pollard, 1986), it has been suggested that they have some role in moving membranes along actin filaments *in vivo*. Although we have known about unconventional myosins for some time, our knowledge of their role in membrane movements has lagged behind the relatively rapid advances in our understanding of the MT-based motors. This delay in some respects is due to the inability to directly visualize single actin filaments by video microscopy. Advances have been made in the understanding of roles of unconventional myosins using gene knockouts in *Acanthamoeba* and *Dictyostelium*, although these efforts have been frustrated somewhat by the presence of multiple isoforms with apparent functional redundancy (see review by Fath and Burgess, 1994). Using immunofluorescence microscopy, unconventional myosins have been found in a punctate, vesicular distribution in liver cells (Coluccio and Conaty, 1993), fibroblasts (Conrad *et al.,* 1993; Wagner *et al.,* 1992), neuronal cells (Espreafico *et al.,* 1992; Wagner *et al.,* 1992), *Dictyostelium* (Fukui *et al.,* 1989), *Acanthamoeba* (Baines *et al.,* 1992; Yonemura and Pollard, 1992),

Drosophila embryos (Kellerman and Miller, 1992; Mermall *et al.,* 1994), and *Saccharomyces* (Lillie and Brown, 1992). Using immunoelectron microscopy, brush border myosin-I has been localized to the surface of cytoplasmic vesicles in intestinal epithelial cells (Drenckhahn and Dermietzel, 1988), myosin-IA is found on small vesicles in *Acanthamoeba* amoeba (Baines *et al.,* 1992), and myosin-V is found on vesicles in growth cones (Evans *et al.,* 1995). These immunolocalization results have been confirmed biochemically by the identification of myosins bound to isolated Golgi membranes from intestinal epithelia (Fath and Burgess, 1993), to small vesicles isolated from *Acanthamoeba* (Adams and Pollard, 1986), and to isolated *Dictyostelium* contractile vacuoles (Zhu and Clarke, 1992).

Although unconventional myosins are associated with intracellular organelles, there has been little evidence more directly showing that myosins are actually translocating vesicles inside cells. Several laboratories have begun using the isolated axoplasm from the squid giant axon to analyze the role of myosins and actin filaments in axonal membrane movements. In the video microscope, vesicles have been observed moving on actin filaments at the periphery of extruded axoplasm (Kuznetsov *et al.,* 1992). In order to show the directionality of these movements, actin filaments of known polarity were polymerized from the ends of acrosomal processes and added to extracts of squid axoplasm. Consistent with a myosin-based movement, axoplasmic vesicles translocated only toward the plus ends of the actin filaments (Bearer *et al.,* 1993; Langford *et al.,* 1994). Although it is not known whether a myosin is driving these movements, myosin-V has been identified in squid axoplasm (Cohen *et al.,* 1994) and an ~235-kDa myosin has been immunolocalized to the surface of some axoplasmic organelles (Bearer *et al.,* 1993). It will be of interest to see if myosin antibodies can be used to block vesicle motility in these assays, as well as in the interior of extruded axoplasm, as has been done for organelles isolated from *Acanthamoeba* (Adams and Pollard, 1986).

Perhaps the most compelling evidence for a role for myosins in cytoplasmic vesicle motility comes from studies of members of the type V class of myosins. Members of this family are two-headed unconventional myosins (reviewed by Mooseker and Cheney, 1995; Titus, 1993a,b). The best characterized member of this family, chicken brain myosin-V (p190, calmodulin), which is associated with vesicles in the Golgi area of cultured neurons and at the tips of growth cones (Evans *et al.,* 1995), has been shown to be a motor in motility assays (Cheney *et al.,* 1993). It is not known, however, if the myosin-V that is on the membranes is active in the movement of these membranes or is simply passive cargo that is being transported to its site of utilization in the growth cone. In *Saccharomyces*, a type V myosin called Myo2p is thought to be involved in the movement of secretory membranes along actin cables from the mother to the bud during cell division (Govindan *et al.,* 1995; Johnston *et al.,* 1991; Lillie and Brown, 1992). Temperature sensitive *myo2–66* mutants, defective in Myo2p, arrest in mitosis with an accumulation of cytoplasmic vesicles in the mother cell and a disorganized actin cytoskeleton. Because *myo2–66* is synthetically lethal in combination with several post-Golgi *sec* mutants, it is thought that Myo2p functions in post-Golgi secretion steps (Govindan *et al.,* 1995). It is not known, however, whether Myo2p functions on the surface of vesicles as a motor or at the PM in organizing the actin cytoskeleton. Another type V myosin, from the mouse *dilute* locus, is thought to be involved in melanosome movement in melanocytes. The *dilute* protein was shown to be associated with melanosomes by immunolocalization and immunoprecipitation (Provance and Mercer, 1995). The melanosomes in wild-type melanocytes are evenly distributed throughout the cytoplasm while melanosomes in *dilute* melanocytes are clustered around the nucleus.

Based largely on conjecture developed from its *in vitro* properties, on the polarity of actin filaments within microvilli, and on the morphology of microvillus growth, myosin-I in the intestinal epithelium has been proposed to move membranous vesicles transporting proteins and lipids from the Golgi complex or from basolateral endosomes through the terminal web of actin filaments to their sites of incorporation into the apical plasma membrane (Collins *et al.,* 1990; Conzelman and Mooseker, 1987, Fath and Burgess, 1993; Shibayama *et al.,* 1987). Electron microscopic immunolocalization of myosin-I on vesicles in the apical cytoplasm of enterocytes (Drenckhahn and Dermietzel, 1988) is consistent with such a membrane translocation role. Further support comes from work from our laboratory in which we have isolated a mixed population of Golgi-enriched cytoplasmic membranes from isolated intestinal epithelial cells and found that a subset of membranes, which bind actin filaments *in vitro* in an ATP-sensitive manner, possessed myosin-I as a cytoplasmically oriented peripheral membrane protein (Fath and Burgess, 1993). Recent genetic evidence for a role of myosin-I in membrane movements to the apical PM in intestinal epithelial cells has come from the laboratory of Louvard *et al.* These workers expressed full length and truncated forms of brush border myosin-I in polarized epithelial Caco-2 cells (Durrbach *et al.,* 1995). They found that a full-length motor domain was required for the correct localization of the myosin-I in microvilli. Expressed forms with truncated motor domains were associated with endogenous myosin-I in punctate structures along the basolateral domain of the cell. The normally apically targeted enzyme alkaline phosphatase was no longer transported apically, but was concentrated in the Golgi complex. Although it is not known how the expression of truncated myosin-I affects the expression and function of other cellular proteins, these data are consistent with a role for myosin-I in the transloca-tion of apically targeted vesicles in polarized epithelial cells.

VI. MULTIPLE MOTOR PROTEINS ARE ASSOCIATED WITH SINGLE CYTOPLASMIC VESICLES

We have outlined earlier some of the evidence of directed membrane movement in cells utilizing either an actin- or a MT-based system. In the past 3 or 4 years, evidence has begun to accumulate that not only do these two systems coexist in the same cells, but they may function as parallel or serial pathways for the movement of post-Golgi secretory vesicles (for review, see Atkinson *et al.,* 1992; Titus, 1993b). In nearly all eukaryotic cells, a carrier vesicle moving from the Golgi complex to the PM must traverse a MT-rich, actin filament–poor cell interior and then an actin filament–rich, MT-poor cell cortex. Thus, newly synthesized membranes may travel on MTs using a dynein or kinesin motor in the deep cytoplasm and then a myosin to traverse the actin-rich cortex to reach the PM. Evidence for several types of motors on one organelle has come from several independent sources. For example, Kuznetsov *et al.* (1992), using video-enhanced and fluorescence microscopy, followed single vesicles moving at the periphery of isolated squid axoplasm that apparently jumped from a MT track to an actin track, suggesting that both motors are on the same vesicle. What could be the role of both motors on the same axonal vesicle? Because actin filaments are clustered in domains surrounding the axonal MTs in the interior of the squid giant axon (Fath and Lasek, 1988), the actin filaments may provide a means of returning a vesicle that detached and drifted away from a MT back to the MT. Alternatively, the motors may function in different regions of the neuron. For example, a vesicle may use kinesin to move out of the cell body and down the axon, but, upon reaching the end of the MTs at the nerve terminal, bind to the actin filaments via its associated myosin motor and then translocate to the synaptic membrane. The presence of both types of motors on the same vesicle is consistent with the localization of 50- to 100-nm vesicles with myosin-V on their surfaces to both MTs and actin filaments in cultured superior cervical ganglion growth cones (Evans *et al.,* 1995).

Further independent evidence that a single vesicle may utilize both MT and actin-based motilities comes from studies in polarized epithelial cells. The MTs in polarized epithelial cells rarely, if at all, extend through the cell cortex to the PM (Drenckhahn and Dermietzel, 1988; Sandoz *et al.,* 1985). The cell cortex, especially that of the apical brush border, is rich in actin and likely has no MT-based movement (Burgess, 1987; Fath *et al.,*

1993; Louvard, 1989). The basolateral cortex of epithelial cells is also MT poor and actin rich, although not as organized as the apical surface. The basolateral membrane, beneath the zonula adherens actin circumferential ring, is composed of an anastomosing actin meshwork containing fodrin, ankyrin, and myosin-I and -II among other cytoskeletal proteins (Burgess, 1987; Mooseker, 1985). Therefore, the final 0.5–1 μm to the epithelial cell apical or basolateral PM must be traversed by a vesicle in the absence of MTs and through an actin meshwork. Thus, MT-based motors alone may not be sufficient to translocate a post-Golgi vesicle to the cell surface. An actin-based motor, perhaps attached to the same vesicle translocated first along MTs, may be required to move vesicles through the actin-rich terminal web to reach the PM (Fath and Burgess, 1993). Consistent with this contention, we have isolated Golgi-derived vesicles from intestinal epithelial cells that have myosin-I, cytoplasmic dynein, and dynactin on their cytoplasmic surfaces. Because myosin-I co-sedimented with dynein-containing Golgi membranes that were bound to MTs, we proposed that some membranes contain both myosin-I and dynein (Fath *et al.,* 1994).

Further evidence for a role of both types of motors working with parallel functions comes from genetics studies with *Saccharomyces*. In the aforementioned Myo2p mutants with defective vesicular transport, the overexpression of the kinesin related protein Smy1p, which co-localizes with Myo2p at sites of active growth in wild-type cells (S. Brown, personal communication), can suppress *myo2* defects (Johnston *et al.,* 1991; Lillie and Brown, 1992). Although Smy1p shares similarity with members of the kinesin protein family, it has not been shown to be a bona fide MT-based motor. In fact, both budding and Smy1p/Myo2p localization require actin and not MTs (S. Brown, personal communication). Furthermore, although these mutants accumulate vesicles in the mother cell, the bulk secretion of many proteins continues in these cells even when MTs are disrupted with nocodazole (Govindan *et al.,* 1995). This lack of additive effect of both actin and MT disruption suggests that Myo2 may not act in parallel with a MT-based transport system. It is thought that Smy1p is involved in a late stage of the secretory pathway, perhaps by helping Myo2p to reach and/or remain at the proper site.

VII. CONCLUSIONS

In the past 10 years we have traveled great distances along airways and highways toward an understanding of the role of molecular motors in the organization and movements of membrane compartments. We are beginning to realize that little is accomplished by simple diffusion and that directed transport is necessary for both the translocation and targeting of membranes to their correct destinations. Our depth of understanding of the mode of transport between membrane compartments still lags behind research coming from many laboratories concerning the mechanisms and components required for membrane budding and fusion. Now, as we understand more about how membranes may move between donor and acceptor compartments, a more complete picture of the synthetic pathways is emerging. In this chapter we have discussed the organization of the cytoskeleton and membrane trafficking in three model systems: fibroblasts, polarized epithelia, and neurons. Although there are differences in the positioning of the Golgi complex, ER, and PM relative to the MTOC in these varied cell types, there is a conservation of the pathways used to move between these compartments; only the identity of the motor is changed. In the coming years, those in the motility field who work on MT- and actin-based motors may find that their once seemingly independent pathways are merging. Future work must focus on how motors are targeted to selected membranes and how multiple motors on the same organelle are regulated in different regions of the cytoplasm.

References

Achler, C., Filmer, D., Merte, C., and Drenckhahn, D. (1989). Role of microtubules in polarized delivery of apical membrane proteins to the brush border of the intestinal epithelium. *J. Cell Biol.* **109,** 179–189.

Adams, R. J., and Pollard, T. D. (1986). Propulsion of organelles isolated from *Acanthamoeba* along actin filaments by myosin-I. *Nature* (*London*) **322,** 754–756.

Adams, R. J., and Pollard, T. D. (1989). Binding of myosin I to membrane lipids. *Nature* (*London*) **340,** 565–568.

Atkinson, S. J., Doberstein, S. K., and Pollard, T. D. (1992). Moving off the beaten track. *Curr. Biol.* **2,** 326–328.

Bacallao, R., Antony, C., Dotti, C., Karsenti, E., Stelzer, E. H. K., and Simons, K. (1989). The subcellular organization of Madin–Darby canine kidney cells during the formation of a polarized epithelium. *J. Cell Biol.* **109,** 2817–2832.

Baines, I. C., Brzeska, H., and Korn, E. D. (1992). Differential localization of *Acanthamoeba* myosin I isoforms. *J. Cell Biol.* **119,** 1193–1203.

Baines, I. C., Corigliano-Murphy, A., and Korn, E. D. (1995). Quantification and localization of phosphorylated myosin I isoforms in *Acanthamoeba castellanii. J. Cell Biol.* **130,** 591–603.

Bartles, J. R., and Hubbard, A. L. (1988). Plasma membrane protein sorting in epithelial cells: Do secretory pathways hold the key? *Trends Biochem. Sci.* **13,** 181–184.

Bearer, E. L., DeGiorgis, J. A., Bodner, R. A., Kao, A. W., and Reese, T. S. (1993). Evidence for myosin motors on organelles in squid axoplasm. *Proc. Natl. Acad. Sci. U.S.A.* **90,** 11252–11256.

Bennett, G., Carlet, E., Wild, G., and Parsons, S. (1984). Influence of colchicine and vinblastine on the intracellular migration of secretory and membrane glycoproteins: III. Inhibition of intracellular migration of membrane glycoproteins in rat intestinal columnar cells and hepatocytes as visualized by light and electron-microscope radioautography after ^{3}H-fucose injection. *Am. J. Anat.* **170,** 545–566.

Breitfeld, P. P., McKinnon, W. C., and Mostov, K. E. (1990). Effect of nocodazole on vesicular traffic to the apical and basolateral surfaces of polarized MDCK cells. *J. Cell Biol.* **111,** 2365–2373.

Burgess, D. R. (1987). The brush border: A model for structure, biochemistry, motility, and assembly of the cytoskeleton. *Adv. Cell Biol.* **1,** 31–58.

Cheney, R. E., O'Shea, M. K., Heuser, J. E., Coelho, M. V., Wolenski, J. S., Espreafico, E. M., Forscher, P., Larson, R. E., and Mooseker, M. S. (1993). Brain myosin-V is a two-headed unconventional myosin with motor activity. *Cell* **75,** 13–23.

Cohen, D. L., Kuznetsov, S. A., and Langford, G. M. (1994). Myosin V in squid axoplasm and optic lobes. *Mol. Biol. Cell* **5,** 278a.

Collins, K., Sellers, J. R., and Matsudaira, P. (1990). Calmodulin dissociation regulates brush border myosin I (110-kD-calmodulin) mechanochemical activity in vitro. *J. Cell Biol.* **110,** 1137–1147.

Coluccio, L. M., and Conaty, C. (1993). Myosin-I in mammalian liver. *Cell Motil. Cytoskeleton* **24,** 189–199.

Conrad, P. A., Giuliano, K. A., Fisher, G., Collins, K., Matsudaira, P. T., and Taylor, D. L. (1993). Relative distribution of actin, myosin I, and myosin II during wound healing response of fibroblasts. *J. Cell Biol.* **120,** 1381–1391.

Conzelman, K. A., and Mooseker, M. S. (1987). The 110-kD protein-calmodulin complex of the intestinal microvillus is an actin-activated MgATPase. *J. Cell Biol.* **105,** 313–324.

Corthésy-Theulaz, I., Pauloin, A., and Pfeffer, S. (1992). Cytoplasmic dynein participates in the centrosomal localization of the Golgi complex. *J. Cell Biol.* **118,** 1333–1345.

Dillman, J. F., and Pfister, K. K. (1994). Differential phosphorylation in vivo of cytoplasmic dynein associated with anterogradely moving organelles. *J. Cell Biol.* **127,** 1671–1681.

Doberstein, S. K., Baines, I. C., Wiegand, G., Korn, E. D., and Pollard, T. P. (1993). Inhibition of contractile vacuole function *in vivo* by antibodies against myosin-I. *Nature* (*London*) **365,** 841–843.

Dotti, C. G., Parton, R. G., and Simons, K. (1991). Polarized sorting of glypiated proteins in hippocampal neurons. *Nature* (*London*) **349,** 158–161.

Drenckhahn, D., and Dermietzel, R. (1988). Organization of the actin filament cytoskeleton in the intestinal brush border: A quantitative and qualitative immunoelectron microscope study. *J. Cell Biol.* **107,** 1037–1048.

Durrbach, A., Louvard, D., and Coudrier, E. (1995). Brush border myosin I is required for the apical translocation of GPI anchor membrane proteins. *Mol. Biol. Cell* **6,** 268a.

Eilers, U., Klumperman, J., and Hauri, H.-P. (1989). Nocodazole, a microtubule-active drug, interferes with apical protein delivery in cultured intestinal epithelial cells (Caco-2). *J. Cell Biol.* **108,** 13–22.

Espreafico, E. M., Cheney, R. E., Matteoli, M., Nascimento, A. A. C., DeCamilli, P. V., Larson, R. E., and Mooseker, M. S. (1992). Primary structure and cellular localization of chicken brain myosin-V (p190), an unconventional myosin with calmodulin light chains. *J. Cell Biol.* **119,** 1541–1558.

Euteneur, U., Koonce, M. P., Pfister, K. K., and Schliwa, M. (1988). An ATPase with properties expected for the organelle motor of the giant amoeba, *Reticulomyxa. Nature* (*London*) **332,** 176–178.

Evans, L. L., Hammer, J., and Bridgman, P. C. (1995). Subcellular localization of myosin V in nerve growth cones. *Mol. Biol. Cell* **6,** 145a.

Fath, K. R., and Burgess, D. R. (1993). Golgi-derived vesicles from developing epithelial cells bind actin filaments and possess myosin-I as a cytoplasmically oriented peripheral membrane protein. *J. Cell Biol.* **120,** 117–127.

Fath, K. R., and Burgess, D. R. (1994). Membrane motility mediated by unconventional myosin. *Curr. Opin. Cell Biol.* **6,** 131–135.

Fath, K. R., and Burgess, D. R. (1995). Not actin alone. *Curr. Biol.* **5,** 591–593.

Fath, K., and Lasek, R. (1988). Two classes of actin microfilaments are associated with the inner cytoskeleton of axons. *J. Cell Biol.* **107,** 613–621.

Fath, K. R., Trimbur, G. M., and Burgess, D. R. (1994). Molecular motors are differentially distributed on Golgi membranes from polarized epithelial cells. *J. Cell Biol.* **126,** 661–675.

Fukui, Y., Lynch, T. J., Brzeska, H., and Korn, E. D. (1989). Myosin I is located at the leading edges of locomoting *Dictyostelium* amoebae. *Nature* (*London*) **341,** 328–331.

Gilbert, T., Le Bivic, A., Quaroni, A., and Rodriguez-Boulan, E. (1991). Microtubular organization and its involvement in the biogenetic pathways of plasma membrane proteins in Caco-2 intestinal epithelial cells. *J. Cell Biol.* **113,** 275–288.

Govindan, B., Bowser, R., and Novick, P. (1995). The role of Myo2, a yeast class V myosin, in vesicular transport. *J. Cell Biol.* **128,** 1055–1068.

Grafstein, B., and Forman, D. S. (1980). Intracellular transport in neurons. *Physiol. Rev.* **60,** 1167–1283.

Hirokawa, N., Sato-Yoshitake, R., Yoshida, T., and Kawashima, T. (1990). Brain dynein (MAP1C) localizes on both anterogradely and retrogradely transported membranous organelles in vivo. *J. Cell Biol.* **111,** 1027–1037.

Hirokawa, N., Sato-Yoshitake, R., Kobayashi, N., Pfister, K. K., Bloom, G. S., and Brady, S. T. (1991). Kinesin associates with anterogradely transported membranous organelles *in vivo. J. Cell Biol.* **114,** 295–302.

Ho, W.-C., Allan, V. J., van Meer, G., Berger, E. G., and Kreis, T. E. (1989). Reclustering of scattered Golgi elements occurs along microtubules. *Eur. J. Cell Biol.* **48,** 250–263.

Holzbaur, E. L. F., and Vallee, R. B. (1994). Dyneins: Molecular structure and cellular function. *Annu. Rev. Cell Biol.* **10,** 339–372.

Hoyt, M. A. (1994). Cellular roles of kinesin and related proteins. *Curr. Biol.* **6,** 63–68.

Hugon, J. S., Bennett, G., Pothier, P., and Ngoma, Z. (1987). Loss of microtubules and alteration of glycoprotein migration in organ cultures of mouse intestine exposed to nocadazole or colchicine. *Cell Tissue Res.* **248,** 653–662.

Hunziker, W., Male, P., and Mellman, I. (1990). Differential microtubule requirements for transcytosis in MDCK cells. *EMBO J.* **9,** 3515–3525.

Johnston, G. C., Prendergast, J. A., and Singer, R. A. (1991). The *Saccharomyces cerevisiae* MYO2 gene encodes an essential myosin for vectorial transport of vesicles. *J. Cell Biol.* **113,** 539–551.

Karki, S., and Holzbaur, E. L. F. (1995). Affinity chromatography demonstrates a direct binding between cytoplasmic dynein and the dynactin complex. *J. Biol. Chem.* **270,** 28806–28811.

Kellerman, K. A., and Miller, K. G. (1992). An unconventional myosin heavy chain gene from *Drosophila melanogaster. J. Cell Biol.* **119,** 823–834.

Kupfer, A. G., Louvard, D., and Singer, S. J. (1983). Polarization of the Golgi apparatus and the microtubule-organizing center in cultured fibroblasts at the edge of an experimental wound. *Proc. Natl. Acad. Sci. U.S.A.* **79,** 2603–2607.

Kuznetsov, S. A., Langford, G. M., and Weiss, D. G. (1992). Actin-dependent organelle movement in squid axoplasm. *Nature* (*London*) **356,** 722–725.

Lacey, M. L., and Haimo, L. T. (1992). Cytoplasmic dynein is a vesicle protein. *J. Biol. Chem.* **267,** 4793–4798.

Lafont, F., Burkhardt, J. K., and Simons, K. (1994). Involvement of microtubule motors in basolateral and apical transport in kidney cells. *Nature* (*London*) **372,** 801–803.

Langford, G. M., Kuznetsov, S. A., Johnson, D., Cohen, D. L., and Weiss, D. G. (1994). Movement of axoplasmic organelles on actin filaments assembled on acrosomal processes: Evidence for a barbed-end-directed organelle motor. *J. Cell Sci.* **107,** 2291–2298.

Lillie, S. H., and Brown, S. S. (1992). Suppression of a myosin defect by a kinesin-related gene. *Nature* (*London*) **356,** 358–361.
Lin, S. X. H., and Collins, C. A. (1992). Immunolocalization of cytoplasmic dynein to lysosomes in cultured cells. *J. Cell Sci.* **101,** 125–137.
Lippincott-Schwartz, J., and Cole, N. B. (1995). Roles for microtubules and kinesin in membrane traffic between the endoplasmic reticulum and the Golgi complex. *Biochem. Soc. Trans.* **23,** 544–548.
Louvard, D. (1989). The function of the major cytoskeletal components of the brush border. *Curr. Opin. Cell Biol.* **1,** 51–57.
Luby-Phelps, K. (1994). Physical properties of cytoplasm. *Curr. Biol.* **6,** 3–9.
Matter, K., Bucher, K., and Hauri, H.-P. (1990). Microtubule perturbation retards both the direct and the indirect apical pathway but does not affect sorting of plasma membrane proteins in intestinal epithelial cells (Caco-2). *EMBO J.* **9,** 3163–3170.
Mays, R. W., Beck, K. A., and Nelson, J. W. (1994). Organization and function of the cytoskeleton in polarized epithelial cells: A component of the protein sorting machinery. *Curr. Opin. Cell Biol.* **6,** 16–24.
Mellman, I., Matter, K., Yamamoto, E., Pollack, N., Roome, J., Felsenstein, K., and Roberts, S. (1995). Mechanisms of molecular sorting in polarized cells: relevance to Alzheimer's disease. *In* "Alzheimer's Disease: Lessons from Cell Biology" (K. S. Kosik, Y. Christen, and D. J. Selkoe, eds.), pp. 14–26. Springer-Verlag, Berlin.
Mermall, V., McNally, J. G., and Miller, K. G. (1994). Transport of cytoplasmic particles catalysed by an unconventional myosin in living *Drosophila* embryos. *Nature* (*London*) **369,** 560–562.
Mooseker, M. S. (1985). Organization, chemistry and assembly of the cytoskeletal apparatus of the intestinal brush border. *Annu. Rev. Cell Biol.* **1,** 209–241.
Mooseker, M. (1993). A multitude of myosins. *Curr. Biol.* **3,** 245–248.
Mooseker, M. S., and Cheney, R. E. (1995). Unconventional myosins. *Annu. Rev. Cell Dev. Biol.* **11,** 633–675.
Morris, J., and Lasek, R. (1984). Monomer-polymer equilibria in the axon: direct measurement of tubulin and actin as polymer and monomer in axoplasm. *J. Cell Biol.* **98,** 2064–2076.
Parczyk, K., Haase, W., and Kondor-Koch, C. (1989). Microtubules are involved in the secretion of proteins at the apical cell surface of the polarized epithelial cell, Madin-Darby kidney. *J. Biol. Chem.* **264,** 16837–16846.
Paschal, B. M., Shpetner, H. S., and Vallee, R. B. (1987). Map 1C is a microtubule-activated ATPase which translocates microtubules in vitro and has dynein-like properties. *Nature* (*London*) **105,** 1273–1282.
Provance, D. W. Jr., and Mercer, J. A. (1995). Association of *dilute* with melanosomes in melanocytes. *Mol. Biol. Cell.* **6,** 146a.
Rayment, I., Holden, H. M., Whittacker, M., Yohn, C. B., Lorenz, M., Holmes, K. C., and Milligand, R. A. (1993a). Structure of the actin-myosin complex and its implications for muscle contraction. *Science* **261,** 58–65.
Rayment, I., Rypniewski, W. R., Schmidt-Base, K., Smith, R., Tomchick, D. R., Benning, M. M., Winkelmann, D. A., Wesenberg, G., and Holden, H. M. (1993b). Three-dimensional structure of myosin subfragment-1: A molecular motor. *Science* **261,** 50–57.
Rindler, M. J., Ivanov, I. E., and Sabatini, D. D. (1987). Microtubule-acting drugs lead to the nonpolarized delivery of the influenza hemagglutinin to the cell surface of polarized Madin-Darby canine kidney cells. *J. Cell Biol.* **104,** 231–241.
Rogalski, A. A., Bergmann, J. E., and Singer, S. J. (1984). Effect of microtubule assembly status on the intracellular processing and surface expression of an integral protein of the plasma membrane. *J. Cell Biol.* **99,** 1101–1109.

Salas, P. J. I., Misek, D. E., Vega-Salas, D. E., Gundersen, D., Cereijido, M., and Rodriguez-Boulan, R. (1986). Microtubules and actin filaments are not critically involved in the biogenesis of epithelial cell surface polarity. *J. Cell Biol.* **102,** 1853–1867.

Sandoz, D., Lainé, M.-C., and Nicolas, G. (1985). Distribution of microtubules within the intestinal terminal web as revealed by quick-freezing and cryosubstitution. *Eur. J. Cell Biol.* **39,** 481–484.

Schnapp, B. J., and Reese, T. S. (1989). Dynein is the motor for retrograde axonal transport of organelles. *Proc. Natl. Acad. Sci. U.S.A.* **86,** 1548–1552.

Schroer, T. A., Steuer, E. R., and Sheetz, M. P. (1989). Cytoplasmic dynein is a minus end–directed motor for membranous organelles. *Cell* **56,** 937–946.

Sharp, D. J., Yu, W., and Baas, P. W. (1995). Transport of dendritic microtubules establishes their nonuniform polarity orientation. *J. Cell Biol.* **130,** 95–103.

Sheetz, M. P., and Martenson, C. H. (1991). Axonal transport: Beyond kinesin and cytoplasmic dynein. *Curr. Opin. Neurobiol.* **1,** 393–398.

Shibayama, T., Carboni, J. M., and Mooseker, M. S. (1987). Assembly of the intestinal brush border: Appearance and redistribution of microvillar core proteins in developing chick enterocytes. *J. Cell Biol.* **105,** 335–344.

Simons, K., and Wandinger-Ness, A. (1990). Polarized sorting in epithelia. *Cell* **62,** 207–210.

Thyberg, J., and Moskalewski, S. (1985). Microtubules and the organization of the Golgi complex. *Exp. Cell Res.* **159,** 1–16.

Titus, M. A. (1993a). From fat yeast and nervous mice to brain myosin-V. *Cell* **75,** 9–11.

Titus, M. A. (1993b). Myosins. *Curr. Opin. Cell Biol.* **5,** 77–81.

Trahair, J. F., Neutra, M. R., and Gordon, J. I. (1989). Use of transgenic mice to study the routing of secretory proteins in intestinal epithelial cells: Analysis of human growth hormone compartmentalization as a function of cell type and differentiation. *J. Cell Biol.* **109,** 3231–3242.

Vale, R. D., Reese, T. S., and Sheetz, M. P. (1985). Identification of a novel force-generating protein, kinesin, involved in microtubule-based motility. *Cell* **42,** 39–50.

van der Sluijs, P., Bennett, M. K., Antony, C., Simons, K., and Kreis, T. E. (1990). Binding of exocytic vesicles from MDCK cells to microtubules *in vitro*. *J. Cell Sci.* **95,** 545–553.

Vaughan, K. T., and Vallee, R. B. (1995). Cytoplasmic dynein binds dynactin through a direct interaction between the intermediate chains and p150^{Glued}. *J. Cell Biol.* **131,** 1507–1516.

Wagner, M. C., Barylko, B., and Albanesi, J. P. (1992). Tissue distribution and subcellular localization of mammalian myosin I. *J. Cell Biol.* **119,** 163–170.

Walker, R. A., and Sheetz, M. P. (1993). Cytoplasmic microtubule-associated motors. *Annu. Rev. Biochem.* **62,** 429–451.

Xu, P., Zott, A. S., and Zott, H. G. (1995). Identification of Acan125 as a myosin-I–binding protein present with myosin-I on cellular organelles of *Acanthamoeba*. *J. Biol. Chem.* **270,** 25316–25319.

Yonemura, S., and Pollard, T. D. (1992). The localization of myosin I and myosin II in *Acanthamoeba* by fluorescence microscopy. *J. Cell Sci.* **102,** 629–642.

Zhu, Q., and Clarke, M. (1992). Association of calmodulin and an unconventional myosin with the contractile vacuole complex of *Dictyostelium discoideum*. *J. Cell Biol.* **118,** 347–358.

CHAPTER 5

Membrane-Cytoskeleton Interaction in Regulated Exocytosis and Apical Insertion of Vesicles in Epithelial Cells

Xuebiao Yao and John G. Forte
Department of Molecular and Cell Biology, University of California, Berkeley, California 94720

I. INTRODUCTION

Most eukaryotic cells have some degree of spatial asymmetry arising ontogenically from the asymmetry of cell division. Spatial asymmetry is highly developed within all epithelial cells, for which their polarization is intrinsically linked to function. Among the many functions of polarized epithelial cells are the regulated and asymmetrical transport processes that involve the turnover of cytoplasmic vesicles with the plasma membrane through exocytosis and endocytosis. There are several categorical types of regulated exocytic processes in epithelial cells:

1. Translocation and fusion of secretory vesicles from the cytoplasm to the apical plasma membrane and subsequent release of vesicular contents (e.g., acetylcholine-stimulated amylase release from pancreatic acinar cells).
2. Translocation and recruitment of transporter proteins to the apical plasma membrane of the cells with consequent functional activation, such as the antidiuretic hormone-(ADH) mediated recruitment of water channels in the amphibian bladder and mammalian renal collecting duct.
3. A regulated exocytic event that incorporates both translocation of the transporter to the apical plasma membrane and release of vesicular contents into the extracellular fluid (e.g., acetylcholine-stimulated recruitment of H,K-ATPase for hydrochloric acid (HCl) secretion and release of intrinsic factor in gastric parietal cells).

All of these processes are a special form of exocytosis, involving the movement, docking, and insertion of vesicular membranes and their constituent proteins into the plasma membrane. When the stimulatory signal is removed, the transporters are withdrawn back to the cytoplasm by endocytic processes that involve elements such as dynamin (Takei *et al.*, 1995; Hinchaw and Schmid, 1995) and the small GTP-binding protein rab5 and its accessory proteins (Stenmark *et al.*, 1995). Every step within these dynamic events is modulated by cytoskeletal components within the cell.

Certainly there are differences among these types of exocytic events, and they may be further contrasted with another type of fast exocytosis observed in nerve terminals, where the trafficking of vesicles and release of vesicular contents occur very rapidly. However, there are several general themes underlying these different forms of exocytosis; these are reviewed by Gayer, Campanelli, and Scheller (see Chapter 12, this volume). In this chapter we give an overview of several specific examples of regulated exocytosis, including the role of the actin-based cytoskeleton and the cooperative actions of actin-binding proteins in these systems.

Actin is a primitive and very abundant protein, representing about 5% of the total protein in most nonmuscle cells, with the remarkable property to self-assemble and disassemble between a monomeric form, called G-actin, and an oligomeric filamentous form (F-actin). Under the conditions of ionic strength and temperature within most cells, the equilibrium for this reaction lies far to the polymerized form (>99%), and yet in virtually all cells mechanisms exist to restrain the steady state for G-actin–F-actin interconversion so that G-actin is frequently as high as 20–60% of the total actin. This steady state is achieved by a variety of proteins known as actin-binding proteins that react with actin monomer or filaments so that they may be rapidly interconverted when necessary. A second and equally important problem is the spatial organization of actin within the cell so that specific localized pools might be altered as conditions dictate; again, this is largely a function of the distribution of specific actin-binding proteins. The diverse actin-binding proteins can be broadly grouped within three categorical functional realms, after the suggestion of Stossel (1989). One group of actin-binding proteins regulates the steady state between polymerization to form F-actin and depolymerization to form G-actin, including those that sequester, nucleate, sever, and cap the actin forms. A second group of actin-binding proteins promote some higher order of complexity by cross-linking and organizing the filaments into tight or loose bundles, or orthogonal nets. The third group of organizing proteins serves to form the associations between actin filaments and membranes, including both the plasma membrane and various organelle surfaces. In addition, new data suggest that a high degree of morphological and functional asymmetry within the actin cytoskeleton, especially in polarized epithelia, may also be related to spatial distribution of actin isoforms within the cell (Yao *et al.,* 1995a; Yeh and Svoboda, 1994).

II. THE BRUSH BORDER CYTOSKELETON

Surface protrusions are found on almost all higher eukaryotes. Microvilli, filopodia, spikes, membrane ruffles, and the like all contain an actin-based cytoskeleton. Since these actin-based structures are dynamic and fragile, it is difficult to study their functionality in many cell types. For this reason, the experimental accessibility of intestinal microvilli has drawn much attention, and we introduce a current model for the cytoskeletal network present in apical microvilli of intestinal epithelial cells as a comparative example for exploring the cortical substructure in cells in which the actin cytoskeleton is less highly organized and more dynamic.

Apical microvilli of intestinal epithelial cells, the so-called brush border, were first isolated by Miller and Crane (1961). The brush border is composed of about 1000 highly organized microvilli, each with a core of F-actin bundles inserting into a second meshlike actin-based structure called the terminal web. As the procedures for isolating and purifying brush borders improved (Bretscher and Weber, 1978), the resolution of cortical structure and the structure of the terminal web became apparent.

The core of each microvillus contains about 20 F-actin bundles, all with uniform polarity and approximately 1–2 μm in length (Mooseker and Tilney, 1975). Most of the actin bundle is confined to the microvillus and surrounded by the plasma membrane, but about one-third projects into the apical cytoplasm as a rootlet (Bretscher, 1991). By subjecting the brush borders to shear force, the microvilli are dissociated from the protruding rootlet and thus purified. After extracting purified brush borders with Triton X100 in the presence of a Ca^{2+} chelator, the remnant microvillar core actin structure consists of five major peptides, including actin, villin, fimbrin, and brush border myosin-I (Bretscher and Weber, 1978; Matsudaira and Burgess, 1979). There are also several minor proteins, among which ezrin has been characterized.

A. Villin

Villin (a 95-kDa polypeptide) is a major F-actin–bundling protein of the microvillar core. Villin has versatile properties; in the presence of Ca^{2+} it severs actin filaments into shorter filaments, whereas in the absence of Ca^{2+} it bundles actin filaments (Bretscher and Weber, 1980; Mooseker *et al.,* 1980; Matsudaira and Burgess, 1982). In addition, villin nucleates the assembly of actin in a Ca^{2+}-dependent pattern (Mooseker *et al.,* 1980; Craig and Powell, 1980). Studies show that villin possesses the activity of F-actin nucleation, capping, and severing. Expression of villin is highly tissue specific, being found primarily in brush borders of the small intestine and kidney proximal tubule, in pancreatic acinar cells, and in the visceral endoderm of the postimplantation embryo (Bretscher *et al.,* 1981; Ezzel *et al.,* 1989). Louvard and his colleagues (Costa de Beauregard *et al.,* 1995) showed that the assembly of brush borders is impaired in polarized epithelial intestinal cells when the expression of villin was suppressed by antisense RNA.

B. Fimbrin

Fimbrin (a 68-kDa polypeptide) was first isolated from intestinal microvilli, and has been shown to arrange F-actin filaments *in vitro* so that all filaments within a bundle have the same polarity (Bretscher, 1981,1982; Glenney *et al.,* 1981). This bundling property is sensitive to ionic strength and millimolar levels of Mg^{2+}, but insensitive to Ca^{2+} in the physiological range (Bretscher, 1981; de Arruda *et al.,* 1990). In contrast to villin, fimbrin is widely distributed in all nonmuscle cells. Fimbrin is concentrated in structures such as microvilli and microspikes that contain compact actin filament bundles (Bretscher and Weber, 1980). The cDNA sequence of chicken fimbrin revealed that fimbrin is highly homologous to the human plastins and appears to represent a plastin isoform (de Arruda *et al.*, 1990). Although the fimbrins/plastins are a family of highly conserved actin-binding proteins, each isoform displays a remarkable tissue specificity. For example, T-plastin is normally present in epithelial and mesenchymal cells, whereas L-plastin is expressed in hematopoietic cells. By overexpressing plastin isoforms in an epithelial cell line (LLC-PK1 cells), Arpin *et al.* (1994) showed that T-plastin binds tightly to the actin-based cytoskeleton and induces shape changes in microvilli. In contrast, L-plastin has no effect on microvillar structure and binds loosely to the actin cytoskeleton. Expression studies further demonstrated that T-plastin and villin had differential organizational effects on microvilli; T-plastin–positive cells had thick microvilli concentrated near cell–cell contacts, whereas cells overexpressing villin had long microvilli over the entire apical surface.

C. Brush Border Myosin-I

A great deal of attention has been given to the lateral bridge that links the core actin bundle to the plasma membrane since the ultrastructure of the brush border was resolved in detail. Based on the initial observation that treatment of isolated microvilli with ATP solubilized a 110-kDa polypeptide with concomitant disappearance of lateral bridges (Matsudaira and Burgess, 1979), it was proposed that this polypeptide was a major component of the lateral bridge structure. Collins and Borysenko (1984) partially purified the 110-kDa polypeptide, noting that this polypeptide had enzymatic properties characteristic of nonmuscle myosin. The studies of Pollard and Korn (1973) offered compelling evidence that the 110-kDa microvillar polypeptide is a single-headed myosin related to the nonconventional myosin-I originally identified in *Acanthamoeba* but since found to be more generally distributed. Further characterization, using monoclonal antibodies against *Acanthamoeba* myosin (Carboni *et al.*, 1988) and cDNA cloning (Halsall and Hammer, 1990), confirmed that the 110-kDa brush border peptide belongs the the myosin-I family (referred as to brush border myosin-I). All the myosins share a similar head region, including an actin-binding domain and an ATP-binding domain. It is the shorter tail region of the myosin-I family that displays the distinctive feature of binding to active membrane regions (e.g., microvilli, microspikes, filopodia, leading lamellae); thus it is thought to represent a motor for interactions and motions between membranes and cytoskeleton. Presently, it is not clear how brush border myosin-I is attached to the plasma membrane, although it has been speculated that the association might be analogous to the binding of *Acanthamoeba* myosin-I to acidic phospholipids (Adams and Pollard, 1989). Unfortunately, there is no direct evidence for the translocation of vesicles by myosin motors within cells. However, *in vitro* assays have demonstrated vesicles prepared from squid axons (Kuznetsov *et al.,* 1994) and algal cells (Kachar, 1985) being driven by ATP-dependent myosin motors along actin cables and reconstituted actin filaments. Another class of unconventional myosins, chicken myosin-V, has also been implicated in vesicular transport by virtue of its localization to Golgi–vicinal vesicles and functional properties. Myosin-V has two heads and a short stalk; it has Mg^{2+}-ATPase activity and is a barbed-end–directed motor along actin filaments (Cheney *et al.,* 1993).

D. Ezrin

Ezrin is an 80-kDa polypeptide initially identified as a minor component of brush borders from small intestine (see Bretscher, 1991) and as a *src* kinase substrate in A431 cells (Hunter and Cooper, 1981). Several studies suggest that ezrin is a membrane–cytoskeleton linker protein that may be coupled to cell activation and signaling pathways via phosphorylation. Stimulation of A431 cells with epidermal growth factor caused the formation of surface microspikes and microvilli as well as the tyrosine phosphorylation of ezrin, which was localized to these same surface structures (Bretscher, 1989). The phosphorylation of ezrin concomitant with cAMP-mediated stimulation of gastric parietal cells (Urushidani *et al.,* 1987) suggested that the putative membrane–cytoskeleton linker might play some role in the membrane transformation and recruitment underlying gastric acid secretion. Based on cDNA analysis, ezrin is classified as a member of the band 4.1 superfamily, supporting the notion that ezrin might link actin filaments to the plasma membrane (Gould *et al.,* 1989). Sequence data now identify several proteins with high homology to ezrin, including radixin, moesin, merlin, and EM10 (Turunen *et al.,* 1994). The most striking homologies exist within the N-terminal 300 amino acids and at the C-terminal putative actin-binding domain(s). Funayama *et al.* (1991) found that radixin binds to the actin filament at its plus end. Moreover, this group has shown that ezrin, radixin, and moesin all have a mutual plasma membrane target, CD44 (Tsükita *et al.,* 1994). In an attempt to reveal the actin-binding site, Turunen *et al.* (1994) expressed a series of ezrin fragments and showed that ezrin has a C-terminal actin-binding site that is conserved among the ezrin–radixin–moesin family. However, there has been debate as to whether native ezrin does bind to actin filaments, based largely on the lack of success in the initial attempts to probe for an association between skeletal muscle actin and intestinal ezrin (Bretscher, 1983). Because of an observed immunocytochemical co-localization of ezrin with the cytoplasmic β-actin isoform (but not the cytoplasmic γ-actin isoform) at the apical membrane of parietal cells, and a co-immunoprecipitation of ezrin and β-actin from parietal cell extracts, Yao *et al.* (1996a) suggested that the interaction of ezrin with F-actin was isoform specific. Using affinity chromatography, Shuster and Herman (1995) found that ezrin was preferentially retained on immobilized actin filaments built from cytoplasmic β-actin monomers, but not from skeletal muscle α-actin, although their *in vitro* reconstitution experiment suggested that the association between β-actin and ezrin was indirect, possibly via a 72-kDa polypeptide.

III. ASYMMETRIC STRUCTURE OF EPITHELIAL CELLS AND POLARIZED EXOCYTOSIS IN EPITHELIAL CELLS

Polarity of epithelial cells exhibits several levels of complexity, including (1) structural polarity involving the organization of intracellular organelles; (2) membrane polarity, the asymmetrical distribution of plasma membrane components between apical and basolateral membrane domains; and (3) the functional manifestation of structural and membrane polarity as a secretory activity that occurs at the apical plasma membrane. The two plasma membrane domains, basolateral and apical membranes, serve very different functions and therefore have quite disparate protein and lipid compositions. Studies suggest that the machinery involved in asymmetric cell division and the establishment cell polarity might be conserved throughout the eukaryotic kingdom (Way *et al.*, 1994). Here we briefly review the operation of regulated exocytic processes in the context of several cell-physiological transport systems.

A. Amylase Release from Pancreatic Acinar Cells

In the pancreatic acinar cell, digestive enzymes stored in secretory (zymogen) granules are released into the glandular lumen by exocytosis upon stimulation. Pancreatic acinar cells have been widely used as a model for studying regulated exocytosis in nonexcitable cells. Although this process occurs rapidly, the exocytotic images have been captured by transmission electron microscopy (Palade, 1975) and by freeze fracture (Tanaka *et al.*, 1980). The stimulation-mediated insertion of the zymogen granule membrane into the apical plasma membrane results in a transient enlargement of the apical membrane (Jamieson and Palade, 1971). Stimulation of zymogen secretion by cholescytokinin (CCK), mediated via the Ca^{2+}-dependent intracellular signaling pathway, involves three steps: (1) movement of the zymogen granule to the apical membrane followed by (2) recognition of the two partners and then (3) fusion of the membranes in an energy-dependent manner (Palade, 1975). A functional actin cytoskeleton is required for normal pancreatic secretion (O'Konski and Pandol, 1990,1993), and it has been proposed that an actin-based motor provides propulsive forces for the migration of zymogen granule (Chang and Jamieson, 1986).

B. *ADH-Induced Water Channel Translocation in Bladder and Collecting Duct*

The discovery of ADH-stimulated exocytic recruitment of water channels originated from early observations on the oxytocin-mediated appearance of particle arrays within the apical cell membrane of the frog urinary bladder (Chevalier *et al.,* 1974). Subsequent studies showed that water channel repartitioning in toad and frog bladder epithelial cells resulted from exocytosis stimulated by ADH, in which large vesicles known as aggrephores fuse with the apical plasma membrane, transferring the water channel from the walls of the aggrephore to the apical membrane (Hays, 1983; Humbert *et al.,* 1977). In nonstimulated cells, the majority of water channel–containing vesicles are located 0.2–1.0 μm from the apical plasma membrane, with some as close as 0.1 μm (Hays *et al.,* 1987). Morphologically similar vesicles have also been identified near the basolateral membrane, but since these "basal aggrephores" were nonresponsive to ADH (Shinowara *et al.,* 1989), it is not certain whether they indeed contain water channels and/or whether there are multiple mechanisms to promote fusion. Information gleaned from amphibian bladder has promoted analogous studies of the exquisitely ADH-sensitive mammalian collecting duct. Vesicles primarily in the subapical region of principal cells and inner medullary collecting duct cells provide the water channels to the apical membrane by exocytosis upon ADH stimulation. Like the amphibian bladder, cyclic AMP and protein kinase A (PKA) drive the hydro-osmotic response to ADH in the collecting duct, but the mobilization of intracellular Ca^{2+} and protein kinase C also appear to play a modulating role (Breyer, 1991; Burnatowska-Hledin and Spielman, 1987).

C. *HCl Secretion and Intrinsic Factor Release from Parietal Cells*

Hormone-regulated secretion in gastric parietal cells represents a somewhat different class of exocytosis in which the insertion of the proton-pumping transporter into the apical plasma membrane is concomitant with vesicular content release (intrinsic factor) into the lumen. The gastric acid–secreting parietal cell has several distinctive morphological features that have functional significance (Forte and Soll, 1989). The resting, or nonsecreting, parietal cell has only a limited area of apical membrane in direct contact with the gland lumen; most of the apical membrane extends into the cytoplasm as a series of invaginated canals, called canaliculi. The canaliculi are covered with numerous short, stubby microvilli with a system of submembrane microfilaments that extends 1–2 μm into the cytoplasm (Foret *et al.,* 1977). Another typical feature of parietal cells is the abundance of tubular and vesicular membranes within the cytoplasm, called tubulovesicles, containing the primary transport pump (H,K-ATPase) for acid secretion. Parietal cells are also relatively unique epithelial cells by virtue of their many mitochondria, which are necessary to provide energy support for proton pumping. The majority of the cytoplasmic volume (60–70%) is occupied by tubulovesicles and mitochondria.

Activation of the parietal cell to secrete HCl via histamine receptors and the cAMP pathway involves one of the most dramatic events of membrane remodeling in a cell system (see Forte and Yao, 1996). Quantitative ultrastructural studies revealed that maximal stimulation of parietal cells elicited a 5- to 10-fold increase of the apical membrane area, which now contains elongated microvilli (Helander and Hirschowitz, 1972; Forte *et al.,* 1977). According to the membrane recycling hypothesis of acid secretion, massive recruitment of H,K-ATPase–containing tubulovesicles into the apical membrane accounts for the microvillar elongation associated with stimulation (Forte *et al.,* 1977,1990). The expanded apical surface and sustained HCl secretion continue until the stimulus is withdrawn, whereupon the cell returns to the resting state. Thus, activation of gastric acid secretion involves the recycling of proton pumps between a cytoplasmic vesicular compartment and the apical plasma membrane. An implied function for the actin-based cytoskeleton in secretagogue-mediated secretion might be predicted on the basis of surface remodeling, and has been supported by the observation that cytochalasins disorganize the cytoskeleton and inhibit the secretory response to histamine stimulation (Black *et al.,* 1982).

IV. FUNCTIONAL ROLE OF CYTOSKELETON IN POLARIZED SECRETION

Functional activity of the actin-based cytoskeleton not only determines cell polarity but also participates in the polarized secretion by epithelial cells. Several modes of action of the actin cytoskeleton underlie regulated secretory processes, including (1) serving as a *barrier* to regulate the release of secretory vesicles; (2) *remodeling* of the membrane–cytoskeleton to facilitate polarized secretion; (3) serving as a track for *motor-mediated translocation* of secretory vesicles (myosin, dynein, actin-related proteins, etc.); and (4) serving as a *scaffold* to buffer the reserve pool or the rapidly releasable pool of transport-containing vesicles (synapsin-I).

A. Barrier to Regulate the Release of Secretory Vesicles

The postulate that a layer of cortical actin acts as a barrier to the exocytic release of secretory granules was first proposed by Orci *et al.* (1972), who observed that insulin release by pancreatic β-cells was facilitated by cytochalasin B. More recently, the hypothesis that cortical actin acts as a physical barrier to granule docking has been supported by the observation, in several secretory systems, that the actin network beneath the plasma membrane transiently depolymerizes during exocytosis (Norman *et al.,* 1994; Vitale *et al.,* 1991). Cortical actin has therefore come to be regarded as part of the clamping apparatus to regulate secretion, which has elements of both stable and dynamic actin cytoskeleton.

In the case of ADH-regulated water transport, the use of actin and microtuble disruptors provided early evidence for the importance of the cytoskeleton for exocytosis (Pearl and Taylor, 1983). Hays and his colleagues found that a rapid depolymerization of F-actin occurred in toad bladder epithelial cells treated with ADH or 8-Br-cAMP (Ding *et al.,* 1991). In unstimulated bladder cells, monomeric G-actin represented 37% (53 μmol/liter) of the total actin, and G-actin was increased to 77 μmol/liter concomitant with exocytosis and increased water transport (Hays and Lindberg, 1991). Under conditions that exist in the epithelial cell (and virtually all cells), the G-actin concentration greatly exceeds its critical concentration for polymerization to F-actin (Hall, 1994), indicating the requirement for G-actin–binding proteins to sustain the steady-state levels. The data further suggest the possibility that regulation of the G-actin sequestration system may be fundamental to the exocytic event. Confocal microscopy was used to demonstrate that the site of actin depolymerization was only at the apical pole of the ADH-treated toad bladder cells (Holmgren *et al.,* 1992). Moreover, using quantitative immunogold technique, Gao *et al.,* (1992) concluded that ADH treatment selectively depolymerized the actin cytoskeleton in the cortex and between microvilli, but not within the microvilli per se. Testing whether actin-depolymerizing factors can mimic ADH stimulation, low levels of cytochalasin D were found to produce a significant net depolymerization of F-actin (Franki *et al.,* 1992), but no increase in aggrephore fusion or water flow occurred (Wade and Kachadorian, 1988). When added together with ADH, cytochalasin D potentiated the rate of ADH-induced aggrephore fusion to the apical membrane, but this occurred in the face of decreased water flow (Franki *et al.,* 1992). While these data are consistent with the hypothesis that cortical F-actin impedes aggrephore access, they also suggest that an alternative pool of F-actin filaments might be required for the translocation of water channels to the apical membrane or for the stabilization of the apical membrane in carrying out osmotic flow.

The adrenal chromaffin cell is a well-studied polarized cell in which the actin cortical network provides a barrier for regulated exocytosis (Perrin *et al.,* 1987; Aunis and Bader, 1988). In stimulated chromaffin cells, disassembly of cortical actin appeared to precede catecholamine secretion and was correlated with increased cytosolic Ca^{2+} levels (Cheek *et al.,* 1989). Comparable actin disassembly was observed in digitonin-permeabilized cells actin-treated with micromolar levels of Ca^{2+}, and the effects were potentiated by phorbol esters and GTPγS, suggesting some regulatory effect by protein kinase C (Burgoyne *et al.,* 1989). Several actin-binding proteins have been implicated in the cytoskeletal rearrangements associated with chromaffin granule secretion. For example, fodrin, which is normally uniformly localized to the cortex, is rearranged into submembranous patches after stimulation, and treatment of permeabilized cells with an antibody against fodrin inhibited the secretory response to Ca^{2+} (Perrin *et al.,* 1987). It has also been suggested that scinderin, a Ca^{2+}-dependent F-actin–severing protein, mediates secretagogue-evoked catecholamine release from chromaffin cells by severing the cortical actin cytoskeleton. In nicotine-stimulated cells, scinderin, but not the related protein gelsolin, was observed to redistribute, leaving patches of the cortex devoid of the protein that correlated with the same regions devoid of assembled actin (Vitale *et al.,* 1991). Moreover, specific protein kinase C inhibitors (calphostin C and chelerythine) blocked all responses to secretagogue-induced exocytosis with the exception of the release of Ca^{2+} from intracellular stores (Zhang *et al.,* 1995). In addition to the implication that protein kinase C is involved in the secretagogue-induced responses, these data show that, in the absence of F-actin disassembly, elevation of intracellular Ca^{2+} concentration by itself is not capable of triggering catecholamine release.

To test if the cortical actin cytoskeleton provides a barrier for CCK-regulated exocytosis in pancreatic acinar cells, Muallem *et al.* (1995) introduced specific monomeric actin-binding proteins (β-thymosins) into streptolysin-*O*–permeabilized cells and measured the release of amylase. Low concentrations of the actin monomer-sequestering peptides triggered a rapid and robust exocytosis, with a profile similar to the initial phase of CCK stimulation but independent of intracellular Ca^{2+} oscillation. Interestingly, high concentrations of these polypeptides inhibited all phases of exocytosis. These studies are consistent with some "degradation" of F-actin as a fundamental event for zymogen secretion and suggest that the cortical actin cytoskeleton might act as a negative clamp in the regulation of exocytosis. However, the data also demonstrate that a minimal actin cytoarchitecture is necessary for exocytosis, possibly acting as a structural support for apical surface activity and/or a pathway for vesicular transport. In any event, it would be of great interest to probe if there are two spatially separated pools of actin that are differentially affected by these depolymerizing peptides. O'Konski and Pandol (1993) offered alternative evidence for a functionally stable F-actin pool, showing that diminished secretory activity caused by hyperstimulating acinar cells with the peptide cerulein was associated with marked changes in the apical cytoskeleton (i.e., ablation of microvilli, the terminal actin web, and intermediate filament bands). Experiments to identify the intracellular signaling mechanism responsible for the cytoskeletal changes suggested that they result from excessive phospholipase C effects on membrane phospholipids (O'Konski and Pandol, 1993).

A proposed model for the role of the cortical actin cytoskeleton as a barrier regulator of exocytosis is shown in Fig. 1. In effect, a layer of actin

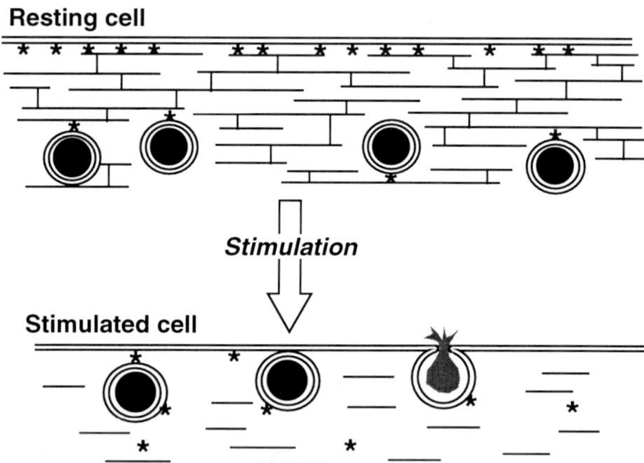

FIGURE 1 A schematic model of the cortical actin cytoskeleton as a barrier regulator of exocytosis. In resting cells, the access of secretory vesicles to the apical surface is limited by the cross-linked F-actin network in the cortex. The continuity of vesicle membranes and the plasma membrane to F-actin filaments is maintained through linker proteins, such as fodrin (asterisks). Stimulation results in the liberation of vesicles from the cytoskeleton, the clearing of some fodrin from the plasma membrane, and the activation of agents that promote disassembly of the actin network, such as scinderin or gelsolin. Secretory vesicles are now free to dock and fuse with the apical plasma membrane for secretion and/or recruitment of functional membrane proteins. (Adapted from Burgoyne and Cheek, 1987.)

filaments interferes with vesicular access to the secretory surface until an appropriate signal is received. Vesicle membranes and the plasma membrane maintain continuity with F-actin filaments through linker proteins. Agents that promote disassembly of the actin network, such as cytochalasin *in vitro* or actin-severing or -sequestering proteins *in vivo,* promote vesicular egress through the regulated docking and fusion machinery. A component of stabilized F-actin is included as an essential microfilamentous framework on which the membrane rearrangements can occur.

B. Remodeling of Membrane–Cytoskeleton to Facilitate Polarized Secretion

Remodeling of membrane–cytoskeleton has been implicated in acid secretion of parietal cells in which drastic growth of microvilli occurs in the apical canalicular plasma membrane upon histamine stimulation (Forte and Soll, 1989). However, we found that parietal cell actin is predominantly in the F-actin form (~90%), and that there were no significant changes in the steady-state levels of either F-actin or G-actin when parietal cells were stimulated (Ogihara *et al.,* 1995). These same studies demonstrated that cytochalasin D promoted a shift toward G-actin in parallel with inhibition of acid secretion, confirming earlier observations of Black *et al.* (1982); however, stabilization of parietal cell F-actin with phalloidin in α-toxin–permeabilized cells did not alter the HCl secretory process (Ogihara *et al.,* 1995). In the case of the T84 intestinal cell line, stabilization of actin filaments with phalloidin did not alter the activation and insertion of Cl^- channels (cystic fibrosis transmembrane conductance regulator) at the apical membrane, although cAMP-induced activation of the $Na^+,K^+,2Cl^-$ cotransporter at the basolateral membrane appeared to be microfilament dependent and responsible for inhibitory effects in the phalloidin-loaded state (Matthews *et al.,* 1992,1993). Thus it appears that disorganization of the apical actin network is not required for regulated exocytic recruitment of transport proteins in all cases. Since a large expansion of the apical surface is clearly associated with stimulation of the parietal cell, we speculate that some membrane–cytoskeleton linking proteins, such as ezrin and myosin, might facilitate the remodeling and elongation of microvilli by invaginating and cinching the newly received H,K-ATPase–rich membrane surface to the microfilamentous substructure. However, a disruption of the apical cytoskeletal architecture would severely limit the secretory response. In fact, the well-formed microfilament bundles within microvilli of resting parietal cells extend 1 μm or more into the cytoplasm and can provide the structural framework upon which the stimulation-associated microvillar extension occurs (Black *et al.,* 1982). A schematic view of how massive surface remodeling might occur over a stable base of microfilaments is presented in Fig. 2.

By virtue of its abundance and co-distribution with apical actin microfilaments, a role for ezrin has been implied in parietal cell secretory surface remodeling. It has been suggested that phosphorylation of ezrin by the secretagogue-elicited PKA pathway might enhance the association of ezrin with the actin-based cytoskeleton in parietal cells (Hanzel *et al.,* 1991). Mandel and his colleagues extended this interesting correlation to the anoxia-induced serine–threonine dephosphorylation of ezrin concomitant with its dissociation from the actin-based cytoskeleton of renal proximal tubular cells (Chen *et al.,* 1995).

Two actin isoforms have been shown to be spatially separated within distinct functional domains of the parietal cells. The cytoplasmic β-actin isoform is prominently localized to the apical and canalicular microvilli of the parietal cell, whereas the cytoplasmic γ-actin isoform is predominantly distributed to the basolateral membrane (Yao *et al.,* 1995a). Confocal microscopy demonstrated that ezrin was almost exclusively co-distributed with

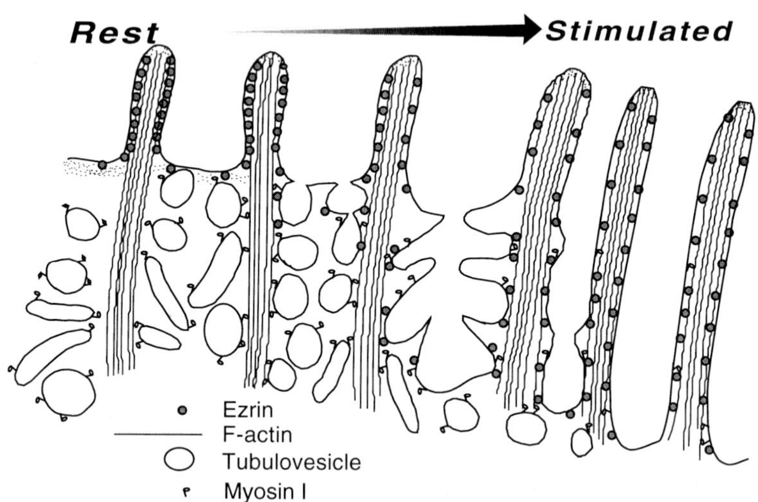

FIGURE 2 Massive membrane recruitment requires apical surface remodeling via cytoskeletal interactions. The apical surface of the gastric parietal cell is schematically shown in going from rest to a maximally stimulated state (left to right). Stimulation results in the recruitment of H,K-ATPase–containing tubulovesicles to the apical surface, possibly facilitated by movement along microfilamentous tracts by motor proteins (e.g., myosin-I). Organization of the expanded membrane surface requires membrane–cytoskeleton interactions via linker proteins (e.g., ezrin) to elongate the microvilli by invagination.

the cytoplasmic β-actin isoform at the apical pole of parietal cells. Selective segregation of β-actin with ezrin was further supported by biochemical studies in which β-actin was predominantly co-immunoprecipitated with ezrin (Yao *et al.,* 1995a). Moreover, *in vitro* reconstitution experiments demonstrate that the ezrin more specifically binds to F-actin assembled from cytoplasmic β-actin as compared to skeletal α-actin (Shuster and Herman, 1995; Yao *et al.,* 1996a). These data suggest that actin isoforms might segregate within different functional domains and exert specific functions by isoform-selective binding proteins.

C. Serving as a Track for Motor-Mediated Translocation of Secretory Vesicles

In the intestinal brush border, the mechanoenzyme myosin-I links the microvillus core actin filaments with the plasma membrane. It has been shown that myosin-I is associated with vesicles in mature enterocytes (Drenckhahn and Dermietzel, 1988), suggesting a potential role in mediating vesicle motility. To test if myosin-I is associated with Golgi-derived vesicles that are involved in the rapid assembly of brush borders in intestinal crypt cells, Fath and Burgess (1993) carried out a detailed study in which they showed that 50- to 100-nm vesicles, a size similar to those observed *in situ,* contain myosin-I, as demonstrated by immunoblotting and immunolabel-negative staining as well as the presence of a *trans*-Golgi marker enzyme. In addition, the bound myosin-I could be extracted from the vesicles using NaCl, KI, and Na_2CO_3, indicating that it is peripheral vesicular protein. Moreover, the myosin-I–containing vesicles were shown to bundle actin filaments in an ATP-dependent manner. Taken together, these results are consistent with a role for myosin-I as an apically targeted motor for vesicle translocation in epithelial cells. The model shown in Fig. 3 has been proposed by Fath and Burgess (1993) as a means for targeting vesicular transport.

A putative role for myosin in regulated secretion comes from recent studies of gastric parietal cells. Dabike *et al.* (1994) showed that myosin was associated with tubulovesicular membranes in resting parietal cells but was dissociated from the secretory canalicular membrane in the secreting state. Preliminary data from our own laboratory show that myosin-I is colocalized with H,K-ATPase in rabbit gastric parietal cells (Yao *et al.,* 1996b). Since the cytochalasins are known to disrupt the actin cytoskeleton and inhibit acid secretion (Black *et al.,* 1982), it is tempting to propose that myosin-I serves as motor to move the tubulovesicle cargo apically by using

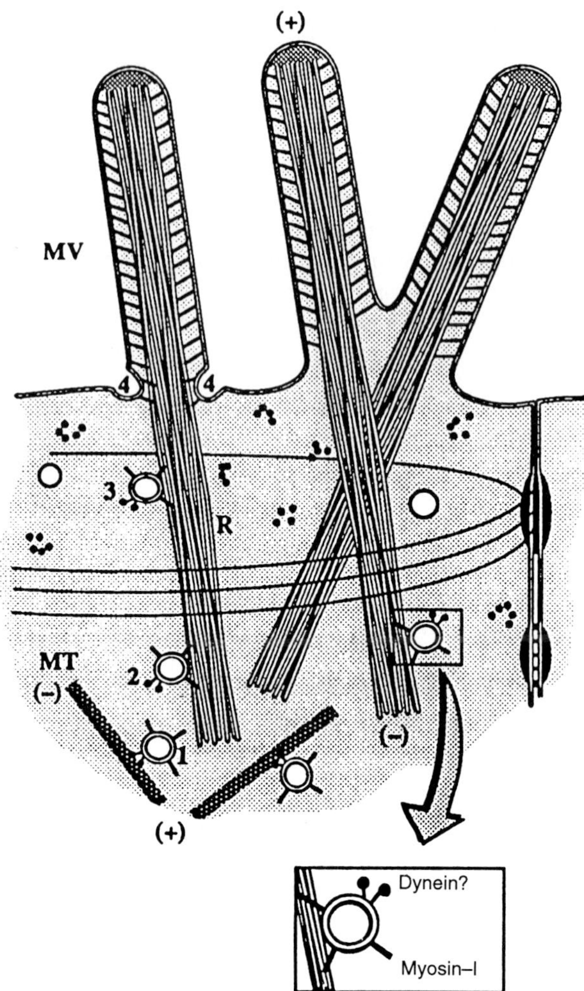

FIGURE 3 Myosin motors may facilitate vesicle movement to the apical surface in this schematic view of the transport of vesicles to the apical plasma membrane in developing intestinal epithelial cells. (1) Vesicles move from their genesis in the Golgi complex to the apical cytoplasm along microtubules (MT) using a minus end–directed motor (dynein?). (2) Upon reaching the end of the MT, the vesicles bind to actin filaments in the microvillus (MV) rootlet (R) through their membrane-bound myosin-I. (3) Vesicles translocate apically through the terminal web toward the plus end of the actin filaments using the myosin-I motor. (4) Upon reaching the the base of the MV, the vesicles fuse with the plasma membrane and link the membrane with the actin core through the attached myosin-I. (From Fath and Burgess, 1993.)

the actin track. The challenge ahead is to understand the detailed mechanism underlying tubulovesicle translocation.

D. Serving as a Scaffold to Buffer the Interchange between a Reserve Pool and an Immediately Releasable Pool of Transport-Containing Vesicles

Neural synapses are the most highly developed cellular secretory systems in which exocytic events can be fired at a high rate, and for sustained periods of time. It has been proposed that the pattern of neurotransmitter release might be explained via the activation of distinct vesicle pools: (1) a vesicle pool in close proximity to the presynaptic membrane available for immediate release; and (2) a larger, more distant, reserve pool whose access to the membrane is limited by a cytoskeletal meshwork, cross-linked via any one of a number of actin-binding proteins. On continued and prolonged stimulation, reserve vesicles would be liberated from the cytoskeleton and become available for exocytosis. One family of neuron-specific proteins, called the synapsins, has been implicated in the distinctive features of neurotransmitter release via phosphorylation-dependent interactions between synaptic vesicles and the actin-based cytoskeleton. Pieribone *et al.* (1995) showed that vesicles located at synaptic release sites are composed of two pools, a proximal pool, located adjacent to the presynaptic membrane, that is devoid of synapsin-I and a distal pool containing synapsin-I. Injection of antibodies against synapsin resulted in the loss of the distal pool, without any apparent effect on the proximal pool. In addition, these authors showed that depletion of the distal pool markedly depressed neurotransmitter release evoked by high-frequency but not by low-frequency stimulation. Thus, sustained neurotransmitter release in response to high-frequency bursts of impulses seems to require the availability of the synapsin-associated pool of vesicles.

Access of chromaffin granules to their exocytic release site in the adrenal chromaffin cell is also under the influence of a cytoskeletal meshwork that is the result of interaction between actin filaments and proteins associated with the cytoplasmic surface of the granules, such as α-actinin (Bader and Aunis, 1983), fodrin (Aunis and Perrin, 1984), and synapsin-II, an isoform of neural synapsin-I. Phosphorylation of synapsin-II was closely correlated with granule release and exocytosis in chromaffin cells (Firestone and Browning, 1992). Aunis and Bader (1988) noted that release of chromaffin granules seemed to have both cytoskeletal-dependent and -independent components that became more or less apparent depending on the nature of the preparation used. Thus they proposed that one pool of granules might be immediately subadjacent to the plasma membrane, not requiring cytoskeletal reorganization, while a second pool is entrapped within an actin network representing a reserve population to be released when the first pool is exhausted and only when released from the cytoskeleton. A schematic representation of the possible role of cross-linked cytoskeletal networks to regulate vesicle access, incorporating a two-step Ca^{2+}-dependent vesicle egress model proposed by Heinemann *et al.* (1993), is shown in Fig. 4.

Studies of ADH-induced water channel translocation in toad bladder and collecting duct cells have suggested that a spectrin–actin complex might provide a scaffold to withhold the water channel–containing vesicles. Using immunoelectron microscopy, Hays and his colleagues found that spectrin (fodrin) is located to aggrephores and in close proximity to actin filaments, consistent with the possibility that actin–spectrin complexes may aggrephore positioning and release (Hays *et al.,* 1994).

V. SUMMARY

The actin cytoskeleton plays a variety of roles in regulated exocytic processes. We have taken the liberty of broadly categorizing these activities.

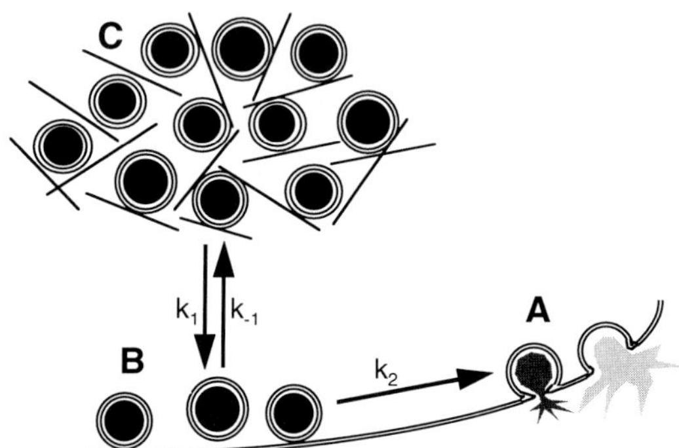

FIGURE 4 The reserve pool of secretory vesicles may be sustained within a modulatable actin cytoskeleton in this schematic representation of the relationship between three pools of secretory vesicles. One immediately activatable pool (B) is close to the plasma membrane and ready for recruitment to the exocytic pool (A) via an irreversible fusion reaction (k_2). Endocytic recovery of membrane and transporters would occur by an independent pathway. A third pool of secretory vesicles is entrained within the subplasmalemmal cytoskeleton (C) and serves as a reservoir to supply repeated or sustained exocytic demand. Mobilization of the reserve pool would occur via reversible destabilization of the subplasmalemmal cytoskeleton (k_1/k_{-1}), which in turn would be related to stimulus-generated signals (e.g., Ca^{2+}, cAMP) and subsequent activation of cytoskeletal mobilizing proteins, such as gelsolin and scinderin. The reserve pool (but not the immediately activatable pool) would also be affected by experimental manipulations such as exposure to antifodrin antibody or actin-destabilizing agents (cytochalasin, DNase I). (Adapted from Aunis and Bader, 1988, and Heinemann *et al.,* 1993.)

Evidence clearly shows that in some systems a layer of cortical actin filaments, and possibly even subcortical actin filaments, represents a barrier that exocytic vesicles must negotiate for access to the apical plasma membrane. In these cases, regulation of the exocytic event is intimately tied to activation of those proteins that control the state and organization of the actin barrier and/or the actin of entrapment. In almost all cases, whether or not a cortical barrier function has been demonstrated, some basic actin cytoskeletal substructure must be maintained for the normal operation of regulated vesicular recruitment to the plasma membrane. This seems to be a requisite foundation for remodeling the apical cell surface and would demand a different set of regulator proteins keyed to membrane–cytoskeleton linking functions. Finally, there is suggestive evidence that functional motor proteins play a role in directing vesicular traffic along actin tracks to the membrane docking sites. Certainly, specific differences do exist among the mechanisms for polarized exocytic secretion, and we might well expect a wide variation in interactions with the actin cytoskeleton; however, it is clear that the future of the problem requires an understanding of those proteins that bind and modulate the functional engagement of the actin cytoskeleton with the secretory membranes.

References

Adams, R. J., and Pollard, T. D. (1989). Binding of myosin I to membrane lipids. *Nature (London)* **340,** 565–588.

Aunis, D., and Bader, M. F. (1988). The cytoskeleton as a barrier to exocytosis in secretory cells. *J. Exp. Biol.* **139,** 253–266.

Aunis, D., and Perrin, D. (1984). Chromaffin granule membrane–F-actin interactions and spectrin-like protein of subcellular organelles: A possible relationship. *J. Neurochem.* **42,** 1558–1569.

Arpin, M., Friederich, E., Algrain, M., Vernel, F., and Louvard, D. (1994). Functional differences between L- and T-plastin isoforms. *J. Cell Biol.* **127,** 1995–2008.

Bader, M. F., and Aunis, D. (1983). The 97-kD alpha-actinin-like protein in chromaffin granule membranes from adrenal medulla: Evidence for localization on the cytoplasmic surface and for binding to actin filaments. *Neuroscience* **8,** 165–181.

Black, J. A., Forte, T. M., and Forte, J. G. (1982). The effects of microfilament disrupting agents on HCl secretion and ultrastructure of piglet oxyntic cells. *Gastroenterology* **83,** 595–604.

Bretscher, A. (1981). Fimbrin is a cytoskeletal protein that cross-links F-actin in vitro. *Proc. Natl. Acad. Sci. U.S.A.* **78,** 6849–6853.

Bretscher, A. (1982). Characterization and ultrastructural role of the major components of the intestinal microvillus cytoskeleton. *Cold Spring Harbor Symp. Quant. Biol.* **46,** 871–879.

Bretscher, A. (1983). Purification of an 80,000-dalton protein that is a component of the isolated microvillus cytoskeleton, and its localization in nonmuscle cells. *J. Cell Biol.* **97,** 425–432.

Bretscher, A. (1989). Rapid phosphorylation and reorganization of ezrin and spectrin accompany morphological changes induced in A-431 cells by epidermal growth factor. *J. Cell Biol.* **108,** 921–930.

Bretscher, A. (1991). Microfilament structure and function in the cortical cytoskeleton. *Annu. Rev. Cell Biol.* **7,** 337–374.

Bretscher, A., and Weber, K. (1978). Localization of actin and microfilament-associated proteins in the microvilli and terminal web of the intestinal brush border by immunofluorescence microscopy. *J. Cell Biol.* **79,** 839–845.

Bretscher, A., and Weber, K. (1980). Fimbrin, a new microfilament-associated protein present in microvilli and other cell surface structures. *J. Cell Biol.* **86,** 335–340.

Bretscher, A., Osborn, M., Wehland, J., and Weber, K. (1981). Villin associates with specific microfilamentous structure as seen by immunofluorescence microscopy on tissue sections and cells microinjected with villin. *Exp. Cell Res.* **135,** 213–219.

Breyer, M. D. (1991). Regulation of water and salt transport in collecting duct through calcium-dependent signaling mechanisms. *Am. J. Physiol.* **260,** F1–F11.

Burgoyne, R. D., and Cheek, T. R. (1987). Role of fodrin in secretion. *Nature* (*London*) **326,** 448.

Burgoyne, R. D., Morgan, A., and O'Sullivan, A. J. (1989). The control of cytoskeletal actin and exocytosis in intact and permeabilized adrenal chromaffin cells: Role of calcium and protein kinase C. *Cell. Signal.* **1,** 323–334.

Burnatowska-Hledin, M. A., and Spielman, W. S. (1987). Vasopressin increases cytosolic free calcium in LLC-PK1 cells through a V1-receptor. *Am. J. Physiol.* **253,** F328–F332.

Carboni, J. M., Conzelman, K., Adams, R. A., and Mooseker, M. S. (1988). Structural and immunological characterization of the myosin-like 110 kD subunit of the intestinal microvillar 110 K-calmodulin complex: Evidence for discrete myosin head and calmodulin-binding domains. *J. Cell Biol.* **107,** 1749–1757.

Chang, A., and Jamieson, J. D. (1986). Stimulus-secretion coupling in the developing exocrine pancreas: Secretory responsiveness to cholecystokinin. *J. Cell Biol.* **103,** 2353–2365.

Cheek, T. R., Jackson, T. R., O'Sullivan, A. J., Moreton, R. B., Berridge, M. J., and Burgoyne, R. D. (1989). Simultaneous measurements of cytosolic calcium and secretion in single bovine adrenal chromaffin cells by fluorescent imaging of fura-2 in cocultured cells. *J. Cell Biol.* **109,** 1219–1227.

Chen, J., Coh, J. A., and Mandel, L. J. (1995). Dephosphorylation of ezrin as an early event in renal microvillar breakdown and anoxic injury. *Proc. Natl. Acad. Sci. U.S.A.* **92,** 7495–7499.

Cheney, R. E., O'Shea, M. K., Heuser, J. E., Coelho, M. V., Wolenski, J. S., Espreafico, E. M., Forscher, P., Larson, R. E., and Mooseker, M. S. (1993). Brain myosin-V is a two-headed unconventional myosin with motor activity. *Cell* **75,** 13–23.

Chevalier, J., Bourguet, J., and Hugon, J. S. (1974). Membrane associated particles: Distribution in frog urinary bladder epithelium at rest and after oxytocin treatment. *Cell Tiss. Res.* **152,** 129–140.

Collins, K., and Borysenko, C. W. (1984). The 110,000 dalton actin- and calmodulin-binding protein from the intestinal brush border is a myosin-like ATPase. *J. Biol. Chem.* **259,** 14128–14135.

Costa de Beauregard, M. A., Pringault, E., Robine, S., and Louvard, D. (1995). Suppression of villin expression by antisense RNA impairs brush border assembly in polarized epithelial intestinal cells. *EMBO J.* **14,** 409–421.

Craig, S. W., and Powell, L. D. (1980). Regulation of actin polymerization by villin, a 95,000 dalton cytoskeleton component of intestinal brush borders. *Cell* **22,** 739–746.

Dabike, M., Munizaga, A., and Koenig, C. S. (1994). Parietal cells contain most of the myosin, filamin and actin present in rat gastric glands. *Biol. Res.* **27,** 29–38.

de Arruda, M. V., Watson, S., Lin, C. S., Leavitt, J., and Matsudaira, P. (1990). Fimbrin is a homologue of the cytoplasmic phosphoprotein plastin and has domains homologous with calmodulin and actin gelation proteins. *J. Cell Biol.* **111,** 1069–1079.

Ding, G. H., Franki, N., Condeelis, J., and Hays, R. M. (1991). Vasopressin depolymerizes F-actin in toad bladder epithelial cells. *Am. J. Physiol.* **260,** C9–C16.

Drenckhahn, D., and Dermietzel, R. (1988). Organization of the actin filament cytoskeleton in the intestinal brush border: A quantitative and qualitative immunoelectron microscope study. *J. Cell Biol.* **107,** 1037–1048.

Ezzel, R. M., Chafel, M., and Matsudaira, P. (1989). Differential localization of villin and fimbrin during development of mouse visceral endoderm and intestinal epithelium. *Development* **106,** 407–409.

Fath, K. R., and Burgess, D. R. (1993). Golgi-derived vesicles from developing epithelial cells bind actin filaments and possess myosin-I as a cytoplasmically oriented peripheral membrane protein. *J. Cell Biol.* **120,** 117–127.

Firestone, J. A., and Browning, M. D. (1992). Synapsin II phosphorylation and catecholamine release in bovine adrenal chromaffin cells: Additive effects of histamine and nicotine. *J. Neurochem.* **58,** 441–447.

Forte, J. G., and Soll, A. (1989). Cell biology of hydrochloric acid secretion. *In* "Handbook of Physiology, Sect. 6: Vol. III: Salivary, Gastric, Pancreatic, and Hepatobiliary Secretion" The Gastrointestinal System, (J. G. Forte, ed.), pp. 207–228. American Physiological Society, Bethesda, MD.

Forte, J. G., and Yao, X. (1996). The membrane-recruiment-and-recycling hypothesis of gastric HCl secretion. *Trends Cell Biol.* **6,** 45–48.

Forte, J. G., Hanzel, D. K., Okamoto, C., Chow, D., and Urushidani, T. (1990). Membrane and protein recycling associated with gastric HCl secretion. *J. Intern. Med.* **732,** 17–26.

Forte, T. M., Machen, T. E., and Forte, J. G. (1977). Ultrastructural changes in oxyntic cells associated with secretory function: A membrane-recycling hypothesis. *Gastroenterology* **73,** 941–955.

Franki, N., Ding, G., Gao, Y., and Hays, R. M. (1992). Effect of cytochalasin D on the actin cytoskeleton of the toad bladder epithelial cell. *Am. J. Physiol.* **263,** C995–C1000.

Funayama, N., Nagafuchi, A., Sato, N., Tsukita, S., and Tsukita, S. (1991). Radixin is a novel member of the band 4.1 family. *J. Cell Biol.* **115,** 1039–1048.

Gao, Y., Franki, N., Macaluso, F., and Hays, R. M. (1992). Vasopressin decreases immunogold labeling of apical actin in the toad bladder granular cell. *Am. J. Physiol.* **263,** C908–C912.

Glenney, J. R., Kaulfus, P., and Weber, K. (1981). F-actin binding and bundling properties of fimbrin, a major cytoskeletal protein of the microvillus core filaments. *J. Biol. Chem.* **256,** 9283–9288.

Gould, K. L., Bretscher, A., Esch, F. S., and Hunter, T. (1989). cDNA cloning and sequencing of the protein-tyrosine kinase substrate, ezrin, reveals homology to band 4.1. *EMBO J.* **8,** 4133–4142.

Hall, A. (1994). Small GTP-binding proteins and the regulation of the actin cytoskeleton. *Annu. Rev. Cell Biol.* **10,** 31–54.

Halsall, D. J., and Hammer, J. A. III. (1990). A second isoform of chicken border myosin I contain a 29 aa-residue inserted sequence that binds calmodulin. *FEBS Lett.* **267,** 126–130.

Hanzel, D., Reggio, H., Bretscher, A., Forte, J. G., and Mangeat, P. (1991). The secretion-stimulated 80K phosphoprotein of parietal cells is ezrin, and has properties of a membrane cytoskeletal linker in the induced apical microvilli. *EMBO J.* **10,** 2363–2373.

Hays, R. M. (1983). Alteration of luminal membrane structure by antidiuretic hormone. *Am. J. Physiol.* **245,** C289–C296.

Hays, R. M., and Lindberg, U. (1991). Actin depolymerization in the cyclic AMP-stimulated toad bladder epithelial cell, determined by the DNAse method. *FEBS Lett.* **280,** 397–399.

Hays, R. M., Franki, N., and Ding, G. (1987). Effects of antidiuretic hormone on the collecting duct. *Kidney Int.* **31,** 530–537.

Hays, R. M., Franki, N., Simon, H., and Gao, Y. (1994). Antidiuretic hormone and exocytosis: Lessons from neurosecretion. *Am. J. Physiol.* **269,** C1507–C1524.

Heinemann, C., von Ruden, L., Chow, R. H,, and Neher, E. (1993). A two-step model of secretion control in neuroendocrine cells. *Pflug. Arch. Eur. J. Physiol.* **424,** 105–112.

Helander, H. F., and Hirschowitz, B. I. (1972). Quantitative ultrastructural studies of microvilli and changes in the tubulovesicular compartment of mouse parietal cells in relation to gastric acid secretion. *J. Cell Biol.* **63,** 951–961.

Hinchaw, J. E., and Schmid, S. L. (1995). Dynamin self-assembles into rings suggesting a mechanism for coated vesicle budding. *Nature* (*London*) **374,** 190–192.

Holmgren K., Magnusson, K. E., Franki, N., and Hays, R. M. (1992). ADH-induced depolymerization of F-actin in the toad bladder granular cell: A confocal microscope study. *Am. J. Physiol.* **262,** C672–C677.

Humbert, F., Montesano, R., Grosso, A., DeSousa, R. C., and Orci, L. (1977). Particle aggregates in plasma and intracellular membranes of toad bladder (granular cell). *Experientia* (*Basel*) **33,** 1364–1367.

Hunter, T., and Cooper, J. A. (1981). Epidermal growth factor induces rapid tyrosine phosphorylation of proteins in A431 human tumor cells. *Cell* **24,** 741–752.

Jamieson, J. D., and Palade, G. E. (1971). Synthesis, intracellular transport and discharge of secretory proteins in stimulated pancreatic exocrine cells. *J. Cell Biol.* **50,** 135–158.

Kachar, B. (1985). Direct visualization of organelle movement along actin filaments dissociated from characean algae. *Science* **227,** 1355–1357.

Kuznetsov, S. A., Rivera, D. T., Severin, F. F., Weiss, D. G., and Langford, G. M. (1994). Movement of axoplasmic organelles on actin filaments from skeletal muscle. *Cell Motil. Cytoskeleton* **28,** 231–242.

Matthews, J. B., Awtrey, C. S., and Madara, J. L. (1992). Microfilament-dependent activation of Na^+-K^+-$2Cl^-$ cotransport by cAMP in intestinal epithelial monolayers. *J. Clin. Invest.* **90,** 1608–1316.

Matthews, J. B., Awtrey, C. S., Thompson, R., Hung, T., Tally, K. J., and Madara, J. L. (1993). Na^+-K^+-$2Cl^-$ cotransport and Cl^- secretion evoked by heat-stable enterotoxin is microfilament dependent in T84 cells. *Am. J. Physiol.* **265,** G370–G378.

Matsudaira, P. T., and Burgess, D. R. (1979). Identification and organization of the components of the isolated microvillus cytoskeleton. *J. Cell Biol.* **83,** 667–673.

Matsudaira, P. T., and Burgess, D. R. (1982). Organization of the cross-filaments in intestinal microvilli. *J. Cell Biol.* **92,** 657–664.

Miller, D., and Crane, R. K. (1961). The digestive function of the epithelium of the small intestine. II. Localization of disaccharide hydrolysis in the isolated brush border portion of intestinal epithelial cells. *Biochim. Biophys. Acta* **52,** 293–298.

Mooseker, M. S., and Tilney, L. G. (1975). Organization of an actin filament-membrane complex: Filament polarity and membrane attachment in the microvilli of intestinal epithelial cells. *J. Cell Biol.* **67,** 725–743.

Mooseker, M. S., Grave, T. A., Wharton, K. A., Falco, N., and Howe, C. L. (1980). Regulation of microvillus structure: Calcium-dependent isolation and cross-linking of actin filaments in the microvilli of intestinal epithelial cells. *J. Cell Biol.* **87,** 809–822.

Muallem, S., Kwiatkowska, K., Xu, X., and Yin, H. L. (1995). Actin filament disassembly is a sufficient final trigger for exocytosis in nonexcitable cells. *J. Cell Biol.* **128,** 589–598.

Norman, J. C., Price, L. S., Ridley, A. J., Hall, A., and Koffer, A. (1994). Actin filament organization in activated mast cells is regulated by heterotrimeric and small GTP-binding proteins. *J. Cell Biol.* **126,** 1005–1015.

Ogihara, S., Yao, X., and Forte, J. G. (1995). Actin cytoskeleton and secretion by gastric parietal cells. *Mol. Biol. Cell.* **6,** 2155A.

O'Konski, M. S., and Pandol, S. J. (1990). Effects of caerulein on the apical cytoskeleton of the pancreatic acinar cell. *J. Clin. Invest.* **86,** 1649–1657.

O'Konski, M. S., and Pandol, S. J. (1993). Cholecystokinin JMV-180 and caerulein effects on the pancreatic acinar cell cytoskeleton. *Pancreas* **8,** 638–646.

Orci, L., Gabbay, K. H. and Malaisse, W. J. (1972). Pancreatic beta-cell web: Its possible role in insulin secretion. *Science* **175,** 1128–1130.

Palade, G. (1975). Intracellular aspects of the process of protein synthesis. *Science* **189,** 347–358.

Pearl, M., and Taylor, A. (1983). Actin filaments and vasopressin-stimulated water flow in toad urinary bladder. *Am. J. Physiol.* **245,** C28–C39.

Perrin, D., Langley, O. K., and Aunis, D. (1987). Anti-alpha-fodrin inhibits secretion from permeabilized chromaffin cells. *Nature* (*London*) **326,** 498–501.

Pieribone, V. A. Shupliakov, O., Brodin, L., Hilfiker-Rothenfluh, S., Czernik, A. J., and Greengard, P. (1995). Distinct pools of synaptic vesicles in neurotransmitter release. *Nature* (*London*) **375,** 493–497.

Pollard, T. D., and Korn, E. D. (1973). Acanthamoeba myosin. I. Isolation from Acanthamoeba castellanii of an enzyme similar to mucle myosin. *J. Biol. Chem.* **248,** 4682–4690.

Shinowara, N. L., Palaia, T. A., and Discala, V. A. (1989). Intramembrane particle structures in epithelial cells of the toad urinary bladder: A quantitative freeze-fracture study. *Biol. Cell* **66,** 65–76.

Shuster, C. B., and Herman, I. M. (1995). Indirect association of ezrin with F-actin: Isoform specificity and calcium sensitivity. *J. Cell Biol.* **128,** 837–848.

Stenmark, H., Vitale, G., Ullrich, O., and Zerial, M. (1995). Rabaptin-5 is a direct effector of the small GTPase rab5 in endocytic membrane fusion. *Cell* **83,** 423–432.

Stossel, T. P. (1989). From signal to pseudopod: How cells control cytoplasmic actin assembly. *J. Biol. Chem.* **264,** 18261–18264.

Takei, K., McPherson, P. S., Schmid, S. L., and De Camilli, P. (1995). Tubular membrane invaginations coated by dynamin rings are induced by GTP-γS in nerve terminals. *Nature* (*London*) **374,** 186–190.

Tanaka, Y., DeCamilli, P., and Meldolesi, J. (1980). Membrane interactions between secretion granules and plasmalemma in the three exocrine glands. *J. Cell Biol.* **84,** 438–453.

Tsukita, S., Oishi, K., Sato, N., Sagara, J., Kawai, A., and Tsukita, S. (1994). ERM family members as molecular linkers between the cell surface glycoprotein CD44 and actin-based cytoskeletons. *J. Cell Biol.* **126,** 391–401.

Turunen, O., Wahlstrom, T., and Vaheri, A. (1994). Ezrin has a COOH-terminal actin-binding site that is conserved in the ezrin protein family. *J. Cell Biol.* **126,** 1445–1453.

Urushidani, T., Hanzel, D. K., and Forte, J. G. (1987). Protein phosphorylation associated with stimulation of rabbit gastric glands. *Biochim. Biophys. Acta.* **930,** 209–219.

Vitale, M. L., Rodriguez Del Castillo, A., Tchakarov, L., and Trifaro, J. M. (1991). Cortical filamentous actin disassembly and scinderin redistribution during chromaffin cell stimulation precede exocytosis, a phenomenon not exhibited by gelsolin. *J. Cell Biol.* **113,** 1057–1067.

Wade, J. B., and Kachadorian, W. A. (1988). Cytochalasin B inhibition of toad bladder apical membrane responses to ADH. *Am. J. Physiol.* **255,** C526–C530.

Way, J. C., Wang, L., Run, J., and Hung, M. (1994). Cell polarity and the mechanism of asymmetric cell division. *BioEssays* **16,** 925–930.

Yao, X., Chaponnier, C., Gabbiani, G., and Forte, J. G. (1995a). Polarized distribution of actin isoform in gastric parietal cells. *Mol. Biol. Cell* **6,** 541–557.

Yao, X., Cheng, L., Rao, L., Han, B. G., and Forte, J. G. (1995b). Biochemical characterization of ezrin-actin interaction. *Mol. Biol. Cell* **6,** 140A.

Yao, X., Cheng, L., and Forte, J. G. (1996a). Biochemical characterization of ezrin-actin interaction. *J. Biol. Chem.* **271,** 7224–7229.

Yao, X., Karam, S., Ramilo, M., Rong, Q., Thibodeau, A., and Forte, J. G. (1996b). Stimulation of gastric acid secrtion by cAMP in a novel α-toxin-permeabilized gland model. *Am. J. Physiol.* In press.

Yeh, B., and Svoboda, K. K. (1994). Intracellular distribution of beta-actin mRNA is polarized in embryonic corneal epithelia. *J. Cell Sci.* **107,** 105–115.

Zhang, L., Rodriguez Del Castillo, A., and Trifaro, J. M. (1995). Histamine-evoked chromaffin cell scinderin redistribution, F-actin disassembly and secretion: In the absence of cortical F-actin disassembly, an increase in intracellular Ca^{2+} fails to trigger exocytosis. *J. Neurochem.* **65,** 1297–1308.

CHAPTER 6

The Spectrin Cytoskeleton and Organization of Polarized Epithelial Cell Membranes

Prasad Devarajan* **and Jon S. Morrow**
Departments of Pathology and *Pediatrics, Yale University School of Medicine, New Haven, Connecticut 06510

I. THE ERYTHROCYTE PARADIGM

Our current understanding of the structure and function of the erythrocyte membrane cytoskeleton is a tribute to three decades of biochemical, biophysical, and morphological studies from laboratories throughout the world. Consisting primarily of spectrin, actin, ankyrin, protein 4.1, protein 4.9, adducin, and tropomyosin, the distribution and interactions of these components in the red cell are now well understood (reviewed in Coleman *et al.*, 1989b; Bennett, 1990a; Bennett and Lambert, 1991; Luna and Hitt, 1992; Bennett and Gilligan, 1993; Morrow *et al.*, 1996). The predominant component is spectrin, an antiparallel flexible rod-shaped heterodimer usually composed of two self-associating subunits (α, β) that interact with F-actin to form a stoichiometric and roughly pentagonal to hexagonal submembranous scaffolding. Spectrin–actin junctions are stabilized by protein 4.1, protein 4.9, adducin, and tropomyosin (reviewed in Bennett, 1990b; Bennett and Gilligan, 1993; Gilligan and Bennett, 1993). The spectrin–actin meshwork is linked by ankyrin to the anion exchanger, by protein 4.1 to glycophorin C and the anion exchanger, and by ankyrin-independent direct linkages to unidentified membrane proteins (Howe *et al.,* 1985; Steiner and Bennett, 1988; Davis and Bennett, 1994; Lombardo *et al.,* 1994b). Additional recently elucidated linkages of the spectrin-based cytoskeleton to the red cell membrane include the interaction of adducin with stomatin (Sinard *et al.,* 1994) and several ligands that bind α-spectrin's SH3 (*src* homology) domain, including erythrocyte dynamin (Cianci *et al.,* 1995). The major role of the spectrin–actin infrastructure in red cells is to provide membrane stability and to ensure deformability, thus enabling erythrocytes to survive the turbulent capillary microcirculation and splenic sinusoids. This task is accomplished by the skeleton's elastic properties and, perhaps more importantly, by its role in maintaining integral membrane proteins in a laterally homogeneous state (Mohandas and Evans, 1994; Svetina *et al.,* 1995; Kralj-Iglic *et al.,* 1996).

II. UNEXPECTED CYTOSKELETAL DIVERSITY IN EPITHELIAL CELLS

With the discovery of a spectrin-based cytoskeleton in probably all eukaryotic cells, several aspects of the erythrocyte paradigm have been challenged. Spectrin in nonerythroid cells often displays a nonhomogeneous spatial and temporal distribution and may not be associated with the peripheral plasma membrane at all. For instance, circulating lymphocytes contain cytoplasmic aggregates of spectrin that redistribute in response to various stimuli (Gregorio *et al.,* 1992,1993,1994). In neurons and skeletal muscle, different isoforms of spectrin are confined to distinct regions of the postsynaptic membrane, such as βIΣ2-spectrin at the postsynaptic density of cerebellar granule cells (Riederer *et al.,* 1986,1987; Malchiodi-Albedi *et al.,* 1993) or a βIΣ* isoform with the acetylcholine receptor of skeletal muscle (Bloch and Morrow, 1989). Epithelial cells present perhaps the biggest challenge to the red cell paradigm. Experimental removal of spectrin from the membrane of cultured epithelial cells by microinjection of anti-spectrin antibodies does not result in altered cell shape or membrane instability (Mangeat and Burridge, 1984). Cultured epithelial cells maintained as contact naive display a cytoplasmic distribution of spectrin (Nelson and Veshnock, 1986,1987; Morrow *et al.,* 1989,1991). Formation of cell–cell contact initiates a remarkable sequence of events that culminates in the polarized assembly of the spectrin and its associated proteins at the lateral plasma membrane. In epithelial cells lining the kidney tubules, certain isoforms of spectrin and ankyrin are even restricted to infoldings of the basolateral plasma membrane domain (Kashgarian *et al.,* 1988; Morrow *et al.,* 1991). This unique spatial organization is disrupted following ischemic injury and restituted during recovery, coincident with the recovery of the polarized distribution of crucial integral membrane proteins (Siegel *et al.,* 1994). Collectively, these observations suggest that the spectrin-based cytoskeleton in these cells serves functions that transcend its traditional view as simply a structural component lending deformation and shear stability to the membrane. Unraveling the mechanisms by which these proteins are directed to a polarized distribution at the epithelial cell membrane, how they interact, and their roles is an intriguing challenge to contemporary cellular and molecular biologists, and is the focus of this chapter. What follows is a detailed description of four major components of the epithelial cell cytoskeleton, an account of the dynamics of their membrane assembly, and a discussion of contemporary questions that follow from the emerging complexity of this process.

A. Spectrin

Proteins closely related to the α- and β-subunits of erythrocyte spectrin have been detected in virtually all vertebrate tissues (Coleman *et al.,* 1989b; Bennett and Gilligan, 1993; Morrow *et al.,* 1996), in invertebrates (Pollard, 1984; Bennett and Condeelis, 1988; Dubreuil *et al.,* 1989; Byers *et al.,* 1992), and even in plants (Michaud *et al.,* 1991; Faraday and Spanswick, 1993; Sikorski *et al.,* 1993; Lorenz *et al.,* 1995). Because spectrin has been discovered by so many investigators, a myriad of names has emerged. To rectify this semantic complexity, a unified nomenclature for spectrins has been proposed (Malchiodi-Albedi *et al.,* 1993; Winkelmann and Forget, 1993; Morrow *et al.,* 1996). This nomenclature is based on the four known human spectrin genes; nonhuman spectrins are denoted by their homologous human spectrin (or by separate notation if not homologous). There are currently two recognized human genes for each spectrin subunit (Table I), these are denoted by Roman numerals in their order of discovery. Thus αI and βI encode red cell spectrin and αII and βII the nonerythroid spectrins. Alternatively spliced transcripts are referred to as "subtypes" and denoted by the symbol Σ followed by Arabic numerals. Unconfirmed transcripts are designated with an asterisk until their identity is ascertained. Thus red

TABLE I

Proposed Nomenclature for Human Spectrins

Old name	Chrom.	Transcripts	Tissues	New name	Reference
α-Spectrin	1	1 (~8 kb)	Red cells (240 kDa)	αIΣ1	Sahr *et al.* (1990)
α-Fodrin; α_G	9	1 (~9 kb)	Nonerythroid (284 kDa)	αIIΣ1	Moon and McMahon (1990)
		2 (alt. splice)[a]	Lung, nonerythroid	αIIΣ2	Moon and McMahon (1990)
		3 (alt. splice)	Nonerythroid	αIIΣ3	Moon and McMahon (1990)
		4 (alt. splice)	Fetal brain, nonerythroid	αIIΣ4	Cianci and Morrow (1995)
β-Spectrin	14	1 (7.8 kb)	Red cells, brain, heart (246 (kDa)	βIΣ1	Prchal *et al.* (1987), Winkelmann *et al.* (1988)
		2 (~8.5 kb)	Muscle, brain (270 kDa)	βIΣ2	Winkelmann *et al.* (1990), Malchiodi-Albedi *et al.* (1993)
		3 (?)	MDCK Golgi (220 kDa)	βIΣ*	Beck *et al.* (1994), Devarajan *et al.* (1996)
β-Fodrin; β_G	2	1 (9.4 kb)	Nonerythroid (275 kDa)	βIIΣ1	Hu *et al.* (1992)

[a] Alt. splice, isoforms arising from alternative splicing of C-terminal sequences.

cell β-spectrin and muscle β-spectrin, both arising as alternative transcripts from the same gene on chromosome 14, are termed βIΣ1 and βIΣ2, respectively. A recently described spectrin associated with the Golgi complex in Madin–Darby cainine kidney (MDCK) cells (Beck *et al.,* 1994; Devarajan *et al.,* 1996) cross-reacts exclusively with antibodies directed against βIΣ1-spectrin, and is thus currently termed βIK* pending sequence availability. The general form of β-spectrin (previously termed fodrin or β_G) arises from chromosome 2 and is designated βII-spectrin, with alternative transcripts subtyped as above.

Each spectrin molecule is an antiparallel heterodimer of two subunits, termed α (M_r 280,000) and β (M_r 245,000–460,000). Each subunit is composed of multiple 106-residue homologous repeat units (collectively termed domain II), flanked by nonhomologous N- and C-terminal sequences (domains I and III, respectively) (reviewed in Winkelmann and Forget, 1993; Hartwig, 1994; Morrow *et al.,* 1996). Typically, domain II of β-spectrin contains 17 structural repeats and that of α-spectrin contains 22 repeats. Heterodimerization is achieved by the initial binding between high-affinity sites involving the N-terminus of β-spectrin (domain I + repeat 1) and the C-terminal region of α-spectrin (repeats 20 and 21) (Speicher *et al.,* 1992; Viel and Branton, 1994; Pradhan *et al.,* 1996). The self-association of α/β-spectrin heterodimers to form tetramers is crucial in the red cell for its role as a membrane stabilizer and organizer, and requires an additional interaction between the last repeat unit in β-spectrin and with the N-terminus (domain I) of α-spectrin (Morrow *et al.,* 1980; Speicher *et al.,* 1993; Kennedy *et al.,* 1994).

Two additional classes of protein interactions are essential for functional assembly of spectrins at the peripheral membrane of erythrocytes and epithelial cells (Table II). First, five to six spectrin molecules are cross-linked to short F-actin filaments, forming a lattice-like scaffolding of five- and six-sided polygons (Byers and Branton, 1985; Liu *et al.,* 1987; Ursitti and Wade, 1993). The actin-binding region resides in a highly conserved 27-residue region of domain I of β-spectrin (Karinch *et al.,* 1990; Frappier *et al.,* 1991) and is conserved between several members of a superfamily of actin–cross-linking proteins, including α-actinin, dystrophin, and fimbrin (for reviews, see Dubreuil, 1991; Hartwig, 1994). In addition, the spectrin–actin junctions are further stabilized by protein 4.1 and adducin.

A second class of critical interactions includes spectrin's direct and indirect linkages to the membrane. The best characterized of these acts via the adapter molecule ankyrin, which links β-spectrin to the integral membrane protein band 3 (Bennett and Stenbuck, 1979). The binding site for ankyrin has been localized to the 15th repeat of β-spectrins (Kennedy *et al.,* 1991). Other indirect linkages include

TABLE II

Spectrin Interactions

Type	Linkages	Spectrin sites	References
Heterodimerization	Lateral α/β spectrins	α domain III/β domain I	Morrow *et al.* (1980), Speicher *et al.* (1992), Viel and Branton (1994)
Self-association	Head–head α/β spectrins	α domain I/β repeat 17	Morrow *et al.* (1980), DeSilva *et al.* (1992), Kennedy *et al.* (1994)
Actin cross-linking	Spectrin/actin	β domain I	Becker *et al.* (1990), Karinch *et al.* (1990)
Membrane association	Spectrin/membrane	MAD1 (β repeat 1)	Lombardo *et al.* (1994b)
Spectrin/membrane	IP3- and $\beta\gamma$-GTP–binding proteins	MAD2 (β domain III) (pleckstrin homology)	Lombardo *et al.* (1994b), Touhara *et al.* (1994), Wang and Shaw (1995)
Spectrin/membrane	Membrane glycoprotein	β repeats 3–8	Davis and Bennett (1994)
Spectrin/ankyrin		β repeat 15	Kennedy *et al.* (1991)
Spectrin/protein 4.1		β domain I	Becker *et al.* (1990, 1993)
Spectrin/adducin		β domain I	Mische *et al.* (1987), Coleman *et al.* (1989a)
Spectrin/α-catenin		Not known	Lombardo *et al.* (1994a)
Regulatory roles	Spectrin/SH3 ligands; α-ENaC; dynamin, etc.	α repeat 10	Rotin *et al.* (1994), Cianci *et al.* (1995)
	Spectrin/calmodulin	α repeat 11	Harris *et al.* (1988)
	Spectrin/calpain	α repeat 11	Harris *et al.* (1988, 1989)

1. The binding of the N-terminal β-domain I with red cell protein 4.1 (Becker *et al.,* 1990,1993), which in turn is linked to glycophorin, band 3, and actin (Hemming *et al.,* 1994; Morris and Lux, 1995).
2. Binding of spectrin to erythrocyte adducin (Gardner and Bennett, 1987; Mische *et al.,* 1987; Coleman *et al.,* 1989a), which in turn interacts with the putative sodium channel regulatory protein stomatin (Sinard *et al.,* 1994).
3. An interaction with α-catenin that has been identified *in vitro* (Lombardo *et al.,* 1994a), and that if confirmed *in vivo* will be of great interest, since it would provide a molecular link between the E-cadherin–mediated adhesion complex and the spectrin skeleton at sites of epithelial cell–cell contact (Morrow *et al.,* 1991; Kemler, 1992; McNeill *et al.,* 1993).
4. Linkage by way of its binding to F-actin, which in turn may be bound to the membrane in several different ways (reviewed in Luna, 1991), including by linkage to α-catenin (Rimm *et al.,* 1995).

In addition, β-spectrin binds directly to at least three ligands in NaOH-stripped membranes (Davis and Bennett, 1994; Lombardo *et al.,* 1994b). Two of these sites, found in βII-spectrin (fodrin) and βIΣ2- (muscle) spectrin, mediate strong binding to membranes and have been termed membrane association domains 1 and 2 (MAD1 and MAD2). MAD1 is located within the first repeat unit of βI- and βII-spectrin (Lombardo *et al.,* 1992), and appears to be the dominant interaction responsible for assembly of spectrin to the membrane in MDCK cells (Stabach *et al.,* 1993, and as discussed later). MAD2 resides within the C-terminal domain (domain III) of βII- and βIΣ2-spectrin (Davis and Bennett, 1994; Lombardo *et al.,* 1994b). While MAD2 mediates strong binding to membranes *in vitro,* it appears to play no role in the assembly of the nascent spectrin skeleton to the membrane in either epithelial (Stabach *et al.,* 1993) or muscle (S. A. Weed and J. S. Morrow, unpublished observations) cells. MAD2 also contains a pleckstrin homology domain with putative β/γ heterotrimeric G-protein– (Macias *et al.,* 1994; Touhara *et al.,* 1994; Wang *et al.,* 1994; D. S. Wang *et al.,* 1995) and inositol-1,4,5-triphosphate– (Wang and Shaw, 1995) binding activities. A third direct membrane association site appears to reside within repeats 3–8 of both βI and βII spectrin, based on the ability of recombinant peptides from this region to bind stripped membranes (Davis and Bennett, 1994). The ligand for this site appears to be a calmodulin-sensitive glycoprotein (Davis and Bennett, 1995).

The α-spectrin subunit displays fewer recognized protein-binding sites than β-spectrin; however, it does contain several putative regulatory regions. For instance, repeat 10 of both αI- and αII-spectrin contains an SH3 (src homology domain 3) motif (Wasenius *et al.*, 1989). This domain is found in a wide variety of signaling and cytoskeletal proteins. Spectrin's SH3 domain has been shown to interact with the epithelial Na^+ channel (α-ENaC) (Rotin *et al.*, 1994) and the GTPase dynamin (αI but not αII) (Cianci *et al.*, 1995), as well as with several as yet unidentified ligands that may include signal transduction molecules and other cytoskeletal control elements (Merilainen *et al.*, 1993; Cianci *et al.*, 1995). In addition, a calmodulin-binding region and an adjacent calpain cleavage site have been identified within repeat 11 of αII-spectrin (Harris *et al.*, 1988). It is postulated that the concerted action of Ca^{2+}, calmodulin, and microcalpain may regulate the stability and plasticity of the spectrin–actin lattice (Harris *et al.*, 1989; Harris and Morrow, 1990), the affinity of heterodimeric αII/βII-spectrin for the membrane (Steiner *et al.*, 1989), and the organization of membrane receptor domains and the response of cells to injury (reviewed in Glantz and Morrow, 1995). Calcium-dependent spectrin proteolysis is also thought to contribute to the cytoskeletal changes accompanying epithelial and neuronal cell hypoxia (Glantz and Morrow, 1995), neuronal degeneration (Seubert *et al.*, 1990), and neonatal brain development (Najm *et al.*, 1991).

B. *Ankyrin*

Ankyrins are a family of large proteins that have emerged as crucial adapter molecules mediating linkages between several integral membrane proteins and the underlying spectrin-based cytoskeleton (reviewed in Bennett and Lambert, 1991; Bennett, 1992). Although first described in erythrocytes, ankyrin-like proteins have been discovered in virtually all vertebrate cells (Lux *et al.*, 1990) and even in invertebrate tissues (Axton *et al.*, 1994; Dubreuil and Yu, 1994; Otsuka *et al.*, 1995). Multiple general and tissue-specific isoforms of ankyrin have arisen both by gene duplication and by alternative mRNA splicing, resulting in a protein family with enormous complexity. In an attempt to rectify the semantic confusion, we propose that the following nomenclature be used for the ankyrins, following conventions similar to those now used for the spectrins. This nomenclature is based on the three known human ankyrin genes; nonhuman ankyrins are denoted by their homologies to the human ankyrins (or by separate notation if nonhomologous). The human ankyrins are denoted by the capital letters ANK, and the mouse ankyrins by the letters Ank (see Kapfhamer *et al.*, 1995; Peters *et al.*, 1995), followed by Roman numerals in their order of discovery. Alternatively spliced transcripts are denoted by the symbol Σ followed by Arabic numerals; unconfirmed transcripts are designated with an asterisk until their identity is ascertained. Thus, as shown in Table III, human erythrocyte ankyrins (Ank_R, arising from chromosome 8) are properly termed ANKIΣ1 (for protein 2.1), ANKIΣ2 (for protein 2.2), etc.; human brain ankyrins (Ank_B, mapped to chromosome 4) are referred to as ANKIIΣ1 (440 kDa) and ANKIIΣ2 (220 kDa); and the general isoforms of ankyrin (Ank_G, located on chromosome 10) are named ANKIIIΣ1 (480 kDa), ANKIIIΣ2 (270 kDa), ANKIIIΣ3 (190 kDa), and ANKIIIΣ4 (119 kDa, Ank_{G119}), etc.

Most ankyrins described to date (Hall and Bennett, 1987; Lambert *et al.,* 1990; Lux *et al.,* 1990; Kunimoto *et al.,* 1991; Otto *et al.,* 1991; Bennett, 1992; Gallagher *et al.,* 1992; Birkenmeier *et al.,* 1993; Kordeli *et al.,* 1995) contain three independently folded domains: (i) a highly conserved N-terminal 89-kDa domain of repeats that associates with several membrane proteins; (ii) a well-conserved central 62- to 67-kDa domain that binds spectrin; and (iii) a variable C-terminal "regulatory" domain that can modulate the activities of the first two domains. Novel ankyrin isoforms that lack all or part of the repetitive or regulatory domain have been elucidated (Peters *et al.,* 1995; Devarajan *et al.,* 1996). A striking feature of the 89-kDa domain is the presence of 24 tandemly arrayed repeats, each consisting of 33 amino acids. Ankyrin-like repeats are present in an intriguing number of apparently unrelated proteins (Lux *et al.,* 1990; Bennett, 1992), including membrane proteins involved in cell differentiation (e.g., *Drosophila* Notch and *Caenorhabditis elegans* Glp-1), cytoplasmic proteins that regulate the cell cycle (e.g., *Saccharomyces cerevisiae* SW16 and SW14), and nuclear proteins (e.g., the transcription factors NF-κB and GABP-β). These findings suggest a role for ankyrin well beyond simply linking cytoskeletal proteins to the membrane.

Biophysical studies reveal the 89-kDa domain to be globular and composed of four independently folded subdomains of 6 repeats each (Michaely and Bennett, 1993). All four sets of 6 repeats appear to be required for proper folding into a membrane-binding globular configuration, with the removal of even a single repeat reducing α-helicity by 40%. All four subdomains may participate in binding to AE1 (band 3), although the most N-terminal 12 repeats alone retain most of the activity of the whole protein (Davis and Bennett, 1990a,1990b). However, the isoform of band 3 (AE3) in kidney does not bind ankyrin (Ding *et al.,* 1994), presumably due to differences in its cytoplasmic domain (C. C. Wang *et al.,* 1995). The voltage-dependent sodium channel in the brain has been shown to associate primarily with the terminal 11 repeats (Srinivasan *et al.,* 1992), and α-Na,K-ATPase requires both the repeats domain and the spectrin-binding domain to achieve the highest affinity interaction with ankyrin in erythrocytes and epithelial cells (Davis and Bennett, 1990a). Other ligands include gastric

TABLE III

Proposed Nomenclature for Human Ankyrins

Old name	Chromosome	Transcript	Tissues	New name	Reference
Ank_R, ANK1	8	1 (7.2 kb)	Red cells (206 kDa), cerebellum, purkinje	ANKIΣ1	Lambert *et al.* (1990), Lux *et al.* (1990)
		2 (6.8 kb)	Red cells	ANKIΣ2	Hall and Bennett (1987), Lux *et al.* (1990)
Ank_B, ANK2	4	1 (~13 kb)	Neurons, glia (440 kDa)	ANKIIΣ1	Otto *et al.* (1991)
		2 (~9 kb)	Unmyelinated axon (220 kDa)	ANKIIΣ2	Kunimoto *et al.* (1991), Otto *et al.* (1991)
Ank_G, ANK3	10	1 (15 kb)	Nodes of Ranvier, axonal initial segments (480 kDa)	ANKIIIΣ1	Kunimoto *et al.* (1991), Kapfhamer *et al.* (1995)
		2 (10 kb)	Nodes of Ranvier, axonal initial segments (220 kDa)	ANKIIIΣ2	Kunimoto *et al.* (1991), Kapfhamer *et al.* (1995)
		3 (~7 kb)	Kidney, other epithelial tissues (190 kDa)	ANKIIIΣ3	Peters *et al.* (1995), Devarajan *et al.* (1996)
		4 (~6 kb)	Kidney, placenta, muscle, MDCK Golgi (116 kDa)	ANKIIIΣ4	Devarajan *et al.* (1996)

H,K-ATPase (Smith *et al.*, 1993), the renal amiloride-sensitive sodium channel (Smith *et al.*, 1991), and the cardiac Na^+/Ca^{2+} exchanger (Li *et al.*, 1993). Besides integral membrane transport proteins and spectrin, ankyrin also binds vimentin (Georgatos and Marchesi, 1985), tubulin (Davis *et al.*, 1992), the inositol triphosphate receptor (Bourguignon *et al.*, 1993; Bourguignon and Jin, 1995), and the cell adhesion molecule (CAM) CD44 (Lokeshwar *et al.*, 1994). The marked diversity among these proteins suggests that a complex evolutionary process has generated individualized binding sites within ankyrin that mediate its attachment to multiple ligands.

The 62- to 67-kDa spectrin-binding domain is conserved in all human isoforms of ankyrin characterized to date. Within this region, the "minimal" spectrin-binding region has recently been identified as encompassing residues 1136–1160 of red cell ANKIΣ1 (Platt *et al.*, 1993) and residues 669–860 of renal ANKIIIΣ4 (Devarajan *et al.*, 1996). However, the N-terminal region of the spectrin-binding domain, which is less well conserved, also appears to play a role (Davis and Bennett, 1990b; Fearon *et al.*, 1992). ANKIΣ1 (red cell) and ANKIIΣ2 (brain) interact with βIΣ1- and βIIΣ*-spectrin with much different affinities, and it has been postulated that the divergent N-terminal sequence within the larger spectrin-binding domain of ankyrin may impart upon different members of the ankyrin family their characteristic spectrin-binding affinities (Platt *et al.*, 1993). Other factors that influence spectrin–ankyrin interactions are the state of spectrin self-association and the phosphorylation state of ankyrin (Weaver *et al.*, 1984; Lu *et al.*, 1985; Cianci *et al.*, 1988). Unphosphorylated ankyrin preferentially binds to spectrin tetramers rather than to dimers with a 10-fold greater affinity; this preferential binding is abolished by phosphorylation, although the sites involved and the physiological significance of this interaction remain unclear. In addition, the affinity of ankyrin for spectrin oligomers is enhanced by concomitant binding of the anion exchanger with the repeats domain (Cianci *et al.*, 1988). Finally, spectrin–ankyrin interactions are influenced by alternative splicing of ANKIΣ1, which deletes 163 residues within the regulatory domain to produce a smaller "activated" ANKIΣ2 (protein 2.2) with an increase in affinity for spectrin, anion exchanger, and tubulin (Hall and Bennett, 1987; Davis *et al.*, 1992). Collectively, these *in vitro* studies suggest that the interaction between ankyrin, spectrin, and their membrane-binding sites is finely tuned and subject to many levels of regulation, the purpose of which remains enigmatic.

C. Protein 4.1

Protein 4.1 is an 80-kDa sulfhydryl-rich phosphoprotein originally described on the basis of its electrophoretic mobility in erythrocyte membranes (reviewed in Davis *et al.*, 1992). It is now recognized as the prototypical member of a larger family of proteins involved with mediating the interaction of actin filaments with membranes (Funayama *et al.*, 1991; Sato *et al.*, 1992; Franck *et al.*, 1993). The other members of this family include ezrin, moesin, and radixin. All members of this protein 4.1 superfamily share a highly conserved N-terminal 30-kDa domain that mediates membrane binding (Lombardo and Low, 1994) but whose biological function is otherwise poorly understood. Immunoreactive forms of protein 4.1 have since been found in most tissues, where they exhibit a complex heterogeneity in molecular weights, subcellular localization, and primary amino acid sequence (Tang *et al.*, 1990; Conboy *et al.*, 1991). A rich isoform diversity has arisen by posttranslational modifications such as differential phosphorylation and glycosylation, as well as by tissue-specific and developmentally regulated alternative splicing of the pre-mRNA derived from the gene on chromosome 1p33-p34.2 (Huang *et al.*, 1993). Immunoreactive 4.1 polypeptides range in size from 30 to 210 kDa, and range in distribution from membrane associated in red cells and polarized epithelial cells to nuclear staining in cultured leukemic and MDCK cells (Tang *et al.*, 1988; Correas, 1991), perinuclear in endothelial cells (Leto and Marchesi, 1984), and on stress fibers in yet other cells (Cohen *et al.*, 1982). A unified nomenclature is not available and must await complete characterization of the gene and definition of intron–exon boundaries.

In its most abundant form, protein 4.1 is an 80-kDa monomer in solution, folded into four distinct domains (Leto and Marchesi, 1984): (i) a hydrophobic 30-kDa N-terminal domain conserved between all members of the protein 4.1 superfamily—this region contains the membrane association site (via the anion exchanger and glycophorin C/D in red cells) (Pasternack *et al.*, 1985; Hemming *et al.*, 1994,1995); (ii) a hydrophilic 16-kDa region containing a protein kinase C phosphorylation site (Horne *et al.*, 1985); (iii) a highly charged 10-kDa domain which mediates the interaction with spectrin and actin (Correas *et al.*, 1986); and (iv) an acidic 12.6-kDa C-terminal domain. Protein 4.1 also binds calmodulin at the junction between the 30-kDa and 16-kDa domains (Kelly *et al.*, 1991; Tanaka *et al.*, 1991; Cianci *et al.*, 1996), with a corresponding reduction in its membrane affinity (Lombardo and Low, 1994).

Since spectrin-actin complexes precede expression of protein 4.1 in maturing erythrocytes (Chasis *et al.*, 1993), and since protein 4.1 directly binds both spectrin and actin (Becker, *et al.*, 1990; Morris and Lux, 1995), protein 4.1 no doubt plays a role in the stabilization of the spectrin-actin network after it has formed. However, a large fraction of protein 4.1 associates with red cell membranes independently of spectrin, suggesting that these two functions of protein 4.1 (i.e., spectrin–actin binding and membrane binding) are distinct. The major integral membrane protein ligands for protein 4.1 in red cells are the anion exchanger and glycophorins C and D (Pasternack *et al.,* 1985; Hemming *et al.,* 1994,1995). Its binding to glycophorin also involves p55, a MAGUK family of guanylate kinase–related proteins (Marfatia *et al.,* 1995). Isoforms of AE1 are present in most human tissues, while the glycophorins represent a typical type I membrane protein. Thus, membrane-binding sites for 4.1 probably exist in all tissues. However, it has been shown that, although the kidney anion exchanger isoform of AE1 is restricted to the basolateral domain and co-localizes with ankyrin, it does not bind renal protein 4.1 or ankyrin (Ding *et al.,* 1994; C. C. Wang *et al.,* 1995a). Protein 4.1 in cultured epithelial cells, as with other members of this gene family, often localizes along zones of cell–cell contact (Bretscher, 1989; Morrow *et al.,* 1991; Franck *et al.,* 1993), raising the possibility that it associates with a component(s) of the cell–cell adhesion apparatus.

D. Adducin

Adducin is a spectrin-binding and actin-bundling protein first identified as a substrate for protein kinase C (Palfrey and Waseem, 1985), and later purified and characterized using its calmodulin-binding activity (Gardner and Bennett, 1986). There are approximately 30, 000 copies per red blood cell (Morrow *et al.,* 1996). The functional unit of adducin consists of α- (103-kDa) and β- (97-kDa) subunits; immunoreactive isoforms have since been discovered in several cells and tissues (Gilligan and Bennett, 1993). Molecular cloning has revealed that the two human erythrocyte subunits are 66% homologous at the amino acid level (Joshi *et al.,* 1991). The α-subunit is expressed as a 4-kb mRNA species in erythrocytes, brain, kidney, and liver; an additional alternatively spliced isoform has been described in brain (Taylor *et al.,* 1992). At least three alternatively spliced tissue-specific isoforms of β-adducin exist (Gilligan and Bennett, 1991). The β-1 (predicted size 80 kDa) isoform is expressed in low abundance in erythroid tissues (~4 kb) and brain (~8 kb) and not in kidney or liver (Joshi *et al.,* 1991); β-2 (63 kDa) is encoded by an ~3.5-kb message in kidney, liver, spleen, and heart (Tripodi *et al.,* 1991); and the β-3 polypeptide (26 kDa) is detected in kidney, heart, and brain (Gilligan and Bennett, 1991).

Both α- and β-adducin polypeptides are folded into three independent domains: (i) a 39-kDa basic protease-resistant N-terminal globular domain that is highly conserved between the subunits; (ii) a 9-kDa neck region; and (iii) a 32- to 35-kDa protease-sensitive C-terminal tail domain configured as a long, flexible random coil (Joshi and Bennett, 1990; Joshi *et al.,* 1991;

Scaramuzzino and Morrow, 1993). The calmodulin-binding site resides in the protease-sensitive tail domain (Scaramuzzino and Morrow, 1993). Adducin exists predominantly as $\alpha2/\beta2$ tetramers, resulting from the four head domains contacting to form a globular core with extending and interacting tails (Hughes and Bennett, 1995). The site for binding to spectrin–actin complexes has been identified as the C-terminal tail domains of both α- and β-adducin (Hughes and Bennett, 1995). Adducin binds in isolation to either β-spectrin (Coleman *et al.,* 1989a) or F-actin, which it bundles with high affinity (Mische *et al.,* 1987). In addition, adducin will also bind simultaneously to spectrin and actin independently of protein 4.1 binding, leading to stabilization of the spectrin-actin complex (Gardner and Bennett, 1987; Mische *et al.,* 1987). The interaction of adducin with actin and spectrin is a true state function, in that the nature of the final complex of the three proteins is independent of their order of assembly; at least *in vitro,* the binding of these three proteins to each other also exists in a relatively rapid equilibrium (Mische *et al.,* 1987). Another novel feature of adducin has been recognized: its ability to bind to the putative ion channel regulatory protein stomatin (Sinard *et al.,* 1995). While the significance of this interaction remains to be explored, it is noteworthy that both adducin and stomatin homologues are present in epithelial cells, and that stomatin is a homologue of MEC-2, a protein in *C. elegans* involved with ion channel regulation and mechanosensory signal transduction (Huang *et al.,* 1995).

The temporal sequence of gene expression in maturing erythrocytes (Nehls *et al.,* 1991) suggests that adducin might play a role in the early assembly of a nascent spectrin–actin erythroid membrane skeleton. Adducin is also localized at the lateral cell borders of cultured epithelial cells (Kaiser *et al.,* 1989) and may be one of the first proteins recruited to sites of cell–cell contact in these cells as well, suggesting an important role in the organization of epithelial cell junctions.

E. Cadherin and Catenin

The biology of cadherins and catenins is detailed elsewhere in this volume (see Chapter 9); only aspects directly pertaining to their interactions with the cytoskeleton at sites of cell–cell contact are summarized here. The cadherin superfamily is a group of proteins that mediate calcium-dependent homophilic adhesion between cells (for other reviews, see Takeichi, 1991; Luna and Hitt, 1992; Gumbiner and McCrea, 1993; Kemler, 1993). Epithelial cells ubiquitously express the classical or E-cadherins (also called uvomorulin, L-CAM, cell CAM 120/80), which are 120-kDa integral membrane glycoproteins comprosed of three domains: (i) an extracellular domain of a 113-residue conserved N-terminal region and three homologous putative calcium-binding repeats; (ii) a single transmembrane domain; and (iii) a highly conserved cytoplasmic domain that interacts with at least three cytosolic proteins, termed α-, β-, and γ-catenins (Ozawa *et al.,* 1989; Moon *et al.,* 1993; Näthke *et al.,* 1993; Peifer *et al.,* 1993; Fleming *et al.,* 1994; Hinck *et al.,* 1994b). In the presence of millimolar concentrations of extracellular calcium, epithelial cells form sites of cell–cell contact (called adherens junctions) where E-cadherin and the catenins concentrate, and initiate a cascade of events that culminates in the polarized distribution of spectrin, ankyrin, and integral membrane proteins such as the Na,K-ATPase. The hierarchy of interactions at the adhesion complex suggests that the cytoplasmic domain of cadherin interacts directly with β-catenin (or plakoglobin), and that β-catenin links cadherin to α-catenin (Jou *et al.,* 1995). Cytoplasmic pools of β-catenin also exist (Hinck *et al.,* 1994b). α-Catenin bundles F-actin and thereby tethers actin filaments to the adhesion complex (Rimm *et al.,* 1994). α-Catenin also interacts directly with spectrin, at least *in vitro* (Lombardo *et al.,* 1994a). If this same reaction occurs *in vivo,* it would provide a second mechanism by which the spectrin–actin skeleton may be linked directly to the adhesion receptor machinery.

III. A MODEL FOR THE CYTOSKELETON IN POLARIZED EPITHELIA

The phenomenon of epithelial cell polarity, whereby the cellular distribution of integral membrane and cytoskeletal proteins is restricted to predetermined domains, plays a fundamental role in the ontogeny and functions of a variety of organs and tissues (Rodriguez-Boulan and Nelson, 1989; Simons *et al.,* 1992). Polarity is especially critical to epithelial cells lining the renal tubules, gastrointestinal tract, and other organs that perform the crucial functions of cell volume regulation and vectoral transport of ions and nutrients. An initial requirement for the generation of cellular polarity appears to be the establishment of cell–cell contacts and tight junctions, which delineate biochemically distinct apical and basolateral membrane compartments (Rodriguez-Boulan and Nelson, 1989; Takeichi, 1991; Nelson, 1992; Simons *et al.,* 1992; Näthke *et al.,* 1993). Transporters, enzymes, and receptors are then targeted from the *trans*-Golgi network to one of these two domains by complex and incompletely understood mechanisms (Nelson *et al.,* 1992; Matter and Mellman, 1994). Subsequently, polarized membrane proteins must be restricted to the targeted domain and protected from lateral diffusion or endocytosis. Several recent findings suggest that the spectrin–actin cytoskeleton assembles just after the establishment of productive cell–cell contact involving the cadherin-based adhesion complex

(Hinck *et al.*, 1994a), and that it may play a key role in the stabilization of the polarized state (reviewed in Rodriguez-Boulan and Nelson, 1989; Morrow *et al.*, 1991,1996). The findings that isoforms of spectrin, ankyrin, Na,K-ATPase, E-cadherin, α- and β-catenin, actin, adducin, and protein 4.1 are co-distributed, co-localized, and complexed in intact and cultured (MDCK) renal tubule epithelial cells allow reasoned speculations on the structure of the epithelial cell spectrin–actin skeleton (Fig. 1).

IV. ROLE OF CELL–CELL CONTACT IN ESTABLISHING SPECTRIN POLARITY

Cultured MDCK cells are a well-established model of a polarized epithelium (Rodriguez-Boulan and Nelson, 1989). When these cells are maintained in micromolar calcium, contact formation and cell polarity are greatly reduced due to inactivation of E-cadherin (Ozawa *et al.*, 1990), and the distribution of spectrin, ankyrin, E-cadherin, and Na,K-ATPase is cyto-

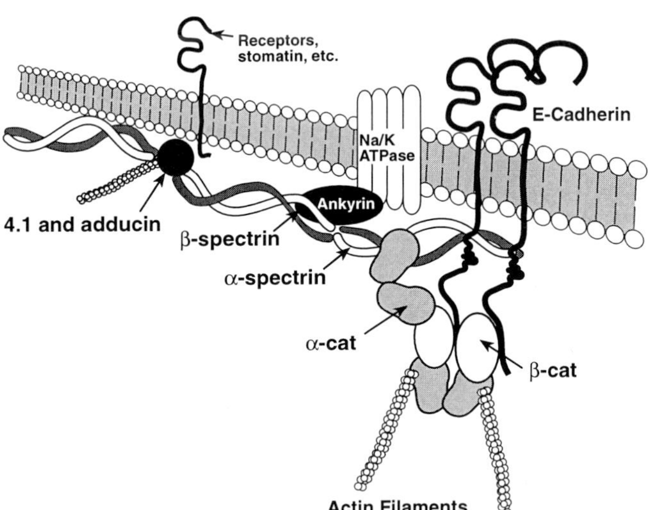

FIGURE 1 Hypothetical structure of the spectrin–actin skeleton and its relationship to the cadherin adhesion complex in epithelial cells. Interactions initiated by the adhesion complex, either indirectly by the binding of α-catenin to F-actin (Rimm *et al.*, 1994) or directly by its interaction with spectrin (Lombardo *et al.*, 1994a), probably serve to initiate the assembly of the nascent spectrin skeleton in the regions of cell–cell contact. Subsequent binding of other integral and cytoplasmic proteins tethers integral proteins to the skeletal lattice, controlling their lateral distribution in the membrane and creating organized mosaics. (Courtesy of Dr. David Rimm, Yale University.)

plasmic (Nelson and Veshnock, 1986). Synchronized assembly of the polarized state can be induced by 2-m*M* calcium and is mediated by E-cadherin (McNeill *et al.,* 1990; Takeichi, 1991). When such cells are sparsely plated, populations at the earliest states of cell–cell contact can be identified and studied by time-lapse microscopy and immunofluorescence techniques. This approach has allowed dissection of the sequence of early cellular events leading to the polarized state (McNeill *et al.,* 1993). Such studies have shown that E-cadherin is the first molecule to be recruited and enriched at stable contact sites. This occurs while cytoskeletal components (spectrin and actin) are still cytoplasmic. This early enrichment upon contact requires homotypic interactions between E-cadherin molecules of adjacent cells, and may involve conformational changes and/or posttranslational modifications (Clayton *et al.,* 1995). Interestingly, subsequent newly synthesized E-cadherin molecules are delivered predominantly to the apical membrane but are rapidly endocytosed and degraded, whereas the E-cadherin targeted to the basolateral membrane retains its polarized distribution (Wollner *et al.,* 1992). It has been postulated that an interaction with the spectrin-based cytoskeleton at the basolateral membrane may serve to stabilize and maintain the distribution of E-cadherin by preventing its internalization and lateral mobility. Indeed, molecules that closely follow E-cadherin assembly at sites of contact include actin, spectrin, adducin, and protein 4.1 (Kaiser *et al.,* 1989; also P. Devarajan and J. S. Morrow, unpublished observations; Morrow *et al.,* 1991; Rimm *et al.,* 1995).

What are the factors responsible for the initial recruitment of spectrin to sites of cell–cell contact? Although the cytoplasmic domain of E-cadherin does impair the binding of βII-spectrin to membranes *in vitro* (Lombardo *et al.,* 1993), direct binding between E-cadherin and spectrin has not been demonstrated. However, there may be a direct link between α-catenin and spectrin (Lombardo *et al.,* 1994a). Since the catenins have been shown to bind newly synthesized E-cadherin even before its arrival at the cell surface (Ozawa and Kemler, 1992), we postulate that the cadherin–catenin complex acts as a "nucleating" center once assembled on the plasma membrane and "triggered" by cell–cell contact.

V. THE SPREAD OF SPECTRIN: A VESICULAR MEMBRANE SKELETON

Coincident with the recognition of many isoforms of spectrin, ankyrin, and other associated proteins has come the question of their role. Recent data suggest a surprising answer. It now appears that homologues of the cortical membrane skeleton, assembled from different and probably unique isoforms of spectrin (Beck *et al.,* 1994; Devarajan *et al.,* 1996), ankyrin

(Devarajan *et al.*, 1996), and probably other components, including protein 4.1 (Cohen *et al.*, 1982; Correas, 1991) and adducin (Pinto *et al.*, 1991), appear to be associated with internal vesicular membranes and perhaps other internal structures as well. Where this has been most carefully studied, it appears that an isoform of $\beta I\Sigma^*$-spectrin and ANKIIIΣ4 ankyrin are associated with the Golgi and *trans*-Golgi apparatus. Their association (at least for spectrin) is sensitive to disruption by Brefeldin A (Beck *et al.*, 1994), and both are co-distributed with β-COP (Fig. 2) (Devarajan *et al.*, 1996), a component of the vesicle coatamer and a marker for vesicles involved in endoplasmic reticulum, Golgi, and *trans*-Golgi shuttling (Matter and Mellman, 1994). These proteins may also be associated with some types of transcytotic vesicles, although probably not those associated with clathrin coats (Whitney *et al.*, 1995). Significantly, in MDCK cells, the components of the cortical (as opposed to the vesicular) skeleton are $\alpha II/\beta II$-spectrin and ANKIIΣ2 and ANKIIIΣ2 (Drenckhahn and Bennett, 1987; Morrow *et al.*, 1989), whereas neither $\beta I\Sigma^*$-spectrin nor ANKIIIΣ4 ankyrin is associated with the plasma membrane.

The implications of these findings are significant. Recent discoveries in the red cell emphasize the dominating contributions of integral membrane proteins to cell shape and membrane stability (Mohandas *et al.*, 1992; Schofield *et al.*, 1992; reviewed in Mohandas, 1992; Mohandas and Evans, 1994; Morrow *et al.*, 1996). Theoretical considerations based on membrane energetics emphasize how changes in integral membrane protein distributions can per se alter membrane shape and stability (Svetina *et al.*, 1990,1995; Kralj-Iglic *et al.*, 1996). One principle emerging from these considerations is that the role of the spectrin cortical membrane skeleton is not so much to directly provide *mechanical* stability or elasticity to the membrane, but rather to provide *organizational* stability by controlling the distribution of integral membrane proteins. Without such control, both experimental and theoretical considerations suggest that biological membranes are intrinsically unstable and prone to vesiculation or other shape change (reviewed in Morrow *et al.*, 1996). Such considerations suggest that the cortical spectrin skeleton is more appropriately thought of as an *organizing center* about which integral and cytoplasmic proteins are congregated, rather than as a *supporting infrastructure*. These concepts are embodied in the linked mosaic model of the membrane skeleton (Morrow *et al.*, 1996), and suggest a reason why the cell has found the need for a vesicular spectrin skeleton. Indeed, the control and sorting of integral membrane proteins through a cacophony of internal membrane compartments is the central and most characteristic feature of membrane biogenesis pathways. The theoretical considerations outlined earlier must apply with equal validity to all membranes. This fact, together with the identification of a vesicular spectrin

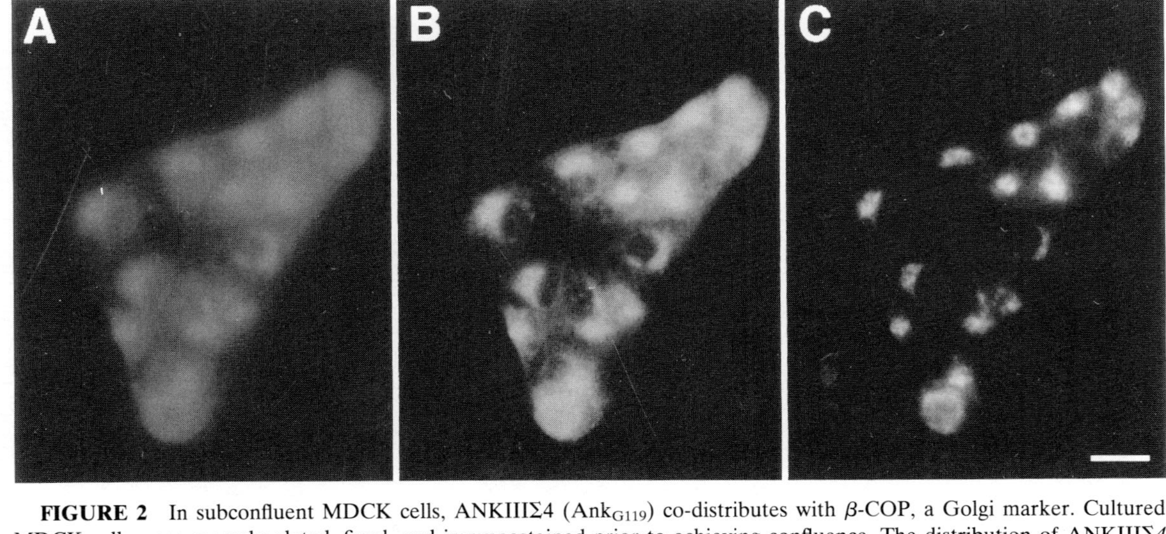

FIGURE 2 In subconfluent MDCK cells, ANKIIIΣ4 (Ank_{G119}) co-distributes with β-COP, a Golgi marker. Cultured MDCK cells were sparsely plated, fixed, and immunostained prior to achieving confluence. The distribution of ANKIIIΣ*4* (A) or β-COP (B) was determined by indirect immunofluorescence. The rightmost micrograph (C) represents the appearance of the preparation when viewed through a filter that passes the emission of both CY-2 (green) and CY-3 (red). Areas of absolute coincidence are revealed here as white areas. Note the strong coincident staining over the Golgi complex. There is no staining at the plasma membrane, which is characterized by a different ankyrin (Devarajan *et al.*, 1996). Bar = 50 μm.

membrane skeleton (Beck *et al.,* 1994; Devarajan *et al.,* 1996), makes it likely that the same mechanisms operating at the plasma membrane to control membrane stabilization, shape, and protein distribution also stabilize the cell's internal membrane compartments. If this holds true, it would explain not only why these proteins are found on internal membrane compartments (which presumably are not subject to shear stresses), but also possibly the sensitivity of sorting pathways to agents that disrupt actin filaments (Kelly, 1991).

VI. ASSEMBLY OF SPECTRIN AND ANKYRIN IN EPITHELIAL CELLS

Since isoforms of spectrin and ankyrin are found to co-localize, coprecipitate, and exist as complexes in epithelial cells (Koob *et al.,* 1988; Morrow *et al.,* 1989; Nelson *et al.,* 1990; Morrow *et al.,* 1991), and to directly bind *in vitro,* we sought to determine whether a hierarchy exists in the membrane localization of these two molecules. Five lines of evidence indicate that αII/βII-spectrin assembles first at the plasma membrane in an ankyrin-independent manner:

1. In semiconfluent MDCK cells, spectrin localizes very early at regions of cell–cell contact and assembles into a Triton-insoluble pool at a time when most ankyrin is not membrane associated (Morrow *et al.,* 1991; Stabach *et al.,* 1993; Devarajan and Morrow, manuscript in preparation).
2. In MDCK cell pulse-chase experiments, the half-life of spectrin is greater than that of ankyrin, indicating that spectrin is stabilized first (C. D. Cianci and J. S. Morrow, unpublished observations).
3. Spectrin possesses direct membrane-association sites (MAD1 and MAD2, see previously) that mediate ankyrin-independent assembly to membranes *in vitro.*
4. MDCK cells transfected with recombinant spectrin peptides encompassing the ankyrin-binding domain, but deleting MAD1 and several associated repeat units, assemble a native spectrin skeleton that does not incorporate the recombinant peptide (Morrow *et al.,* 1991).
5. Transfected recombinant constructs containing sequences encompassing the N-terminal MAD1 domain and other flanking repeat units of βII-spectrin, but which lack the ankyrin-binding domain, still assemble into the cortical skeleton (Stabach *et al.,* 1993).

Collectively, these observations suggest that the proteins that interact with ankyrin do not appear to mediate the assembly of the nascent cortical spectrin skeleton. We hypothesize that these interactions (i.e., those that involve ankyrin) guide the formation of organized protein domains at points defined by cell–cell contact, or by other proteins responsible for the assem-

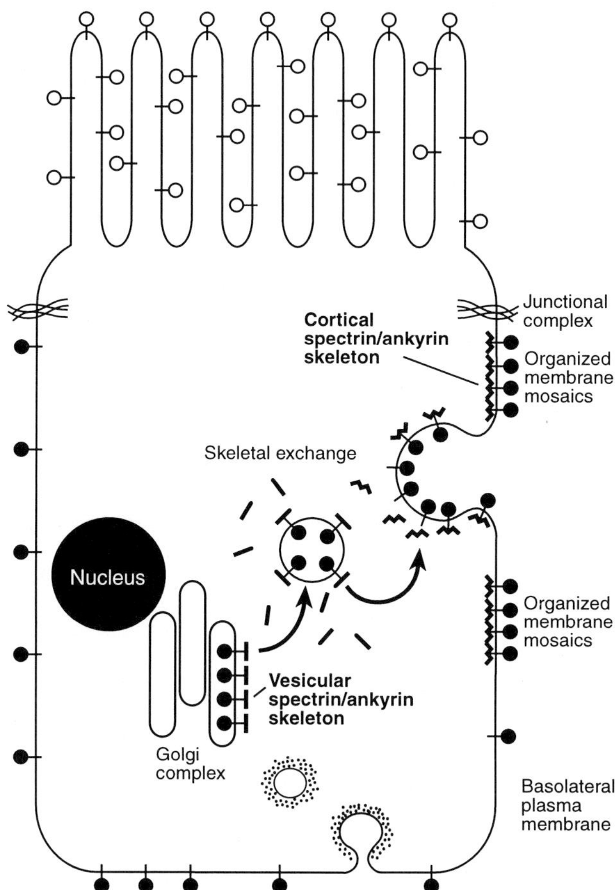

FIGURE 3 A hypothesis concerning the role of the vesicular and cortical spectrin–ankyrin skeletons. Molecular sorting of integral membrane proteins is carried out by complex vesicular trafficking pathways (reviewed in Matter and Mellman, 1994). Two spectrin and ankyrin compartments have recently been identified (Beck *et al.,* 1994; Devarajan *et al.,* 1996). The vesicular spectrin–ankyrin skeleton, composed of βIΣ*-spectrin and ANKIIIΣ4 ankyrin, is associated with Golgi and *trans*-Golgi vesicles but not with the cortical skeleton. Conversely, ANKIIΣ* and ANKIIIΣ2 forms of ankyrin, and βIIΣ*-spectrin, characterize the cortical skeletal compartment. We hypothesize that vesicles destined for the basolateral membrane carry a vesicular skeleton with them until docking with the plasma membrane, at which point the vesicular skeletal elements exchange with cortical skeletal elements, and the targeted vesicle is incorporated into the membrane.

bly of a nascent spectrin skeleton. The consequence of this process is to form an array of surface microdomains in defined region of the cell, a feature termed linked mosaics (Morrow *et al.,* 1996).

Finding of unique subsets of spectrin and ankyrin associated with vesicular membranes, and their absence at the plasma membrane compartment, suggests that the linkages of integral membrane proteins to the vesicular skeleton are exchanged when these vesicles reach their targets at the basolateral membrane. Fig. 3 schematically outlines this hypothetical pathway for polarized membrane assembly and the putative participation of the spectrin skeleton. While this model must still be considered highly speculative, it presents many testable hypotheses. These are the subject of ongoing investigations.

Acknowledgments

The authors thank Drs. David Rimm, Ira Mellman, Scott Weed, and John Sinard for advice and critical comments. The studies described herein that were carried out by the authors were funded in part by grants from the National Institutes of Health to J. S. Morrow and P. Devarajan.

References

Axton, J. M., Shamanski, F. L., Young, L. M., Henderson, D. S. , Boyd J. B., and Orr, W. T. (1994). The inhibitor of DNA replication encoded by the Drosophila gene plutonium is a small, ankyrin repeat protein. *EMBO J.* **13,** 462–470.

Beck, K. A., Buchanan, J. A., Malhotra V., and Nelson, W. J. (1994). Golgi spectrin: Identification of an erythroid beta-spectrin homolog associated with the Golgi complex. *J. Cell Biol.* **127,** 707–723.

Becker, P. S., Schwartz, M. A., Morrow, J. S., and Lux, S. E. (1990). Radiolabel-transfer cross-linking demonstrates that protein 4.1 binds to the N-terminal region of beta spectrin and to actin in binary interactions. *Eur. J. Biochem.* **193,** 827–836.

Becker, P. S., Tse, W. T., Lux, S. E., and Forget, B. G. (1993). Beta spectrin kissimmee: A spectrin variant associated with autosomal dominant hereditary spherocytosis and defective binding to protein 4.1. *J. Clin. Invest.* **92,** 612–616.

Bennett, H., and Condeelis, J. (1988). Isolation of an immunoreactive analogue of brain fodrin that is associated with the cell cortex of Dictiostelium amoebae. *Cell Motil. Cytoskeleton* **11,** 303–307.

Bennett, V. (1990a). Spectrin-based membrane skeleton: A multipotential adaptor between plasma membrane and cytoplasm. *Physiol. Rev.* **70,** 1029–1065.

Bennett, V. (1990b). Spectrin: A structural mediator between diverse plasma membrane proteins and the cytoplasm. *Curr. Opin. Cell Biol.* **2,** 51–56.

Bennett, V. (1992). Ankyrins. Adaptors between diverse plasma membrane proteins and the cytoplasm. *J. Biol. Chem.* **267,** 8703–8706.

Bennett, V., and Gilligan, D. M. (1993). The spectrin-based membrane skeleton and micron-scale organization of the plasma membrane. *Annu. Rev. Cell Biol.* **9,** 27–66.

Bennett, V., and Lambert, S. (1991). The spectrin skeleton: From red cells to brain. *J. Clin. Invest.* **87,** 1483–1489.

Bennett, V., and Stenbuck, P. J. (1979). The membrane attachment protein for spectrin is associated with band 3 in human erythrocyte membranes. *Nature* (*London*) **280,** 468–473.

Birkenmeier, C. S., White, R. A., Peters, L. L., Hall, E. J., Lux, S. E., and Barker, J. E. (1993). Complex patterns of sequence variation and multiple 5′ and 3′ ends are found among transcripts of the erythroid ankyrin gene. *J. Biol. Chem.* **268,** 9533–9540.

Bloch, R. J., and Morrow, J. S. (1989). An unusual β-spectrin associated with clustered acetylcholine receptors. *J. Cell Biol.* **108,** 481–493.

Bourguignon, L. Y., and Jin, H. (1995). Identification of the ankyrin-binding domain of the mouse T-lymphoma cell inositol 1, 4, 5-trisphosphate (IP3) receptor and its role in the regulation of IP3-mediated internal Ca2+ release. *J. Biol. Chem.* **270,** 7257–7260.

Bourguignon, L. Y., Jin, H., Iida, N., Brandt, N. R., and Zhang, S. H. (1993). The involvement of ankyrin in the regulation of inositol 1, 4, 5-trisphosphate receptor-mediated internal Ca2+ release from Ca2+ storage vesicles in mouse T-lymphoma cells. *J. Biol. Chem.* **268,** 7290–7297.

Bretscher, A. (1989). Rapid phosphorylation and reorganization of ezrin and spectrin accompany morphological changes induced in A-431 cells by epidermal growth factor. *J. Cell Biol.* **108,** 921–930.

Byers, T. J., and Branton, D. (1985). Visualization of the protein associations in the erythrocyte membrane skeleton. *Proc. Natl. Acad. Sci. U.S.A.* **82,** 6153–6157.

Byers, T., Brandin, E., Lue, R., Winograd, E., and Branton, D. (1992). The complete sequence of Drosophila beta-spectrin reveals supra-motifs. *Proc. Natl. Acad. Sci. U.S.A.* **89,** 6187–6191.

Chasis, J. A., Coulombel, L., Conboy, J., McGee, S., Andrews, K., Kan, Y. W., and Mohandas, N. (1993). Differentiation-associated switches in protein 4.1 expression. Synthesis of multiple structural isoforms during normal human erythropoiesis. *J. Clin. Invest.* **91,** 329–338.

Cianci, C. D., and Morrow, J. S. (1995). Cloning and characterization of human fetal brain alpha II spectrin. *GenBank* Accession number U26396.

Cianci, C. D., Giorgi, M., and Morrow, J. S. (1988). Phosphorylation of ankyrin down-regulates its cooperative interaction with spectrin and protein 3. *J. Cell Biochem.* **37,** 301–315.

Cianci, C. D., Gallagher, P. G., Forget, B. G., and Morrow, J. S. (1995). Unique subsets of proteins bind the SH3 domain of αI spectrin and Grb2 in differentiating mouse erythroleukemia (MEL) cells. *Mol. Cell Biol.* **5,** 271a.

Cianci, C. D., Harris, A. S., Mische, S., and Morrow, J. S. (1996). The calmodulin binding site of protein 4.1. Submitted.

Clayton, L., McConnell, J. M., and Johnson, M. H. (1995). Control of the surface expression of uvomorulin after activation of mouse oocytes. *Zygote* **3,** 177–189.

Cohen, C. M., Foley, S. F., and Korsgren, C. (1982). A protein immunologically related to erythrocyte band 4.1 is found on stress fibres on non-erythroid cells. *Nature* (*London*) **299,** 648–650.

Coleman, T. R., Fishkind, D. J., Mooseker, M. S., and Morrow, J. S. (1989a). Contributions of the beta-subunit to spectrin structure and function. *Cell Motil. Cytoskeleton* **12,** 248–263.

Coleman, T. R., Fishkind, D. J., Mooseker, M. S., and Morrow, J. S. (1989b). Functional diversity among spectrin isoforms. *Cell Motil. Cytoskeleton* **12,** 225–247.

Conboy, J. G., Chan, J. Y., Chasis, J. A., Kan, Y. W., and Mohandas, N. (1991). Tissue- and development-specific alternative RNA splicing regulates expression of multiple isoforms of erythroid membrane protein 4.1. *J. Biol. Chem.* **266,** 8273–8280.

Correas, I. (1991). Characterization of isoforms of protein 4.1 present in the nucleus. *Biochem. J.* **279,** 581–585.

Correas, I., Leto, T. L., Speicher, D. W., and Marchesi, V. T. (1986). Identification of the functional site of erythrocyte protein 4.1 involved in spectrin-actin associations. *J. Biol. Chem.* **261,** 3310–3315.

Davis, J. Q., and Bennett, V. (1990a). The anion exchanger and Na+K(+)-ATPase interact with distinct sites on ankyrin in in vitro assays. *J. Biol. Chem.* **265,** 17252–17256.

Davis, L. H., and Bennett, V. (1990b). Mapping the binding sites of human erythrocyte ankyrin for the anion exchanger and spectrin. *J. Biol. Chem.* **265,** 10589–10596.

Davis, L. H., and Bennett, V. (1994). Identification of two regions of beta G spectrin that bind to distinct sites in brain membranes. *J. Biol. Chem.* **269,** 4409–4416.

Davis, L., and Bennett, V. (1995). Purification of a calmodulin-sensitive spectrin-binding membrane glycoprotein from brain. *Mol. Biol. Cell* **6,** 1566A.

Davis, L. H., Davis, J. Q., and Bennett, V. (1992). Ankyrin regulation: An alternatively spliced segment of the regulatory domain functions as an intramolecular modulator. *J. Biol. Chem.* **267,** 18966–18972.

DeSilva, T. M., Peng, K. C., Speicher, K. D., and Speicher, D. W. (1992). Analysis of human red cell spectrin tetramer (head-to-head) assembly using complementary univalent peptides. *Biochemistry* **31,** 10872–10878.

Devarajan, P., Stabach, P. R., Mann, A. S., Ardito, T., Kashgarian, M., and Morrow, J. S. (1996). Identification of a small cytoplasmic ankyrin (Ank_{G119}) in kidney and muscle that binds $\beta I\Sigma^*$ spectrin and associates with the Golgi apparatus. *J. Cell Biol.* **133,** 819–830.

Ding, Y., Casey, J. R., and Kopito, R. R. (1994). The major kidney AE1 isoform does not bind ankyrin (Ank1) in vitro: An essential role for the 79 NH2-terminal amino acid residues of band 3. *J. Biol. Chem.* **269,** 32201–32208.

Drenckhahn, D., and Bennett, V. (1987). Polarized distribution of Mr 210, 000 and 190, 000 analogs of erythrocyte ankyrin along the plasma membrane of transporting epithelia, neurons and photoreceptors. *Eur. J. Cell Biol.* **43,** 479–486.

Dubreuil, R. R. (1991). Structure and evolution of the actin crosslinking proteins. *Bioessays* **13,** 219–226.

Dubreuil, R. R., and Yu, J. (1994). Ankyrin and beta-spectrin accumulate independently of alpha-spectrin in Drosophila. *Proc. Natl. Acad. Sci. U.S.A.* **91,** 10285–10289.

Dubreuil, R. R., Byers, T. J., Sillman, A. L., Bar-Zvi, D., Goldstein, L. S., and Branton, D. (1989). The complete sequence of Drosophila alpha-spectrin: Conservation of structural domains between alpha-spectrins and alpha-actinin. *J. Cell Biol.* **109,** 2197–2205.

Faraday, C. D., and Spanswick, R. M. (1993). Evidence for a membrane skeleton in higher plants: A spectrin-like polypeptide co-isolates with rice root plasma membranes. *FEBS Lett.* **318,** 313–316.

Fearon, E. R., Finkel, T., Gillison, M. L., Kennedy, S. P., Casella, J. F., Tomaselli, G. F., Morrow, J. S., and Dang, C. V. (1992). Karyoplasmic interaction selection strategy (KISS): A general strategy to detect protein-protein interactions in mammalian cells. *Proc. Natl. Acad. Sci. U.S.A.* **89,** 7958–7962.

Fleming, T. P., Butler, L., Lei, X., Collins, J., Javed, Q., Sheth, B., Stoddart, N., Wild, A., and Hay, M. (1994). Molecular maturation of cell adhesion systems during mouse early development. *Histochemistry* **101,** 1–7.

Franck, Z., Gary, R., and Bretscher, A. (1993). Moesin, like ezrin, colocalizes with actin in the cortical cytoskeleton in cultured cells, but its expression is more variable. *J. Cell Sci.* **105,** 219–231.

Frappier, T., Stetzkowski-Marden, F., and Pradel, L. A. (1991). Interaction domains of neurofilament light chain and brain spectrin. *Biochem. J.* **275,** 521–527.

Funayama, N., Nagafuchi, A., Sato, N., Tsukita, S., and Tsukita, S. (1991). Radixin is a novel member of the band 4.1 family. *J. Cell Biol.* **115,** 1039–1048.

Gallagher, P. G., Tse, W. T., Scarpa, A. L., Lux, S. E., and Forget, B. G. (1992). Large numbers of alternatively spliced isoforms of the regulatory region of human erythrocyte ankyrin. *Trans. Assoc. Am. Physicians* **105,** 268–277.

Gardner, K., and Bennett, V. (1986). A new erythrocyte membrane-associated protein with calmodulin binding activity. Identification and purification. *J. Biol. Chem.* **261,** 1339–1348.

Gardner, K., and Bennett, V. (1987). Modulation of spectrin-actin assembly by erythrocyte adducin. *Nature* **328,** 359–362.

Georgatos, S. D., and Marchesi, V. T. (1985). The binding of vimentin to human erythrocyte membranes: A model system for the study of intermediate filament-membrane interactions. *J. Cell Biol.* **100,** 1955–1961.

Gilligan, D., and Bennett, V. (1991). Alternative splicing of adducin beta subunit. *J. Cell Biol.* **115,** 42a.

Gilligan, D. M., and Bennett, V. (1993). The junctional complex of the membrane skeleton. *Semin. Hematol.* **30,** 74–83.

Glantz, S. B., and Morrow, J. S. (1995). The spectrin-actin cytoskeleton and membrane integrity in hypoxia. *In* "Tissue Oxygen Deprivation: Developmental, Molecular and Integrated Function" (G. G. Haddad and G. Lister, eds.). Marcel Dekker, Inc., New York.

Gregorio, C. C., Kubo, R. T., Bankert, R. B., and Repasky, E. A. (1992). Translocation of spectrin and protein kinase C to a cytoplasmic aggregate upon lymphocyte activation. *Proc. Natl. Acad. Sci. U.S.A.* **89,** 4947–4951.

Gregorio, C. C., Black, J. D., and Repasky, E. A. (1993). Dynamic aspects of cytoskeletal protein distribution in T lymphocytes: Involvement of calcium in spectrin reorganization. *Blood Cells* **19,** 361–371.

Gregorio, C. C., Repasky, E. A., Fowler, V. M., and Black, J. D. (1994). Dynamic properties of ankyrin in T lymphocytes: colocalization with spectrin and protein kinase C beta. *J. Cell Biol.* **125,** 345–358.

Gumbiner, B. M., and McCrea, P. D. (1993). Catenins as mediators of the cytoplasmic functions of cadherins. *J. Cell Sci. Suppl.* **17,** 155–158.

Hall, T. G., and Bennett, V. (1987). Regulatory domains of erythrocyte ankyrin. *J. Biol. Chem.* **262,** 10537–10545.

Harris, A. S., and Morrow, J. S. (1990). Calmodulin and calcium-dependent protease I coordinately regulate the interaction of fodrin with actin. *Proc. Natl. Acad. Sci. U.S.A.* **87,** 3009–3013.

Harris, A. S., Croall, D., and Morrow, J. S. (1988). The calmodulin-binding site in fodrin is near the site of calcium-dependent protease-I cleavage. *J. Biol. Chem.* **263,** 15754–15761.

Harris, A. S., Croall, D. E., and Morrow, J. S. (1989). Calmodulin regulates fodrin susceptibility to cleavage by calcium-dependent protease I. *J. Biol. Chem.* **264,** 17401–17408.

Hartwig, J. H. (1994). Spectrin superfamily, subfamily 1: The spectrin family. *Protein Profile* **1,** 715–749.

Hemming, N. J., Anstee, D. J., Mawby, W. J., Reid, M. E., and Tanner, M. J. (1994). Localization of the protein 4.1-binding site on human erythrocyte glycophorins C and D. *Biochem. J.* **299,** 191–196. [published erratum appears in *Biochem. J.* (1994). **300,** 920].

Hemming, N. J., Anstee, D. J., Staricoff, M. A., Tanner, M. J. A., and Mohandas, N. (1995). Identification of the membrane attachment sites for protein 4.1 in the human erythrocyte. *J. Biol. Chem.* **270,** 5360–5366.

Hinck, L., Nathke, I. S., Papkoff, J., and Nelson, W. J. (1994a.) Dynamics of cadherin/catenin complex formation: Novel protein interactions and pathways of complex assembly. *J. Cell Biol.* **125,** 1327–1340.

Hinck, L., Näthke, I. S., Papkoff, J., and Nelson, W. J. (1994b). Beta-catenin: A common target for the regulation of cell adhesion by Wnt-1 and Src signaling pathways. *Trends Biochem.* **19,** 538–542.

Horne, W. C., Leto, T. L., and Marchesi, V. T. (1985). Differential phosphorylation of multiple sites in protein 4.1 and protein 4.9 by phorbol ester-activated and cyclic AMP-dependent protein kinases. *J. Biol. Chem.* **260,** 9073–9076.

Howe, C. L., Sacramone, L. M., Mooseker, M. S., and Morrow, J. S. (1985). Mechanisms of cytoskeletal regulation: Modulation of membrane affinity in avian brush border and erythrocyte spectrins. *J. Cell Biol.* **101,** 1379–1385.

Hu, R.-J., Watanabe, M., and Bennett, V. (1992). Characterization of human brain cDNA encoding the general isoform of β-spectrin. *J. Biol. Chem.* **267,** 18715–18722.

Huang, J. P., Tang, C. J., Kou, G. H., Marchesi, V. T., Benz, E. J., Jr. and Tang, T. K. (1993). Genomic structure of the locus encoding protein 4.1: Structural basis for complex combinational patterns of tissue-specific alternative RNA splicing. *J. Biol. Chem.* **268,** 3758–3766.

Huang, M., Gu, G., Ferguson, E. L., and Chalfie, M. (1995). A stomatin-like protein necessary for mechanosensation in C. elegans. *Nature (London)* **378,** 292–295.

Hughes, C. A., and Bennett, V. (1995). Adducin: A physical model with implications for function in assembly of spectrin-actin complexes. *J. Biol. Chem.* **270,** 18990–18996.

Joshi, R., and Bennett, V. (1990). Mapping the domain structure of human erythrocyte adducin. *J. Biol. Chem.* **265,** 13130–13136.

Joshi, R., Gilligan, D. M., Otto, E., McLaughlin, T., and Bennett, V. (1991). Primary structure and domain organization of human alpha and beta adducin. *J. Cell Biol.* **115,** 665–675.

Jou, T. S., Stewart, D. B., Stappert, J., Nelson, W. J., and Marrs, J. A. (1995). Genetic and biochemical dissection of protein linkages in the cadherin-catenin complex. *Proc. Natl. Acad. Sci. U.S.A.* **92,** 5067–5071.

Kaiser, H. W., O'Keefe, E., and Bennett, V. (1989). Adducin: Ca^{++}-dependent association with sites of cell-cell contact. *J. Cell Biol.* **109,** 557–569.

Kapfhamer, D., Miller, D. E., Lambert, S., Bennett, V., Glover, T. W., and Burmeister, M. (1995). Chromosomal localization of the ankyrinG gene (ANK3/Ank3) to human 10q21 and mouse 10. *Genomics* **27,** 189–191.

Karinch, A. M., Zimmer, W. E., and Goodman, S. R. (1990). The identification and sequence of the actin-binding domain of human red blood cell beta-spectrin. *J. Biol. Chem.* **265,** 11833–11840.

Kashgarian, M., Morrow, J. S., Foellmer, H. G., Mann, A. S., Cianci, C., and Ardito, T. (1988). Na,K-ATPase co-distributes with ankyrin and spectrin in renal tubular epithelial cells. *In* "The NA+, K+-Pump, Part B: Cellular Aspects," (J. C. Skou, J. G. Norby, A. B. Manusbach, and M. Esmann, eds.), pp. 245–250. Alan R. Liss, Inc., New York.

Kelly, G. M., Zelus, B. D., and Moon, R. T. (1991). Identification of a calcium-dependent calmodulin-binding domain in Xenopus membrane skeleton protein 4.1. *J. Biol. Chem.* **266,** 12469–12473.

Kelly, R. B. (1991). Secretory granule and synaptic vesicle formation. *Curr. Opin. Cell Biol.* **3,** 654–660.

Kemler, R. (1992). Classical cadherins. *Semin. Cell Biol.* **3,** 149–155.

Kemler, R. (1993). From cadherins to catenins: Cytoplasmic protein interactions and regulation of cell adhesion. *Trends Genet.* **9,** 317–321.

Kennedy, S. P., Warren, S. L., Forget, B. G., and Morrow, J. S. (1991). Ankyrin binds to the 15th repetitive unit of erythroid and non-erythroid β-spectrin. *J. Cell Biol.* **115,** 267–277.

Kennedy, S. P., Weed, S. A., Forget, B. G., and Morrow, J. S. (1994). A partial structural repeat forms the heterodimer self-association site of all β-spectrins. *J. Biol. Chem.* **269,** 11400–11408.

Koob, R., Zimmermann, M., Schoner, W., and Drenckhahn, D. (1988). Colocalization and coprecipitation of ankyrin and Na+,K+-ATPase in kidney epithelial cells. *Eur. J. Cell Biol.* **45,** 230–237.

Kordeli, E., Lambert, S., and Bennett, V. (1995). AnkyrinG: A new ankyrin gene with neural-specific isoforms localized at the axonal initial segment and node of Ranvier. *J. Biol. Chem.* **270,** 2352–2359.

Kralj-Iglic, V., Svetina. S., and Zeks, B. (1996). Lateral distribution of membrane constituents and cellular shapes. *Eur. Biophys. J.* In press.

Kunimoto, M., Otto, E., and Bennett, V. (1991). A new 440-kD isoform is the major ankyrin in neonatal rat brain. *J. Cell Biol.* **115,** 1319–1331.

Lambert, S., Yu, H., Prchal, J. T., Lawler, J., Ruff, P., Speicher, D., Cheung, M. C., Kan, Y. W., and Palek, J. (1990). cDNA sequence for human erythrocyte ankyrin. *Proc. Natl. Acad. Sci. U.S.A.* **87,** 1730–1734.

Leto, T. L., and Marchesi, V. T. (1984). A structural model of human erythrocyte protein 4.1. *J. Biol. Chem.* **259,** 4603–4608.

Li, Z. P., Burke, E. P., Frank, J. S., Bennett, V., and Philipson, K. D. (1993). The cardiac Na+-Ca2+ exchanger binds to the cytoskeletal protein ankyrin. *J. Biol. Chem.* **268,** 11489–11491.

Liu, S. C., Derick, L. H., and Palek, J. (1987). Visualization of the hexagonal lattice in the erythrocyte membrane skeleton. *J. Cell Biol.* **104,** 527–536.

Lokeshwar, V. B., Fregien, N., and Bourguignon, L. Y. (1994). Ankyrin-binding domain of CD44(GP85) is required for the expression of hyaluronic acid-mediated adhesion function. *J. Cell Biol.* **126,** 1099–1109.

Lombardo, C. R., and Low, P. S. (1994). Calmodulin modulates protein 4.1 binding to human erythrocyte membranes. *Biochim. Biophys. Acta* **1196,** 139–144.

Lombardo, C. R., Willardson, B. M., and Low, P. S. (1992). Localization of the protein 4.1-binding site on the cytoplasmic domain of erythrocyte membrane band 3. *J. Biol. Chem.* **267,** 9540–9546.

Lombardo, C. R., Rimm, D. L., Kennedy, S. P., Forget, B. G., and Morrow, J. S. (1993). Ankyrin independent membrane sites for non-erythroid spectrin. *Mol. Biol. Cell* **4**(Suppl.), 57a.

Lombardo, C. R., Rimm, D. L., Koslov, E., and Morrow, J. S. (1994a). Human recombinant alpha-catenin binds to spectrin. *Mol. Biol. Cell* **5**(Suppl.), 47a.

Lombardo, C. R., Weed, S. A., Kennedy, S. P., Forget, B. G., and Morrow, J. S. (1994b). βII-spectrin (fodrin) and βIΣ2-spectrin (muscle) contain NH2- and COOH-terminal membrane association domains (MAD1 & MAD2). *J. Biol. Chem.* **269,** 29212–29219.

Lorenz, M., Bisikirska, B., Hanus-Lorenz, B., Strzalka, K., and Sikorski, A. F. (1995). Proteins reacting with anti-spectrin antibodies are present in Chlamydomonas cells. *Cell Biol. Int.* **19,** 83–90.

Lu, P. W., Soong, C. J., and Tao, M. (1985). Phosphorylation of ankyrin decreases its affinity for spectrin tetramer. *J. Biol. Chem.* **260,** 14958–14964.

Luna, E. J. (1991). Molecular links between the cytoskeleton and membranes. *Curr. Opin. Cell Biol.* **3,** 120–126.

Luna, E. J., and Hitt, A. L. (1992). Cytoskeleton–plasma membrane interactions. *Science* **258,** 955–964.

Lux, S. E., John, K. M., and Bennett, V. (1990). Analysis of cDNA for human erythrocyte ankyrin indicates a repeated structure with homology to tissue-differentiation and cell-cycle control proteins. *Nature* (*London*) **344,** 36–42.

Macias, M. J., Musacchio, A., Ponstingl, H., Nilges, M., Saraste, M., and Oschkinat, H. (1994). Structure of the pleckstrin homology domain from beta-spectrin. *Nature* (*London*) **369,** 675–677.

Malchiodi-Albedi, F., Ceccarini, M., Winkelmann, J. C., Morrow, J. S., and Petrucci, T. C. (1993). The 270 kDa splice variant of erythrocyte β-spectrin (βIΣ2) segregates *in vivo* and *in vitro* to specific domains of cerebellar neurons. *J. Cell Sci.* **106,** 67–78.

Mangeat, P. H., and Burridge, K. (1984). Immunoprecipitation of nonerythrocyte spectrin within live cells following microinjection of specific antibodies: Relation to cytoskeletal structures. *J. Cell Biol.* **98,** 1363–1377.

Marfatia, S. M., Leu, R. A., Branton, D., and Chishti, A. H. (1995). Identification of the protein 4.1 binding interface on glycophorin C and p55, a homologue of the Drosophila discs-large tumor suppressor protein. *J. Biol. Chem.* **270,** 715–719.

Matter, K., and Mellman, I. (1994). Mechanisms of cell polarity: Sorting and transport in epithelial cells. *Curr. Opin. Cell Biol.* **6,** 545–554.

McNeill, H., Ozawa, M., Kemler, R., and Nelson, W. J. (1990). Novel function of the cell adhesion molecule uvomorulin as an inducer of cell surface polarity. *Cell* **62,** 309–316.

McNeill, H., Ryan, T. A., Smith, S. J., and Nelson, W. J. (1993). Spatial and temporal dissection of immediate and early events following cadherin-mediated epithelial cell adhesion. *J. Cell Biol.* **120,** 1217–1226.

Merilainen, J., Palovuori, R., Sormunen, R., Wasenius, V. M., and Lehto, V. P. (1993). Binding of the alpha-fodrin SH3 domain to the leading lamellae of locomoting chicken fibroblasts. *J. Cell Sci.* **105,** 647–654.

Michaely, P., and Bennett, V. (1993). The membrane-binding domain of ankyrin contains four independently folded subdomains, each comprised of six ankyrin repeats. *J. Biol. Chem.* **268,** 22703–22709.

Michaud, D., Guillet, G., Rogers, P. A., and Charest, P. M. (1991). Identification of a 220 kDa membrane-associated plant cell protein immunologically related to human beta-spectrin. *FEBS Lett.* **294,** 77–80.

Mische, S. M., Mooseker, M. S., and Morrow, J. S. (1987). Erythrocyte adducin: A calmodulin regulated actin bundling protein that stimulates spectrin-actin binding. *J. Cell Biol.* **105,** 2837–2845.

Mohandas, N. (1992). Molecular basis for red cell membrane viscoelastic properties. *Biochem. Soc. Trans.* **20,** 776–782.

Mohandas, N., and Evans, E. (1994). Mechanical properties of the red cell membrane in relation to molecular structure and genetic defects. *Annu. Rev. Biophys. Biomol. Struct.* **23,** 787–818.

Mohandas, N., Winardi, R., Knowles, D., Leung, A., Parra, M., George, E., Conboy, J., and Chasis, J. (1992). Molecular basis for membrane rigidity of hereditary ovalocytosis: A novel mechanism involving the cytoplasmic domain of band 3. *J. Clin. Invest.* **89,** 686–692.

Moon, R. T., and McMahon, A. P. (1990). Generation of diversity in nonerythroid spectrins. Multiple polypeptides are predicted by sequence analysis of cDNAs encompassing the coding region of human nonerythroid alpha-spectrin. *J. Biol. Chem.* **265,** 4427–4433.

Moon, R. T., DeMarais, A., and Olson, D. J. (1993). Responses to Wnt signals in vertebrate embryos may involve changes in cell adhesion and cell movement. *J. Cell Sci. Suppl.* **17,** 183–188.

Morris, M. B., and Lux, S. E. (1995). Characterization of the binary interaction between human erythrocyte protein 4.1 and actin. *Eur. J. Biochem.* **231,** 644–650.

Morrow, J. S., Speicher, D. W., Knowles, W. J., Hsu, C. J., and Marchesi, V. T. (1980). Identification of functional domains of human erythrocyte spectrin. *Proc. Natl. Acad. Sci. U.S.A.* **77,** 6592–6596.

Morrow, J. S., Cianci, C., Ardito, T., Mann, A., and Kashgarian, M. T. (1989). Ankyrin links fodrin to alpha Na/K ATPase in Madin-Darby canine kidney cells and in renal tubule cells. *J. Cell Biol.* **108,** 455–465.

Morrow, J. S., Cianci, C. D., Kennedy, S. P., and Warren, S. L. (1991). Polarized assembly of spectrin and ankyrin in epithelial cells. *In* "Ordering the Membrane Cytoskeleton Trilayer" (M. S. Mooseker and J. S. Morrow, eds.), pp. 227–244. Academic Press, New York.

Morrow, J. S., Cianci, C. D., Kennedy, S. P., Sinard, J. H., Rimm, D. L., and Weed, S. A. (1996). Of membrane stability and linked mosaics: The spectrin cytoskeleton. *In* "Handbook of Physiology" (J. Hoffman and J. Jamieson, eds.). Oxford University Press, New York.

Najm, I., Vanderklish, P., Lynch, G., and Baudry, M. (1991). Effect of treatment with difluoromethylornithine on polyamine and spectrin breakdown levels in neonatal rat brain. *Brain Res. Dev. Brain Res.* **63,** 287–289.

Näthke, I. S., Hinck, L. E., and Nelson, W. J. (1993). Epithelial cell adhesion and development of cell surface polarity: Possible mechanisms for modulation of cadherin function, organization and distribution. *J. Cell Sci. Suppl.* **17,** 139–145.

Nehls, V., Drenckhahn, D., Joshi, R., and Bennett, V. (1991). Adducin in erythrocyte precursor cells of rats and humans: Expression and compartmentalization. *Blood* **78,** 1692–1696.

Nelson, W. J. (1992). Renal epithelial cell polarity. *Curr. Opin. Nephrol. Hypertens.* **1,** 59–67.

Nelson, W. J., and Veshnock, P. J. (1986). Dynamics of membrane-skeleton (fodrin) organization during development of polarity in Madin-Darby canine kidney epithelial cells. *J. Cell Biol.* **103,** 1751–1765.

Nelson, W. J., and Veshnock, P. J. (1987). Modulation of fodrin (membrane skeleton) stability by cell-cell contact in Madin-Darby canine kidney epithelial cells. *J. Cell Biol.* **104,** 1527–1537.

Nelson, W. J., Shore, E. M., Wang, A. Z., and Hammerton, R. W. (1990). Identification of a membrane-cytoskeletal complex containing the cell adhesion molecule uvomorulin (E-cadherin), ankyrin, and fodrin in Madin-Darby canine kidney epithelial cells. *J. Cell Biol.* **110,** 349–357.

Nelson, W. J., Wilson, R., Wollner, D., Mays, R., McNeill, H., and Siemers, K. (1992). Regulation of epithelial cell polarity: A view from the cell surface. *Cold Spring Harbor Symp. Quant. Biol.* **57,** 621–630.

Otsuka, A. J., Franco, R., Yang, B., Shim, K. H., Tang, L. Z., Zhang, Y. Y., Boontrakulpoontawee, P., Jeyaprakash, A., Hedgecock, E., Wheaton, V. I., et al. 1995. An ankyrin-related gene (unc-44) is necessary for proper axonal guidance in Caenorhabditis elegans. *J. Cell Biol.* **129,** 1081–1092.

Otto, E., Kunimoto, M., McLaughlin, T., and Bennett, V. (1991). Isolation and characterizatoin of cDNAs encoding human brain ankyrins reveal a family of alternatively spliced genes. *J. Cell Biol.* **114,** 241–253.

Ozawa, M., and Kemler, R. (1992). Molecular organization of the uvomorulin-catenin complex. *J. Cell Biol.* **116,** 989–996.

Ozawa, M., Baribault, H., and Kemler, R. (1989). The cytoplasmic domain of the cell adhesion molecule uvomorulin associates with three independent proteins structurally related in different species. *EMBO J.* **8,** 1711–1717.

Ozawa, M., Engel, J., and Kemler, R. (1990). Single amino acid substitutions in one Ca^{++} binding site of uvomorulin abolish the adhesive function. *Cell* **63,** 1033–1038.

Palfrey, H. C., and Waseem, A. (1985). Protein kinase C in the human erythrocyte: Translocation to the plasma membrane and phosphorylation of bands 4.1 and 4.9 and other membrane proteins. *J. Biol. Chem.* **260,** 16021–16029.

Pasternack, G. R., Anderson, R. A., Leto, T. L., and Marchesi, V. T. (1985). Interactions between protein 4.1 and band 3: An alternative binding site for an element of the membrane skeleton. *J. Biol. Chem.* **260,** 3676–3683.

Peifer, M., Orsulic, S., Pai, L. M., and Loureiro, J. (1993). A model system for cell adhesion and signal transduction in Drosophila. *Dev. Suppl.* 163–176.

Peters, L. L., John, K. M., Lu, F. M., Eicher, E. M., Higgins, A., Yialamas, M., Turtzo, L. C., Otsuka, A. J., and Lux, S. E. (1995). Ank3 (epithelial ankyrin), a widely distributed new member of the ankyrin gene family and the major ankyrin in kidney, is expressed in alternatively spliced forms, including forms that lack the repeat domain. *J. Cell Biol.* **130,** 313–330.

Pinto, C. C., Goldstein, E. G., Bennett, V., and Sobel, J. S. (1991). Immunofluorescence localization of an adducin-like protein in the chromosomes of mouse oocytes. *Dev. Biol.* **146,** 301–311.

Platt, O. S., Lux, S. E., and Falcone, J. F. (1993). A highly conserved region of human erythrocyte ankyrin contains the capacity to bind spectrin. *J. Biol. Chem.* **268,** 24421–24426.

Pollard, T. D. (1984). Purification of a high molecular weight actin filament gelation protein from Acanthamoeba that shares antigenic determinants with vertebrate spectrins. *J. Cell Biol.* **99,** 1970–1980.

Pradhan, D., Lombardo, C., Stabach, P. R., and Morrow, J. S. (1996). Surface plasmon resonance detects a strong heterodimer nucleation site in domain I of beta-spectrin. In preparation.

Prchal, J. T., Morley, B. J., Yoon, S. H., Coetzer, T. L., Palek, J., Conboy, J. G., and Kan, Y. W. (1987). Isolation and characterization of cDNA clones for human erythrocyte beta-spectrin. *Proc. Natl. Acad. Sci. U.S.A.* **84,** 7468–7472.

Riederer, B. M., Zagon, I. S., and Goodman, S. R. (1986). Brain spectrin (240/235) and brain spectrin (240/235E): Two distinct spectrin subtypes with different locations within mammalian neural cells. *J. Cell Biol.* **102,** 2088–2096.

Riederer, B. M., Zagon, I. S., and Goodman, S. R. (1987). Brain spectrin (240/235) and brain spectrin (240/235E): Differential expression during mouse brain development. *J. Neurosci.* **7,** 864–874.

Rimm, D. L., Kabriaei, P., and Morrow, J. S. (1994). a(E)-catenin is an actin binding and bundling protein. *Mol. Biol. Cell* **5**(Suppl.), 5a.

Rimm, D. L., Koslov, E. R., Kebriaei, P., Cianci, C. D., and Morrow, J. S. (1995). α_1(E)-catenin is a novel actin binding and bundling protein mediating the attachment of F-actin to the membrane adhesion complex. *Proc. Natl. Acad. Sci. U.S.A.* **92,** 8813–8817.

Rodriguez-Boulan, E., and Nelson, W. J. (1989). Morphogenesis of the polarized epithelial cell phenotype. *Science* **245,** 718–725.

Rotin, D., Bar-Sagi, D., O'Brodovich, H., Merilainen, J., Lehto, V. P., Canessa, C. M., Rossier, B. C., and Downey, G. P. (1994). An SH3 binding region in the epithelial Na+ channel (alpha rENaC) mediates its localization at the apical membrane. *EMBO J.* **13,** 4440–4450.

Sahr, K. E., Laurila, P., Kotula, L., Scarpa, A. L., Coupal, E., Leto, T. L., Linnenbach, A. J., Winkelmann, J. C., Speicher, D. W., Marchesi, V. T., Curtis, P. J., and Forget, B. G. (1990). The complete cDNA and polypeptide sequences of human erythroid α-spectrin. *J. Biol. Chem.* **265,** 4434–4443.

Sato, N., Funayama, N., Nagafuchi, A., Yonemura, S., Tsukita, S., and Tsukita, S. (1992). A gene family consisting of ezrin, radixin and moesin: Its specific localization at actin filament/plasma membrane association sites. *J. Cell Sci.* **103,** 131–143.

Scaramuzzino, D. A., and Morrow, J. S. (1993). Calmodulin-binding domain of recombinant erythrocyte beta-adducin. *Proc. Natl. Acad. Sci. U.S.A.* **90,** 3398–3402. [published erratum appears in *Proc. Natl. Acad. Sci. U.S.A.* (1993.) **90**(16), 7908].

Schofield, A. E., Tanner, M. J., Pinder, J. C., Clough, B., Bayley, P. M., Nash, G. B., Dluzewski, A. R., Reardon, D. M., Cox, T. M., Wilson, R. J., et. al. (1992). Basis of unique red cell membrane properties in hereditary ovalocytosis. *J. Mol. Biol.* **223,** 949–958.

Seubert, P., Peterson, C., Vanderklish, P., Cotman, C., and Lynch, G. (1990). Increased spectrin proteolysis in the brindled mouse brain. *Neurosci. Lett.* **108,** 303–308.

Siegel, N. J., Devarajan, P., and Van Why, S. (1994). Renal cell injury: Metabolic and structural alterations. *Pediatr. Res.* **36,** 129–136.

Sikorski, A. F., Swat, W., Brzezinska, M., Wroblewski, Z., and Bisikirska, B. (1993). A protein cross-reacting with anti-spectrin antibodies is present in higher plant cells. *Z. Naturforsch. [C]* **48,** 580–583.

Simons, K., Dupree, P., Fiedler, K., Huber, L. A., Kobayashi, T., Kurzchalia, T., Olkkonen, V., Pimplikar, S., Parton, R., and Dotti, C. (1992). Biogenesis of cell-surface polarity in epithelial cells and neurons. *Cold Spring Harbor Symp. Quant. Biol.* **57,** 611–619.

Sinard, J. H., Stewart, G. W., Argent, A. C., and Morrow, J. S. (1994). Stomatin binding to adducin: A novel link between transmembrane ion transport and the cytoskeleton. *Mol. Biol. Cell* **5**(Suppl.), 421a.

Sinard, J. H., Stewart, G. W., Argent, A. C., Gilligan, D. M., and Morrow, J. S. (1995). A novel isoform of beta adducin utilizes an alternatively spliced exon near the C-terminus. *Mol. Biol. Cell* **6,** 269a.

Smith, P. R., Saccomani, G., Joe, E. H., Angelides, K. J., and Benos, D. J. (1991). Amiloride-sensitive sodium channel is linked to the cytoskeleton in renal epithelial cells. *Proc. Natl. Acad. Sci. U.S.A.* **88,** 6971–6975.

Smith, P. R., Bradford, A. L., Joe, E. H., Angelides, K. J., Benos, D. J., and Saccomani, G. (1993). Gastric parietal cell H(+)-K(+)-ATPase microsomes are associated with isoforms of ankyrin and spectrin. *Am. J. Physiol.* **264,** C63–C70.

Speicher, D. W., Weglarz, L., and DeSilva, T. M. (1992). Properties of human red cell spectrin heterodimer (side-to-side) assembly and identification of an essential nucleation site. *J. Biol. Chem.* **267,** 14775–14782.

Speicher, D. W., DeSilva, T. M., Speicher, K. D., Ursitti, J. A., Hembach, P., and Weglarz, L. (1993). Location of the human red cell spectrin tetramer binding site and detection of a related "closed" hairpin loop dimer using proteolytic footprinting. *J. Biol. Chem.* **268,** 4227–4235.

Srinivasan, Y., Lewallen, M., and Angelides, K. J. (1992). Mapping the binding site on ankyrin for the voltage-dependent sodium channel from brain. *J. Biol. Chem.* **267,** 7483–7489.

Stabach, P. R., Kennedy, S. P., and Morrow, J. S. (1993). Polarized assembly of the spectrin skeleton is ankyrin-independent in MDCK cells. *Mol. Biol. Cell* **4,** 58a.

Steiner, J. P., and Bennett, V. (1988). Ankyrin-independent membrane protein-binding sites for brain and erythrocyte spectrin. *J. Biol. Chem.* **263,** 14417–14425.

Steiner, J. P., Walke, H. T. J., and Bennett, V. (1989). Calcium/calmodulin inhibits direct binding of spectrin to synaptosomal membranes. *J. Biol. Chem.* **264,** 2783–2791.

Svetina, S., Kralj-Iglic, V., and Zeks, B. (1990). Cell shape and lateral distribution of mobile membrane constituents. *In* "Tenth School on Biophysics of Membrane Transport, School Proceedings," pp. 139–155. (J. Kuczera and S. Przestalski, eds.). Univ. Wroclaw, Wroclaw.

Svetina, S., Iglic, A., Kralj-Iglic, V., and Zeks, B. (1995). Cytoskeleton and red cell shape. *Cell Mol. Biol. Lett.* **1,** 67–78.

Takeichi, M. (1991). Cadherin cell adhesion receptors as a morphogenetic regulator. *Science* **251,** 1451–1455.

Tanaka, T., Kadowaki, K., Lazarides, E., and Sobue, K. (1991). Ca^{2+}-dependent regulation of the spectrin/actin interaction by calmodulin and protein 4.1. *J. Biol. Chem.* **266,** 1134–1140.

Tang, T. K., Leto, T. L., Marchesi, V. T., and Benz, E. J. (1988). Expression of specific isoforms of protein 4.1 in erythroid and non-erythroid tissues. *Adv. Exp. Med. Biol.* **241,** 81–95.

Tang, T. K., Qin, Z., Leto, T., Marchesi, V. T., and Benz, E. J., Jr. (1990). Heterogeneity of mRNA and protein products arising from the protein 4.1 gene in erythroid and nonerythroid tissues. *J. Cell Biol.* **110,** 617–624.

Taylor, S. A., Snell, R. G., Buckler, A., Ambrose, C., Duyao, M., Church, D., Lin, C. S., Altherr, M., Bates, G. P., Groot, N., et al. (1992). Cloning of the alpha-adducin gene from the Huntington's disease candidate region of chromosome 4 by exon amplification. *Nature Genet.* **2,** 223–227.

Touhara, K., Inglese, J., Pitcher, J. A., Shaw, G., and Lefkowitz, R. J. (1994). Binding of G protein beta gamma-subunits to pleckstrin homology domains. *J. Biol. Chem.* **269,** 10217–10220.

Tripodi, G., Piscone, A., Borsani, G., Tisminetzky, S., Salardi, S., Sidoli, A., James, P., Pongor, S., Bianchi, G., and Baralle, F. E. (1991). Molecular cloning of an adducin-like protein: Evidence of a polymorphism in the normotensive and hypertensive rats of the milan strain. *Biochem. Biophys. Res. Commun.* **177,** 939–947.

Ursitti, J. A., and Wade, J. B. (1993). Ultrastructure and immunocytochemistry of the isolated human erythrocyte membrane skeleton. *Cell Motil. Cytoskeleton* **25,** 30–42.

Viel, A., and Branton, D. (1994). Interchain binding at the tail end of the Drosophila spectrin molecule. *Proc. Natl. Acad. Sci. U.S.A.* **91,** 10839–10843.

Wang, C. C., Moriyama, R., Lombardo, C. R., and Low, P. S. (1995). Partial characterization of the cytoplasmic domain of human kidney band 3. *J. Biol. Chem.* **270,** 17892–17897.

Wang, D. S., and Shaw, G. (1995). The association of the C-terminal region of beta I sigma II spectrin to brain membranes is mediated by a PH domain, does not require membrane proteins, and coincides with a inositol-1, 4, 5 triphosphate binding site. *Biochem. Biophys. Res. Commun.* **217,** 608–615.

Wang, D. S., Shaw, R., Winkelmann, J. C., and Shaw, G. (1994). Binding of PH domains of beta-adrenergic receptor kinase and beta-spectrin to WD40/beta-transducin repeat containing regions of the beta-subunit of trimeric G-proteins. *Biochem. Biophys. Res. Commun.* **203,** 29–35.

Wang, D. S., Shaw, R., Hattori, M., Arai, H., Inoue, K., and Shaw, G. (1995). Binding of pleckstrin homology domains to WD40/beta-transducin repeat containing segments of the protein product of the Lis-1 gene. *Biochem. Biophys. Res. Commun.* **209,** 622–629.

Wasenius, V. M., Saraste, M., Salven, P., Eramaa, M., Holm, L., and Lehto, V. P. (1989). Primary structure of the brain alpha-spectrin. *J. Cell Biol.* **108,** 79–93.

Weaver, D. C., Pasternack, G. R., and Marchesi, V. T. (1984). The structural basis of ankyrin function II. Identification of two functional domains. *J. Biol. Chem.* **259,** 6170–6175.

Whitney, J. A., Gomez, M., Sheff, D., Kreis, T. E., and Mellman, I. (1995). Cytoplasmic coat proteins involved in endosome function. *Cell* **83,** 703–713.

Winkelmann, J. C., and Forget, B. G. (1993). Erythroid and nonerythroid spectrins. *Blood* **81,** 3173–3185.

Winkelmann, J. C., Leto, T. L., Watkins, P. C., Eddy, R., Shows, T. B., Linnenbach, A. J., Sahr, K. E., Kathuria, N., Marchesi, V. T., and Forget, B. G. (1988). Molecular cloning of the cDNA for human erythrocyte beta-spectrin. *Blood* **72,** 328–334.

Winkelmann, J. C., Costa, F. F., Linzie, B. L., and Forget, B. G. (1990). Beta spectrin in human skeletal muscle. Tissue-specific differential processing of 3′ β spectrin pre-mRNA generates a β spectrin isoform with a unique carboxyl terminus. *J. Biol. Chem.* **265,** 20449–20454.

Wollner, D. A., Krzeminski, K. A., and Nelson, W. J. (1992). Remodeling the cell surface distribution of membrane proteins during the development of epithelial cell polarity. *J. Cell Biol.* **116,** 889–899.

CHAPTER 7

Axonal Ankyrins and Ankyrin-Binding Proteins: Potential Participants in Lateral Membrane Domains and Transcellular Connections at the Node of Ranvier

Stephen Lambert and Vann Bennett
Howard Hughes Medical Institute and Departments of Cell Biology and Biochemistry, Duke University Medical Center, Durham, North Carolina 27710

I. INTRODUCTION

Molecular interactions at the interface between extracellular and cytoplasmic compartments are believed to be of basic importnce for establishment of lateral membrane domains and formation of specialized sites of cell–cell contact. This interface is especially important for nerve axons, which require mechanisms for appropriate organization of ion channels involved in initiation and conductance of action potentials, and recognition systems for synaptogenesis during early development as well as to form specific contacts with glial cells along the length of axons. Localization of ion channels and axoglial contacts have evolved to a striking degree in myelinated axons, which are a vertebrate adaptation that permits rapid propagation of action potentials without increasing the diameter of axons. Myelinated axons are enveloped by multiple insulating layers of glial cell membrane interrupted at specialized regions known as nodes of Ranvier, which are responsible for ion fluxes of action potentials. Nodes of Ranvier are flanked by intricate glial cell processes and contain high concentrations of voltage-dependent sodium channels. Nodes of Ranvier of myelinated nerves are of considerable clinical interest due to their involvement in pathological conditions, including diabetic peripheral neuropathy (Sima *et al.,* 1993) and trauma (Maxwell *et al.,* 1991); they are also the sites of regeneration of damaged peripheral nerve axons (Fawcett and Keynes, 1990). Myelinated axons and nodes of Ranvier also exemplify several basic issues for cell biologists: formation of polarized cell domains, assembly of integral proteins into lateral membrane domains, and formation of morphological structures that require cooperation between distinct types of cells.

This chapter focuses on specialized isoforms of the membrane–skeletal protein ankyrin, recently discovered to be localized in axons and targeted to axon initial segments and nodes of Ranvier in the nervous system. Ankyrin was initially characterized in human erythrocytes, where it forms a high-affinity link between the cytoplasmic domain of the anion exchanger and the spectrin–actin network (reviewed in Bennett and Gilligan, 1993). Ankyrin has a general role in other tissues as an adaptor between a variety of integral membrane proteins and the spectrin skeleton. Ankyrin-binding proteins relevant to axons include the voltage-dependent sodium channel (Srinivasan *et al.,* 1988), Na,K-ATPase (Ariyasu *et al.,* 1985; Nelson and Veshnock, 1987), and the L1–neurofascin–NrCAM family of cell adhesion molecules (CAMs) (Davis *et al.,* 1993; Davis and Bennett, 1994). Ankyrin and ankyrin-binding membrane proteins thus potentially are capable of mediating key interactions involved in localization of ion channels and coordinated interactions between axons and glial cells.

II. OVERVIEW OF ANKYRIN GENES, FUNCTIONAL DOMAINS, AND SPLICEOFORMS

The ankyrin gene family of mammals currently includes three members defined by cDNA sequence: (1) ankyrin$_R$ (ANK1) (Lux *et al.,* 1990; Lambert *et al.,* 1990), expressed in erythrocytes and restricted to a subset of neurons in the central nervous system (Lambert and Bennett, 1993a,b); (2) ankyrin$_B$ (ANK2) (Otto *et al.,* 1991), the major form of ankyrin in brain; and (3) ankyrin$_G$ (ANK3) (Kordeli *et al.,* 1995; Peters *et al.,* 1995), which is found in most tissues and includes alternatively spliced forms targeted to nodes of Ranvier (see Section VI). Diseases attributed to ankyrins in humans include cases of hereditary spherocytosis resulting from decreased expression and/or expression of mutated forms of ankyrin$_R$ (Lux and Palek, 1995). The *nb/nb* mutation in mice results in a nearly complete deficiency of ankyrin$_R$ with a phenotype of severe anemia and degeneration of a subset of Purkinje cell neurons accompanied by signs of cerebellar dysfunction (Peters *et al.*, 1991). Ankyrin also is expressed in *Drosophila* (Dubreiul and Yu, 1994) and in the nematode *Caenorhabditis elegans* (Otsuka *et al.*, 1995). Mutations in the ankyrin gene in *C. elegans* result in the unc44 phenotype due to abnormal axonal guidance during development (Otsuka *et al.*, 1995).

Ankyrins are modular proteins composed of three domains conserved among the different family members as well as specialized domains found in alternatively spliced isoforms (see Fig. 1). The conserved domains are an N-terminal 89- to 95-kDa membrane-binding domain, a 62-kDa spectrin-binding domain, and a 12-kDa "death domain." The membrane-binding domains are composed of 24 copies of a 33-residue repeat present in many types of proteins (Bork, 1993). The 33-residue repeats form four subdomains each composed of the basic folding unit of six repeats (Michaely and Bennett, 1993). The role of six-repeat subdomains as protein-binding sites will be discussed later. The death domain is followed by a regulatory domain subject to alternative splicing in the case of ankyrin$_R$ that modulates binding of both the anion exchanger and spectrin (Hall and Bennett, 1987; Davis *et al.*, 1992). Death domains were first reported in proteins such as Fas and the tumor necrosis factor receptor, which participate in apoptosis pathways (Cleveland and Ihle, 1995). These domains can associate with related death domains in other proteins. The protein interactions of the ankyrin death domain could involve self-association and/or interactions with other proteins and are not yet resolved.

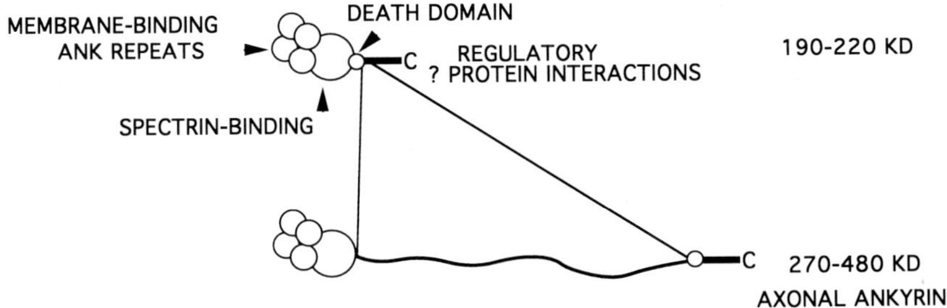

FIGURE 1 Schematic model for domain organization of the ankyrin family. The predicted inserted tail domain in axonal ankyrins is not drawn to scale.

Ankyrin genes are subject to tissue-dependent alternative splicing resulting in deletion as well as insertion of functional domains. $Ankyrin_B$ and $ankyrin_G$ include isoforms missing the membrane-binding domain (Kunimoto *et al.,* 1991; Kordeli *et al.,* 1995; Peters *et al.,* 1995). The functions of these truncated ankyrins are not known but are likely to involve an intracellular as opposed to a plasma membrane role. $Ankyrin_G$ transcripts include at least three other small forms that remain to be defined in terms of sequence.

The $ankyrin_B$ gene also encodes an alternate transcript with an insertion of 6 kb placed between the spectrin-binding domain and death domain. The large transcript of $ankyrin_B$ encodes a 440-kDa polypeptide with a 220-kDa sequence inserted between the membrane–spectrin-binding domains and the C-terminal domain (Kunimoto *et al.,* 1991; Chan *et al.,* 1993). Much of the inserted sequence has the configuration of an extended random coil based on physical properties of expressed polypeptides (Chan *et al.,* 1993). This sequence is likely to be highly phosphorylated and has 220 predicted sites for phosphorylation by protein kinases (casein kinase 2, protein kinase C, and proline-directed protein kinase). Many of the predicted phosphorylation sites for protein kinase C are located in a series of fifteen 12-residue serine-rich repeats. The properties of the inserted sequence suggest a model for 440-kDa $ankyrin_B$ in which the globular membrane-associated head domain is separated from the death domain by an extended filamentous tail domain encoded by the inserted sequence. The length of the tail domain could be up to 0.5 μ if fully extended, which is a distance that in principle could be resolved in the light microscope. Functions of the inserted sequence are not yet known, but one obvious idea is that it connects molecules interacting with the death domain and membrane-binding domains. Given the potential distance spanned by the inserted sequence, these molecules could be in distinct membrane domains or even different membrane compartments.

$Ankyrin_G$ also includes transcripts with large sequences inserted between spectrin-binding and death domains (Kordeli *et al.,* 1995). The encoded sequences all begin with a serine–threonine-rich stretch of about 400 residues, and then have an additional sequence with similarity to that of the inserted sequence of 440-kDa $ankyrin_B$. Two transcripts containing a serine-rich sequence are expressed in brain and heart that encode polypeptides of 270 and 480 kDa. The 270-kDa form lacks a 190-kDa stretch of tail domain. Functions of the serine-rich domains are not known, but these domains may have a special role related to axons (see Section VI).

III. ANKYRIN-BINDING MEMBRANE PROTEINS

A notable feature of ankyrins is their ability to interact through the membrane-binding domain with multiple membrane proteins with apparently unrelated primary sequence. Currently known ankyrin-binding proteins fall into three categories: ion channels, calcium release channels, and CAMs (reviewed in Bennett and Gilligan, 1993). Several of these ankyrin-binding proteins are localized in axons and in particular at nodes of Ranvier. This section presents experimental evidence that these proteins indeed associate with ankyrin, and discusses the basis for such diversity in recognition of target proteins by the ankyrin family. Finally, new *in vitro* evidence is discussed indicating that ankyrin is multivalent and has the capacity to simultaneously associate with two or more membrane proteins.

The voltage-dependent sodium channel was first discovered to have ankyrin-binding activity due to co-purification of ankyrin and spectrin during isolation of the sodium channel (Srinivasan *et al.,* 1988). Pure ankyrin was subsequently found to associate with the purified sodium channel reconstituted into liposomes with a K_d of 20–50 nmol/liter. The voltage-dependent sodium channel co-localizes with ankyrin at the ultrastructural level in postsynaptic membrane infoldings of the neuromuscular junction (Flucher and Daniels, 1989), and both proteins have been independently localized at nodes of Ranvier and axon initial segments (Kordeli *et al.,* 1990). The voltage-dependent sodium channel has three membrane-spanning subunits (Catterall, 1995), one containing the voltage-gated channel and the other two with undefined functions. It is not known which subunit binds to ankyrin. Another unknown is which isoform of the sodium channel is localized at the node of Ranvier.

Na,K-ATPase was first discovered to associate with ankyrin using enzyme derived from kidney (Nelson and Veshnock, 1987), and has subsequently been demonstrated to be co-localized with ankyrin in lateral cell domains of kidney and other epithelial tissues (Morrow *et al.,* 1989; Koob *et al.,* 1987). Complexes of ankyrin and the Na,K-ATPase have been isolated from Madin–Darby canine kidney cells (Hammerton *et al.,* 1991), providing further evidence that these proteins interact *in vivo.* The Na,K-ATPase has two subunits, an α-subunit containing the ATPase and channel activities and a β-subunit. The α-subunit has been implicated in ankyrin binding in blot overlay experiments (Morrow *et al.,* 1989; Deverajan *et al.,* 1994; Jordon *et al.,* 1995). The Na,K-ATPase α- and β-subunits are both members of gene families. Activities of different α-subunits in binding to ankyrin have not been compared. One of the Na,K-ATPase α-subunits is located at the nodes of Ranvier (Ariyasu *et al.,* 1985), although it is not clearly established which gene encodes this protein or which member of the β-subunit is associated.

The major class of ankyrin-binding proteins in brain is a group of CAMs in the immunoglobulin/fibronectin type III (Ig/FnIII) superfamily, which together comprise over 1% of the membrane protein in the adult nervous system. These proteins include neurofascin, L1, NrCAM, and NgCAM in vertebrates and neuroglian in *Drosophila* (Sonderegger and Rathgen, 1992; Rathgen and Jessel, 1991; Grumet, 1991; Hortsch and Goodman, 1991), and have in common a relatively conserved cytoplasmic domain now known to contain the ankyrin-binding site (Davis and Bennett, 1994). Functions of these proteins are believed to include involvement in neurite outgrowth, axonal fasiculation and targeting, cell migration, and synaptogenesis during embryonic and postnatal development. Other activities reported for members of the superfamily include participation in signal transduction pathways across membranes (Schuch *et al.,* 1989; Williams *et al.,* 1992; Ignelzi *et al.,* 1994). Associations of the extracellular domains of these proteins occur as a result of self-association, as well as interactions with other members of the Ig/FnIII superfamily and with extracellular molecules such as restrictin (Sonderegger and Rathjen, 1992; Morales *et al.,* 1993; Felsenfeld *et al.,* 1994).

Neurofascin was the first member of the Ig/FnIII family to be identified as an ankyrin-binding protein (Davis *et al.,* 1993). Neurofascin was isolated from rat brain using an ankyrin affinity column, and identified as having ability to bind directly to ankyrin in a blot overlay assay. Neurofascin was subsequently isolated, an antibody was raised, and cDNA encoding the protein was cloned. The cDNA sequence revealed the close relationship to chicken neurofascin, with nearly 80% sequence identity. Pure neurofascin associated with ankyrin in solution with a K_d of 70 nmol/liter and in a 1:1 molar stoichiometry, thus verifying results of the blot overlay experiment (Davis *et al.,* 1993). Subsequent analysis of proteins eluted from the ankyrin affinity column revealed the presence of L1 and the rat form of NrCAM (Davis and Bennett, 1994). A critical region of the binding site in the cytoplasmic domain was identified using antibody competition, and is highly conserved among vertebrate members of the family as well as neuroglian from *Drosophila* (Davis and Bennett, 1994). A data base search with this sequence also reveals a predicted protein in *C. elegans* (not shown), indicating that the ankyrin-binding site is ancient.

The genes encoding neurofascin, L1, NrCAM, and NgCAM are subject to additional diversity due to alternative exon usage. A complete understanding of this family will require definition of polypeptides in terms of actual exons. This level of understanding is some years away but has begun with neurofascin. Neurofascin has five potential splice sites (Davis *et al.,* 1993,1995; Volkmer *et al.,* 1992). Two of the major forms of neurofascin, termed 186-kDa and 155-kDa neurofascin, have been defined in terms of exon usage by characterization of full-length cDNAs, and exhibit difference at each of the five splice sites (Davis *et al.,* 1995). The 186-kDa neurofascin, is missing the third FnIII domain but has a mucin-like domain inserted between the FnIII domains and the plasma membrane. The 155-kDa neurofascin, in contrast, has a complete set of four FnIII domains but is lacking the mucin-like sequence.

Antibodies to 186-kD neurofascin strongly stain axon initial segments in the central nervous system and nodes of Ranvier in both central and peripheral neurons in a pattern similar to that of the nodal form of ankyrin (see Section VI). Antibodies to 155-kDa neurofascin stain neuronal plasma membranes in a more uniform fashion, do not stain mature nodes of Ranvier in central or peripheral nerves, but stain other structures that are consistent with bundles of unmyelinated axons. The localization of 186-kDa neurofascin at nodes of Ranvier and axon initial segments suggests the possibility of a specific role for the mucin-like domain in lateral and/or transcellular interactions involving voltage-sensitive Na^+ channels in these excitable membrane domains.

The 186-kDa neurofascin is the first member of the family of ankyrin-binding CAMs to be assigned to the node of Ranvier and axon initial segments and to be defined in terms of actual exons. L1, another member of the family, is expressed in unmyelinated axons and disappears with the onset of myelination (Martini, 1994). Neurofascin is not the only CAM at these sites, however. Preliminary results from our laboratory have revealed the presence of NrCAM at nodes of Ranvier as well as unmyelinated axons. The precise exons present in the nodal and unmyelinated forms of NrCAM have not been defined.

IV. ANKYRIN CONTAINS MULTIPLE BINDING SITES FOR MEMBRANE PROTEINS

The diversity of potential binding partners for ankyrin raises the question of whether these proteins all bind to a common site, as is the case for calmodulin and its binding proteins, or whether ankyrin contains more than one and perhaps multiple distinct binding sites. The membrane-binding domain of ankyrin, which is responsible for most of the binding interactions, is composed of 24 consecutive 33-residue repeats termed Ank repeats. Ank repeats are present in many types of proteins, including transcription factors and molecules such as Notch that determine cell fate (Michaely and Bennett, 1992). Ank repeats in these proteins have been implicated in protein–protein interactions. No common feature has yet been attributed to proteins that bind to Ank repeats, which is consistent with the diversity displayed by ankyrin.

The first step in elucidating the role of Ank repeats of ankyrin in forming binding sites was to understand their folding properties. Analysis of native ankyrin and recombinant polypeptides by protease sensitivity and circular dicroism spectroscopy has demonstrated that the 24 Ank repeats are organized into four 6-repeat folding domains. The role of these 6-repeat domains in protein recognition has been addressed by detailed studies of association of ankyrin with the anion exchanger cytoplasmic domain and neurofascin (Michaely and Bennett, 1995a,b). Conclusions from these studies are that binding sites require intact 6-repeat subdomains, and that two 6-repeat subdomains can cooperate to form a binding site. Diversity in binding sites on ankyrin results from several factors: utilization of different subdomains, different combinations of subdomains, and distinct sites within the same subdomain pair.

An important conclusion from analysis of association of the anion exchanger and neurofascin with ankyrin is that ankyrin has multiple binding sites and can accommodate more than one membrane protein at the same time (Michaely and Bennett, 1995a,b). The anion exchanger associates with two sites that are cooperatively coupled, one composed of subdomains 3 plus 4, and the other of subdomain 2. The corresponding ankyrin-binding sites on the anion exchanger also are likely to be distinct, raising the possibility that under certain conditions ankyrin can form linear arrays of anion exchanger in the plane of the plasma membrane. Ankyrin also has two sites for neurofascin, although these do not exhibit cooperativity. One site is located on subdomains 3 plus 4, and the other is on subdomains 2 plus 3. Ankyrin can simultaneously bind the anion exchanger and neurofascin based on lack of competition between these proteins. The ability of ankyrin to form heterocomplexes with CAMs and an ion channel has important physiological implications for assembly of ion channels at sites defined by extracellular contacts, and is discussed later in the context of the node of Ranvier.

V. A SPECIALIZED FORM OF ANKYRIN$_B$ TARGETED TO AXONS

The first axonal ankyrin to be discovered was the 440-kDa alternate form of ankyrin$_B$ with the 2100-residue random coil sequence inserted between spectrin-binding and death domains (see Section II) (Kunimoto *et al.*, 1991; Chan *et al.*, 1993; Kunimoto, 1995). This form of ankyrin is predominantly confined to unmyelinated and premyelinated fibers in the postnatal period, and is also present in growth cones and at the ends of neurite processes in early development (Kunimoto, 1995). Levels of 440-kDa ankyrin$_B$ peak around postnatal day 10 in rats and then decline to about 30% of day 10 levels as myelination progresses (Kunimoto *et al.,* 1991).

Analysis of 440-kDa ankyrin$_B$ in optic nerve, which has a much more uniform program of myelination than total brain, revealed a complementary relationship between 440-kDa ankyrin$_B$ and myelin basic protein (Chan *et al.,* 1993). In optic nerve, levels of 440-kDa ankyrin fall to less than 5% of maximal postnatal values at the same time that expression of myelin basic protein increases. Immunofluorescence staining of premyelinated optic nerve revealed staining consistent with fasciculated bundles of axons. Targeting of 440-kDa ankyrin to axons is essentially complete since the ganglion cell neurons reveal no detectable 440-kDa ankyrin by immunoblot or by immunofluorescence. The decline in expression of 440-kDa ankyrin requires participation of glial cells and is not an intrinsic program of the neurons since 440-kDa ankyrin remains elevated in mice with the hypomyelinating *shiverer* mutation (Chan *et al.,* 1993).

Two members of the Ank CAM family are expressed in unmyelinated axons with patterns of expression similar to 440-kDa ankyrin$_B$. L1 is downregulated at early stages of myelination in the optic nerve (Bartsch *et al.,* 1989) and in the peripheral nervous system (Martini, 1994). The 155-kDa alternatively spliced form of neurofascin also is expressed in peripheral unmyelinated axons in adults (Davis *et al.,* 1995). These proteins are both candidates to participate in membrane complexes with 440-kDa ankyrin$_B$.

These initial studies raise a number of questions regarding the role of 440-kDa ankyrin in unmyelinated axons. Function(s) of the extended tail domain of 440-kDa ankyrin$_B$ are not known but are likely to have relevance to specialized requirements of axons. The full extended length of the tail domain of 0.5 μ could reach deep into the axoplasm, permitting interaction with cytoskeletal proteins or internal membranes. Preliminary experiments have not yet revealed association of C-terminal regions of 440-kDa ankyrin with neurofilaments or microtubules, but these negative results certainly are not definitive. An important and currently unknown bit of information is identification of protein neighbors of the death domain of ankyrin. It also would be important to define by microscopy the location of the N-terminal and C-terminal domains of 440-kDa ankyrin$_B$. Again, given the potential length of the tail, these domains may be resolvable even by light microscopy.

Another set of questions arise from the striking downregulation of 440-kDa ankyrin$_B$ that occurs coincident with myelination. Presumably, the function of 440-kDa ankyrin is not compatible with enveloping of the axon by myelin. One hypothesis is that ankyrin, through L1, is involved in stabilizing axon–axon contacts in fasciculated bundles of axons, and these contacts must be disrupted to allow axoglial contacts. In any case, the signaling pathway between glial cells and neurons leading to coordinate downregulation of both L1 and 440-kDa ankyrin$_B$ remains to be elucidated.

The 440-kDa ankyrin$_B$ is one of a relatively few proteins known to be selectively targeted to axons. Other axonal proteins include GAP43 (Goslin *et al.*, 1988) and the microtubule-associated protein tau (Binder *et al.*, 1985). While no obvious primary structural features are shared by these proteins, they do have in common multiple predicted sites of phosphorylation. It will be of interest to determine whether 440-kDa ankyrin is transported by fast and/or slow transport mechanisms. It also will be important to determine which features of the primary structure are responsible for directing 440-kDa ankyrin to axons.

VI. SPECIALIZED ISOFORMS OF ANKYRIN$_G$ AT THE NODE OF RANVIER AND AXON INITIAL SEGMENTS

The discovery that the voltage-dependent sodium channel associated with ankyrin prompted a search for ankyrin with a distribution at sites known to have high local concentrations of the channel, such as the node of Ranvier and axon initial segments. An immunoreactive form of ankyrin distinct from ankyrin$_B$ was indeed found to be located at these sites by immunofluorescence (Kordeli *et al.*, 1990). Immunogold labeling confirmed that ankyrin at the node of Ranvier was located on the axonal plasma membrane (Kordeli *et al.*, 1990). The gene encoding the nodal form of ankyrin was distinct from ankyrin$_R$ as well as ankyrin$_B$, since mice deficient in ankyrin$_R$ still exhibited staining at nodes of Ranvier (Kordeli and Bennett, 1991).

The gene encoding nodal ankyrin was identified as ankyrin$_G$ (G for general expression and giant) based on characterization of a third ankyrin cDNA distinct from the known ankyrins (Kordeli *et al.*, 1995). Ankyrin$_G$ actually includes multiple transcripts expressed in most tissues as well as the isoform present at nodes of Ranvier, and was independently discovered in a search for kidney ankyrins (Peters *et al.*, 1995). Two transcripts encoding ankyrin isoforms targeted to the node of Ranvier have distinguishing features of a predicted extended tail domain as well as a serine–threonine-rich stretch of sequence contiguous to the spectrin-binding domain. The predicted polypeptides are 270 and 480 kDa, with the difference in size resulting from a deletion of 190 kDa of tail domain sequence. Antibodies raised against the serine-rich domain and spectrin-binding domain that react with both 270- and 480-kDa polypeptides label nodes of Ranvier in peripheral and central axons as well as axon initial segments. In addition, antibodies specific for the 480-kDa polypeptide also label these structures.

The 480-kDa form of ankyrin$_G$ thus is a component of nodes of Ranvier and axon initial segments, and the 270-kDa polypeptide may also be present. The 270-kDa polypeptide is expressed later in development than the 480-kDa form and is likely to have a distinct function (S. Lambert and V. Bennett, unpublished data).

Ankyrin$_G$ isoforms are the first example of a cytoplasmic protein selectively targeted to nodes of Ranvier and axon initial segments, and could eventually provide useful clues to pathways for assembly of other components at these domains. Spectrin has previously been reported to be concentrated at nodes of Ranvier (Koenig and Repasky, 1985), but spectrin also is located between nodes by immunofluorescence and immunogold electron microscopy (Trapp *et al.,* 1987; Levine and Willard, 1981). Ankyrins, perhaps together with spectrin, may be a component of the amorphous submembrane coat observed in transmission electron micrographs (Robertson, 1959). Ankyrin and associated proteins may also be responsible for fine transmembrane and transcellular filaments visualized at the node of Ranvier (Ichimura and Ellisman, 1991). Such structures could be explained by the extended extracellular domain of neurofascin connected to ankyrin with an extended tail domain passing into the axoplasm and attached to cytoskeletal components.

The serine–threonine-rich domain is a striking feature of nodal ankyrins, and has recently been demonstrated to be *O*-glycosylated (X. Zhang and V. Bennett, unpublished data). Other axonal proteins modified by addition of a monosaccharide on serine–threoinine residues include neurofilament subunits (Dong *et al.,* 1993). Consequences of *O*-glycosylation are not known, but it could represent a mechanism to cap potential phosphorylation sites (Haltiwanger *et al.,* 1992). It is of interest in this regard that phosphorylation of neurofilaments is decreased at nodes of Ranvier (Mata *et al.,* 1992). One hypothesis to explain this observation is that a sugar transferase is localized at the node of Ranvier and blocks phosphorylation, thus participating in modifications that distinguish the node from other areas along the axon. It will be of interest to evaluate whether other proteins are *O*-glycosylated at the node of Ranvier using a protein–GlucNAC-specific antibody. Other potential roles of *O*-glycosylation include promotion of self-association of serine-rich domains or association of the serine-rich domain with an intracellular "lectin."

VII. COORDINATE RECRUITMENT OF ANKYRIN$_G$ AND ANKYRIN-BINDING PROTEINS DURING MORPHOGENESIS OF THE NODE OF RANVIER

The role of ankyrin and associated proteins in assembly and/or maintenance of the node of Ranvier has not yet been directly evaluated. However, developmental studies suggest that ankyrin, neurofascin, and the voltage-dependent sodium channel associate into microdomains of axons as early events in morphogenesis of the node of Ranvier. Ankyrin is present in premyelinated axons at early stages of axonal growth during embryonic life (Lambert *et al.,* 1995). During myelination in the sciatic nerve in the postnatal period, ankyrin starts to cluster in concentrated zones, flanking the site of presumed node formation. As myelination proceeds, these "cluster zones" appear to fuse to form mature nodes of Ranvier. Concomitant with ankyrin clustering, antibodies against the sodium channel and neurofascin also stain cluster zones, suggesting that these molecules are associated with ankyrin during nodal development. The ankyrin clusters can appear before accumulation of myelin-associated glycoprotein, which is a Schwann cell protein expressed early in the formation of compact myelin. Antibodies against ankyrin$_G$ and the nodal form of neurofascin thus provide probes for early intermediates in development of the node of Ranvier. In support of this conclusion, ankyrin and neurofascin co-clusters also are evident in the hypomyelinating mutant mice *trembler* and *jimpy,* which do not form compact myelin (S. Lambert and V. Bennett, unpublished data). Interestingly, similar clusters of sodium channels have been reported in regenerating peripheral nerve (Dugandzija-Novakovic *et al.,* 1995), which may represent a reversion to an earlier developmental stage of the axon as a response to injury.

Co-localization of ankyrin with ankyrin-binding proteins early in development is a gratifying result that supports *in vitro* evidence for association between ankyrin and its target proteins. *In vitro* studies further indicate that the ankyrin membrane-binding domain has at least two independent binding sites, and has the potential to form heterocomplexes between neurofascin and at least one ion channel (Michaely and Bennett, 1995a,b). Hence, a multivalent ankyrin could play a key role in the coordinate recruitment of adhesion molecules and ion channels to the node of Ranvier (Fig. 2). A potential benefit of coupling the sodium channel to a CAM would be to provide transcellular (neuron–Schwann cell) signaling to directly affect the polarized distribution of ion channels to sites of axoglial contact in the axonal membrane. The molecular nature of the axoglial contact is currently unknown but could be contributed in the peripheral nervous system by specialized Schwann cell processes that extend into the nodal cleft, and by astrocytes in the central nervous system (Black and Waxman, 1988).

VIII. FUTURE DIRECTIONS

These studies have revealed a set of interacting proteins involved in membrane–cytoskeletal connections that are targeted to nodes of Ranvier

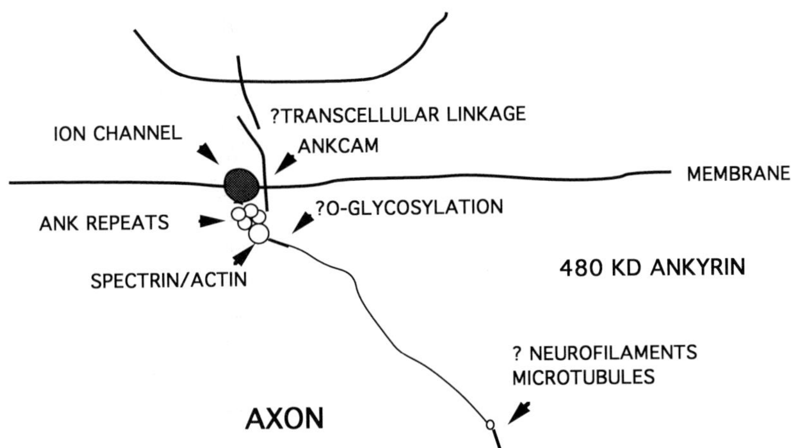

FIGURE 2 Speculative model for role of ankyrin$_G$ in forming lateral association between ion channels and neurofascin at sites of glia–axon contact. The predicted tail domain is not drawn to scale.

and are defined in terms of actual exon usage in the case of ankyrin$_G$ and 186-kDa neurofascin. Utilization of these molecules as probes for early stages of nodal development is just beginning, and has already led to identification of an early intermediate in the assembly pathway. It will be of interest to extend the light microscopy to an ultrastructural level. Other studies of mechanisms of axonal transport and targeting of these molecules to nodes are possible, and may provide clues as to the molecular basis for formation and maintenance of these specialized membrane domains. One major challenge for future work will be to extend predictions of ankyrin activity in promoting lateral and transcellular protein interactions based on *in vitro* biochemical analysis to actual biology of neurons. Experimental systems have been developed for formation of nodes of Ranvier utilizing co-cultures of neurons and Schwann cells (Bunge and Wood, 1987). In principle, this type of tissue culture system in conjunction with expression of antisense or dominant-negative constructs could be used to evaluate contributions of individual proteins and interactions. Other strategies could involve gene disruption by homologous recombination in mice.

A second, more fundamental challenge will be to fully elucidate the signaling pathways initiated by contact between axons and glial cells, which may be mediated at least in part by members of the family of ankyrin-binding CAMs. These molecules or their neighbors could be responsible for axonal stimulation of the myelination program by Schwann cells, and Schwann cell regulation of local organization of neurofilaments (deWaegh *et al.*, 1992). Walsh, Schachner, and their colleagues have presented evidence that L1 participates in signaling pathways involving pertussis-sensitive G-proteins and calcium channels (Schuch *et al.,* 1989; Williams *et al.,* 1992; Ignelzi *et al.,* 1994). Missing information includes the ligand system that activates L1 and other members of the family, and details of the intracellular components. Potential clinical benefits of these studies could result from insight into mechanisms of nerve regeneration and repair following injury.

Acknowledgments

Stimulating discussions with Jonathan Davis and Peter Michaely are gratefully acknowledged.

References

Ariyasu, R. G., Nichol, J. A., and Ellisman, M. H. (1985). Localization of sodium/potassium adenosine triphosphatase in multiple cell types of the murine nervous system with antibodies raised against the enzyme from kidney. *J. Neurosci.* **5,** 2581–2596.

Bartsch, U., Kerchief, F., and Stationer, M. (1989). Immunohistological localization of the adhesion molecules L1, N-CAM, and MAG in the developing and adult optic nerve of mice. *J. Comp. Neurol.* **284,** 451–462.

Bennett, V., and Gilligan, D. M. (1993). The spectrin-based membrane skeleton and micron-scale organization of the plasma membrane. *Annu. Rev. Cell Biol.* **9,** 27–66.

Binder, L. I., Frankfurter, A., and Rebhun, L. I. (1985). The distribution of tau in the mammalian central nervous system. *J. Cell Biol.* **101,** 1371–1378.

Black, J. A., and Waxman, S. (1988). The perinodal astrocyte. *Glia* **1,** 1169–1183.

Bork, P. (1993). Hundreds of ankyrin-like repeats in functionally diverse proteins: Mobile modules that cross phyla horizontally? *Proteins* **17,** 363–374.

Bunge, R. P., and Wood, P. M. (1987). Tissue culture studies of interactions between axons and myelinating cells of the central and peripheral nervous system. *Prog. Brain Res.* **71,** 143–152.

Catterall, W. A. (1995). Structure and function of voltage-gated ion channels. *Annu. Rev. Biochem.* **64,** 493–531.

Chan, W., Kordeli, E., and Bennett, V. (1993). 440-kD ankyrin$_B$: Structure of the major developmentally regulated domain and selective localization in unmyelinated axons. *J. Cell Biol.* **123,** 1463–1473.

Cleveland, J. L., and Ihle, J. N. (1995). Contenders in Fas/TNF death signaling. *Cell* **81,** 479–482.

Davis, J., and Bennett, V. (1994). Ankyrin-binding activity shared by the neurofascin/L1/NrCam family of nervous system cell adhesion molecules. *J. Biol. Chem.* **269,** 27163–27166.

Davis, J. Q., McLaughlin, T., and Bennett, V. (1993). Ankyrin-binding proteins related to nervous system cell adhesion molecules: Candidates to provide transmembrane and intercellular connections in adult brain. *J. Cell Biol.* **232,** 121–133.

Davis, J., Zhang, F., Ren, Q., and Bennett, V. (1995). Neurofascin: Correlation of molecular splicing with targeting to nodes of Ranvier and axon initial segments. *Mol. Cell. Biol.* **6,** 177a.

Davis, L. H., Davis, J. Q., and Bennett, V. (1992). Ankyrin regulation: An alternatively spliced segment of the regulatory domain functions as an intramolecular modulator. *J. Biol. Chem.* **267,** 18966–18972.

Deverajan, P., Scaramuzzino, D., and Morrow, J. S. (1994). Ankyrin binds to two distinct cytoplasmic domains of Na,K-ATPase alpha subunit. *Proc. Natl. Acad. Sci. U.S.A.* **91,** 2965–2969.

deWaegh, S., Lee, V. M.-Y., and Brady, S. T. (1992). Local modulation of neurofilament phosphorylation, axonal caliber, and slow axonal transport by myelinateing Schwann cells. *Cell* **68,** 451–463.

Dong, D., Xu, Z.-S., Chevrier, M., Cotter, R., Cleveland, D., and Hart, G. W. (1993). Glycosylation of mammalian neurofilaments. *J. Biol. Chem.* **268,** 16679–16687.

Dubreuil, R., and Yu, J. (1994). Ankyrin and beta-spectrin accumulate independently of alpha-spectrin in Drosophila. *Proc. Natl. Acad. Sci. U.S.A.* **91,** 10285–10289.

Dugandzija-Novakovic, S., Koszowski, A. G., Levinson, S. R., and Shrager, P. (1995). Clustering of Na^+ channels and node of Ranvier formation in remyelinating axons. *J. Neurosci.* **15,** 492–503.

Fawcett, J. W., and Keynes, R. J. (1990). Peripheral nerve regeneration. *Annu. Rev. Neurosci.* **13,** 43–60.

Felsenfeld, D., Hynes, M., Skoler, K., Furley, A., and Jessel, T. M. (1994). TAG-1 can mediate homophilic binding but neurite outgrowth on TAG-1 requires an L1-like molecule and beta1 integrins. *Neuron* **12,** 675–690.

Flucher, B., and Daniels, M. (1989). Membrane proteins in the neuromuscular junction: Distribution of Na channels and ankyrin is complementary to acetylcholine receptors and the 43-KD protein. *Neuron* **3,** 163–175.

Goslin, K., Schreyer, J. Skene, P., and Banker, G. (1988). Development of neuronal polarity: GAP43 distinguishes axonal from dendritic growth cones. *Nature* (*London*) **336,** 672–674.

Grumet, M. (1991). Cell adhesion molecules and their subgroups in the nervous system. *Curr. Opin. Neurobiol.* **1,** 370–376.

Hall, T. G., and Bennett, V. (1987). Regulatory domains of erythrocyte ankyrin. *J. Biol. Chem.* **262,** 10537–10545.

Haltiwanger, R. S., Blomberg, M., and Hart, G. W. (1992). Glycosylation of nuclear and cytoplasmic fractions. *J. Biol. Chem.* **267,** 9005–9013.

Hammerton, R., Krzeminski, K., Mays, R., Ryan, T., Wollner, D., and Nelson, J. (1991). Mechanism for regulating cell surface distribution of Na+, K+-ATPase in polarized epithelial cells. *Science* **254,** 847–850.

Hortsch, M., and Goodman, C. (1991). Cell and substrate adhesion molecules. *Annu. Rev. Cell Biol.* **7,** 505–557.

Ichimura, T., and Ellisman, M. H. (1991). Three-dimensional fine structure of cytoskeletal-membrane interactions at nodes of Ranvier. *J. Neurocytol.* **20,** 667–681.

Ignelzi, M., Miller, D., Soriano, P., and Maness, P. F. (1994). Impaired neurite outgrowth of src-minus cerebellar neurons on the cell adhesion molecule L1. *Neuron* **12,** 873–884.

Jordon, C., Puschel, B., Koob, R., and Drenckhahn, D. (1995). Identification of a binding motif for ankyrin on the α-subunit of Na,K-ATPase. *J. Biol. Chem.* **270,** 29971–29975.

Koenig, E., and Repasky, E. (1985). A regional analysis of alpha-spectrin in the isolated Mauthner neuron and in isolated axons of the goldfish and rabbit. *J. Neurosci.* **5,** 705–714.

Koob, R., Zimmermann, M., Schone, W., and Drenckhahn, D. (1987). Colocalization and coprecipitation of ankyrin and Na+,K+-ATPase in kidney epithelial cells. *Eur. J. Cell Biol.* **45,** 230–237.

Kordeli, E., and Bennett, V. (1991). Distinct ankyrin isoforms at neuron cell bodies and nodes of Ranvier resolved using erythrocyte ankyrin-deficient mice. *J. Cell Biol.* **114,** 1243–1259.

Kordeli, E., Davis, J., Trapp, B., and Bennett, V. (1990). An isoform of ankyrin is localized at nodes of Ranvier in myelinated axons of central and peripheral nerves. *J. Cell Biol.* **110,** 1341–1352.

Kordeli, E., Lambert, S., and Bennett, V. (1995). $Ankyrin_G$: A new ankyrin gene with neural-specific isoforms localized at the axonal initial segment and node of Ranvier. *J. Biol. Chem.* **270,** 2352–2359.

Kunimoto, M. (1995). A neuron-specific isoform of brain ankyrin, 440 kD ankyrin$_B$, is targeted to axons of rat cerebellar neurons. *J. Cell Biol.* **131,** 1821–1830.
Kunimoto, M., Otto, E., and Bennett, V. (1991). A new 440-kDa isoform is the major ankyrin in neonatal rat brain. *J. Cell Biol.* **115,** 1319–1331.
Lambert, S., and Bennett, V. (1993a). From anemia to cerebellar dysfunction: A review of the ankyrin gene family. *Eur. J. Biochem.* **211,** 1–6.
Lambert, S., and Bennett, V. (1993b). Post-mitotic expression of ankyrin$_R$ and β_R-spectrin in discrete neuronal populations of the rat brain. *J. Neurosci.* **13,** 3725–3735.
Lambert, S., Yu, H., Prchal, J., Lawler, J., Ruff, P., Speicher, D., Cheung, M., Kan, Y., and Palek, J. (1990). cDNA sequence for human erythrocyte ankyrin. *Proc. Natl. Acad. Sci. U.S.A.* **87,** 1730–1734.
Lambert, S., Davis, J., Michaely, P., and Bennett, V. (1995). Ankyrin clustering in the coordinate recruitment of ion channels and adhesion molecules during morphogenesis of the node of Ranvier. *Mol. Biol. Cell* **6,** 98a.
Levine, J., and Willard, M. (1981). Fodrin: Axonally transported polypeptides associated with the internal periphery of many cells. *J. Cell. Biol.* **90,** 631–643.
Lux, S. E., and Palek, J. (1995). Disorders of the red cell membrane. *In* "Blood: Principles and Practice of Hematology" (R. I. Handin, S. E. Lux, and T. P. Stossel, eds.), pp. 1701–1818. J. B. Lippincott Co., Philadelphia.
Lux, S. E., John, K. M., and Bennett, V. (1990). Analysis of cDNA for human erythrocyte ankyrin indicates a repeated structure with homology to tissue-differentiation and cell-cycle control proteins. *Nature* (*London*) **344,** 36–42.
Martini, R. (1994). Expression and functional roles of neuronal cell adhesion molecules and extracellular matrix components during development and regeneration of peripheral nerves. *J. Neurocytol.* **23,** 1–28.
Mata, M., Kupina, N., and Fink, D. J. (1992). Phosphorylation-dependent neurofilament epitopes are reduced at the node of Ranvier. *J. Neurocytol.* **21,** 199–210.
Maxwell, W. L., Irvine, A., Graham, J., Adams, T., Gennarelli, R., Tipperman, R., and Sturatis, M. (1991). Focal axonal injury: The early axonal response to stretch. *J. Neurocytol.* **20,** 157–164.
Michaely, P., and Bennett, V. (1992). ANK repeats: A ubiquitous motif involved in macromolecular recognition. *Trends Cell Biol.* **2,** 127–129.
Michaely, P., and Bennett, V. (1993). The membrane-binding domain of ankyrin contains four independently-folded subdomains each comprised of six ankyrin repeats. *J. Biol. Chem.* **268,** 22703–22709.
Michaely, P., and Bennett, V. (1995a). The ANK repeats of erythrocyte ankyrin form two distinct but cooperative binding sites for the erythrocyte anion exchanger. *J. Biol. Chem.* **270,** 22050–22057.
Michaely, P., and Bennett, V. (1995b). Mechanism for binding site diversity on ankyrin: Comparison of binding sites on ankyrin for neurofascin and the Cl^-/HCO_3^- anion exchanger. *J. Biol. Chem.* **270,** 31298–31302.
Morales, G., Hubert, M., Brummendorf, T., Treubert, U., Tarnok, A., Schwarz, U., and Rathgen, F. (1993). Induction of axonal growth by helerophilic interactions between the cell surface recognition proteins F11 and NrCAM/Bravo. *Neuron* **11,** 1113–1122.
Morrow, J. S., Cianci, C. D., Ardito, T., Mann, A. S., and Kashgarian, M. (1989). Ankyrin links fodrin to the alpha subunit of Na,K-ATPase in Madin-Darby canine kidney cells and in intact renal tubule cells. *J. Cell Biol.* **108,** 455–465.
Nelson, W. J., and Veshnock, P. J. (1987). Ankyrin binding to the (Na+ + K+) ATPase and implications for the organization of membrane domains in polarized cells. *Nature* (*London*) **328,** 533–535.

Otsuka, A., Franco, R., Yang, B., Shim, K., Tang, L., Zhang, Y., Boontrakulpoontawee, P., Jeyaprakash, A., Hedgecock, E., Wheaton, V., and Sobery, A. (1995). An ankyrin-related gene (*unc-44*) is necessary for proper axonal guidance in *Caenorhabditis elegans*. *J. Cell Biol.* **129,** 1081–1092.

Otto, E., Kunimoto, M., McLaughlin, T., and Bennett, V. (1991). Isolation and characterization of cDNAs encoding human brain ankyrins reveal a family of alternatively-spliced genes. *J. Cell Biol.* **114,** 241–253.

Peters, L. L., Birkenmeirer, C. S., Bronson, R. T., White, R. A., Lux, S. E., Otto, E., Bennett, V., Higgins, A., and Barker, J. E. (1991). Purkinje cell degeneration associated with erythroid ankyrin deficiency in NB/NB mice. *J. Cell Biol.* **114,** 1233–1241.

Peters, L. L., John, K. M., Lu, F. M., Eicher, E. M., Higgins, A., Yialamas, M., Turtzo, L. C., Otsuka, A., and Lux, S. E. (1995). Ank3 (epithelial ankyrin), a widely distributed new member of the ankyrin gene family and the major ankyrin in kidney, is expressed in alternatively spliced forms, including forms that lack the repeat domain. *J. Cell Biol.* **130,** 313–321.

Rathjen, F. G., and Jessel, T. M. (1991). Glycoproteins that regulate the growth and guidance of vertebrate axons: Domains and dynamics of the immunoglobulin/fibronectin type III subfamily. *Semin. Neurosci.* **3,** 297–307.

Robertson, J. D. (1959). Preliminary observations on the ultrastructure of nodes of Ranvier. *Z. Zellforsch. Mikrosk. Anat.* **50,** 553–560.

Schuch, U., Lohse, M. J., and Schachner, M. (1989). Neural cell adhesion molecules influence second messenger systems. *Neuron* **3,** 13–20.

Sima, A. A., Prasher, A., Nathaniel, V., Werb, M. R., and Greene, D. A. (1993). Overt diabetic neuropathy: Repair of axo-glial dysjunction and axonal atrophy by aldose reductase inhibition and its correlation to improvement in nerve conduction velocity. *Diabetic Med.* **10,** 115–121.

Sonderegger, P., and Rathgen, F. G. (1992). Regulation of axonal growth in the vertebrate nervous system by interactions between glycoproteins belonging to two subgroups of the immunoglobulin superfamily. *J. Cell Biol.* **119,** 1387–1394.

Srinivasan, Y., Elmer, L., Davis, J., Bennett, V., and Angelides, K. (1988). Ankyrin and spectrin associate with voltage-dependent sodium channels in brain. *Nature* (*London*) **333,** 177–180.

Trapp, B. D., Andrews, S. B., Wong, A., O'Connell, M., and Griffin, J. W. (1987). Co-localization of the myelin-associated glycoprotein and the microfilaments, F-actin and spectrin, in Schwann cells of myelinated nerve fibers. *J. Neurocytol.* **18,** 47–60.

Volkmer, H., Hassel, B., Wolff, J. M., Frank, R., and Rathjen, F. G. (1992). Structure of the axonal surface recognition molecule neurofascin and its relationship to a neural subgroup of the immunoglobulin superfamily. *J. Cell Biol.* **118,** 149–161.

Williams, E. J., Doherty, P., Turner, G., Reid, R. A., Hemperly, J. J., and Walsh, F. S. (1992). Calcium influx into neurons can solely account for cell contact-dependent neurite outgrowth stimulated by transfected L1. *J. Cell Biol.* **119,** 883–892.

CHAPTER 8

Molecular and Genetic Dissection of the Membrane Skeleton in *Drosophila*

Ronald R. Dubreuil

Committee on Cell Physiology and Department of Pharmacological and Physiological Sciences, University of Chicago, Chicago, Illinois 60637

I. INTRODUCTION

The plasma membrane is an important boundary that defines cell size and shape, and it is the site of interactions between cells and their environment. A wealth of evidence indicates that the plasma membrane actively contributes to cell form and function, rather than behaving as a passive boundary. The premier model system for these studies has been the human erythrocyte membrane, because of its accessibility to a combination of biochemical, structural, molecular, and genetic experimental approaches.

An important principle to emerge from the erythrocyte model is that the lipid bilayer is fortified by a submembrane network of peripheral membrane proteins. These proteins, commonly refered to as the membrane skeleton, are essential for the normal shape and stability of the erythrocyte membrane.

The membrane skeleton proteins first described in the erythrocyte also have homologues associated with the plasma membrane of other cell types. Biochemical studies have revealed that these homologues participate in the same molecular interactions as their erythrocyte counterparts. However, technical limitations have prevented a direct electron microscopic (EM) analysis of the membrane skeleton in nonerythroid cells. Genetic studies of the nonerythroid membrane skeleton have also been lacking. Consequently, much of the conceptual framework for membrane skeleton function is borrowed from the erythrocyte model.

Recent genetic studies using *Drosophila* as a model system (Lee *et al.*, 1993; Deng *et al.*, 1995) have provided some of the first glimpses of membrane skeleton function in cells other than erythrocytes. The unique genetic tools that are available in *Drosophila* have made it possible to identify and characterize mutations in the α-spectrin gene, which encodes one of the major components of the membrane skeleton. The phenotype of these mutants supports a role for spectrin in the maintenance of cell shape. However, a concomitant defect in cell adhesion raises the interesting possibility that the membrane skeleton contributes to cell shape by a different mechanism than in the specialized case of the erythrocyte. This chapter summarizes some of the questions that have emerged from studies of the membrane skeleton in vertebrate systems and recent answers that have come from work in *Drosophila*.

II. THE ERYTHROCYTE MEMBRANE SKELETON

A. Biochemical and Structural Studies

The contribution of peripheral membrane proteins to plasma membrane structure and function was first appreciated in the human erythrocyte. Biochemical studies of purified red cell membranes revealed a relatively simple complement of major membrane-associated proteins. Topological studies revealed that most of these proteins were not integrally associated with the lipid bilayer, but instead were found at the cytoplasmic face of the membrane (reviewed by Steck, 1974). The spatial organization of these peripheral proteins was deduced largely through biochemical studies (reviewed by Branton *et al.*, 1981) that suggested the presence of a 2-

dimensional protein network beneath the plasma membrane. The presence of this network (schematized in Fig. 1a, inset) was eventually confirmed by EM of artificially spread membrane preparations (Shen *et al.,* 1984; Byers and Branton, 1985). Since there are several reviews of the membrane skeleton (e.g., Bennett, 1990; Bennett and Gilligan, 1993; Lux and Palek, 1995), only a brief overview is provided here.

The most conspicuous component of the membrane skeleton is spectrin. Spectrin is a tetramer made from two heterodimeric complexes of α- and β-subunits (Fig. 1a). Both subunits are made up of many copies of a 106-amino-acid repetitive sequence motif (ellipses) and other nonrepetitive sequences (rectangles). Dimers are joined head to head so that two identical tails are displayed at the termini of the molecule. The tail region contains an actin-binding site (solid rectangle) that allows spectrin to form a 2-

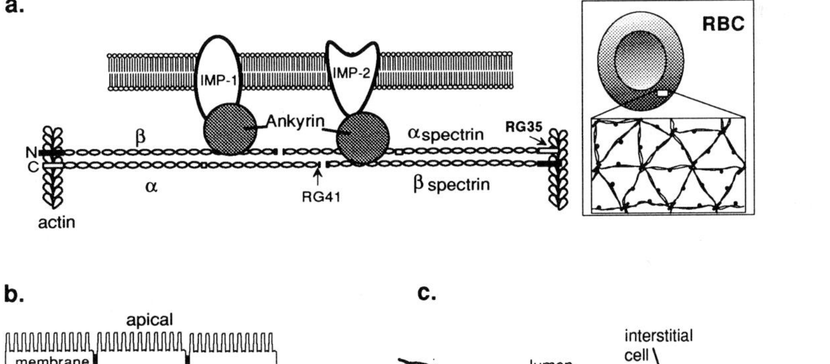

FIGURE 1 The membrane skeleton in erythroid and nonerythroid cells. (a) The membrane skeleton consists primarily of spectrin tetramers, actin, and ankyrin, and is attached to the plasma membrane via integral membrane proteins (e.g., IMP-1 and IMP-2). The C-terminus of α-spectrin and the N-terminus of β-spectrin are indicated to the left. The relative positions of two *Drosophila* frameshift mutations (RG41 and RG35) are shown on α-spectrin. (*inset*) The erythrocyte membrane skeleton is organized as a uniform 2-dimensional network beneath the plasma membrane. About six spectrin tetramers converge at each actin filament junction. (b) The membrane skeleton is polarized in many cell types, such as epithelia, where it is typically (but not always) associated with lateral membrane domains. (c) The cuprophilic cells of the *Drosophila* larval midgut are a polarized epithelial cell type. They can be divided into apical and basolateral cell surface domains, like their more typical mammalian counterparts. The apical domain is studded with microvilli and, although it is invaginated at the cell apex, it is exposed to the lumen of the gut.

dimensional protein network with actin. About six spectrins converge at each actin filament junction (Fig. 1a, inset).

Actin in most cells is a dynamic polymer that can reach micron-scale lengths through the assembly of hundreds of subunits. However, in the red cell actin exists as short, stable oligomers that provide little more than a binding surface for the formation of spectrin network junctions (Pinder and Gratzer, 1983; Byers and Branton, 1985). Several additional proteins (not shown) are associated with the spectrin–actin junction (e.g., protein 4.1, adducin, tropomyosin, dematin, p55), and these may contribute to development of the unusual state of actin in the membrane skeleton.

The membrane skeleton is attached to the lipid bilayer through interactions with ankyrin. Ankyrin forms a bridge that connects the β-subunit of spectrin to the cytoplasmic domain of the major integral membrane protein (IMP) of the erythrocyte, band 3. Other attachments between the membrane skeleton and the lipid bilayer may also have important implications for membrane skeleton assembly and function (Lux and Palek, 1995), but are beyond the scope of this chapter.

B. Human Genetic Studies

A major breakthrough in understanding the function of the red cell membrane skeleton came from the discovery that certain inherited anemias in humans are caused by molecular defects in the membrane skeleton (e.g., hereditary spherocytosis and hereditary elliptocytosis; reviewed by Lux and Palek, 1995). Defects in the shape and stability of the erythrocyte membrane cause affected cells to be lost from the circulation prematurely, resulting in anemia.

With the advent of modern molecular methods, it became possible to identify the molecular lesions that cause anemia. Defects in a number of structural components of the membrane skeleton have been associated with clinically similar phenotypes. It appears that a defect in any one link compromises the entire membrane skeleton network to some degree and produces the disease phenotype. For example, several independent point mutations at codon 28 of α-spectrin are known to produce hereditary elliptocytosis (Coetzer *et al.,* 1991). The conserved arginine normally found at this position is replaced with either a cysteine, serine, histidine, or leucine residue. Each mutation affects the ability of spectrin dimers to form tetramers and thereby prevents spectrin network formation. The network can also be perturbed at the spectrin–actin junction, as is the case with defects in protein 4.1, which normally stabilizes the interaction between spectrin and actin (reviewed by Conboy, 1993).

III. THE MEMBRANE SKELETON IN VERTEBRATE NONERYTHROID CELLS

The simplicity of the erythrocyte made it possible to precisely define the major components of the membrane skeleton; however, in most systems the plasma membrane is far more complex and it is difficult to determine exactly which proteins make up the membrane skeleton. It seems likely that spectrin and ankyrin, together with actin, form the basic structural framework in nonerythroid cells. The remainder of this chapter focuses on these two proteins and associated IMPs.

A. Biochemical and Structural Studies

Spectrin and ankyrin are expressed in a variety of vertebrate cell types, although in most cases the erythroid and nonerythroid proteins are encoded by distinct genes (Kordeli *et al.,* 1995; Peters *et al.,* 1995; reviewed by Bennett and Gilligan, 1993; Winkelmann and Forget, 1993). The purified nonerythroid proteins have *in vitro* properties (e.g., physical properties, protein interactions) that are strikingly similar to those of their red cell counterparts (Glenney and Glenney, 1984; Davis and Bennett, 1984). The appearance of spectrin and ankyrin by EM is also conserved (Bennett *et al.,* 1985), although it has not been possible to examine the structure of the nonerythroid membrane skeleton *in situ*. For now it is assumed that at least some (if not most) of the properties of the erythrocyte membrane skeleton hold true for nonerythroid cells.

While many features of the nonerythoid membrane skeleton are conserved, there are important differences. First, the membrane skeleton is uniformly associated with the plasma membrane in the erythrocyte. However, it was recognized early on that the membrane skeleton is polarized in nonerythroid cells and that its polarity may contribute to the establishment of membrane domains (Lazarides and Nelson, 1983; Drenckhahn *et al.,* 1985; Nelson and Veshnock, 1986). For example, the membrane skeleton is often (but not always) associated with the lateral domain of epithelial cells (Fig. 1b). A second difference from the erythrocyte is that many different IMPs bind to the membrane skeleton through ankyrin in nonerythroid cells. Nonerythroid ankyrins (products of at least three distinct genes and their splice variants; Bennett, 1992; Kordeli *et al.,* 1995; Peters *et al.,* 1995) interact with several physiologically important molecules such as the sodium pump (Nelson and Veshnock, 1987), an anion exchanger from brain (Morgans and Kopito, 1993), and voltage-dependent sodium channels (Srinivasan *et al.,* 1992) and with cell–cell adhesion molecules such as members of the L1 family (Davis *et al.,* 1993) and CD44 (Lokeshwar and

Bourguignon, 1992). Spectrin can also interact with the plasma membrane directly through ankyrin-independent sites (Bourguignon *et al.,* 1985; Steiner and Bennett, 1988; Lombardo *et al.,* 1994).

B. Cell Surface Polarity and the Membrane Skeleton

Do all of the interactions between ankyrin, spectrin, and IMPs simply provide a way to attach the membrane skeleton to the lipid bilayer? An alternative view is that the membrane skeleton contributes to the organization of the plasma membrane by interacting with different membrane proteins in different ways. This view has been incorporated into a model for the establishment of cell surface polarity (summarized by Nelson, 1992). The proposed mechanism is based on two classes of membrane proteins that interact with the membrane skeleton: (1) binding sites that recruit the membrane skeleton from the cytoplasm to a specific plasma membrane domain (e.g. IMP-1; Fig. 1a); and (2) other interacting membrane proteins (e.g., IMP-2) that associate with the preassembled membrane skeleton scaffold, causing them to become segregated within the domain of the membrane skeleton.

Experimental evidence supporting a role for the membrane skeleton in establishment of cell polarity was obtained in cultured fibroblasts (McNeill *et al.,* 1990). The sodium pump was used as a marker in these studies since it had previously been shown to interact with ankyrin (Nelson and Veshnock, 1987). Immunofluorescent staining with appropriate antibodies revealed that spectrin, ankyrin, and the sodium pump were randomly distributed over the surface of control fibroblasts. When fibroblasts were induced to express the homophilic cell adhesion molecule E-cadherin, the membrane skeleton and the sodium pump became enriched at zones of E-cadherin–mediated cell adhesion. In this example, E-cadherin behaved as an inducer of membrane skeleton polarity and the sodium pump, in turn, responded to the position of the membrane skeleton. Studies of the membrane skeleton and the sodium pump in polarized Madin–Darby canine kidney cells showed that the membrane skeleton acts to selectively stabilize the sodium pump in lateral domains of cell–cell contact, even though the nascent molecule is delivered to basolateral and apical domains by the secretory pathway (Hammerton *et al.,* 1991).

The proposed role of the membrane skeleton in establishment of membrane polarity is based on its multivalent nature. However, many questions remain unanswered. In the above example, the mode of interaction between E-cadherin and the membrane skeleton is unknown. Also, the schematic (Fig. 1a) depicts two IMPs interacting with two ankyrin molecules that are bound to one spectrin tetramer. IMP-1 is an inducer of membrane skeleton polarity and IMP-2 responds by co-distributing in the domain of IMP-1. However, any pair of binding sites on the membrane skeleton scaffold could conceivably contribute to membrane polarity (e.g., a site on spectrin and a site on ankyrin, or even two sites on a single ankyrin molecule) as long as one site establishes polarity and the other site passes that information on to a second molecule.

C. Molecular Studies

Much of what we know about membrane skeleton diversity has come from cDNA sequence comparisons (reviewed by Bennett and Gilligan, 1993). Complete sequence data are now available for several isoforms of spectrin and ankyrin. An interesting evolutionary consideration emerged from comparisons of erythroid and nonerythroid spectrins. There is greater sequence similarity between nonerythroid spectrins from organisms as diverse as humans, chickens, and fruit flies than there is between human erythroid and human nonerythroid gene products (reviewed in Dubreuil, 1991). The erythroid gene products first appeared during mammalian evolution, and they rapidly diverged from their ancestors (which are now referred to as "nonerythroid" gene products). Perhaps the observed differences in cellular distribution and protein interactions of the membrane skeleton arose during this period of amino acid sequence divergence.

D. Genetic Studies

Genetic studies were important in establishing the role of the membrane skeleton in erythrocytes, but so far only limited information is available for nonerythroid systems. One reason is that, while nonfatal hereditary anemia occurs quite frequently in the population, it has proven difficult to associate other human diseases with defects of the membrane skeleton. It may be that such mutations are usually lethal very early in life, and therefore escape detection.

1. A Requirement for Ankyrin in Mouse Brain

Mice defective in Ank1, the ankyrin isoform expressed in mammalian erythrocytes (Lambert *et al.,* 1990; Lux *et al.,* 1990), were originally discovered on the basis of an inherited anemic condition (normoblastosis; Bernstein, 1969). Subsequently it was found that the Ank1 gene is also expressed in the brain and that mutant mice exhibit an age-related neurological deficit

(Peters *et al.,* 1991). Interestingly, a population of cerebellar Purkinje neurons that normally express Ank1 degenerates with time in the mutants. These results provided the first evidence that ankyrin function is essential in a nonerythroid cell.

2. Duchenne Muscular Dystrophy Is Caused by a Defect in the Spectrin Homologue Dystrophin

An important breakthrough in understanding the significance of the nonerythroid membrane skeleton came from studies of muscular dystrophy. Kunkel and colleagues cloned the gene responsible for Duchenne muscular dystrophy (Kunkel *et al.,* 1985) and found that its protein product exhibits substantial similarity to spectrin (Koenig *et al.,* 1988). Its domain structure is also quite similar to that of spectrin, which supports the speculation that dystrophin has a role in muscle that is analogous to the role of spectrin in the erythrocyte. Through a subsequent series of biochemical and molecular studies, Campbell and colleagues have shown that one of the functions of dystrophin is to establish a transmembrane linkage from a muscle laminin homologue (merosin) in the extracellular matrix to cytoplasmic actin (reviewed in Campbell, 1995). Evidence that phenotypically similar disease states are produced by genetic defects in dystrophin, or a transmembrane component of the complex (Matsumura *et al.,* 1992), or in merosin (Sunada *et al.,* 1994) support this view of dystrophin as a link in a chain (Campbell, 1995). Extracellular interactions are probably not relevant to the function of the membrane skeleton in erythrocytes. However, nonerythroid spectrins participate in interactions that are intriguingly similar to those of dystrophin (extracellular adhesion molecules → transmembrane linkage → actin). The dystrophin- and spectrin-based membrane skeletons of nonerythroid cells may serve similar roles in transmembrane organization (e.g., in the establishment of cell surface polarity).

IV. THE MEMBRANE SKELETON IN *DROSOPHILA*

A. Nongenetic Approaches

Drosophila spectrin was initially identified by Kiehart and colleagues as a high-molecular-weight actin-binding protein from *Drosophila* tissue culture cells (Dubreuil *et al.,* 1987). At the time, relatively little was known about the degree of similarity between mammalian proteins and their *Drosophila* counterparts (in either sequence or function). Therefore, it was important to establish the degree of similarity between mammalian and

Drosophila spectrins so that ultimately genetic studies could be interpreted in an appropriate context.

1. Biochemical and Structural Studies

Spectrin was purified from *Drosophila* tissue culture cells by standard chromatographic methods (Dubreuil *et al.,* 1987). The purified protein was found to resemble the nonerythroid forms of spectrin that had previously been characterized in vertebrates by several criteria, including the subunit composition (high-molecular-weight α- and β-subunits), the appearance of the native molecule by EM (a double-stranded molecule ~200 nm in length), and its co-sedimentation with F-actin. Calcium- and calmodulin-binding sites are present on the α-subunit, as in vertebrates (Dubreuil *et al.,* 1991). Antibodies produced against recombinant fragments of *Drosophila* α- and β-spectrin were found to cross-react with mammalian subunits (Byers *et al.,* 1987,1989). Interestingly, protein interactions between α- and β-spectrin subunits were also found to be conserved between *Drosophila* and mammals. This conserved interaction was used in an expression cloning strategy to recover β-spectrin cDNAs in *Drosophila* by using α-spectrin as a probe (Byers *et al.,* 1989).

A second isoform of *Drosophila* spectrin was also found in *Drosophila* cultured cells. This isoform shares the same α-subunit as the first but differs in its β-subunit, which is unusually large and was therefore named β_H (for its "high" molecular weight; Dubreuil *et al.,* 1990). The two isoforms of spectrin are now refered to simply as $\alpha\beta$ and $\alpha\beta_H$. Apart from its difference in molecular weight, and a slight difference in length by EM, the $\alpha\beta_H$ molecule appears to be a bona fide spectrin with significant similarities to other spectrin family members (Dubreuil *et al.,* 1990). Both *Drosophila* spectrin isoforms are associated with the plasma membranes of a number of cell types (Pesacreta *et al.,* 1989; Thomas and Kiehart, 1994). The $\alpha\beta_H$ isoform is associated with the apical surface of some polarized cells (Thomas and Kiehart, 1994).

2. Molecular Analysis of the *Drosophila* Membrane Skeleton

The conserved nature of *Drosophila* spectrin was clearly established by cDNA sequence analysis. The domain structure of both subunits (repetitive and nonrepetitive domains) is identical between vertebrates and invertebrates, with 63% identity between fly and chicken α-subunits and 49% identity between β-spectrins from fly and human erythrocytes (Dubreuil *et al.,* 1989; Byers *et al.,* 1992). The spectrin gene family appears to be an ancient one, and the present-day subunit classes found in vertebrates and invertebrates are likely to have arisen through the same ancient gene multiplication events (Dubreuil, 1991).

A homologue of ankyrin has also been identified in *Drosophila* (Dubreuil and Yu, 1994). *Drosophila* ankyrin was initially cloned as a polymerase chain reaction (PCR) product using primers based on the sequence of human brain ankyrin-2 (Otto *et al.,* 1991). The PCR product was then used to recover full-length cDNA clones for sequence analysis. All of the evidence so far indicates that there is a single ankyrin gene in *Drosophila* located on the fourth chromosome. The predicted amino acid sequence is 53% identical to human brain ankyrin and, as was the case with spectrin, the domain structure of fly ankyrin is conserved relative to other known ankyrins. The N-terminal domain of ankyrin repeats and the central spectrin-binding domain exhibit significant sequence identity between fly and human. However, the C-terminal domain, whose function is not yet known, exhibits little sequence identity to human ankyrins. Interestingly, human ankyrins have diverged from one another to a similar extent in this region.

Homologues of some of the other membrane skeleton components originally found in the erythrocyte have been identified in *Drosophila.* Actin and tropomyosin are abundant proteins in a variety of *Drosophila* cells and tissues (Fyrberg and Goldstein, 1990). D4.1 resembles erythrocyte protein 4.1 and is associated with sites of cell–cell adhesion (Fehon *et al.,* 1994). Defects in D4.1 are lethal during embryonic development. The *Drosophila hts* protein is a homologue of adducin, an actin-bundling factor found at the spectrin–actin junction of erythrocytes. Defects in *hts* interfere with normal egg development (Yue and Spradling, 1992). Erythrocyte p55 was found to resemble hdlg, the human homologue of the *Drosophila* discs large tumor suppressor gene (Marfatia *et al.,* 1995). A comprehensive picture is emerging from these studies that will illuminate our understanding of how the different components of the membrane skeleton contribute to plasma membrane structure and function in developing systems.

B. Genetic Studies of α-Spectrin

1. Recovery of α-Spectrin Mutants

The reverse genetic approach in *Drosophila* entails identification of a gene product of interest (in this case spectrin), cloning of cDNA or genomic DNA representing that gene, and subsequent identification of the chromosomal address of that gene. Once the chromosomal address is known, one can follow established genetic methods to recover either new or existing mutations in the gene (Rubin, 1988). Recovery of mutants is greatly accelerated when the address happens to be within a well-studied chromosomal region, and is hindered when the gene resides in a relatively unexplored region. The three *Drosophila* spectrin subunits were localized to regions 62B (α), 16C1-4 (β), and 63CD (β_H). The region of the α-spectrin gene was the best characterized at the time these studies were initiated, and it was the first *Drosophila* spectrin locus for which mutants were recovered.

The α-spectrin region of chromosome 3 (62B) was mutagenized to saturation in the course of studies aimed at neighboring genes (Sliter *et al.,* 1989). Fly stocks carrying representative alleles of the lethal complementation groups from the region were screened for defects in α-spectrin in two ways (Lee *et al.,* 1993). Western blotting with α-spectrin antibodies was used to identify candidates in which the gene product was affected by the mutation (e.g., nonsense mutations yield protein products with altered electrophoretic mobility). A limitation of this approach is its inability to detect missense point mutations that affect the function but not the mobility of the protein. Nevertheless, several spectrin mutant alleles were identified by this strategy and all of them fell within a single complementation group representing a gene in the 62B region. A second approach employed an α-spectrin minigene to rescue the mutant phenotype (lethality) of animals lacking α-spectrin function. A minigene was built from α-spectrin cDNA and a promoter from the ubiquitin gene, which is expressed in all cells throughout *Drosophila* development (Lee *et al.,* 1988). The gene was introduced into germline chromosomal DNA by standard methods. While the minigene rescue approach has its own disadvantages, one important advantage was its capacity to detect all classes of lethal mutations, whether or not they affect a protein's electrophoretic mobility. The α-spectrin transgene was found to specifically rescue mutants from the same complementation group that presented alterations in electrophoretic mobility of α-spectrin.

2. Phenotype Analysis of α-Spectrin Mutants

A number of lethal mutant alleles were produced at the α-spectrin locus, and two of these (RG35 and RG41) have been best characterized (Lee *et al.,* 1993). These recessive lethal alleles were produced by gamma ray mutagenesis. Sequence analysis revealed that both mutations were due to small deletions that destroyed the open reading frame. RG41 introduces a nonsense codon near the beginning of the coding sequence and fails to produce any significant amount of mutant protein. Therefore, it is a null allele. RG35 terminates near the end of the open reading frame and therefore produces a nearly intact protein, except for a calcium-binding domain found near the C-terminus. Interestingly, both of these alleles have the same lethal phase. Homozygous mutants successfully develop as embryos; they hatch as larvae, initially crawl, and respond to tactile stimuli, but they die during the first day of larval development. It appears that some stages of development require zygotic expression of α-spectrin, and others do not.

The interpretation is complicated somewhat by the presence of wild-type maternal spectrin, which may provide α-spectrin function during early embryonic development. However, in the late embryo and early larva, α-spectrin falls to undetectable levels, suggesting that these animals survive for a time in the virtual absence of α-spectrin but invariably die a short time later.

While it is clear that defects in α-spectrin result in larval death, it has been difficult to address the exact cause of death. Nevertheless, important information on the function of spectrin in cells can be obtained without knowing which defect was fatal to the organism. Cellular defects have been most extensively characterized in the larval gut, since it is one of the most conspicuous structures in the first instar larva. The gut is divided into three major regions. The foregut is proximal to the feeding apparatus of the larva. It feeds into the midgut, which is itself compartmentalized into a number of morphologically distinct regions. The hindgut, like the foregut, is a relatively small terminal region of the digestive tract. There are a variety of differentiated cell types in the midgut that can easily be dissected and processed for immunofluorescent staining.

Spectrin can be detected at the plasma membrane of most of the epithelial cell types found in the midgut. A particularly striking pattern of spectrin staining was observed in the proximal "cuprophilic" region. These cells are polarized (Fig. 1c) with a distinct actin-rich apical domain that faces the gut lumen and a basolateral domain facing away from the lumen. The apical domain invaginates to form a cavity that is lined with microvilli. The cavity is continuous with the lumen of the gut, but in some planes of focus it appears to be surrounded by cytoplasm. When stained with fluorescent phalloidin to detect apical microvilli, the cuprophilic cells produce a characteristic pattern corresponding to the actin-rich cavity. Cuprophilic cells alternate with interstitial cells along the length of the gut, so that there is a regular space between the bright zones of actin staining (Fig. 2a). The normally organized pattern of actin staining is markedly disrupted in α-spectrin mutants (Fig. 2d). Even by light microscopy (Lee *et al.*, 1993), the cells appear distended and the normally smooth lumen of the gut appears disorganized. An antibody against α-spectrin stains the basolateral surface of the cuprophilic cells as well as the microvillus-rich apical domain (Fig. 2b). As expected, there is little if any detectable α-spectrin in mutant cuprophilic cells (Fig. 2e). Thus loss of α-spectrin is associated with disruption of the normal cellular organization of the midgut.

The α-spectrin mutation also affects the appearance (shape) of individual cells. When wild-type cells are stained for the sodium pump, a basolateral distribution is detected (Fig. 2c) with little or no staining of the apical microvillar surface. There is no apparent change in the level or distribution

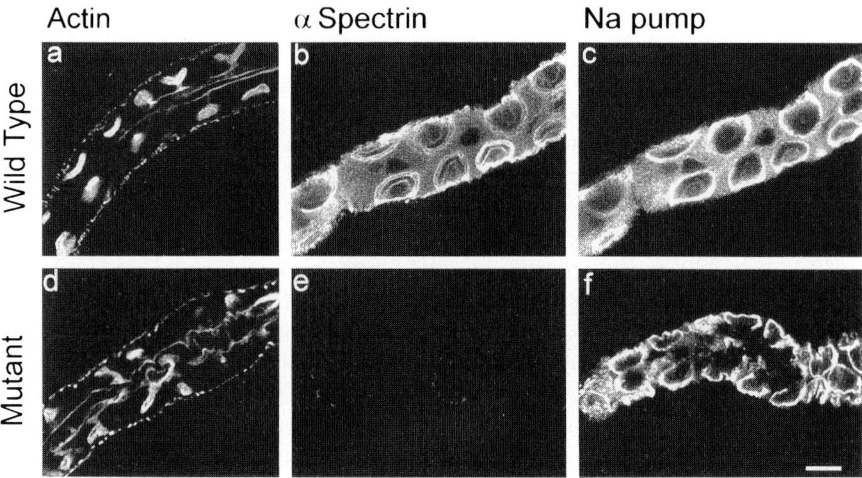

FIGURE 2 Mutations of *Drosophila* α-spectrin affect cell shape and organization in the midgut epithelium. The cuprophilic region of mutant and wild-type first instar larval gut was stained with rhodamine phalloidin (a,d), α-spectrin antibody (b,e), and sodium pump antibody (c,f). See text for details. (Modified from Lee *et al.* (1993). *J. Cell Biol.* **123,** 1797–1809, by copyright permission of the Rockefeller University Press.) Bar = 10 μm.

of sodium pump in α-spectrin mutants (Fig. 2f). However, the normally smooth surface contour of the cuprophilic cell is conspicuously altered so that the cell margins appear scalloped. Thus, it appears that spectrin contributes to the normal shape of cells.

Mutant cells exhibit a further defect in cell adhesion. EM analysis of the junctions between wild-type cuprophilic cells and neighboring interstitial cells reveals that they are zipped up tightly against one another, presumably because of the action of cell adhesion molecules. However, there are prominent gaps visible between cells in EM views of α-spectrin mutants (Lee *et al.,* 1993). Thus defects in α-spectrin also appear to affect cell adhesion, although the mechanism is not yet known. Interestingly, some of the plasma membrane proteins that have been shown to interact with ankyrin in vertebrate systems are cell adhesion molecules (see Section III,A). It is not known if the adhesive function of these molecules is dependent upon their interaction with the membrane skeleton.

Despite the alteration in cell shape, the α-spectrin defect did not detectably alter the polarity of the sodium pump in the midgut epithelium. This result is at odds with results of several studies in vertebrate systems that implicate the spectrin membrane skeleton, and ankyrin in particular, in the establishment of sodium pump polarity (see Section III,D). Yet it does not rule out a role for the membrane skeleton in the establishment of polarity, since there are other important factors to consider. First, it remains to be demonstrated that the *Drosophila* sodium pump interacts with ankyrin. Some cells appear to sort sodium pumps at the level of the secretory pathway instead of by selective stabilization via the membrane skeleton (Mays *et al.*, 1995). Second, some functions of the membrane skeleton may simply not require the presence of α-spectrin.

Does a mutation in α-spectrin constitute a complete knockout of the membrane skeleton? Interestingly, the answer appears to be no (Dubreuil and Yu, 1994). Western blot analysis was used to examine the null RG41 mutation in homozygous larvae shortly before their death. α-Spectrin was readily detectable in wild-type larvae but was below the level of detection in RG41 mutants (Fig. 3). In contrast, similar blots did not reveal a significant effect of the α-spectrin mutation on the accumulation of a number of other relevant molecules, such as the two β-subunits of spectrin and ankyrin. A putative membrane anchor for the membrane skeleton (neuroglian) was also unaffected by the α-spectrin mutation. Neuroglian belongs to a family of cell–cell adhesion molecules that interact with ankyrin in rat brain (Davis *et al.*, 1993) and in *Drosophila* (Dubreuil *et al.*, 1996).

The observed accumulation of membrane skeleton proteins in α-spectrin mutants is probably not a trivial persistence of nonfunctional protein. Immunolocalization of ankyrin in midgut epithelial cells revealed that the wild-type pattern of bright staining at the lateral margins of these cells (Fig. 4a) was apparently unaffected in RG41 mutant larvae (Fig. 4b). The β-subunit

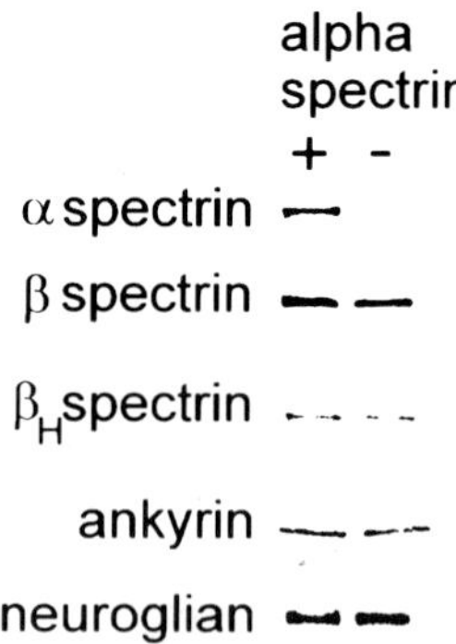

FIGURE 3 A mutation in α-spectrin does not constitute a complete knockout of the membrane skeleton. Western blots of wild-type and α-spectrin mutant larvae reveal that, whereas α-spectrin is undetectable in mutants, other membrane skeleton components such as β-spectrins and ankyrin continue to accumulate at wild-type levels. An integral membrane protein that is associated with ankyrin (neuroglian) also persists at wild-type levels. (Modified from Dubreuil and Yu, 1994.)

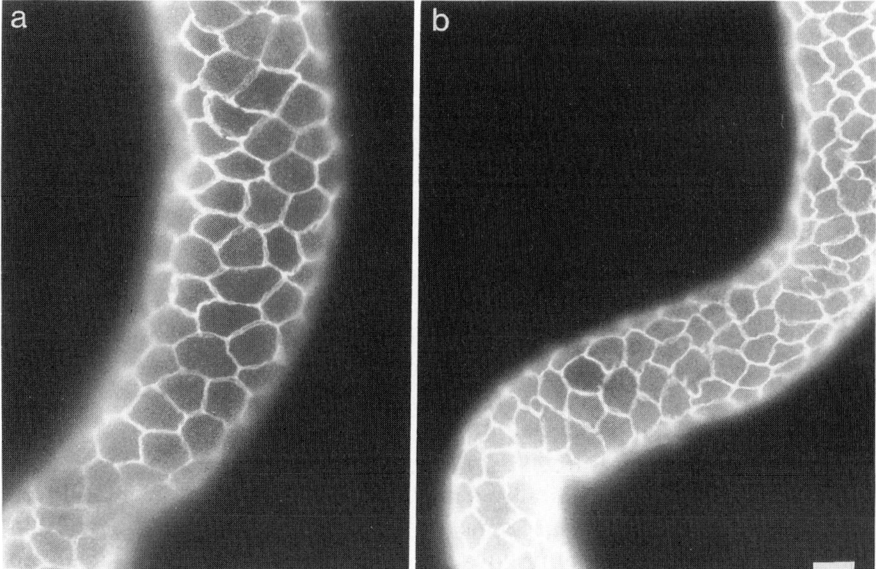

FIGURE 4 The association of ankyrin with the plasma membrane does not require α-spectrin. Epithelial cells of the middle midgut display a characteristic "snakeskin" pattern of ankyrin staining at the thin lateral margins of each cell. The pattern is indistinguishable in wild-type (a) and null α-spectrin mutant (b) midguts. It is not yet known if membrane-associated ankyrin and β-spectrin (not shown) retain partial membrane skeleton function in α-spectrin mutants. Bar = 10 μm.

of spectrin was also detected at the surface of many cells in the absence of detectable α-spectrin (not shown). The observed accumulation and membrane targeting of these membrane skeleton proteins raises the possibility that, while normal cell shape and some aspects of cell adhesion depend on α-spectrin, perhaps some functions of the membrane skeleton persist in the absence of α-spectrin.

3. *Drosophila* Mutants as an Experimental Tool

As described earlier, the phenotype of α-spectrin mutants provides information on the role of the membrane skeleton in cells. Minigene rescue also provides a strategy with which to ask questions about how spectrin carries out its function. Selected properties of the rescue transgene can easily be altered by standard molecular methods, and then the ability of the gene to rescue normal function can be tested in a spectrin mutant genetic background.

One such test examined a region near the N-terminus of α-spectrin that has been implicated in spectrin tetramer formation (Deng *et al.,* 1995). The experiment was based on the finding by Coetzer and colleagues (1991) that four different point mutations in codon 28 of human erythroid α-spectrin each caused hereditary elliptocytosis, probably by inhibition of spectrin tetramer formation (see Section II,B). The corresponding position (codon 22, arginine) is also conserved in *Drosophila* (Dubreuil *et al.,* 1989). First, it was shown that bacterially produced fragments of wild-type *Drosophila* α-spectrin interact with fragments of β-spectrin in an *in vitro* binding assay that models spectrin tetramer formation. The interaction was detected by co-sedimentation of appropriate labeled spectrin fragments and by the formation of a folded protease-resistent structure. Mutations in codon 22 of the α-spectrin fragment blocked the interaction between α- and β-spectrin fragments *in vitro,* suggesting that the conserved arginine is necessary for spectrin tetramer formation in *Drosophila.* The mutation had its greatest effect at elevated temperature (29°C), with only a slight effect on interchain binding at low temperature (19°C).

Next the codon 28 mutation was introduced into an α-spectrin minigene that was then used to produce transformed flies. The ability of the mutant transgene to rescue a homozygous spectrin mutant (RG41) was assessed through appropriate genetic crosses. Interestingly, the mutant transgene was able to rescue spectrin function, but in a temperature-dependent manner. Flies reared at 19°C fared much better than flies reared at 29°C. A transformed fly line expressing a wild-type α-spectrin transgene was not diminished in its ability to rescue mutants at elevated temperature. Thus it appears that the wild-type function of α-spectrin (measured as viability) depends on its ability to form tetramers. Presumably, spectrin tetramers are required to form a 2-dimensional protein network analogous to that found in the erythrocyte.

V. CONCLUSIONS

A. Assembly of the Membrane Skeleton

The effects of the α-spectrin mutation on other membrane skeleton components shed light on the order of events in assembly of the membrane skeleton. This subject was previously addressed in the avian erythrocyte membrane, wherein assembly was proposed to occur by a receptor-mediated stabilization mechanism (Lazarides and Moon, 1984). In this scheme, membrane-proximal components (i.e., band 3) of the complex are both necessary and limiting for assembly of downstream components (i.e., spectrin and ankyrin). Indeed, it was shown that some downstream components are synthesized in excess and that unassembled protein is rapidly degraded

(Lazarides and Moon, 1984). However, the assembly process appears to be more complex than originally envisioned since band 3 is actually one of the last components to become stably associated with the membrane skeleton (Cox *et al.*, 1987).

The genetic approach is well suited to the dissection of biological pathways such as the order of events in membrane skeleton assemby (e.g., Wassarman *et al.*, 1995). Results so far indicate that α-spectrin does not limit the accumulation and membrane assembly of ankyrin and β-spectrin. These results are consistent with a pathway in which membrane skeleton assembly begins with an IMP anchor (e.g., IMP-1 or IMP-2 in Fig. 1a) that recruits ankyrin and ultimately α- and β-spectrin to the plasma membrane. An alternate view is that the cascade begins with an association between β-spectrin and an IMP, and that the recruitment of ankyrin to the membrane skeleton is a subsequent event (Lombardo *et al.*, 1994). There are not enough data yet to distinguish between these possibilities, but genetic methods are likely to be useful in approaching the problem.

B. Function of the Membrane Skeleton

What is the basis for altered cell shape and adhesion in α-spectrin mutants? To answer this question, it is necessary to consider the relative contribution of the α-subunit to the function of intact spectrin and the membrane skeleton. Most of the well-characterized binding sites of spectrin are located on the β-subunit (e.g., actin, ankyrin, other membrane attachment sites; reviewed by Coleman *et al.*, 1989). One important function of spectrin that probably is dependent on the α-subunit is tetramer formation, and therefore the formation of a spectrin–actin network. Three lines of evidence suggest that the phenotype of α-spectrin mutants is due to the failure of spectrin to form a network. First, the codon 22 mutant of α-spectrin is unable to provide normal function, presumably because of its inability to interact with β-spectrin to form tetramers. Second, the RG35 mutant allele encodes a nearly intact α-spectrin chain that only lacks the C-terminal calcium-binding domain. Based on biochemical and sequence similarities between spectrin and α-actinin, it seems likely that this domain regulates the activity of the adjacent actin-binding domain on β-spectrin (Dubreuil *et al.*, 1991). If indeed the RG35 mutation inhibits the actin-binding activity of β-spectrin, then once again the mutant phenotype is due to a failure of spectrin network formation. Third, naturally a null mutant that altogether lacks α-spectrin (RG41) must also fail to produce spectrin tetramers and networks.

The similar requirements for spectrin function in erythrocytes and *Drosophila* cells are intriguing. Both systems appear to depend on spectrin network formation, and network defects cause alterations in cell shape. However, it is important to consider that cell shape is an intrinsic property of the freely circulating erythrocyte. In contrast, adhesive interactions between neighboring cells contribute significantly to the shape and organization of many cell types in higher organisms. Thus, even though membrane skeleton defects in erythroid and nonerythroid cells both alter cell shape, they may do so by different mechanisms.

The membrane skeleton is also thought to influence the composition of membrane domains, which may in turn have broad effects on cell shape and organization. The genetic tests carried out so far have not detected a role for the membrane skeleton in the establishment of membrane polarity. However, it is now clear that further experiments will be necessary to examine the residual functions of the membrane skeleton in α-spectrin mutants. Genetic studies of some of the other components of the membrane skeleton are underway to address the new questions that have emerged through studies of α-spectrin.

References

Bennett, V. (1990). Spectrin based membrane skeleton—a multipotential adaptor between membrane and cytoplasm. *Physiol. Rev.* **70,** 1029–1065.

Bennett, V. (1992). Ankyrins. *J. Biol. Chem.* **267,** 8703–8706.

Bennett, V., and Gilligan, D. M. (1993). The spectrin-based membrane skeleton and micron-scale organization of the plasma membrane. *Annu. Rev. Cell Biol.* **9,** 27–66.

Bennett, V., Baines, A. J., and Davis, J. Q. (1985). Ankyrin and synapsin: Spectrin-binding proteins associated with brain membranes. *J. Cell. Biochem.* **29,** 157–169.

Bernstein, S. E. (1969). Hereditary disorders of the rodent erythron. *In* "Genetics in Laboratory Medicine," pp. 9–33. National Academy of Sciences, Washington DC.

Bourguignon, L. Y. W., Suchard, S. J., Nagpal, M. D., and Glenney, J. R. (1985). T-lymphoma transmembrane glycoprotein (gp180) is linked to the cytoskeleton protein, fodrin. *J. Cell Biol.* **101,** 477–487.

Branton, D., Cohen, C. M., and Tyler, J. (1981). Interaction of cytoskeletal proteins on the human erythrocyte membrane. *Cell* **24,** 24–32.

Byers, T. J., and Branton, D. (1985). Visualization of the protein associations in the erythrocyte membrane skeleton. *Proc. Natl. Acad. Sci. U.S.A.* **82,** 6153–6157.

Byers, T. J., Dubreuil, R. R., Branton, D., Kiehart, D. P., and Goldstein, L. S. B. (1987). Drosophila spectrin II. Conserved features of the alpha subunit are revealed by analysis of cDNA clones and fusion proteins. *J. Cell Biol.* **105,** 2103–2110.

Byers, T. J., Husain-Chishti, A., Dubreuil, R. R., Branton, D., and Goldstein, L. S. B. (1989). Drosophila beta-spectrin: Sequence similarity to the amino-terminal domain of alpha-actinin and dystrophin. *J. Cell Biol.* **109,** 1633–1641.

Byers, T. J., Brandin, E., Winograd, E., Lue, R., and Branton, D. (1992). The complete sequence of Drosophila beta spectrin reveals supra-motifs comprising eight 106-residue segments. *Proc. Natl. Acad. Sci. U.S.A.* **89,** 6187–6191.

Campbell, K. P. (1995). Three muscular dystrophies: Loss of cytoskeleton-extracellular matrix linkage. *Cell* **80,** 675–679.

Coetzer, T. L., Sahr, K., Prchal, J., Blacklock, H., Peterson, L., Koler, R., Doyle, J., Manaster, J., and Palek, J. (1991). Four different mutations in codon 28 of alpha spectrin are associated with structurally and functionally abnormal spectrin alphaI-74 in hereditary elliptocytosis. *J. Clin. Invest.* **88,** 743–749.

Coleman, T. R., Fishkind, D. J., Mooseker, M. S., and Morrow, J. S. (1989). Contributions of the beta-subunit to spectrin structure and function. *Cell Motil. Cytoskeleton* **12,** 248–263.

Conboy, J. (1993). Structure, function, and molecular genetics of erythroid membrane skeletal protein 4.1 in normal and abnormal red blood cells. *Semin. Hematol.* **30,** 58–73.

Cox, J. V., Stack, J. H., and Lazarides, E. (1987). Erythroid anion transporter assembly is mediated by a developmentally regulated recruitment onto a preassembled membrane skeleton. *J. Cell Biol.* **105,** 1405–1416.

Davis, J., and Bennett, V. (1984). Brain ankyrin. *J. Biol. Chem.* **259,** 13550–13559.

Davis, J. Q., McLaughlin, T., and Bennett, V. (1993). Ankyrin-binding proteins related to nervous system cell adhesion molecules: Candidates to provide transmembrane and intercellular connections in adult brain. *J. Cell Biol.* **121,** 121–133.

Deng, H., Lee, J. K., Goldstein, L. S. B., and Branton, D. (1995). Drosophila development requires spectrin network formation. *J. Cell Biol.* **128,** 71–79.

Drenckhahn, D., Schluter, K., Allen, D. P., and Bennett, V. (1985). Colocalization of band 3 with ankyrin and spectrin at the basal membrane of intercalated cells in the rat kidney. *Science* **230,** 1287–1289.

Dubreuil, R. R. (1991). Structure and evolution of the actin crosslinking proteins. *BioEssays* **13,** 219–226.

Dubreuil, R. R., and Yu, J. (1994). Ankyrin and beta spectrin accumulate independently of alpha spectrin in Drosophila. *Proc. Natl. Acad. Sci. U.S.A.* **91,** 10285–10289.

Dubreuil, R., Byers, T. J., Branton, D., Goldstein, L. S. B., and Kiehart, D. P. (1987). Drosophila spectrin I. Characterization of the purified protein. *J. Cell Biol.* **105,** 2095–2102.

Dubreuil, R. R., Byers, T. J., Sillman, A. L., Bar-Zvi, D., Goldstein, L. S. B., and Branton, D. (1989). The complete sequence of Drosophila alpha spectrin: Conservation of structural domains between alpha spectrins and alpha actinin. *J. Cell Biol.* **109,** 2197–2206.

Dubreuil, R. R., Byers, T. J., Stewart, C. T., and Kiehart, D. P. (1990). A beta spectrin isoform from Drosophila (betaH) is similar in size to vertebrate dystrophin. *J. Cell Biol.* **111,** 1849–1858.

Dubreuil, R. R., Brandin, E., Reisberg, J. H. S., Goldstein, L. S. B., and Branton, D. (1991). Structure, calmodulin-binding, and calcium-binding properties of recombinant alpha spectrin polypeptides. *J. Biol. Chem.* **266,** 7189–7193.

Dubreuil, R. R., McVicker, G., Dissanayake, S., Liu, C., Homer, D., and Hortsch, M. (1996). Neuroglian-mediated adhesion induces assembly of the membrane skeleton at sites of cell-cell contact. *J. Cell Biol.* **133,** 647–655.

Fehon, R. G., Dawson, I. A., and Artavanis-Tsakonas, S. (1994). A Drosophila homologue of membrane-skeleton protein 4.1 is associated with septate junctions and is encoded by the coracle gene. *Development* **120,** 545–557.

Fyrberg, E. A., and Goldstein, L. S. B. (1990). The Drosophila cytoskeleton. *Annu. Rev. Cell Biol.* **6,** 559–596.

Glenney, J. R., and Glenney, P. (1984). Comparison of spectrin isolated from erythroid and non-erythroid sources. *Eur. J. Biochem.* **144,** 529–539.

Hammerton, R. W., Krzeminski, K. A., Mays, R. W., Ryan, T. A., Wollner, D. A., and Nelson, W. J. (1991). Mechanism for regulating cell surface distribution of Na+, K+-ATPase in polarized epithelial cells. *Science* **254,** 847–850.

Koenig, M., Monaco, A. P., and Kunkel, L. M. (1988). The complete sequence of dystrophin predicts a rod-shaped cytoskeletal protein. *Cell* **53,** 219–228.

Kordeli, E., Lambert, S., and Bennett, V. (1995). Ankyrin-G. *J. Biol. Chem.* **270,** 2352–2359.

Kunkel, L. M., Monaco, A. P., Middlesworth, W., Ochs, H. D., and Latt, S. A. (1985). Specific cloning of DNA fragments absent from the DNA of a male patient with a chromosomal deletion. *Proc. Natl. Acad. Sci. U.S.A.* **82,** 4778–4782.

Lambert, S., Yu, H., Prchal, J. T., Lawler, J., Ruff, P., Speicher, D., Cheung, M. C., Kan, Y. W., and Palek, J. (1990). cDNA sequence for human erythrocyte ankyrin. *Proc. Natl. Acad. Sci. U.S.A.* **87,** 1730–1734.

Lazarides, E., and Moon, R. T. (1984). Assembly and topogenesis of the spectrin-based membrane skeleton in erythroid development. *Cell* **37,** 354–356.

Lazarides, E., and Nelson, W. J. (1983). Erythrocyte and brain forms of spectrin in cerebellum: Distinct membrane-cytoskeleton domains in neurons. *Science* **220,** 1295–1297.

Lee, H., Simon, J. A., and Lis, J. T. (1988). Structure and expression of ubiquitin genes of Drosophila melanogaster. *Mol. Cell. Biol.* **8,** 4727–4735.

Lee, J., Coyne, R., Dubreuil, R. R., Goldstein, L. S. B., and Branton, D. (1993). Cell shape and interaction defects in alpha-spectrin mutants of Drosophila melanogaster. *J. Cell Biol.* **123,** 1797–1809.

Lokeshwar, V. B., and Bourguignon, L. Y. W. (1992). The lymphoma transmembrane glycoprotein GP85 (CD44) is a novel guanine nucleotide-binding protein which regulates GP85 (CD44)-ankyrin interaction. *J. Biol. Chem.* **267,** 22073–22078.

Lombardo, C. R., Weed, S. A., Kennedy, S. P., Forget, B. G., and Morrow, J. S. (1994). BetaII-spectrin (fodrin) and Beta IE2-spectrin (muscle) contain NH_2- and COOH-terminal membrane association domains (MAD1 and MAD2). *J. Biol. Chem.* **269,** 29212–29219.

Lux, S. E., and Palek, J. (1995). Disorders of the red cell membrane. *In* "Blood: Principles and Practice of Hematology" (R. I. Handin, S. E. Lux, and T. P. Stossel, eds.), pp. 1701–1818. J. B. Lippincott Co., Philadelphia.

Lux, S. E., John, K. M., and Bennett, V. (1990). Analysis of cDNA for human erythrocyte ankyrin indicates a repeated structure with homology to tissue-differentiation and cell-cycle control proteins. *Nature (London)* **344,** 36–42.

Marfatia, S. M., Leu, R. A., Branton, D., and Chishti, A. H. (1995). Identification of the protein 4.1 binding interface on glycophorin C and p55, a homologue of the Drosophila discs-large tumor suppressor protein. *J. Biol. Chem.* **270,** 715–719.

Matsumura, K., Tome, F. M. S., Collin, H., Azibi, K., Chaouch, M., Kaplan, J.-C., Fardeau, M., and Campbell, K. P. (1992). Deficiency of the 50K dystrophin-associated glycoprotein in severe childhood autosomal recessive muscular dystrophy. *Nature* **359,** 320–322.

Mays, R. W., Siemers, K. A., Fritz, B. A., Lowe, A. W., Meer, G. v., and Nelson, W. J. (1995). Hierarchy of mechanisms involved in generating Na/K-ATPase polarity in MDCK epithelial cells. *J. Cell Biol.* **130,** 1105–1115.

McNeill, H., Ozawa, M., Kemler, R., and Nelson, W. J. (1990). Novel function of the cell adhesion molecule uvomorulin as an inducer of cell surface polarity. *Cell* **62,** 309–316.

Morgans, C. W., and Kopito, R. R. (1993). Association of the brain anion exchanger, AE3, with the repeat domain of ankyrin. *J. Cell Sci.* **105,** 1137–1142.

Nelson, W. J. (1992). Regulation of cell surface polarity from bacteria to mammals. *Science* **258,** 948–955.

Nelson, W. J., and Veshnock, P. J. (1986). Dynamics of membrane skeleton (fodrin) organization during development of polarity in Madin-Darby canine kidney epithelial cells. *J. Cell Biol.* **103,** 1751–1765.

Nelson, W. J., and Veshnock, P. J. (1987). Ankyrin binding to (Na+ & K+) ATPase and implications for the organization of membrane domains in polarized cells. *Nature (London)* **328,** 533–536.

Otto, E., Kunimoto, M., McLaughlin, T., and Bennett, V. (1991). Isolation and characterization of cDNAs encoding human brain ankyrins reveal a family of alternatively spliced genes. *J. Cell Biol.* **114,** 241–253.

Pesacreta, T. C., Byers, T. J., Dubreuil, R. R., Keihart, D. P., and Branton, D. (1989). Drosophila spectrin: The membrane skeleton during embryogenesis. *J. Cell Biol.* **108,** 1697–1709.

Peters, L. L., Birkenmeier, C. S., Bronson, R. T., White, R. A., Lux, S. E., Otto, E., Bennett, V., Higgins, A., and Barker, J. E. (1991). Purkinje cell degeneration associated with erythroid ankyrin deficiency in nb/nb mice. *J. Cell Biol.* **114,** 1233–1241.

Peters, L. L., John, K. M., Lu, F. M., Eicher, E. M., Higgins, A., Yialamas, M., Turtzo, L. C., Otsuka, A. J., and Lux, S. E. (1995). Ank3 (epithelial ankyrin), a widely distributed new member of the ankyrin gene family and the major ankyrin in kidney, is expressed in alternatively spliced forms, including forms that lack the repeat domain. *J. Cell Biol.* **130,** 313–330.

Pinder, J. C., and Gratzer, W. B. (1983). Structural and dynamic states of actin in the erythrocyte. *J. Cell Biol.* **96,** 768–775.

Rubin, G. M. (1988). Drosophila melanogaster as an experimental organism. *Science* **240,** 1453–1459.

Shen, B. W., Josephs, R., and Steck, T. L. (1984). Ultrastructure of unit fragments of the skeleton of the human erythrocyte membrane. *J. Cell Biol.* **99,** 810–821.

Sliter, T. J., Henrich, V. C., Tucker, R. L., and Gilbert, L. I. (1989). The genetics of the Dras3-Roughened-ecdysoneless chromosomal region (62B3–4 to 62D3–4) in Drosophila melanogaster: Analysis of recessive lethal mutations. *Genetics* **123,** 327–336.

Srinivasan, Y., Lewallen, M., and Angelides, K. J. (1992). Mapping the binding site on ankyrin for the voltage-dependent sodium channel from brain. *J. Biol. Chem.* **267,** 7483–7489.

Steck, T. L. (1974). The organization of proteins in the human red blood cell membrane. *J. Cell Biol.* **62,** 1–19.

Steiner, J. P., and Bennett, V. (1988). Ankyrin-independent membrane protein-binding sites for brain and erythrocyte spectrin. *J. Biol. Chem.* **263,** 14417–14425.

Sunada, Y., Bernier, S. M., Kozak, C. A., Yamada, Y., and Campbell, K. P. (1994). Deficiency of merosin in dystrophic dy mice and genetic linkage of laminin M chain gene to dy locus. *J. Biol. Chem.* **269,** 13729–13732.

Thomas, G. H., and Kiehart, D. P. (1994). Beta-heavy spectrin has a restricted tissue and sub cellular distribution during Drosophila embryogenesis. *Development* **120,** 2039–2050.

Wassarman, D. A., Therrien, M., and Rubin, G. M. (1995). The Ras signalling pathway in Drosophila. *Curr. Opin. Gen. Dev.* **5,** 44–50.

Winkelman, J. C., and Forget, B. G. (1993). Erythroid and non-erythroid spectrins. *Blood* **81,** 3173–3185.

Yue, L., and Spradling, A. C. (1992). hy-li tai shao, A gene required for ring canal formation during Drosophila oogenesis, encodes a homolog of adducin. *Genes Dev.* **6,** 2443–2454.

CHAPTER 9

Membrane–Cytoskeleton Interactions with Cadherin Cell Adhesion Proteins: Roles of Catenins as Linker Proteins

Margaret J. Wheelock, Karen A. Knudsen,* and Keith R. Johnson
Department of Biology, University of Toledo, Toledo, Ohio 43606 and *Department of Cell Biology, Lankenau Medical Research Center, Philadelphia, Pennsylvania 19096

I. INTRODUCTION

The cadherins are members of a large family of transmembrane glycoproteins that mediate calcium-dependent homotypic cell–cell adhesion. Members of the cadherin family play important roles during tissue morphogenesis and in the maintenance of normal tissues. One important aspect of the cadherins is their binding specificity. With some few exceptions (Volk *et al.,* 1987; Inuzuka *et al.,* 1991; Cepek *et al.,* 1994; Murphy-Erdosh *et al.,* 1995), each cadherin binds to an identical cadherin on an adjacent cell, allowing cells expressing different complements of cadherins to sort out from one another (reviewed in Takeichi, 1990). Thus, in the developing embryo, increased or decreased function of different cadherins promotes segregation of differentiating cell types from one another to form various tissues of the organism (reviewed in Takeichi, 1991). Inappropriate expression or downregulation of cadherins has been associated with tumor cell invasion and may contribute to the detachment of individual cells from the tumor mass (reviewed in Takeichi, 1993; Hülsken *et al.* 1994a).

The cadherin superfamily is composed of diverse proteins, including (1) the classical cadherins [N-, E-, and P-cadherins (reviewed in Takeichi, 1987,1990)], which are the transmembrane components of intercellular junctions known as adherens junctions; (2) the desmogleins and desmocollins, which are the transmembrane proteins of another intercellular junction, the desmosome (reviewed in Buxton and Magee, 1992; Koch and Franke, 1994); (3) the atypical cadherins (reviewed in Grunwald, 1993; Blaschuk *et al.,* 1995); and (4) the protocadherins (Sano *et al.,* 1993). This chapter concentrates on the classical cadherins, and in particular summarizes the current understanding of how they interact with cytoplasmic proteins to form an adherens junction.

Adherens junctions are particularly prominent in epithelia and the myocardium, where they appear, at regions of close cell–cell apposition, as two parallel intracellular plaques into which actin filaments insert (Geiger *et al.,* 1990). Classical cadherins are central molecules in the regulation and function of adherens junctions. As transmembrane molecules, they are composed of three segments: (1) an extracellular domain responsible for cadherin–cadherin interaction; (2) a single-pass transmembrane domain; and (3) a highly conserved cytoplasmic domain that associates with actin filaments and thus serves to connect the outside of the cell to the cytoskeleton. Cadherins are not bound directly to the actin cytoskeleton, but rather are connected indirectly via a group of proteins known as the catenins.

The catenins were identified as proteins co-immunoprecipitating with the classical cadherins and were named α-catenin, β-catenin, and γ-catenin according to their mobility on sodium dodecyl sulfate–polyacrylamide gel electrophoresis (Nagafuchi and Takeichi, 1988; Ozawa *et al.,* 1989; Wheelock and Knudsen, 1991; McCrea *et al.,* 1991; McCrea and Gumbiner, 1991). α-Catenin is a 102-kDa protein that shares some homology with vinculin (Nagafuchi *et al.,* 1991; Herrenknecht *et al.,* 1991). It is associated with the cadherins indirectly through its interaction with β-catenin or γ-catenin. β-Catenin is a 95-kDa protein that shares about 65% identity with γ-catenin

(Fouquet *et al.*, 1992), an 82-kDa protein also named plakoglobin (Knudsen and Wheelock, 1992; Peifer *et al.*, 1992). β-Catenin and plakoglobin associate directly with the cadherins and can substitute for one another in the cadherin–catenin complex (Butz and Kemler, 1994; Hinck *et al.*, 1994; Näthke *et al.*, 1994; Sacco *et al.*, 1995). Whether or not β-catenin and plakoglobin play identical functional roles in the adherens junction is not yet known. Thus, the cell–cell adherens junction is currently thought of as a structure composed of transmembrane cadherin molecules each of which is associated directly with either β-catenin or plakoglobin, which in turn associates directly with α-catenin. α-Catenin then mediates the interaction between the cadherin–catenin complex and the actin cytoskeleton.

Work published in the past few years has provided information on the mechanisms by which the cadherins associate with the cytoskeleton. In this chapter we discuss the biochemical evidence implicating the catenins as linkers between the transmembrane cadherin and the actin cytoskeleton.

II. THE CADHERIN MOLECULE AND ITS ASSOCIATION WITH PLAKOGLOBIN AND β-CATENIN

A. Cadherin Extracellular Domain

The typical vertebrate classical cadherin is a type I transmembrane glycoprotein with a molecular mass of between 110 and 150 kDa. Figure 1

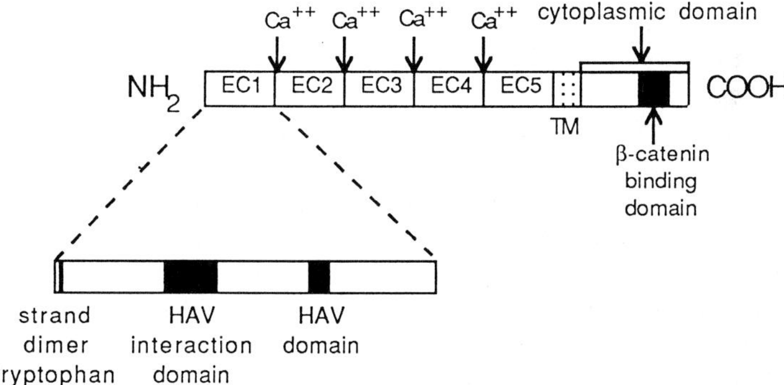

FIGURE 1 Classical cadherin structure. The classical cadherin is a single-pass transmembrane protein. The membrane-spanning domain is labeled TM, the cytoplasmic domain is indicated, and the extracellular portion is labeled EC1–EC5. Calcium ion–binding sites between adjacent EC domains are indicated. EC1 is enlarged to show the strand-dimer tryptophan, the HAV domain, and the domain in which HAV interacts with its homotypic partner cadherin. The domain within the cytoplasmic portion of the cadherin that is necessary for catenin binding is indicated.

presents a diagram of a classical cadherin indicating functional domains. The extracellular (EC) portion can be divided into five domains of approximately 110 amino acids each that are designated EC1–EC5, with EC1 being the extreme N-terminal domain. EC1 is the most highly conserved of the EC domains (Hatta *et al.,* 1988) and is involved in self-association. A tripeptide (HAV), and residues immediately surrounding it, within EC1 mediates cadherin binding to itself (Nose *et al.,* 1990; Blaschuk *et al.,* 1990) by associating with a separate set of amino acids located within the EC1 domain of the interacting partner (Shapiro *et al.,* 1995a). Takeichi and co-workers showed that residues adjacent to the HAV tripeptide are particularly important in conferring homotypic specificity to cadherin interactions; that is, mutations in these residues allow E-cadherin and P-cadherin to interact (Nose *et al.,* 1990). In addition, calcium association with the extracellular domain of cadherin is necessary for its adhesive function (Hatta *et al.,* 1988; Ozawa *et al.,* 1990a). Recent structural studies have shown that it is likely that calcium ions act as bridges between adjacent EC domains (Shapiro *et al.,* 1995a; Pokutta *et al.,* 1994) and that, in the absence of calcium, the cadherin EC domains do not form a structure capable of cadherin–cadherin association (Shapiro *et al.,* 1995a; Overduin *et al.,* 1995). Studies of the crystal structure of EC1 suggest that, in addition to the antiparallel self-association of cadherins expressed on two different cells, cadherins on the surface of the same cell self-associate in a parallel fashion to form a dimer (Shapiro *et al.,* 1995a,b). A tryptophan near the N-terminus of EC1 plays a role in the formation of the strand dimer (see Fig. 1).

B. Cadherin Cytoplasmic Domain

Early studies illustrated the functional importance of the cytoplasmic domain of the classical cadherins. A truncated cadherin retaining the entire extracellular domain, the transmembrane domain, and all but 72 amino acids of the cytoplasmic domain was not able to support cell–cell adhesion in an aggregation assay even though this cadherin was expressed on the surface of transfected cells and retained its ability to bind calcium (Nagafuchi and Takeichi, 1988; Ozawa *et al.,* 1990b). Interestingly, the truncated cadherin did not bind the catenins and hence it was suggested that the catenins are also required for cadherin function. Furthermore, it was hypothesized that the catenins serve to link the cadherin to the cytoskeleton and that this linkage is necessary for adhesive activity.

Kemler and co-workers recently identified a core region of 30 amino acids (832–862) within the C-terminal 72 amino acids that may represent the catenin-binding domain in E-cadherin (Stappert and Kemler, 1994).

Further analysis of this domain using the yeast two-hybrid system has demonstrated that it must be present for cadherin to interact with β-catenin (Jou *et al.*, 1995). Since this domain is serine rich and highly phosphorylated, it has been suggested that phosphorylation of cadherin between amino acids 832 and 862 could be responsible for regulation of both catenin binding and association with the cytoskeleton. This then could regulate the EC adhesive properties of the cadherin (Stappert and Kemler, 1994). However, it must be kept in mind that phosphorylation of serine residues in this region is not essential for cadherin–catenin association since complexes of bacterial recombinant proteins can be assembled *in vitro* (Aberle *et al.*, 1994). The cadherin is thought to interact in the same way with either β-catenin or plakoglobin. Figure 1 presents a diagram of a typical classical cadherin and points out the domains discussed earlier.

III. PLAKOGLOBIN/β-CATENIN AND THEIR ASSOCIATIONS WITH CADHERIN AND α-CATENIN

A. Plakoglobin/β-Catenin Structure

Sequence analysis has revealed that plakoglobin and β-catenin are members of the Armadillo family of proteins (Butz *et al.*, 1992; Peifer and Weischaus, 1990; McCrea *et al.*, 1991). Armadillo is a *Drosophila* protein that was originally identified as one of a group of proteins that regulates segment polarity (Riggleman *et al.*, 1989). Recent studies identifying *Drosophila* cadherins and catenins have made it clear that Armadillo is not only a signaling molecule but is also a structural component of the *Drosophila* junctional complex (Peifer, 1993). Developmental studies in *Xenopus* have shown that plakoglobin and β-catenin play important roles in axis formation, suggesting that they, like Armadillo, play signaling roles during development (Heasman *et al.*, 1994; Funayama *et al.*, 1995; Karnovsky and Klymkowsky, 1995). Thus the current thinking is that members of the Armadillo family of proteins can serve both as signaling molecules and as structural proteins. How and if these two functions are related to one another is not yet known. This chapter concentrates on the roles of plakoglobin and β-catenin as structural components of the adherens junction.

One characteristic of Armadillo family members is a central domain composed of Armadillo repeats (Peifer, 1995). Each repeat is about 42 amino acids in length, and together the repeats form a domain that is thought to be involved in interactions with other proteins in forming protein complexes (Peifer *et al.*, 1994). Plakoglobin and β-catenin each have 13 Armadillo repeats. These two proteins share 65% identity, the majority of which is in the Armadillo repeats. It has been proposed that plakoglobin and β-catenin perform the same function in the adherens junction and that cadherin–catenin complexes contain either plakoglobin or β-catenin (Hinck *et al.*, 1994; Näthke *et al.*, 1994; Butz and Kemler, 1994; Aberle *et al.*, 1994). Plakoglobin, but not β-catenin, is also a component of a separate cellular junction, the desmosome (Cowin *et al.*, 1986). Thus, it is likely that the nonidentical N- and C-terminal domains of these two proteins confer upon them the ability to perform functions that differ substantially from one another. In this chapter we concentrate on the similarities between these proteins.

B. Plakoglobin/β-Catenin Domains That Interact with Cadherin

Our laboratories have studied the domains of plakoglobin that promote association with cadherin and α-catenin. We employed the HT1080 cell line. These cells do not express plakoglobin but do express N-cadherin, β-catenin, and α-catenin. They assemble a functional cadherin complex and form adherens junctions. When plakoglobin was transfected into these cells, they assembled N-cadherin complexes containing plakoglobin in addition to those containing β-catenin. We then made deletion mutations of plakoglobin and transfected these into HT1080 cells in order to map domains on plakoglobin responsible for interactions with both N-cadherin and α-catenin. Using both N-terminal and C-terminal deletions, we found three separate domains to be involved in plakoglobin interactions with N-cadherin. These domains are labeled A–C in Fig. 2. Domains A and B are essential for plakoglobin to interact with N-cadherin. Domain C, while not essential, appears to strengthen the cadherin–plakoglobin interaction. C-terminal deletions placed domains B and C within the C-terminal half of plakoglobin. Domain B resides within Armadillo repeats 7 and 8. As we deleted further into this domain from the C-terminus, we gradually lost the ability of the truncated plakoglobin molecules to remain associated with N-cadherin during immunoprecipitation reactions. The immunoprecipitation reactions were washed with a mild buffer containing Tween-20 as the only detergent. Use of a harsher buffer (RIPA buffer, containing NP-40, sodium dodecyl sulfate, and deoxycholate) to wash the immunoprecipitations revealed that domain C serves to strengthen the plakoglobin–cadherin interaction. Domain C spans amino acids 632–727. N-terminal deletions identified a third interaction domain (domain A) between amino acids 192 and 233 within Armadillo repeats 2 and 3. This domain is essential for plakoglobin to interact with N-cadherin. It is difficult to speculate on the importance of Armadillo repeats 4–6 at this time; however, they may

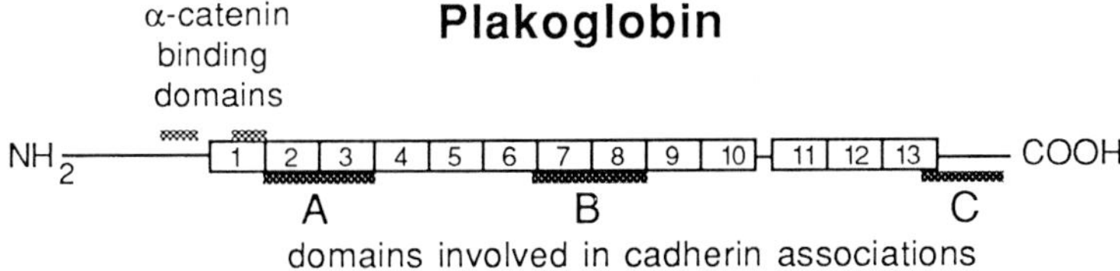

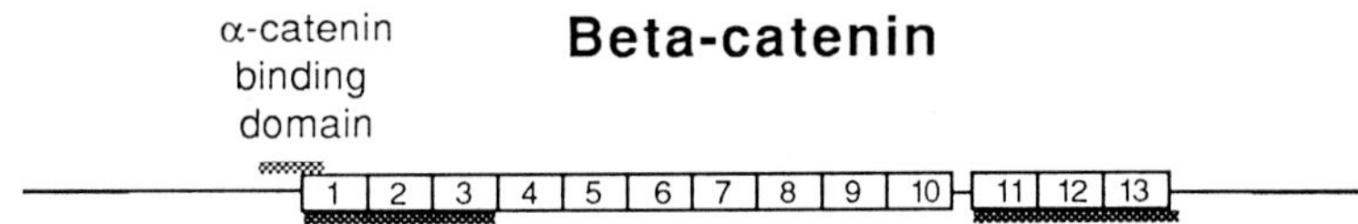

FIGURE 2 Plakoglobin and β-catenin structure. Plakoglobin and β-catenin are drawn with the 13 Armadillo repeats indicated. The domains in plakoglobin that interact with α-catenin are shown above the plakoglobin map and the domains that interact with cadherin are shown below the map. A similar map of β-catenin indicates the domains that interact with cadherin and α-catenin.

also be necessary for plakoglobin to interact with cadherin *in vivo*. Thus we have identified two domains that are essential for plakoglobin interaction with cadherin and one domain that appears to strengthen the interaction.

Birchmeier and co-workers employed similar co-immunoprecipitation experiments using truncated β-catenin transfected along with E-cadherin into Neuro2A cells (Hülsken *et al.*, 1994b). They showed that deleting repeats 1–7 or 8–13 abolished the ability of β-catenin to associate with E-cadherin. Interestingly, they showed that, when β-catenin deletion mutants lacking either repeats 1–7 or 8–13 (neither of these truncated forms of β-catenin binds cadherin) were co-transfected into cells, the two pieces were able to interact with E-cadherin in combination. They also showed that these two mutants interacted with one another independently of their interaction with E-cadherin. This led them to propose a model in which the internal Armadillo repeats interact with the cadherin molecule, with β-catenin wrapping around the cadherin and self-associating through the more N-terminal and C-terminal Armadillo repeats. Our data with plakoglobin support this model. Figure 2 compares the cadherin association sites we identified on plakoglobin with those the Birchmeier laboratory identified on β-catenin. Figure 3 presents a model of the cadherin–catenin complex combining our data with those of the Birchmeier lab, showing β-catenin–plakoglobin in the junction. Although β-catenin and plakoglobin are thought to play homologous roles in the adherens junction, they are not

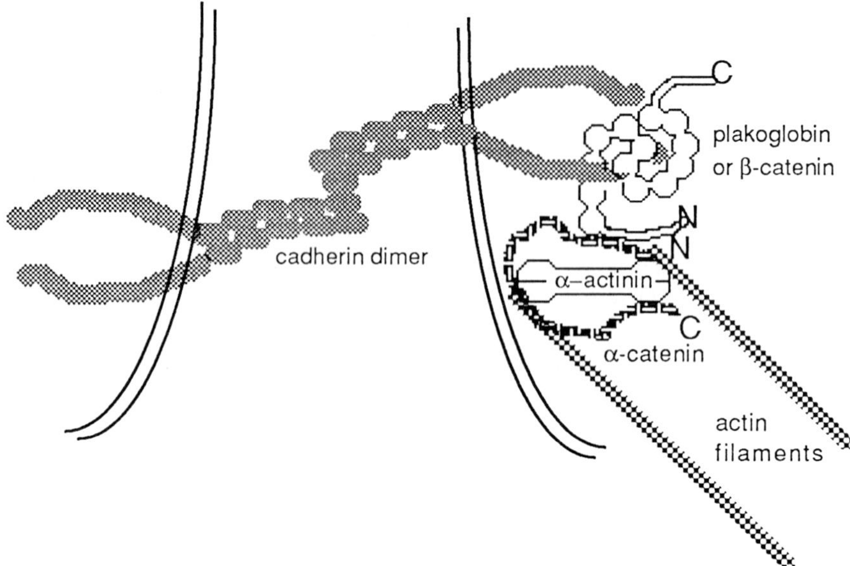

FIGURE 3 Adherens junction structure. The cadherin–catenin complex and its association with the cytoskeleton is depicted. The cadherin molecule is shown as a dimer. To simplify the figure, the cytoplasmic domain of only one partner of the dimer is shown interacting with plakoglobin/β-catenin. The N-terminus of plakoglobin/β-catenin is shown interacting with the N-terminus of α-catenin. α-Catenin is shown interacting with both α-actinin and actin filaments. Key: ▆ cadherin; ▭ β-catenin; ⊟ α-actinin; ▦ α-catenin; ▩ actin.

identical and, therefore, their interactions with α-catenin and cadherin are likely to be very similar but not necessarily identical.

C. Plakoglobin/β-Catenin Domains That Interact with α-Catenin

We used the same strategy described above to map the domains on plakoglobin that interact with α-catenin. Using C-terminal deletions, we found that a region spanning amino acids 133–161 within the first Armadillo repeat of plakoglobin is required for association with α-catenin. N-terminal deletions showed that an additional domain of plakoglobin between amino acids 75 and 113 is also required for interaction with α-catenin. These amino acids reside N-terminal of the first Armadillo repeat. The two domains of plakoglobin that interact with α-catenin are indicated in Fig. 2.

A similar site on α-catenin has been shown to be essential for interaction with α-catenin (Aberle *et al.,* 1994; Hülsken *et al.,* 1994b; Jou *et al.,* 1995;

Rubinfeld *et al.,* 1995). Kemler and co-workers used bacterially synthesized GST-fusion proteins and gel overlay experiments to narrow the site of α-catenin association down to a 30-amino-acid domain within the first Armadillo repeat of β-catenin (Aberle *et al.,* 1994). In support of the biochemical data presented above, a human gastric carcinoma cell line (HSC-39) with a deletion in β-catenin has been identified (Oyama *et al.,* 1994; Kawanishi *et al.,* 1995). This cell line produces β-catenin with an in-frame deletion from amino acids 28 to 134. The deleted β-catenin retains the ability to bind to cadherin but not to α-catenin. The domains of β-catenin identified as sites of α-catenin interaction are indicated in Fig. 2. Thus it appears that plakoglobin and β-catenin use nearly homologous domains to associate with both α-catenin and the cadherins.

The literature reviewed thus far points out how the transmembrane cadherin molecule interacts with plakoglobin/β-catenin and how plakoglobin/β-catenin serve to connect the cadherin to a-catenin. Next we discuss data that implicate α-catenin as the molecule that links the cadherin–catenin complex to the cytoskeleton.

IV. α-CATENIN AND ITS ASSOCIATION WITH PLAKOGLOBIN–β-CATENIN AND THE ACTIN CYTOSKELETON

A. α-Catenin Structure

To date, two α-catenins have been identified, αN-catenin and αE-catenin. In addition, splicing isoforms of both α-catenins exist (Rimm *et al.,* 1994; Uchida *et al.,* 1994). αN-catenin is restricted to nervous tissue and may play an important role in its development (Nagafuchi and Tsukita, 1994), whereas αE-catenin is ubiquitous. Both α-catenins have the ability to associate with both E- and N-cadherin; thus the biological significance of two α-catenins has not been determined. The deduced amino acid sequence of α-catenin revealed that it shares homology with vinculin (Nagafuchi *et al.,* 1991; Herrenknecht *et al.,* 1991), a component of both adherens junctions and focal contacts. Vinculin has been divided into five functional domains: a talin-binding domain, a repetitive domain, a central domain, a proline-rich domain, and a domain that is responsible for self-association. The sequence of α-catenin suggests that it, too, can be divided into five domains and that it shares 26%, 32%, and 34% identity with the first, third, and fifth domains of vinculin, respectively (Nagafuchi *et al.,* 1991). It has been suggested that α-catenin may play a role similar to that of vinculin in connecting the cadherins to the actin cytoskeleton (Nagafuchi *et al.,* 1991; Herrenknecht *et al.,* 1991).

B. α-Catenin Association with Plakoglobin/β-Catenin

To act as a linker molecule between the cadherin complex and the cytoskeleton, α-catenin interacts with plakoglobin/β-catenin and with elements of the cytoskeleton. Nelson and co-workers used the yeast two-hybrid system to show that the N-terminal two-thirds (606 amino acids) of α-catenin is necessary for it to associate with β-catenin (Jou *et al.,* 1995). Our results indicate that, when C-terminal truncations of α-catenin are transfected into mouse NIH 3T3 cells, a deletion mutant retaining only the most N-terminal 220 amino acids retains its ability to co-localize with β-catenin in immunofluorescence light microscopy (M. Nieman, M. J. Wheelock, and K. R. Johnson, unpublished data).

C. α-Catenin as a Linker of Cadherin to the Cytoskeleton

There is compelling evidence that α-catenin is necessary for connecting the cadherin–catenin complex to the actin cytoskeleton. The first clue suggesting that α-catenin was responsible for linking the cadherin–catenin complex to actin came from Kemler's lab (Ozawa *et al.,* 1990b) when they found that α-catenin was required for the cadherin–catenin complex to bind immobilized actin. Further evidence was provided by Takeichi and co-workers and involved studies on a lung carcinoma cell line, PC9 (Hirano *et al.,* 1992; Watabe *et al.,* 1994). These cells express E-cadherin, β-catenin, and plakoglobin but do not express α-catenin. PC9 cells do not aggregate and grow as poorly attached single cells. When PC9 cells were transfected with a cDNA encoding α-catenin, they formed tight aggregates that differentiated into cysts of polarized simple epithelial cells. Since a link between the transmembrane cadherin and the cytoskeleton is essential for cell aggregation, these authors concluded that α-catenin served as the linker protein. Further evidence for the association of α-catenin with the cytoskeleton came from Nagafuchi and Tsukita (Nagafuchi *et al.,* 1994). They constructed a fusion protein between E-cadherin (missing its β-catenin–binding site) and α-catenin. When this fusion protein was transfected into mouse L-cells (which do not express a cadherin and thus do not aggregate) the cells were able to aggregate in a calcium-dependent manner. This suggested that the fusion between E-cadherin and α-catenin could bypass the need for β-catenin. Nagafuchi and Tsukita further showed that a fusion protein of truncated cadherin and the C-terminal half of α-catenin supported full adhesive activity, while a fusion protein of truncated cadherin and the N-terminal half of α-catenin only partially supported cell adhesion. These data suggested that both halves of α-catenin play a role in association with the cytoskeleton but that the C-terminal half is more crucial.

Two other lines of evidence support an association of α-catenin with the actin cytoskeleton: (1) we have shown that, *in vivo*, α-catenin associates directly with α-actinin, which is an actin filament cross-linking protein (Knudsen *et al.*, 1995); and (2) Morrow and co-workers have demonstrated that recombinant αE-catenin binds to and bundles actin filaments in a sedimentation assay (Rimm *et al.*, 1995).

D. α-Catenin Association with α-Actinin

In our studies, fibroblasts were used to begin to unravel the cadherin–cytoskeleton connection. Fibroblasts express N-cadherin, but not E-cadherin, and form a very extensive actin cytoskeleton. They appear to make tenuous cadherin-mediated contacts with one another and form minimal adherens junctions, but do not form tight junctions or desmosomes. Using immunofluorescence light microscopy, we co-localized the cadherin–catenin complex with α-actinin. In agreement with the immunofluorescence studies, α-actinin co-immunoprecipitated with the cadherin–catenin complex in fibroblasts and other cells, suggesting that α-actinin associates directly with the cadherin–catenin complex. We took advantage of the PC9 cells, which do not express α-catenin, to demonstrate that the association of α-actinin with the cadherin complex is dependent upon the presence of α-catenin. In these studies, we showed that α-actinin did not co-immunoprecipitate with the cadherin complex in the absence of α-catenin expression. We further showed that the association of α-catenin with α-actinin was dependent upon the presence of cadherin and independent of actin. These studies led us to propose that the cadherin–catenin complex associates with the actin cytoskeleton through a direct association between α-catenin and α-actinin (Knudsen *et al.*, 1995).

E. α-Catenin Association with Actin Filaments

Recently, Morrow and his collaborators showed that α-catenin binds directly to actin filaments by co-sedimentation assays. In these studies, they added recombinant α-catenin to actin filaments and showed that it bound with an affinity in the micromolar range. Similarly produced β-catenin did not bind to actin filaments unless α-catenin was present. They also showed that α-catenin enhanced the cross-linking of filamentous actin. Electron microscopic data further supported the idea that α-catenin cross-links actin filaments. Two binding sites were identified in α-catenin, one in the C-terminus and one in the N-terminus. Both binding sites were necessary for cross-linking actin filaments (Rimm *et al.,* 1995).

There are several ways α-catenin could mediate actin cross-linking: (1) the C-terminal binding site on α-catenin could interact with one actin filament and the N-terminal site with a different actin filament, thus cross-linking two filaments; (2) α-catenin could mediate actin cross-linking by dimerization; or (3) α-catenin could mediate cross-linking of actin filaments by association with other actin-binding proteins such as α-actinin. Moreover, a combination of these scenarios is possible.

It is presently unclear whether or not α-catenin forms dimers. Data both supporting and refuting this possibility have been presented. Preliminary yeast two-hybrid studies from our laboratory suggest that α-catenin can dimerize (J. Nieset, M. J. Wheelock, and K. R. Johnson, unpublished data); however, there is evidence in the literature that it does not. Birchmeier and colleagues showed that immunoprecipitation of myc-tagged α-catenin with anti-myc antibodies does not co-immunoprecipitate untagged α-catenin (Hülsken *et al.,* 1994b). In addition, our laboratory showed that, when cells express two different cadherins, the two cadherins are immunoprecipitated in separate complexes (Johnson *et al.,* 1993). If α-catenin forms dimers, one might expect to find the two cadherins in the same large complex. However, it should be kept in mind that both Birchmeier's experiments (Hülsken *et al.,* 1994b) and ours (Johnson *et al.,* 1993) used experimental conditions that might be harsh enough to break α-catenin dimers if they exist.

Our data suggesting that α-catenin mediates linkage of the cadherin complex to the cytoskeleton through α-actinin (Knudsen *et al.,* 1995) and those of the Morrow lab suggesting that α-catenin cross-links actin filaments (Rimm *et al.,* 1995) are not inconsistent. It is feasible that both actin and α-actinin interact directly with α-catenin. This would not be surprising considering that α-catenin shares homology with vinculin (Nagafuchi *et al.,* 1991; Herrenknecht *et al.,* 1991), which is known to bind both α-actinin and actin (Belkin and Koteliansky, 1987; Wachsstock *et al.,* 1987; McGregor *et al.,* 1994; Menkel *et al.,* 1994). Moreover, as mentioned earlier, Nagafuchi and Tsukita reported that α-catenin interacts with the cytoskeleton by two different mechanisms, one via the carboxyl half, which results in full cadherin adhesive activity, and the other via the amino half, which does not support full cadherin activity (Nagafuchi *et al.,* 1994).

V. FUTURE DIRECTIONS

There are many questions left to be answered concerning α-catenin and its molecular interactions. These include a more precise mapping of the domain(s) that associate with β-catenin and plakoglobin, and mapping the domain(s) that associate with α-actinin and filamentous actin. Figure 3 presents a model of an adherens junction, taking into consideration the domains of each protein that have been identified to date as protein–protein interaction sites.

Although beyond the scope of this chapter, an equally interesting question still to be answered is how cells regulate associations with their neighbors. It is well established that the classical cadherins are responsible for regulating other junctional complexes in cells, including gap junctions, desmosomes, and tight junctions. For normal tissue integrity, cellular junctional complexes must be tightly regulated. For example, in a tissue such as the intestine, each cell involved in forming the epithelial barrier has a limited life span. As these cells age and undergo apoptosis, they must leave the epithelial sheet without disrupting the barrier function of the tissue. Similar problems are faced by all tissues in which close cell–cell interactions occur. Even more complex are the cell–cell associations that must be broken and formed during development of a complex multicellular organism to ensure proper cell sorting in order to form tissues. These sorting activities are thought to be mediated by cadherins and their associated proteins. Finally, many of the processes that regulate cell–cell interactions may not function properly in disease states characterized by altered cell–cell interactions, such as metastasis. Further studies into the molecular interactions involved in cadherin–catenin–cytoskeleton interactions will increase our understanding of these complex cellular activities, which will eventually lead to a better understanding of disease states resulting from altered cellular interactions.

References

Aberle, H., Butz, S., Stappert, J., Weissig, H., Kemler, R., and Hoschuetzky, H. (1994). Assembly of the cadherin-catenin complex in vitro with recombinant proteins. *J. Cell Sci.* **107,** 3655–3663.

Belkin, A. M., and Koteliansky, V. E. (1987). Interaction of iodinated vinculin, metavinculin and α-actinin with cytoskeletal proteins. *FEBS Lett.* **220,** 291–294.

Blaschuk, O. W., Sullivan, R., David, S., and Pouliot, Y. (1990). Identification of a cadherin cell adhesion recognition sequence. *Dev. Biol.* **139,** 227–229.

Blaschuk, O. W., Munro, S. B., and Farookhi, R. (1995). Cadherins, steroids and cancer. *Endocrine* **3,** 83–89.

Butz, S., and Kemler, R. (1994). Distinct cadherin-catenin complexes in Ca^{++}-dependent cell-cell adhesion. *FEBS Lett.* **355,** 195–200.

Butz, S., Stappert, J., Weissig, H., and Kemler, R. (1992). Plakoglobin and β-catenin: Distinct but closely related. *Science* **257,** 1142–1144.

Buxton, R. S., and Magee, A. I. (1992). Structure and interactions of desmosomal and other cadherins. *Semin. Cell Biol.* **3,** 157–167.

Cepek, K. L., Shaw, S. K., Parker, C. M., Russel, G. J., Morrow, J. S., Rimm, D. L., and Brenner, M. B. (1994). Adhesion between epithelial cells and T lymphocytes mediated by E-cadherin and the $a^E b_7$ integrin. *Nature (London)* **372,** 190–193.

Cowin, P., Kapprell, H.-P., Franke, W. W., Tamkun, J., and Hynes, R. O. (1986). Plakoglobin: A protein common to different kinds of intercellular junctions. *Cell* **46,** 1063–1073.

Fouquet, B., Zimbelmann, R., and Franke, W. W. (1992). Identification of plakoglobin in oocytes and early embryos of *Xenopus laevis:* Maternal expression of a gene encoding a junctional plaque protein. *Differentiation* **51,** 187–194.

Funayama, N., Fagotto, F., McCrea, P., and Gumbiner, B. M. (1995). Embryonic axis induction by the Armadillo repeat domain of β-catenin: Evidence for intracellular signaling. *J. Cell Biol.* **128,** 959–968.

Geiger, B., Ginsberg, D., Salomon, D., and Volberg, T. (1990). The molecular basis for the assembly and modulation of adherens-type junctions. *Cell Diff. Dev.* **32,** 343–353.

Grunwald, G. B. (1993). The structural and functional analysis of cadherin calcium-dependent cell adhesion molecules. *Curr. Opin. Cell Biol.* **5,** 797–805.

Hatta, K., Nose, A., Nagafuchi, A., and Takeichi, M. (1988). Cloning and expression of cDNA encoding a neural calcium-dependent cell adhesion molecule: Its identity in the cadherin gene family. *J. Cell Biol.* **106,** 873–881.

Heasman, J., Crawford, A., Goldstone, K., Garner-Hamrick, P., Gumbiner, B., McCrea, P., Kintner, C., Noro, C. Y., and Wylie, C. (1994). Overexpression of cadherins and underexpression of β-catenin inhibit dorsal mesoderm induction in early *Xenopus* embryos. *Cell* **79,** 791–803.

Herrenknecht, K., Ozawa, M., Eckerskorn, C., Lottspeich, F., Lenter, M., and Kemler, R. (1991). The uvomorulin-anchorage protein alpha-catenin is a vinculin homologue. *Proc. Natl. Acad. Sci. U.S.A.* **88,** 9156–9160.

Hinck, L., Näthke, I. S., Papkoff, J., and Nelson, W. J. (1994). Dynamics of cadherin/catenin complex formation: Novel protein interactions and pathways of complex assembly. *J. Cell Biol.* **125,** 1327–1340.

Hirano, S., Kimoto, N., Shimoyama, Y., Hirohashi, S., and Takeichi, M. (1992). Identification of a neural a-catenin as a key regulator of cadherin function and multicellular organization. *Cell* **70,** 293–301.

Hülsken, J., Behrens, J., and Birchmeier, W. (1994a). Tumor-suppressor gene products in cell contacts: The cadherin-APC-*armadillo* connection. *Curr. Opin. Cell Biol.* **6,** 711–716.

Hülsken, J., Birchmeier, W., and Behrens, J. (1994b). E-cadherin and APC compete for the interaction with β-catenin and the cytoskeleton. *J. Cell Biol.* **127,** 2061–2069.

Inuzuka, H., Miyatani, S., and Takeichi, M. (1991). R-cadherin: A novel Ca^{2+}-dependent cell-cell adhesion molecule expressed in the retina. *Neuron* **7,** 69–79.

Johnson, K. R., Lewis, J. E., Li, D., Wahl, J., Soler, A. P., Knudsen, K. A., and Wheelock, M. J. (1993). P- and E-cadherin are in separate complexes in cells expressing both cadherins. *Exp. Cell Res.* **207,** 252–260.

Jou, T.-S., Stewart, D. B., Stappert, J., Nelson, W. J., and Marrs, J. A. (1995). Genetic and biochemical dissection of protein linkages in the cadherin-catenin complex. *Proc. Natl. Acad. Sci. U.S.A.* **92,** 5067–5071.

Karnovsky, A., and Klymkowsky, M. W. (1995). Anterior axis duplication in *Xenopus* induced by the over-expression of the cadherin-binding protein plakoglobin. *Proc. Natl. Acad. Sci. U.S.A.* **92,** 4522–4526.

Kawanishi, J., Kato, J., Sasaki, K., Fujii, S., Watanabe, N., and Niitsu, Y. (1995). Loss of E-cadherin-dependent cell-cell adhesion due to mutation of the β-catenin gene in a human cancer cell line, HSC-39. *Mol. Cell. Biol.* **15,** 1175–1181.

Knudsen, K. A., and Wheelock, M. J. (1992). Plakoglobin, or an 83-kD homologue distinct from β-catenin, interacts with E-cadherin and N-cadherin. *J. Cell Biol.* **118,** 671–679.

Knudsen, K. A., Soler, A. P., Johnson, K. R., and Wheelock, M. J. (1995). Interaction of alpha-actinin with the N-cadherin/catenin cell-cell adhesion complex via alpha-catenin. *J. Cell Biol.* **130,** 67–77.

Koch, P. J., and Franke, W. W. (1994). Desmosomal cadherins: Another growing multigene family of adhesion molecules. *Curr. Opin. Cell Biol.* **6,** 682–687.

McCrea, P. D., and Gumbiner, B. M. (1991). Purification of a 92-kDa cytoplasmic protein tightly associated with the cell-cell adhesion molecule E-cadherin (uvomorulin): Characterization and extractability of the protein complex from the cell cytostructure. *J. Biol. Chem.* **266,** 4514–4520.

McCrea, P. D., Turck, C. W., and Gumbiner, B. (1991). A homologue of the Armadillo protein in *Drosophila* (plakoglobin) associates with E-cadherin. *Science* **254,** 1359–1361.

McGregor, A., Blanchard, A. D., Rowe, A. J., and Critchley, D. R. (1994). Identification of the vinculin-binding site in the cytoskeletal protein α-actinin. *Biochem. J.* **301,** 225–233.

Menkel, A. R., Kroemker, M., Bubeck, P., Ronsiek, M., Nikoliai, G., and Jockusch, B. M. (1994). Characterization of an F-actin-binding domain in the cytoskeletal protein vinculin. *J. Cell Biol.* **126,** 1231–1240.

Murphy-Erdosh, C., Yoshida, C. K., Paradies, N., and Reichardt, L. F. (1995). The cadherin-binding specificities of B-cadherin and LCAM. *J. Cell Biol.* **129,** 1379–1390.

Nagafuchi, A., and Takeichi, M. (1988). Cell binding function of E-cadherin is regulated by the cytoplasmic domain. *EMBO J.* **7,** 3679–3684.

Nagafuchi, A., and Tsukita, S. (1994). The loss of the expression of α-catenin, the 102 kD cadherin associated protein, in central nervous tissue during development. *Dev. Growth Differ.* **36,** 59–71.

Nagafuchi, A., Takeichi, M., and Tsukita, S. (1991). The 102 kd cadherin-associated protein: Similarity to vinculin and posttranscriptional regulation of expression. *Cell* **65,** 849–857.

Nagafuchi, A., Ishihara, S., and Tsukita, S. (1994). The roles of catenins in the cadherin-mediated cell adhesion: Functional analysis of E-cadherin-α-catenin fusion molecules. *J. Cell Biol.* **127,** 235–245.

Näthke, I. S., Hinck, L., Swedlow, J. R., Papkoff, J., and Nelson, W. J. (1994). Defining interactions and distributions of cadherin and catenin complexes in polarized epithelial cells. *J. Cell Biol.* **125,** 1341–1352.

Nose, A., Tsuji, K., and Takeichi, M. (1990). Localization of specificity determining sites in cadherin cell adhesion molecules. *Cell* **61,** 147–155.

Oyama, T., Kanai, Y., Ochiai, A., Akimoto, S., Oda, T., Yanagihara, K., Nagafuchi, A., Tsukita, S., Shibamoto, S., Ito, F., Takeichi, M., Matsuda, H., and Hirohashi, S. (1994). A truncated β-catenin disrupts the interaction between E-cadherin and α-catenin: A cause of loss of intercellular adhesiveness in human cancer cell lines. *Cancer Res.* **54,** 6282–6287.

Ozawa, M., Baribault, H., and Kemler, R. (1989). The cytoplasmic domain of the cell adhesion molecule uvomorulin associates with three independent proteins structurally related in different species. *EMBO J.* **8,** 1711–1717.

Ozawa, M., Engel, J., and Kemler, R. (1990a). Single amino acid substutitions in one Ca^{2+} binding site of uvomorulin abolish the adhesive function. *Cell* **63,** 1033–1038.

Ozawa, M., Ringwald, M., and Kemler, R. (1990b). Uvomorulin-catenin complex formation is regulated by a specific domain in the cytoplasmic region of the cell adhesion molecule. *Proc. Natl. Acad. Sci. U.S.A.* **87,** 4246–4250.

Overduin, M., Harvey, T. S., Bagby, S., Tong, K. I., Yau, P., Takeichi, M., and Ikura, M. (1995). Solution structure of the epithelial cadherin domain responsible for selective cell adhesion. *Science* **267,** 386–389.

Peifer, M. (1993). The product of the *Drosophila* segment polarity gene *armadillo* is part of a multi-protein complex resembling the vertebrate adherens junction. *J. Cell Sci.* **105,** 993–1000.

Peifer, M. (1995). Cell adhesion and signal transduction: The *armadillo* connection. *Trends Cell Biol.* **5,** 224–229.

Peifer, M., and Weischaus, E. (1990). The segment polarity gene *armadillo* encodes a functionally modular protein that is the *Drosophila* homolog of plakoglobin. *Cell* **63,** 1167–1176.

Peifer, M., McCrea, P. D., Green, K. J., Weischaus, E., and Gumbiner, B. M. (1992). The vertebrate adhesive junction proteins β-catenin and plakoglobin and the *Drosophila* segment polarity gene *armadillo* form a multigene family with similar properties. *J. Cell Biol.* **118,** 681–691.

Peifer, M., Berg, S., and Reynolds, A. B. (1994). A repeating amino acid motif shared by proteins with diverse cellular roles. *Cell* **76,** 789–791.

Pokutta, S., Herrenknecht, K., Kemler, R., and Engel, J. (1994). Conformational changes of the recombinant extracellular domain of E-cadherin upon calcium binding. *Eur. J. Biochem.* **223,** 1019–1026.

Riggleman, B., Wieschaus, E., and Schedl, P. (1989). Molecular analysis of the *armadillo* locus: Uniformly distributed transcripts and a protein with novel internal repeats are associated with a *Drosophila* segment polarity gene. *Genes Dev.* **3,** 96–113.

Rimm, D. L., Kebriaei, P., and Morrow, J. S. (1994). Molecular cloning reveals alternative splice forms of human a(E)-catenin. *Biochem. Biophys. Res. Commun.* **203,** 1691–1699.

Rimm, D. L., Koslov, E. R., Kebriaei, P., Cianci, C. D., and Morrow, J. S. (1995). α_1(E)-catenin is a novel actin binding and bundling protein mediating the attachment of F-actin to the membrane adhesive complex. *Proc. Natl. Acad. Sci. U.S.A.* **92,** 8813–8817.

Rubinfeld, B., Souza, B., Albert, I., Munemitsu, S., and Polakis, P. (1995). The APC protein and E-cadherin form similar but independent complexes with α-catenin, β-catenin and plakoglobin. *J. Biol. Chem.* **270,** 5549–5555.

Sacco, P. A., McGranahan, T. M., Wheelock, M. J., and Johnson, K. R. (1995). Identification of plakoglobin domains required for association with N-cadherin and α-catenin. *J. Biol. Chem.* **270,** 20201–20205.

Sano, K., Tanihara, H., Heimark, R. L., Obata, S., Davidson, M., St. John, T., Taketani, S., and Suzuki, S. (1993). Protocadherins: A large family of cadherin-related molecules in central nervous system. *EMBO J.* **12,** 2249–2256.

Shapiro, L., Fannon, A. M., Kwong, P. D., Thompson, A., Lehmann, M. S., Grübel, G., Legrand, J.-F., Als-Nielson, J., Colman, D. R., and Hendrickson, W. A. (1995a). Structural basis of cell-cell adhesion by cadherins. *Nature* **347,** 327–337.

Shapiro, L., Kwong, P. D., Fannon, A. M., Colman, D. R., and Hendrickson, W. A. (1995b). Considerations on the folding topology and evolutionary origin of cadherin domains. *Proc. Natl. Acad. Sci. U.S.A.* **92,** 6793–6797.

Stappert, J., and Kemler, R. (1994). A short core region of E-cadherin is essential for catenin binding and is highly phosphorylated. *Cell Adhes. Commun.* **2,** 319–327.

Takeichi, M. (1987). Cadherins: A molecular family essential for selective cell-cell adhesion and animal morphogenesis. *Trends Genet.* **3,** 213–217.

Takeichi, M. (1990). Cadherins: A molecular family important in selective cell-cell adhesion. *Annu. Rev. Biochem.* **59,** 237–252.

Takeichi, M. (1991). Cadherin cell adhesion receptors as a morphogenetic regulator. *Science* **251,** 1451–1455.

Takeichi, M. (1993). Cadherins in cancer: Implications for invasion and metastasis. *Curr. Opin. Cell Biol.* **5,** 806–811.

Uchida, N., Shimamura, K., Miyatani, S., Copeland, N. G., Gilbert, D. J., Jenkins, N. A., and Takeichi, M. (1994). Mouse αN-catenin: Two isoforms, specific expression in the nervous system, and chromosomal localization of the gene. *Dev. Biol.* **163,** 75–85.

Volk, T., Cohen, O., and Geiger, B. (1987). Formation of heterotypic adherens-type junctions between L-CAM-containing liver cells and A-CAM containing lens cells. *Cell* **50,** 987–994.

Wachsstock, D. H., Wilkins, J. A., and Lin, S. (1987). Specific interaction of vinculin with α-actinin. *Biochem. Biophys. Res. Commun.* **146,** 554–560.

Watabe, M., Nagafuchi, A., Tsukita, S., and Takeichi, M. (1994). Induction of polarized cell-cell association and retardation of growth by activation of the E-cadherin-catenin adhesion system in a dispersed carcinoma line. *J. Cell Biol.* **127,** 247–256.

Wheelock, M. J., and Knudsen, K. A. (1991). N-cadherin-associated proteins in chicken muscle. *Differentiation* **46,** 35–42.

CHAPTER 10

The Desmosome: A Component System for Adhesion and Intermediate Filament Attachment

Andrew P. Kowalczyk and Kathleen J. Green
Departments of Pathology and Dermatology and the R. H. Lurie Cancer Center, Northwestern University Medical School, Chicago, Illinois 60611

"There is no single building block—there are only complexes of complex systems."

R. Buckminster Fuller, *Critical Path* (1981)

I. INTRODUCTION

A major function of adhesive cell–cell and cell–matrix junctions is to couple the forces of adhesion to the cytoskeleton. In addition to this structural role, it is becoming increasingly apparent that these macromolecular complexes serve to integrate mechanical and chemical signaling pathways. The purpose of this chapter is to review what is currently known about the molecular components of the desmosome, an intercellular junction that mediates cell–cell adhesion and couples this adhesion to the intermediate filament (IF) cytoskeleton. In addition, emerging information on desmosomes as targets of intracellular signaling pathways and as sites of signal initiation is discussed.

II. DESMOSOME STRUCTURE AND MOLECULAR COMPOSITION

A. Desmosome Structure

A high degree of structural analogy is apparent when desmosomes are compared to other adhesive junctions, such as adherens junctions, focal contacts, and hemidesmosomes. In each of these structures, cytoskeletal networks are anchored to transmembrane glycoproteins through a complex of cytoplasmic proteins. Ultrastructurally, desmosomes appear as pairs of highly organized, electron-dense discs that sandwich cell membranes at sites of close intercellular contact (Fig. 1) (see also Farquhar and Palade, 1963; Staehelin, 1974; Green and Stappenbeck, 1994; Collins and Garrod, 1994). The cytoplasmic "plaque" consists of an electron-dense outer plaque subjacent to the plasma membrane and a less dense fibrillar inner plaque through which IFs appear to loop. The "core" region of the desmosome comprises the plasma membranes of the opposing cells and an electron-dense midline, or central dense stratum. The molecular components of the desmosome and the protein interactions that are known to occur in these structures are discussed next.

B. Molecular Components of the Desmosome

Isolated fractions of epidermal extracts enriched for desmosomes contain a number of prominent polypeptides (Skerrow and Matoltsy, 1974; Schwarz *et al.,* 1990). The cDNAs encoding these proteins have been sequenced and many of the structural properties of the proteins have been analyzed. The most abundant components of the desmosomal plaque are the desmoplakins, I and II. Based on sequence analysis of cDNA clones, desmoplakins I and II are predicted to be large (332- and 260-kDa, respectively) homodimeric proteins comprising two globular ends joined by an α-helical coiled-coil rod domain (Green *et al.,* 1990). Alternative splicing of des-

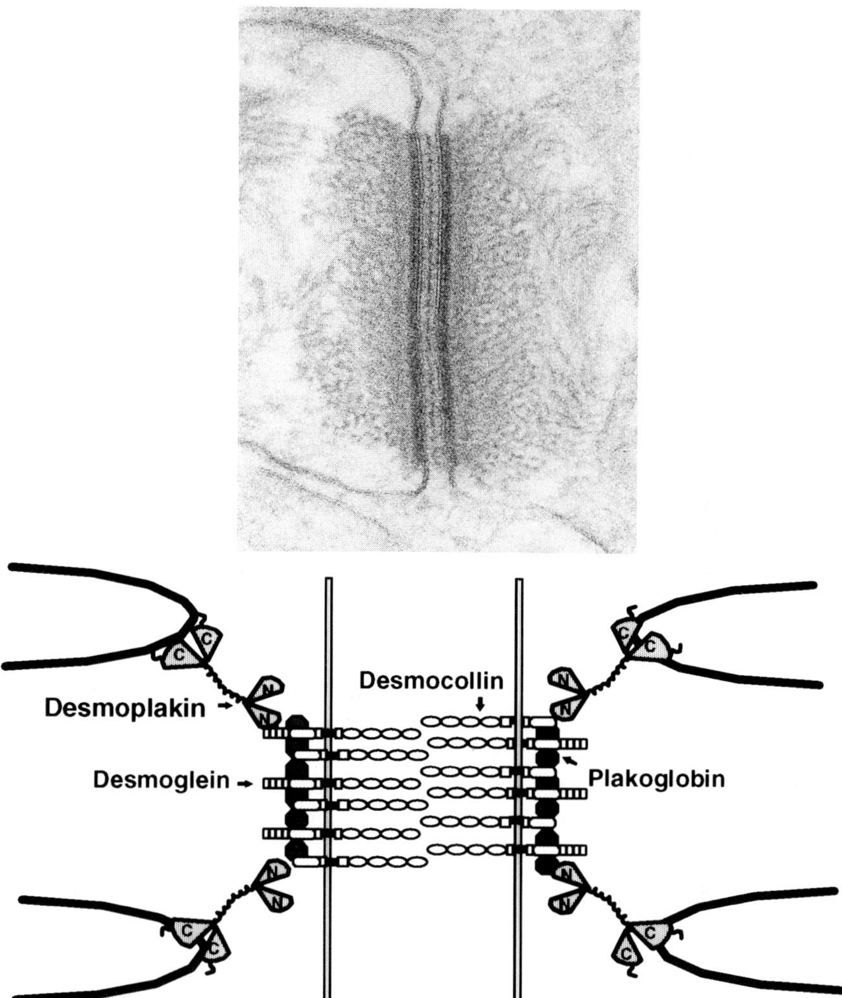

FIGURE 1 Electron micrograph and schematic representation of a desmosome. (Top) On the ultrastructural level, desmosomes consist of mirror-image tripartite plaques that sandwich the opposed lipid bilayers of adjacent cells. The plasma membranes are separated by an approximately 30-nm space filled with electron-dense material that forms the central dense stratum of the desmosomal core. (Bottom) The major molecular components shown schematically. The desmosomal cadherins and the plasma membranes of the adjacent cells are located in the core region while plakoglobin is present in the outer plaque subjacent to the plasma membrane. IFs are shown to loop through the inner plaque and interact with desmoplakin, which may link the IFs to the plakoglobin–cadherin complex. (Refer to text for a detailed discussion of possible interactions between individual components and for a description of other molecules that may contribute to desmosome assembly and function.)

moplakin RNA is thought to result in the generation of the two desmoplakin variants, with the central rod domain being shorter in desmoplakin II (Virata *et al.,* 1992). Desmoplakins are now considered to be part of a family of proteins that includes the IF-associated protein plectin and the 230-kDa bullous pemphigoid antigen (BP230), which is found in the plaque region of hemidesmosomes (Green *et al.,* 1992).

Although desmosomes associate with the IF cytoskeleton and adherens junctions are linked to actin filaments, a plaque protein called plakoglobin is common to both of these adhesive intercellular junctions (Cowin *et al.,* 1986). Plakoglobin is a member of a growing family of proteins that includes the adherens junction protein β-catenin, the *Drosophila* segment polarity protein Armadillo, the protein product of the tumor suppressor gene APC, and the cadherin-associated protein p120 (Peifer *et al.,* 1994). The defining structural feature of these proteins is a 42-amino-acid repeated motif termed the Arm repeat, after the *Drosophila* Armadillo protein. Although the precise function of the repeats in each protein may be different, the Arm motif is predicted to mediate protein–protein interactions. The desmosomal plaque protein band 6, or plakophilin, has also been identified as a member of the plakoglobin–Armadillo family (Hatzfeld *et al.,* 1994; Heid *et al.,* 1994). This protein was originally thought to be restricted to stratified epithelium, but further analysis of cultured epithelial cells indicates that it is more widely expressed. Plakophilin may play a role in linking IFs to desmosomes, as it was demonstrated using *in vitro* binding assays to associate with cytokeratins (Kapprell *et al.,* 1988; Hatzfeld *et al.,* 1994; Heid *et al.,* 1994). However, the protein appeared to be predominantly soluble in cultured epithelial cells, and further studies will be required to determine its role in desmosome assembly and IF attachment (Heid *et al.,* 1994).

The two classes of desmosomal glycoproteins, the desmocollins and desmogleins, are members of the cadherin family of cell adhesion molecules and are thought to be the major transmembrane adhesive components of the desmosome (Koch *et al.,* 1990; Nilles *et al.,* 1991; Wheeler *et al.,* 1991; Collins *et al.,* 1991; Parker *et al.,* 1991; Holton *et al.,* 1990; Mechanic *et al.,* 1991). Like other members of the cadherin superfamily, the amino-terminal extracellular domain of the desmosomal cadherins comprises four copies of a 110-amino-acid motif. The transmembrane domain is flanked by a loosely conserved extracellular and intracellular anchor region. The intracellular anchor is followed by a cytoplasmic tail that includes a highly conserved intracellular cadherin segment. Desmogleins are unusual in that they exhibit an extended cytoplasmic tail that includes variable numbers of a 29 $\pm$ 1-amino-acid repeated motif specific to members of the desmoglein subfamily.

A number of other proteins that localize to desmosomes have been described. A 140-kDa protein localized to the innermost region of the plaque was identified by Ouyang and Sugrue (1992) and may be related to a 140-kDa lamin-B–like protein identified by Cartaud and colleagues (1990). Human and bovine homologues of a 240-kDa calmodulin-binding protein, called keratocalmin and desmocalmin, respectively, are expressed in stratified epithelium and are localized to desmosomes (Fairley *et al.*, 1991; Tsukita, 1985). A 20-kDa GPI-linked protein implicated in keratinocyte cell–cell adhesion, called the E48 antigen, was cloned and reported to be the human homologue of the 20-kDa protein observed in bovine desmosome preparations (Brakenhoff *et al.*, 1995). Other proteins, such as IFAP 300 and plectin, that may also play a role in desmosome function are discussed later.

III. INTERMEDIATE FILAMENT ATTACHMENT TO THE DESMOSOMAL PLAQUE

A. Desmoplakin

The plaque region of the desmosome is thought to link IFs to the cell surface at sites of desmosome-mediated cell–cell adhesion. Candidates for mediating such a function would presumably be localized to the inner plaque region of the desmosome and possess the ability to interact with both the IF cytoskeleton and components of the outer plaque. One protein that appears to meet these requirements is desmoplakin. When cDNA constructs encoding the carboxyl-terminal region of desmoplakin (DP.CT) were expressed transiently in cultured cells, a striking alignment of the protein along IFs was observed (Stappenbeck and Green, 1992). These results indicated that the carboxyl-terminal region of desmoplakin could associate with IF networks. In addition, high levels of DP.CT expression resulted in the disruption of the cellular IF network and the accumulation of electron-dense perinuclear aggregates containing both IF and desmoplakin carboxyl-terminal polypeptides. A construct encoding the desmoplakin carboxyl terminus and rod domains (DP.ΔN) also disrupted the endogenous IF network. However, in this case, the resulting aggregates contained a meshwork of fine-caliber filamentous material that bore a striking resemblance to the inner desmosomal plaque. These results raise the possibility that IFs are not attached to desmosomes as full-caliber 10-nm filaments, but instead are interwoven into the plaque by co-assembling with desmoplakin into a distinct higher order structure.

A study by Kouklis and colleagues (1994) using *in vitro* assays demonstrated that the desmoplakin carboxyl terminus can interact directly with the amino-terminal head region of type II epidermal keratins (K1, K2, K5, K6). Interactions between the desmoplakin carboxyl terminus and the simple epithelial type II keratins, vimentin, or the type I keratins were not detected. However, desmoplakin does interact with IF networks composed of simple epithelial keratins in cultured cells (Stappenbeck and Green, 1992). Preliminary experiments using the yeast two-hybrid system suggest that desmoplakin can interact directly with both simple keratins and vimentin (Meng *et al.*, 1995). It is possible that interactions between desmoplakin and the simple epithelial keratins were not detected *in vitro* due to the stringency of the conditions used in these assays. In addition, accessory proteins that are present in cells but absent from purified protein preparations may augment desmoplakin–IF interactions *in vivo*. However, there also appear to be differences in desmoplakin interactions with the keratins and vimentin. For example, removal of a 68-residue carboxyl-terminal region of desmoplakin inhibits association with keratin but not vimentin networks (Stappenbeck *et al.*, 1993). In addition, desmoplakin interactions with epidermal keratins *in vitro* appear to involve an amino acid sequence, GSRS, that is present in the epidermal type II keratins but not in keratin-8 (Kouklis *et al.*, 1994). These data suggest that a number of distinct mechanisms may mediate desmoplakin interactions with various IF proteins.

In order for a putative linker protein to mediate IF attachment to the desmosome, the protein must interact both with the IF network and with other desmosomal components, such as the desmosomal cadherins or the cadherin-associated protein plakoglobin. Previous studies from our laboratory indicated that desmoplakin polypeptides that included amino-terminal sequences localized to desmosomes when transiently expressed in cultured epithelial cells (Stappenbeck *et al.*, 1993). Deletion constructs lacking the first 194 amino acids of the amino terminus failed to localize to desmosomes. These findings provided the basis for a model predicting that desmoplakin is a functionally modular protein in which the carboxyl-terminal domain of desmoplakin interacts with IF networks and the amino-terminal domain of the protein associates with components of the desmosomal outer plaque. In a study from our laboratory, an amino-terminal polypeptide of desmoplakin lacking the carboxyl terminus was stably expressed in A431 epithelial cells and was found to displace endogenous desmoplakin from the plasma membrane. Interestingly, IF bundles failed to attach to the cell membrane in areas where endogenous desmoplakin was displaced, suggesting a central role for desmoplakin in the attachment of IFs to desmosomes (Bornslaeger *et al.*, 1996). These findings suggest that desmoplakin fulfills the basic requirements for a protein that could link IFs to the desmosome.

B. Other Proteins Involved in IF–Desmosome Interactions

IFAP 300 and plectin are among other candidate molecules that may play a role in linking IFs to desmosomes. In epithelial cells, IFAP 300 is concentrated at desmosomes and hemidesmosomes and can be detected in protein preparations enriched for these structures. IFAP 300 is localized deep in the desmosome plaque and binds to keratin IFs *in vitro,* suggesting that it may play a role in attaching IFs to desmosomes (Skalli *et al.,* 1994). Plectin is an IF-binding protein and, similar to desmoplakin, polypeptides containing the carboxyl terminus of plectin align along IFs and cause disruption of filament networks when transiently expressed in cells (Wiche *et al.,* 1993). However, because it also interacts with α-spectrin and fodrin, and thus exhibits a broad range of cytoskeletal interactions, plectin has been proposed to be a universal linking protein (Foisner and Wiche, 1991). It is possible that desmoplakin, due to its proximity to the desmosomal core, may function to recruit IF specifically to the desmosome plaque, whereas IFAP 300 and plectin may play more generalized roles in the organization of IF filaments near the plaque region of both desmosomes and hemidesmosomes.

IV. DESMOSOMAL CADHERINS

A. Desmosomal Cadherin Genes and Expression Patterns

While the plaque region of the desmosome mediates attachment of IF to the cell surface, the core region of the desmosome is thought to mediate cell–cell adhesion. The core proteins likely to be responsible for this adhesion are the desmosomal cadherins, the desmogleins and desmocollins. To date, three desmoglein and three desmocollin genes have been identified (Buxton *et al.,* 1994) and a common nomenclature has been established for both the human and bovine forms of the desmosomal cadherins (Buxton *et al.,* 1993). In addition to each of the desmosomal cadherin genes that have been identified, each desmocollin gene generates two alternatively spliced mRNA transcripts to yield an "a" and a "b" form. These splice variants differ only in the cytoplasmic domains, with the shorter b form containing 11 amino acids not included in the a form. The functions of these different forms of desmocollin are currently unknown.

The desmosomal cadherins exhibit tissue- and differentiation-specific patterns of expression. Recent evidence indicates that the desmosomal cadherins Dsg2 (Schafer *et al.,* 1994) and Dsc2 (Theis *et al.,* 1993; Nuber *et al.,* 1995) are expressed in all desmosome-bearing tissues that have been examined. Dsg3 is expressed predominantly in the suprabasal layer of the epidermis, whereas Dsc3 expression is detected throughout the living layers of the epidermis (Arnemann *et al.*, 1993; Amagai *et al.*, 1995; Collins *et al.*, 1991; King *et al.*, 1995; Legan *et al.*, 1994; Yue *et al.*, 1995). In contrast, Dsg1 and Dsc1 are expressed primarily in the granular layer of the epidermis (Arnemann *et al.*, 1993; Amagai *et al.*, 1996; King *et al.*, 1995; Theis *et al.*, 1993). Interestingly, the desmoglein genes are ordered in a closely linked tandem array on chromosome 18 that mirrors their expression pattern in the epidermis (Simrak *et al.*, 1995). Although the large number of desmosomal cadherin isoforms has hampered homology comparisons between bovine and human sequences, it is now generally agreed that what was originally termed human Dsc3 (Buxton *et al.*, 1993) should be referred to as Dsc2 (Yue *et al.*, 1995; Nuber *et al.*, 1995). In addition, the most recently identified desmosomal cadherin, also referred to as HT-CP, is most homologous to bovine Dsc3 (Kawamura *et al.*, 1994; King *et al.*, 1995). The current nomenclature and tissue distribution of the human desmosomal cadherins is summarized in Table I. (For information on the bovine desmosomal cadherin genes, the reader is referred to Buxton *et al.*, 1994; Collins *et al.*, 1991; Koch *et al.*, 1990,1991b,1992; Mechanic *et al.*, 1991; Legan *et al.*, 1994; Goodwin *et al.*, 1990.)

TABLE I

Nomenclature and Tissue Distribution of Human Desmosomal Cadherins

Nomenclature	Synonyms (references)	Tissue distribution (references)
Dsc1a,b	DgIV,V (Theis *et al.*, 1993; King *et al.*, 1993)	Upper layers of epidermis (Arnemann *et al.*, 1993; Theis *et al.*, 1993; King *et al.*, 1995; Nuber *et al.*, 1995)
Dsc2a,b	DgII,III (Parker *et al.*, 1991)	Widespread in simple epithelium and epidermis (Arnemann *et al.*, 1993; Theis *et al.*, 1993; Nuber *et al.*, 1995)
Dsc3a,b	HT-CP (Kawamura *et al.*, 1994; King *et al.*, 1995)	Basal and suprabasal layers of epidermis (King *et al.*, 1995)
Dsg1	DGI (Wheeler *et al.*, 1991; Nilles *et al.*, 1991)	Upper layers of epidermis (Arnemann *et al.*, 1993; Amagai *et al.*, 1996)
Dsg2	HDGC (Koch *et al.*, 1991a; Schafer *et al.*, 1994)	Widespread in simple epithelium and lower layers of epidermis (Shafer *et al.*, 1994)
Dsg3	PVA (Amagai *et al.*, 1991)	Suprabasal layers of epidermis (Arnemann *et al.*, 1993; Amagai *et al.*, 1996)

B. The Role of Desmosomal Cadherins in Adhesion

Although most cadherins identified to date function as calcium-dependent adhesion molecules and engage in homophilic adhesion, the role of the desmosomal cadherins in adhesion is not well understood. A chimeric molecule consisting of the Dsg3 extracellular domain and the E-cadherin cytoplasmic domain was found to engage in weak homophilic adhesion (Amagai *et al.*, 1994b). Other evidence suggesting that desmosomal cadherins play a role in desmosome-mediated adhesion is derived from antibody inhibition experiments in which Fab fragments directed against desmocollins were found to inhibit desmosome formation in cultured cells (Cowin *et al.*, 1984). In addition, patients suffering from the autoimmune blistering diseases pemphigus vulgaris and pemphigus foliaceus exhibit circulating antibodies directed against Dsg3 and Dsg1, respectively (Stanley, 1995; Amagai, 1994). These diseases are characterized by severe epidermal blistering. In the case of pemphigus vulgaris, blistering occurs deep in the epidermis, at the interface between the basal and spinous layers, whereas pemphigus foliaceus patients exhibit blisters in the granular layer of the epidermis. The autoantibodies present in pemphigus vulgaris patients recognize primarily Dsg3 and pemphigus foliaceus antibodies react with Dsg1 (Stanley *et al.*, 1984,1986; Amagai *et al.*, 1991), although both types of pemphigus autoantibodies recognize amino-terminal epitopes of the desmogleins (Dmochowski *et al.*, 1994; Kowalczyk *et al.*, 1995; Emery *et al.*, 1995). Furthermore, pemphigus vulgaris and foliaceus antibodies that recognize the extracellular region of Dsg3 or Dsg1, respectively, cause epidermal blistering in a mouse model of pemphigus (Amagai *et al.*, 1992,1994a,1995). Although it remains to be demonstrated directly that the desmosomal cadherins are cell adhesion molecules, the results from studies of the pemphigus diseases strongly suggest that the desmosomal cadherins are important for cell–cell adhesion and that this adhesion is critical to the structural integrity of the epidermis.

C. Cytoplasmic Interactions of the Desmosomal Cadherins

The transmembrane domains and the cytoplasmic tails of the desmosomal cadherins provide the primary link between the extracellular and intracellular regions of the desmosome. It is important, therefore, to understand the interactions that occur between the cadherin cytoplasmic domains and desmosomal plaque components. Using pemphigus sera to immunoprecipitate extracts from human epidermis, Stanley and colleagues demonstrated that plakoglobin associates with both Dsg1 and Dsg3 (Korman *et al.*, 1989). Recent studies have demonstrated that desmocollin also interacts with plakoglobin (Kowalczyk *et al.*, 1994; Troyanovsky *et al.*, 1994b). Plakoglobin appears to bind to the conserved intracellular cadherin segment of both desmogleins and desmocollins (Mathur *et al.*, 1994; Roh and Stanley, 1995a; Troyanovsky *et al.*, 1994a,b). In the case of desmoglein, this binding is now known to be the result of a direct interaction (Mathur *et al.*, 1994; Roh and Stanley, 1995a).

In order to determine the function of the cadherin cytoplasmic tails in desmosome assembly, Troyanovsky and colleagues (1993) expressed chimeric molecules consisting of the gap junction protein connexin 32 and the cytoplasmic tail of either desmoglein or desmocollin. Expression of these chimeric proteins in cultured epithelial cells resulted in the formation of functional gap junction structures. In addition, the desmocollin connexin chimera recruited plakoglobin, desmoplakin, and IFs to the cell surface. Surprisingly, the connexin chimeras containing the desmoglein cytoplasmic domain failed to recruit IFs and, instead, disrupted the assembly of endogenous desmosomes. The underlying basis for the differences in behavior of these chimeric molecules is unknown, but it may reflect differences in the structure of the cytoplasmic tails of the cadherins or differential abilities to interact with cytoplasmic plaque components of the desmosome.

In addition to the plakoglobin-binding domain of the desmosomal cadherins, another region of the desmosomal cadherin tails that appears to be functionally important is the intracellular anchor (IA) domain, a cytoplasmic region juxtaposed to the plasma membrane. When linked to the gap junction protein connexin 32, the IA domain of desmocollin was shown to play a role in recruiting desmoplakin to the plasma membrane (Troyanovsky *et al.*, 1994b). In addition, a chimeric molecule containing the IA domain of Dsg3 and the extracellular domain of E-cadherin was found to function in adhesion (Roh and Stanley, 1995b). This result was surprising since the chimeric protein lacked the plakoglobin-binding domain. A number of previous reports indicated that E-cadherin required interaction with the catenins and the actin cytoskeleton in order to function in adhesion (Ozawa *et al.*, 1990; Hirano *et al.*, 1992). It remains to be determined how the desmoglein IA domain supports the adhesive function of the E-cadherin extracellular domain without binding to the catenins. However, these studies suggest that the IA domains of both the desmogleins and desmocollins are functionally important in desmosome assembly and possibly adhesion.

V. DESMOSOME ASSEMBLY AND REGULATION

A. The Calcium Switch Model of Desmosome Assembly

Desmosome assembly is induced when cells are transferred from low to normal levels of extracellular calcium. This calcium-dependent induction of desmosome formation may be due to the fact that the desmosomal cadherins require calcium in order to mediate adhesion and to cluster into organized membrane domains, a process thought to be an important early step in junction assembly (Gingell and Owens, 1992; Yamada and Miyamoto, 1995). Early studies demonstrated that calcium induced a redistribution of desmosomal components to the plasma membrane and an increase in metabolic stability of the desmosomal proteins (Penn *et al.,* 1987a,b; Pasdar and Nelson, 1988a,b). These changes were also accompanied by a transfer of the desmosomal molecules to a detergent-insoluble pool, presumably resulting from the formation of protein complexes and/or association with the cytoskeleton. More recent studies, however, indicate that at least some aspects of desmosome assembly can occur in the absence of calcium (Demlehner *et al.,* 1995). Interestingly, focal contact assembly appears to be a multistep process that involves both ligation and clustering of integrin receptors (Yamada and Miyamoto, 1995). With respect to the calcium switch models of desmosome assembly, it will be important to determine how cadherin binding (i.e., ligation) and cadherin clustering lead to mature desmosome assembly and to the attachment of IFs at sites of desmosomal adhesion.

B. Plakoglobin: A Cadherin-Binding and Intracellular Signaling Protein

The process of calcium-induced desmosome assembly represents an example of spatial and temporal coordination of protein–protein interactions. To analyze further the interactions between the desmosomal cadherins and plakoglobin, these proteins were expressed in L-cells, a mouse fibroblast cell line that does not express detectable levels of the desmosomal proteins (Kowalczyk *et al.,* 1994). Plakoglobin was rapidly degraded when expressed in L-cells in the absence of the desmosomal cadherins. However, in the presence of either desmoglein or desmocollin, plakoglobin half-life increased from 10–15 min to approximately 3–4 hr (Fig. 2). Previous studies have demonstrated that α- and β-catenin fail to accumulate in L-cells in the absence of classical cadherins (Nagafuchi *et al.,* 1991; Tanihara *et al.,* 1994), presumably due to rapid protein turnover in the absence of a cadherin-binding partner. The stabilization of plakoglobin upon binding to a desmosomal cadherin may promote the local accumulation of plakoglobin specifically at membrane domains occupied by the desmosomal cadherins.

A number of studies have demonstrated that a pool of non-cadherin-associated plakoglobin and β-catenin is present in epithelial cells (Hinck *et al.,* 1994; Näthke *et al.,* 1994). Based on the studies in fibroblasts mentioned earlier, this pool of plakoglobin–β-catenin may be metabolically unstable. However, it is also possible that other binding partners or posttranslational

A

L-Pg
0 5 10 15 20 30 60
min

B

L-Dsg1/Pg
0 0.5 1 3 6 9
h

C

L-Dsc3a/Pg
0 0.5 1 3 6 9
h

D

L-Dsg/Pg
L-Dsc3a/Pg
L-Pg
Arbitrary Units
0
20
40
60
80
100
0
2
4
6
8
10
12
Time (h)

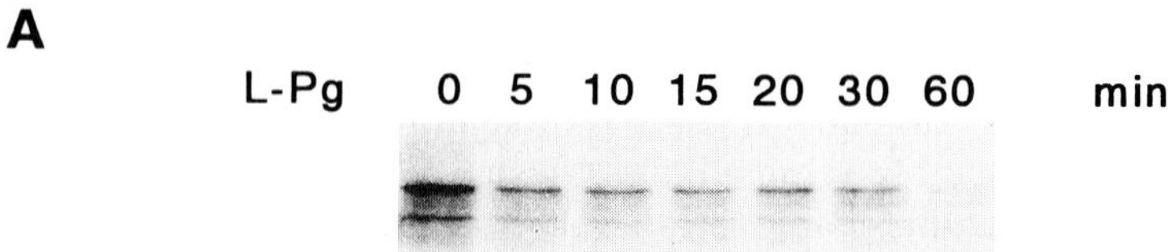

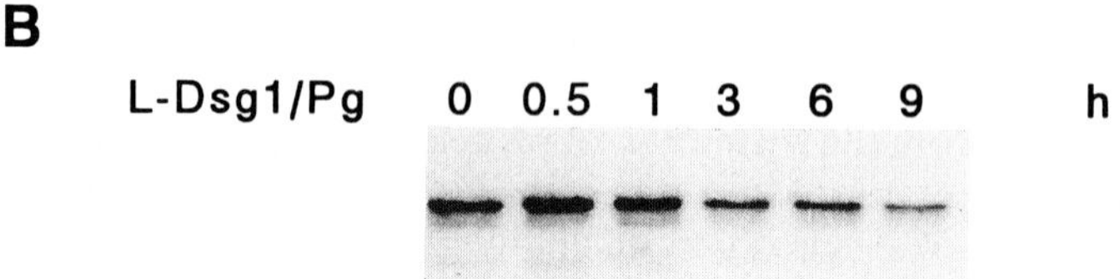

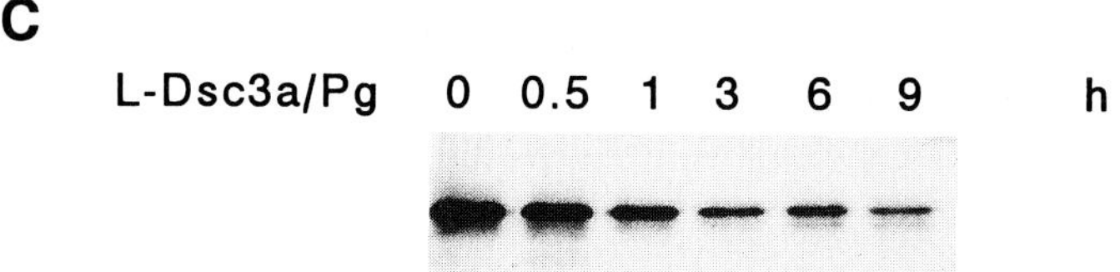

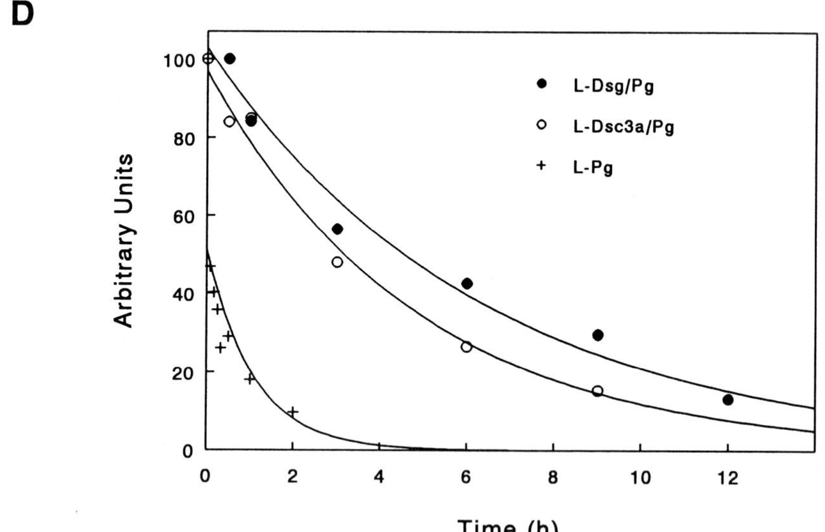

modifications could allow non-cadherin-associated plakoglobin to accumulate in epithelial cells. Several studies have suggested that this non-cadherin-associated pool of plakoglobin–β-catenin may be important for signal transduction. This is supported by the observation that overexpression of plakoglobin (Karnovsky and Klymkowsky, 1995) or β-catenin (Funayama *et al.*, 1995) in *Xenopus* embryos leads to anterior axis duplication. In the case of plakoglobin, this inductive effect was inhibited by coexpression of the desmoglein cytoplasmic tail. Although the regulation of embryonic pattern formation is a complex, multistep process, these findings suggest that the balance between the cadherin-bound and unbound pools of plakoglobin–β-catenin may be crucially important for normal embryonic development (Klymkowsky and Parr, 1995; Gumbiner, 1995). In addition, the extremely rapid degradation of non-cadherin-associated plakoglobin that is observed in fibroblasts (Kowalczyk *et al.*, 1994) (Fig. 2) may represent a mechanism for tightly controlling the accumulation of plakoglobin that is not associated with a cadherin.

C. Regulation of Desmosome Assembly by Protein Phosphorylation

Numerous studies have demonstrated that both adherens junctions and desmosomes are modulated by various growth factors (Thiery and Boyer, 1992). In addition, plakoglobin is tyrosine phosphorylated in some tumor cells and in growth factor–treated cells (reviewed by Cowin, 1994). Although it is currently difficult to reconcile all of the existing literature from different cell systems, desmosome assembly appears to be dictated by a balance between kinase and phosphatase activity. Desmosome assembly in Madin–Darby canine kidney cells was inhibited by okadaic acid, suggesting that phosphatase activity is required for normal desmosome formation. In addition, desmosome disassembly in response to lowered extracellular calcium levels was inhibited by the kinase inhibitor H7 (Pasdar *et al.*, 1995).

FIGURE 2 Rapid degradation of plakoglobin in cells lacking expression of the desmosomal cadherins. Stable mouse L-cell fibroblast lines expressing plakoglobin in the presence and absence of the desmosomal cadherins were generated. Cells were pulse labeled with [^{35}]methionine and plakoglobin was immunoprecipitated from cell lysates after various chase times. Immunoprecipitates from cell lines expressing (A) plakoglobin (Pg), (B) Dsg1 and plakoglobin, or (C) Dsc3a and plakoglobin were analyzed by sodium dodecyl sulfate–polyacrylamide gel electrophoresis (SDS-PAGE) and fluorography. In addition, SDS-PAGE gels were analyzed using a Fuji BAS 2000 Bioimaging analyzer and the amount of radioactivity present in the plakoglobin band at each chase time point was determined. The points were then plotted and fitted to an exponential (D). (Adapted from Kowalczyk *et al.*, 1994.)

However, Sheu *et al.* (1989) reported that activation of protein kinase C could induce the translocation of desmoplakin to the plasma membrane in cells cultured in low calcium, suggesting a role for protein kinase C in desmosome assembly.

While the overall role of protein kinase and phosphatase activity in desmosome assembly is not well understood, phosphorylation of desmoplakin at a serine residue 23 amino acids from the carboxyl terminus of the protein (SerC23) appears to play an important role in modulating the interaction between desmoplakin and IF networks (Stappenbeck *et al.*, 1994). This serine residue is nested within a cAMP-dependent protein kinase consensus sequence. Deletion of the last 28 amino acids of the protein, and therefore SerC23, enhanced alignment of desmoplakin polypeptides along keratin networks when transiently expressed in cultured cells. Furthermore, desmoplakin polypeptides comprising the rod and carboxyl terminus and containing a glycine (GlyC23) substituted for SerC23 exhibited increased ability to align along keratin networks compared to those containing SerC23. Treatment of cells with forskolin to activate cAMP-dependent kinases resulted in the displacement of DPSerC23 from filaments but had no effect on the DPGlyC23 (Fig. 3). In addition, 2-dimensional phosphopeptide analysis demonstrated the absence of a specific phosphopeptide in the DPGlyC23 variant compared to DPSerC23 or endogenous desmoplakin. These results strongly suggest that the phosphorylation of SerC23 inhibits the interaction of the desmoplakin carboxyl terminus with keratin networks and that this event could be regulated by agents that influence intracellular cAMP levels. The molecular mechanisms of this regulation and the role of SerC23 phosphorylation in desmosome assembly are currently under investigation.

VI. ISSUES ARISING

Many of the major proteins originally identified in bovine desmosome preparations have been isolated and cDNA clones encoding these proteins have been sequenced. These important contributions have provided some of the tools necessary to begin an analysis of how the various desmosomal components interact to form a structure that functions in cell adhesion, IF attachment, and, perhaps, signal transduction. In order to gain a more complete understanding of how individual desmosomal molecules interact and how these protein complexes lead to desmosome formation, several key questions must be addressed. For example, it will be important to elucidate the mechanisms by which the desmosomal cadherins contribute to intercellular adhesion. Unlike the classical cadherins, such as E-cadherin,

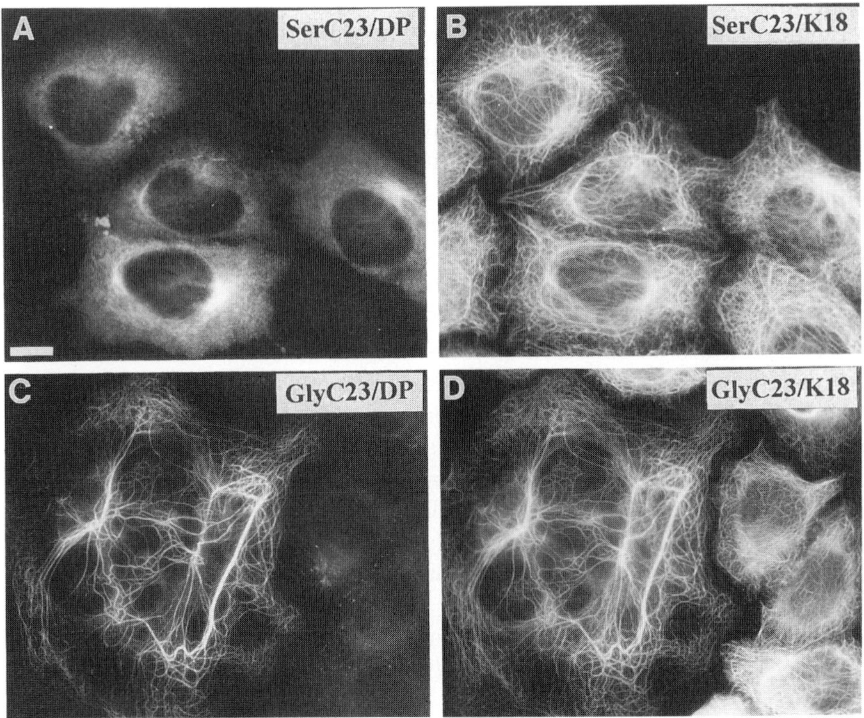

FIGURE 3 Forskolin treatment disrupts DPSerC23 but not DPGlyC23 alignment along keratin networks in HeLa cells. HeLa cells were transfected with desmoplakin cDNA constructs encoding the carboxyl terminus and rod domains and a serine (SerC23; A and B) or glycine (GlyC23; C and D) 23 residues from the carboxyl terminus. Dual-label immunofluorescence was performed using rabbit polyclonal antibody NW38 to detect desmoplakin (A and C) and mouse monoclonal antibody KS-B17.2 to detect keratin 18 (B and D). The GlyC23 polypeptide aligns along keratin networks in forskolin-treated cells, whereas the control SerC23 polypeptide does not. (Adapted from Stappenbeck *et al.,* 1994.)

the desmosomal cadherins do not appear to engage in simple homophilic binding events (Amagai *et al.,* 1994b; A. P. Kowalczyk and K. J. Green, unpublished observations). It is currently unknown how, or even if, the desmosomal cadherins function as adhesion molecules. In addition, although the desmoplakin carboxyl terminus has been demonstrated to interact with IFs, it remains unclear precisely how the desmoplakin–IF complex is attached to the desmosome. In addition, the role of other proteins that may be involved in linking filament networks to the desmosome must be characterized before comprehensive models of desmosome assembly can be constructed and tested.

Similarly, it is virtually certain that a number of regulatory molecules are present in desmosomes at substoichiometric levels. In the case of focal contacts, a growing number of regulatory proteins have been identified in addition to the structural components of these junctions (Yamada and Miyamoto, 1995). The receptor protein tyrosine phosphatase PTPμ (Brady-Kalnay *et al.*, 1995) and the epidermal growth factor receptor (Hoschuetzky *et al.*, 1994) have been found to interact with adherens junction components and c-*erbB*-2 has been shown to associate with plakoglobin (Ochiaj *et al.*, 1994). The identification of regulatory components that interact with desmosomal proteins will open new avenues to explore the regulation of desmosome assembly and should facilitate our understanding of the molecular mechanisms that control desmosome formation. In addition, desmosomes may not be just the recipients of regulatory cues, but may also be a site at which intracellular signals are transduced from the extracellular environment. Although this idea is only starting to receive attention from investigators in the field, there is evidence that would support this notion. For example, antisera from pemphigus patients bind to the surface of cultured keratinocytes and cause a transient increase in intracellular calcium levels (Seishima *et al.*, 1995; Esaki *et al.*, 1995), raising the possibility that binding of pemphigus sera to cell surface desmosomal cadherins transduces an intracellular signal. In addition, pemphigus antibodies appear to cause increases in keratinocyte urokinase activity (Hashimoto *et al.*, 1983; Lazarus and Jensen, 1991). These observations raise the exciting possibility that desmosomes convey specific information to intracellular signaling pathways and may even influence gene expression.

Finally, a number of studies have revealed that the genetic basis for the inherited skin diseases epidermolysis bullosa simplex and epidermal hyperkeratosis involve mutations in the epidermal keratins (Steinert and Bale, 1993; Fuchs, 1994; Roop, 1995; Epstein, 1992; McLean, 1995). However, the underlying basis for other skin diseases that might affect the IF–desmosome complex, such as Hailey–Hailey's and Darier's diseases, is not known. Given the potential role of desmosomes in maintaining tissue integrity, it is possible that these diseases are caused by mutations in desmosomal molecules or proteins that modulate desmosome function. The identification of disease states that involve desmosomal components and the generation of transgenic animals with disrupted desmosomal genes should further our understanding of the function of these proteins on the tissue level.

Acknowledgments

We would like to thank our colleagues who shared their work with us prior to publication. In addition, we extend our appreciation to Dr. Elayne Bornslaeger for critical reading of this manuscript and to members of the Green lab for contributing to many of the ideas and interpretations that were discussed in this chapter. Work in K. J. Green's laboratory is supported by the National Institutes of Health (1-RO1-AR41836 and RO1-AR43380), the March of Dimes and the Council for Tobacco Research, USA (2432B). A. P. Kowalczyk is supported by a postdoctoral fellowship from the Dermatology Foundation. K. J. Green is a recipient of an American Cancer Society Faculty Research Award.

References

Amagai, M. (1994). Molecular mechanisms of pemphigus deseases. *In* "Molecular Mechanisms of Epithelial Cell Junctions: From Development to Disease" (S. Citi, ed.), pp. 245–257. R. G. Landes Co., Austin, TX.

Amagai, M., Klaus-Kovtun, V., and Stanley, J. R. (1991). Autoantibodies against a novel epithelial cadherin in pemphigus vulgaris, a disease of cell adhesion. *Cell* **67,** 869–877.

Amagai, M., Karpati, S., Prussick, R., Klaus-Kovtun, V., and Stanley, J. R. (1992). Autoantibodies against the amino-terminal cadherin-like binding domain of pemphigus vulgaris antigen are pathogenic. *J. Clin. Invest.* **90,** 919–926.

Amagai, M., Hashimoto, T., Shimizu, N., and Nishikawa, T. (1994a). Absorption of pathogenic autoantibodies by the extracellular domain of pemphigus vulgaris antigen (Dsg3) produced by baculovirus. *J. Clin. Invest.* **94,** 59–67.

Amagai, M., Karpati, S., Klaus-Kovtun, V., Udey, M. C., and Stanley, J. R. (1994b). The extracellular domain of pemphigus vulgaris antigen (desmoglein 3) mediates weak homophilic adhesion. *J. Invest. Dermatol.* **102,** 402–408.

Amagai, M., Hashimoto, T., Green, K. J., Shimizu, N., and Nishikawa, T. (1995). Antigen-specific immunoadsorption of pathogenic autoantibodies in pemphigus foliaceus. *J. Invest. Dermatol.* **104,** 895–901.

Amagai, M., Koch, P. J., Nishikawa, T., and Stanley, J. R. (1996). Pemphigus vulgaris antigen (desmoglein 3) is localized in the lower epidermis, the site of blister formation in patients. *J. Invest. Dermatol.* **106,** 351–355.

Arnemann, J., Sullivan, K. H., Magee, A. I., King, I. A., and Buxton, R. S. (1993). Stratification-related expression of isoforms of the desmosomal cadherins in human epidermis. *J. Cell Sci.* **104,** 741–750.

Bornslaeger, E. A., Corcoran, C. M., Stappenbeck, T. S., and Green, K. J. (1996). Breaking the connection: Displacement of the desmosomal Plaque Protein Desmoplakin from cell–cell interfaces disrupts Anchorage of intermediate filament bundles and alters intercellular junction assembly. *J. Cell Biol.* In press.

Brady-Kalnay, S. M., Rimm, D. L., and Tonks, N. K. (1995). Receptor protein tyrosine phosphatase PTPμ associates with cadherins and catenins in vivo. *J. Cell Biol.* **130,** 977–986.

Brakenhoff, R. H., Gerretsen, M., Knippels, E. M. C., van Dijk, M., van Essen, H., Weghuis, D. O., Sinke, R. J., Snow, G. B., and van Dongen, G. A. M. S. (1995). The human E48 antigen, highly homologous to the murine Ly-6 antigen ThB, is a GPI-anchored molecule apparently involved in keratinocyte cell-cell adhesion. *J. Cell Biol.* **129,** 1677–1689.

Buxton, R. S., Cowin, P., Franke, W. W., Garrod, D. R., Green, K. J., King, I. A., Koch, P. J., Magee, A. I., Rees, D. A., Stanley, J. R., and Steinberg, M. S. (1993). Nomenclature of the desmosomal cadherins. *J. Cell Biol.* **121,** 481–483.

Buxton, R. S., Magee, A. I., King, I. A., and Arnemann, J. (1994). Desmosomal genes. *In* "Molecular Biology of Desmosomes and Hemidesmosomes" (J. E. Collins and D. R. Garrod, eds.), pp. 1–17. R. G. Landes Co., Austin, TX.

Cartaud, A., Ludosky, M. A., Courvalin, J. C., and Cartaud, J. (1990). A protein antigenically related to nuclear lamin B mediates the associatation of intermediate filaments with desmosomes. *J. Cell Biol.* **111,** 581–588.

Collins, J. E., and Garrod, D. R. (eds.). (1994). "Molecular Biology of Desmosomes and Hemidesmosomes." R. G. Landes Co., Austin, TX.

Collins, J. E., Legan, P. K., Kenny, T. P., MacGarvie, J., and Holton, J. L. (1991). Cloning and sequence analysis of desmosomal glycoproteins 2 and 3 (desmocollins): Cadherin-like desmosomal adhesion molecules with heterogeneous cytoplasmic domains. *J. Cell Biol.* **113,** 381–391.

Cowin, P. (1994). Plakoglobin. *In* "Molecular Biology of Desmosomes and Hemidesmosomes" (J. E. Collins and D. R. Garrod, eds.), pp. 53–68. R. G. Landes Co., Austin, TX.

Cowin, P., Mattey, D., and Garrod, D. (1984). Identification of desmosomal surface components (desmocollins) and inhibition of desmosome formation by specific FAB'. *J. Cell Sci.* **70,** 41–60.

Cowin, P., Kapprell, H., Franke, W. W., Tamkun, J. W., and Hynes, R. O. (1986). Plakoglobin: A protein common to different kinds of intercellular adhering junctions. *Cell* **46,** 1063–1073.

Demlehner, M. P., Schafer, S., Grund, C., and Franke, W. W. (1995). Continual assembly of half-desmosomal structures in the absence of cell contacts and their frustrated endocytosis: A coordinated Sisyphus cycle. *J. Cell Biol.* **131,** 745–760.

Dmochowski, M., Hashimoto, T., Amagai, M., Kudoh, J., Shimizu, N., Koch, P. J., Franke, W. W., and Nishikawa, T. (1994). The extracellular aminoterminal domain of bovine desmoglein 1 (Dsg1) is recognized only by certain pemphigus foliaceus sera, whereas its intracellular domain is recognized by both pemphigus vulgaris and pemphigus foliaceus sera. *J. Invest. Dermatol.* **103,** 173–177.

Emery, D. J., Diaz, L. A., Fairley, J. A., Lopez, A., Taylor, A. F., and Giudice, G. J. (1995). Pemphigus foliaceus and pemphigus vulgaris autoantibodies react with the extracellular domain of desmoglein-1. *J. Invest. Dermatol.* **104,** 323–328.

Epstein, E. H. (1992). Molecular genetics of epidermolysis bullosa. *Science* **256,** 799–804.

Esaki, C., Seishima, M., Yamada, T., Osada, K., and Kitajima, Y. (1995). Pharmacologic evidence for involvement of phospholipase C in pemphigus IgG-induced inositol 1,4,5-trisphsphate generation, intracellular calcium increase, and plasminogen activator secretion in DJM-1 cells, a squamous cell carcinoma line. *J. Invest. Dermatol.* **105,** 329–333.

Fairley, J. A., Scott, G. A., Jensen, K. D., Goldsmith, L. A., and Diaz, L. A. (1991). Characterization of keratocalmin, a calmodulin-binding protein from human epidermis. *J. Clin. Invest.* **88,** 315–322.

Farquhar, M. G., and Palade, G. E. (1963). Junctional complexes in various epithelia. *J. Cell Biol.* **17,** 375–412.

Foisner, R., and Wiche, G. (1991). Intermediate filament-associated proteins. *Curr. Opin. Cell Biol.* **3,** 75–81.

Fuchs, E. (1994). Intermediate filaments and disease: Mutations that cripple cell strength. *J. Cell Biol.* **125,** 511–516.

Funayama, N., Fagotto, F., McCrea, P., and Gumbiner, B. M. (1995). Embryonic axis induction by the armadillo repeat domain of β-catenin: Evidence for intracellular signaling. *J. Cell Biol.* **128,** 959–968.

Gingell, D., and Owens, N. (1992). How do cells sense and respond to adhesive contacts? Diffusion-trapping of laterally mobile membrane proteins at maturing adhesions may initiate signals leading to local cytoskeletal assembly response and lamella formation. *J. Cell Sci.* **101,** 255–266.

Goodwin, L., Hill, J. E., Raynor, K., Raszi, L., Manabe, M., and Cowin, P. (1990). Desmoglein shows extensive homology to the cadherin family of cell adhesion molecules. *Biochem. Biophys. Res. Commun.* **173,** 1224–1230.

Green, K. J., and Stappenbeck, T. S. (1994). The desmosomal plaque: Role in attachment of intermediate filaments to the cell surface. *In* "Molecular Mechanisms of Epithelial Junctions: From Development to Disease" (S. Citi, ed.), pp. 157–171. R. G. Landes Co., Austin, TX.

Green, K. J., Parry, D. A. D., Steinert, P. M., Virata, M. L. A., Wagner, R. M., Angst, B. D., and Nilles, L. A. (1990). Structure of the human desmoplakins. Implications for function in the desmosomal plaque. *J. Biol. Chem.* **265,** 2603–2612.

Green, K. J., Virata, M. L. A., Elgart, G. W., Stanley, J. R., and Parry, D. A. D. (1992). Comparative structural analysis of desmoplakin, bullous pemphigoid antigen and plectin: Members of a new gene family involved in organization of intermediate filaments. *Int. J. Biol. Macromol.* **14,** 145–153.

Gumbiner, B. M. (1995). Signal transduction by β-catenin. *Curr. Opin. Cell Biol.* **7,** 634–640.

Hashimoto, K., Shafran, K. M., Webber, P. S., Lazarus, G. S., and Singer, K. H. (1983). Anti-cell surface pemphigus autoantibody stimulates plasminogen activator activity of human epidermal cells. *J. Exp. Med.* **157,** 259–272.

Hatzfeld, M., Gunnar, I. K., Plessmann, U., and Weber, K. (1994). Band 6 protein, a major constituent of desmosomes from stratified epithelia, is a novel member of the armadillo multigene family. *J. Cell Sci.* **107,** 2259–2270.

Heid, H. W., Schmidt, A., Zimbelmann, R., Schafer, S., Winter-Simanowski, S., Stumpp, S., Keith, M., Figge, U., Schnolzer, M., and Franke, W. W. (1994). Cell type-specific desmosomal plaque proteins of the plakoglobin family: Plakophilin 1 (band 6 protein). *Differentiation* **58,** 113–131.

Hinck, L., Näthke, I. S., Papkoff, J., and Nelson, W. J. (1994). Dynamics of cadherin/catenin complex formation: novel protein interactions and pathways of complex assembly. *J. Cell Biol.* **125,** 1327–1340.

Hirano, S., Kimoto, N., Shimoyama, Y., Hirohashi, S., and Takeichi, M. (1992). Identification of a neural α-catenin as a key regulator of cadherin function and multicellular organization. *Cell* **70,** 293–301.

Holton, J. L., Kenny, T. P., Legan, P. K., Collins, J. E., Keen, J. N., Sharma, R., and Garrod, D. R. (1990). Desmosomal glycoproteins 2 and 3 (desmocollins) show N-terminal similarity to calcium-dependent cell-cell adhesion molecules. *J. Cell Sci.* **97,** 239–246.

Hoschuetzky, H., Aberle, H., and Kemler, R. (1994). β-Catenin mediates the interaction of the cadherin-catenin complex with epidermal growth factor receptor. *J. Cell Biol.* **127,** 1375–1380.

Kapprell, H.-P., Owaribe, K., and Franke, W. W. (1988). Identification of a basic protein of Mr 75,000 as an accessory desmosomal plaque protein in stratified and complex epithelia. *J. Cell Biol.* **106,** 1679–1691.

Karnovsky, A., and Klymkowsky, M. W. (1995). Anterior axis duplication in Xenopus induced by the over-expression of the cadherin-binding protein plakoglobin. *Proc. Natl. Acad. Sci. U.S.A.* **92,** 4522–4526.

Kawamura, K., Watanabe, K., Suzuki, T., Yamakawa, T., Kamiyama, T., Nakagawa, H., and Tsurufuji, S. (1994). cDNA cloning and expression of a novel human desmocollin. *J. Biol. Chem.* **269,** 26295–26302.

King, I. A., Arnemann, J., Spurr, N. K., and Buxton, R. S. (1993). Cloning of the cDNA (DSC1) coding for human type 1 desmocollin and its assignment to chromosome 18. *Genomics* **18,** 185–194.

King, I. A., Sullivan, K. H., Bennett, R. Jr., and Buxton, R. S. (1995). The desmocollins of human foreskin epidermis: Identification and chromosomal assignment of a third gene and expression patterns of the three isoforms. *J. Invest. Dermatol.* **105,** 314–321.

Klymkowsky, M. W., and Parr, B. (1995). The body language of cells: The intimate connection between cell adhesion and behavior. *Cell* **83,** 5–8.

Koch, P. J., Walsh, M. J., Schmelz, M., Goldschmidt, M. D., Zimbelmann, R., and Franke, W. W. (1990). Identification of desmoglein, a constitutive desmosomal glycoprotein, as a member of the cadherin family of cell adhesion receptors. *Eur. J. Cell Biol.* **53,** 1–12.

Koch, P. J., Goldschmidt, M. D., Walsh, M. J., Zimbelmann, R., and Franke, W. W. (1991a). Complete amino acid sequence of the epidermal desmoglein precursor polypeptide and identification of a second type of desmoglein gene. *Eur. J. Cell Biol.* **55,** 200–208.

Koch, P. J., Goldschmidt, M. D., Walsh, M. J., Zimbelmann, R., Schmelz, M., and Franke, W. W. (1991b). Amino acid sequence of bovine muzzle epithelial desmocollin derived from cloned cDNA: A novel subtype of desmosomal cadherins. *Differentiation* **47,** 29–36.

Koch, P. J., Goldschmidt, M. D., Zimbelmann, R., Troyanovsky, R., and Franke, W. W. (1992). Complexity and expression patterns of the desmosomal cadherins. *Proc. Natl. Acad. Sci. U.S.A.* **89,** 353–357.

Korman, N. J., Eyre, R. W., Klaus-Kovtun, V., and Stanley, J. R. (1989). Demonstration of an adhering-junction molecule (plakoglobin) in the autoantigens of pemphigus foliaceus and pemphigus vulgaris. *N. Engl. J. Med.* **321,** 631–635.

Kouklis, P. D., Hutton, E., and Fuchs, E. (1994). Making a connection: Direct binding between keratin intermediate filaments and desmosomal proteins. *J. Cell Biol.* **127,** 1049–1060.

Kowalczyk, A. P., Palka, H. L., Luu, H. H., Nilles, L. A., Anderson, J. E., Wheelock, M. J., and Green, K. J. (1994). Posttranslational regulation of plakoglobin expression: Influence of the desmosomal cadherins on plakoglobin metabolic stability. *J. Biol. Chem.* **269,** 31214–31223.

Kowalczyk, A. P., Anderson, J. E., Borgwardt, J. E., Hashimoto, T., Stanley, J. R., and Green, K. J. (1995). Pemphigus sera recognize conformationally sensitive epitopes in the amino-terminal region of desmoglein-1. *J. Invest. Dermatol.* **105,** 147–152.

Lazarus, G. S., and Jensen, P. J. (1991). Plasminogen activators in epithelial biology. *Semin. Thromb. Hemost.* **17,** 210–216.

Legan, P. K., Yue, K. K. M., Chidgey, M. A. J., Holton, J. L., Wilkinson, R. W., and Garrod, D. R. (1994). The bovine desmocollin family: A new gene and expression patterns reflecting epithelial cell proliferation and differentiation. *J. Cell Biol.* **126,** 507–518.

Mathur, M., Goodwin, L., and Cowin, P. (1994). Interactions of the cytoplasmic domain of the desmosomal cadherin Dsg1 with plakoglobin. *J. Biol. Chem.* **269,** 14075–14080.

McLean, W. H. I. (1995). Intermediate filaments in disease. *Curr. Opin. Cell Biol.* **7,** 118–125.

Mechanic, S., Raynor, K., Hill, J. E., and Cowin, P. (1991). Desmocollins form a subset of the cadherin family of cell adhesion molecules. *Proc. Natl. Acad. Sci. U.S.A.* **88,** 4476–4480.

Meng, J.-J., Green, K. J., and Ip, W. (1995). The molecular basis for the interaction between intermediate filament (IF) proteins and desmoplakin. *Mol. Biol. Cell* **6,** 376a.

Nagafuchi, A., Takeichi, M., and Tsukita, S. (1991). The 102kd cadherin-associated protein: Similarity to vinculin and postranscriptional regulation of expression. *Cell* **65,** 849–857.

Näthke, I. S., Hinck, L., Swedlow, J. R., Papkoff, J., and Nelson, W. J. (1994). Defining interactions and distributions of cadherin and catenin complexes in polarized epithelial cells. *J. Cell Biol.* **125,** 1341–1352.

Nilles, L. A., Parry, D. A. D., Powers, E. E., Angst, B. D., Wagner, R. M., and Green, K. J. (1991). Structural analysis and expression of human desmoglein: Cadherin-like component of the desmosome. *J. Cell Sci.* **99,** 809–821.

Nuber, U. A., Schafer, S., Schmidt, A., Koch, P. J., and Franke, W. W. (1995). The widespread human desmocollin Dsc2 and tissue-specific patterns of synthesis of various desmocollin subtypes. *Eur. J. Cell Biol.* **66,** 69–74.

Ochiaj, A., Akimoto, S., Kanai, Y., Shibata, T., Oyama, T., and Hirohashi, S. (1994). c-erbB-2 gene product associates with catenins in human cancer cells. *Biochem. Biophys. Res. Commun.* **205,** 73–78.

Ouyang, P., and Sugrue, S. P. (1992). Identification of an epithelial protein related to the desmosome and intermediate filament network. *J. Cell Biol.* **118,** 1477–1488.

Ozawa, M., Ringwald, M., and Kemler, R. (1990). Uvomorulin-catenin complex formation is regulated by a specific domain in the cytoplasmic region of the cell adhesion molecule. *Proc. Natl. Acad. Sci. U.S.A.* **87,** 4246–4250.

Parker, A. E., Wheeler, G. N., Arnemann, J., Pidsley, S. C., Ataliotis, P., Thomas, C. L., Rees, D. A., Magee, A. I., and Buxton, R. S. (1991). Desmosomal glycoproteins II and III: Cadherin-like junctional molecules generated by alternative splicing. *J. Biol. Chem.* **266,** 10438–10445.

Pasdar, M., and Nelson, W. J. (1988a). Kinetics of desmosome assembly in Madin-Darby canine kidney epithelial cells: Temporal and spatial regulation of desmoplakin and stabilization upon cell-cell contact. I. Biochemical analysis. *J. Cell Biol.* **106,** 677–685.

Pasdar, M., and Nelson, W. J. (1988b). Kinetics of desmosome assembly in Madin-Darby canine kidney epithelial cells: Temporal and spatial regulation of desmoplakin organization and stabilization upon cell-cell contact. II. Morphological analysis. *J. Cell Biol.* **106,** 687–695.

Pasdar, M., Li, Z., and Chan, H. (1995). Desmosome assembly and disassembly are regulated by reversible protein phosphorylation in cultured epithelial cells. *Cell Motil. Cytoskeleton* **30,** 108–121.

Peifer, M., Berg, S., and Reynolds, A. B. (1994). A repeating amino acid motif shared by proteins with diverse cellular roles. *Cell* **76,** 789–791.

Penn, E. J., Burdett, I. D. J., Hobson, C., Magee, A. I., and Rees, D. A. (1987a). Structure and assembly of desmosome junctions: Biosynthesis and turnover of the major desmosome components of Madin-Darby canine kidney cells in low calcium medium. *J. Cell Biol.* **105,** 2327–2334.

Penn, E. J., Hobson, C., Rees, D. A., and Magee, A. I. (1987b). Structure and assembly of desmosome junctions: Biosynthesis, processing, and transport of the major protein and glycoprotein components in cultured epithelial cells. *J. Cell Biol.* **105,** 57–68.

Roh, J.-Y., and Stanley, J. R. (1995a). Plakoglobin binding by human Dsg3 (pemphigus vulgaris antigen) in keratinocytes requires the cadherin-like intracytoplasmic segment. *J. Invest. Dermatol.* **104,** 720–724.

Roh, J.-Y., and Stanley, J. R. (1995b). Intracellular domain of desmoglein 3 (pemphigus vulgaris antigen) confers adhesive function on the extracellular domain of E-cadherin without binding catenins. *J. Cell Biol.* **128,** 939–947.

Roop, D. (1995). Defects in the barrier. *Science* **267,** 474–475.

Schafer, S., Koch, P. J., and Franke, W. W. (1994). Identification of the ubiquitous human desmoglein, Dsg2, and the expression catalogue of the desmoglein subfamily of desmosomal cadherins. *Exp. Cell Res.* **211,** 391–399.

Schwarz, M. A., Owaribe, K., Kartenbeck, J., and Franke, W. W. (1990). Desmosomes and hemidesmosomes: Constitutive molecular components. *Annu. Rev. Cell Biol.* **6,** 461–491.

Seishima, M., Esaki, C., Osada, K., Mori, S., Hashimoto, T., and Kitajima, Y. (1995). Pemphigus IgG, but not bullous pemphigoid IgG, causes a transient increase in intracellular calcium and inositol 1, 4, 5-triphosphate in DJM-1 cells, a squamous cell carcinoma. *J. Invest. Dermatol.* **104,** 33–37.

Sheu, H.-M., Kitajima, Y., and Yaoita, H. (1989). Involvement of protein kinase C in translocation of desmoplakins from cytosol to plasma membrane during desmosome formation in human squamous cell carcinoma cells grown in low to normal calcium concentration. *Exp. Cell Res.* **185,** 176–190.

Simrak, D., Cowley, C. M. E., Buxton, R. S., and Arnemann, J. (1995). Tandem arrangement of the closely linked desmoglein genes on human chromosome 18. *Genomics* **25,** 591–594.

Skalli, O., Jones, J. C. R., Gagescu, R., and Goldman, R. D. (1994). IFAP 300 is common to desmosomes and hemidesmosomes and is a possible linker of intermediate filaments to these junctions. *J. Cell Biol.* **125,** 159–170.

Skerrow, C. J., and Matoltsy, A. G. (1974). Chemical characterization of isolated epidermal desmosomes. *J. Cell Biol.* **63,** 524–530.

Staehelin, L. A. (1974). Structure and function of intercellular junctions. *Int. Rev. Cytol.* **39,** 191–283.

Stanley, J. R. (1995). Autoantibodies against adhesion molecules and structures in blistering skin diseases. *J. Exp. Med.* **181,** 1–4.

Stanley, J. R., Koulu, L., and Thivolet, C. (1984). Distinction between epidermal antigens binding pemphigus vulgaris and pemphigus foliaceus autoantibodies. *J. Clin. Invest.* **74,** 313–320.

Stanley, J. R., Klaus Kovtun, V., and Sampaio, S. A. (1986). Antigenic specificity of fogo selvagem autoantibodies is similar to North American pemphigus foliaceus and distinct from pemphigus vulgaris autoantibodies. *J. Invest. Dermatol.* **87,** 197–201.

Stappenbeck, T. S., and Green, K. J. (1992). The desmoplakin carboxyl terminus coaligns with and specifically disrupts intermediate filament networks when expressed in cultured cells. *J. Cell Biol.* **116,** 1197–1209.

Stappenbeck, T. S., Bornslaeger, E. A., Corcoran, C. M., Luu, H. H., Virata, M. L. A., and Green, K. J. (1993). Functional analysis of desmoplakin domains: Specification of the interaction with keratin versus vimentin intermediate filament networks. *J. Cell Biol.* **123,** 691–705.

Stappenbeck, T. S., Lamb, J. A., Corcoran, C. M., and Green, K. J. (1994). Phosphorylation of the desmoplakin COOH terminus negatively regulates its interaction with keratin intermediate filament networks. *J. Biol. Chem.* **269,** 29351–29354.

Steinert, P. M., and Bale, S. J. (1993). Genetic skin diseases caused by mutations in keratin intermediate filaments. *Trends Genet.* **9,** 280–284.

Tanihara, H., Kido, M., Obata, S., Heimark, R. L., Davidson, M., St. John, T., and Suzuki, S. (1994). Characterization of cadherin-4 and cadherin-5 reveals new aspects of cadherins. *J. Cell Sci.* **107,** 1694–1704.

Theis, D. G., Koch, P. J., and Franke, W. W. (1993). Differential synthesis of type 1 and type 2 desmocollin mRNAs in human stratified epithelia. *Int. J. Dev. Biol.* **37,** 101–110.

Thiery, J. P., and Boyer, B. (1992). The junction between cytokines and cell adhesion. *Curr. Opin. Cell Biol.* **4,** 782–792.

Troyanovsky, S. M., Eshkind, L. G., Troyanovsky, R. B., Leube, R. E., and Franke, W. W. (1993). Contributions of cytoplasmic domains of desmosomal cadherins to desmosome assembly and intermediate filament anchorage. *Cell* **72,** 561–574.

Troyanovsky, S. M., Troyanovsky, R. B., Eshkind, L. G., Krutovshikh, V. A., Leube, R. E., and Franke, W. W. (1994a). Identification of the plakolgobin-binding domain in desmoglein and its role in plaque assembly and intermediate filament anchorage. *J. Cell Biol.* **127,** 151–160.

Troyanovsky, S. M., Troyanovsky, R. B., Eshkind, L. G., Leube, R. E., and Franke, W. W. (1994b). Identification of amino acid motifs in desmocollin, a desmosomal glycoprotein, that are required for plakoglobin binding and plaque formation. *Proc. Natl. Acad. Sci. U.S.A.* **91,** 10790–10794.

Tsukita, S. (1985). Desmocalmin: A calmodulin-binding high molecular weight protein isolated from desmosomes. *J. Cell Biol.* **101,** 2070–2080.

Virata, M. L. A., Wagner, R. M., Parry, D. A. D., and Green, K. J. (1992). Molecular structure of the human desmoplakin I and II amino terminus. *Proc. Natl. Acad. Sci. U.S.A.* **89,** 544–548.

Wheeler, G. N., Parker, A. E., Thomas, C. L., Ataliotis, P., Poynter, D., Arnemann, J., Rutman, A. J., Pidsley, S. C., Watt, F. M., Rees, D. A., Buxton, R. S., and Magee, A. I. (1991). Desmosomal glycoprotein DGI, a component of intercellular desmosome junctions, is related to the cadherin family of cell adhesion molecules. *Proc. Natl. Acad. Sci. U.S.A.* **88,** 4796–4800.

Wiche, G., Gromov, D., Donovan, A., Castanon, M. J., and Fuchs, E. (1993). Expression of plectin cDNA in cultured cells indicates a role of COOH-terminal domain in intermediate filament association. *J. Cell Biol.* **121,** 607–619.

Yamada, K. M., and Miyamoto, S. (1995). Integrin transmembrane signaling and cytoskeletal control. *Curr. Opin. Cell Biol.* **7,** 681–689.

Yue, K. K. M., Holton, J. L., Clarke, J. P., Hyam, J. L. M., Hashimoto, T., Chidgey, M. A. J., and Garrod, D. R. (1995). Characterisation of a desmocollin isoform (bovine DSC3) exclusively expressed in lower layers of stratified epithelia. *J. Cell Sci.* **108,** 2163–2173.

CHAPTER 11

Protein Interactions in the Tight Junction: The Role of MAGUK Proteins in Regulating Tight Junction Organization and Function

Alan S. Fanning,[*] Lynne A. Lapierre,[*] Alexandra R. Brecher,[†] Christina M. Van Itallie,[*] and James Melvin Anderson[*,†]

[*]Departments of Internal Medicine and [†]Cell Biology, Yale University School of Medicine, New Haven, Connecticut 06520

I. PERSPECTIVE

Cell junctions serve the interrelated functions of transferring information across the plasma membrane and organizing links to the cortical cytoskeleton. While likely to adhere to this paradigm, tight junctions are presently better known for the specialized properties of creating a physiological intercellular barrier used to maintain distinct tissue spaces and to separate the apical from the lateral plasma membranes (Cereijido, 1992). However, expanding knowledge of the molecular components of the tight junction has begun to provide insight into its potential role in membrane signaling and in organizing the subjunctional cortical cytoskeleton. While knowledge about regulation of junction assembly and the interaction between proteins of the junction remains limited, it is clear that protein kinases and actin play important roles. In this chapter, we present some of what is known about the proteins of the tight junction and their interactions. A very important insight is offered by recognition that both ZO-1 and ZO-2, cytoplasmic plaque proteins of the tight junction, are members of the membrane-associated guanylate kinase (MAGUK) homologue protein family (Kim, 1995). Other members of this family are located on the cytoplasmic surface of a variety of cell–cell contacts ranging from the septate junctions in *Drosophila* to vertebrate synapses and the lateral surface of epithelial cells. Growing evidence suggests MAGUK proteins are involved in organizing the structural and functional links between transmembrane proteins, signaling pathways, and the cortical cytoskeleton. Review of the experimental evidence addressing the functional role of MAGUK proteins leads us to speculate that the cytoplasmic surface of the tight junction can be considered a specialized case of protein interactions that are used to connect membrane events at many different regions of the cell cortex.

II. FUNCTION AND MOLECULAR ORGANIZATION OF TIGHT JUNCTIONS

The separation of tissue compartments by sheets of epithelial cells is a hallmark of metazoans (Powell, 1981, Cereijido, 1992). In vertebrates the intercellular seal is created by the tight junction, while in invertebrates this function is mediated by the septate junction and other less well described structures (Lane, 1991). A century ago, histologists noted a thickening between polarized mammalian epithelial cells at the extreme apical end of the lateral cell interface. The paracellular barrier created at this so-called terminal bar was thought to result from secretion of a nonvital and unregulated extracellular cement. With the advent of electron microscopy, the terminal bar was shown to be a complex of three morphologically distinct junctions now known to be biochemically and functionally distinct. The most apical is the continuous zonula occludens (tight junction), which actually creates the paracellular seal. The tight junction is invariably flanked by the zonula adherens (adherens junction), which serves in adhesion and induction of cell organization (Fig. 1). While one can exist without the other under experimental manipulation (e.g., Warren *et al.,* 1988), they are

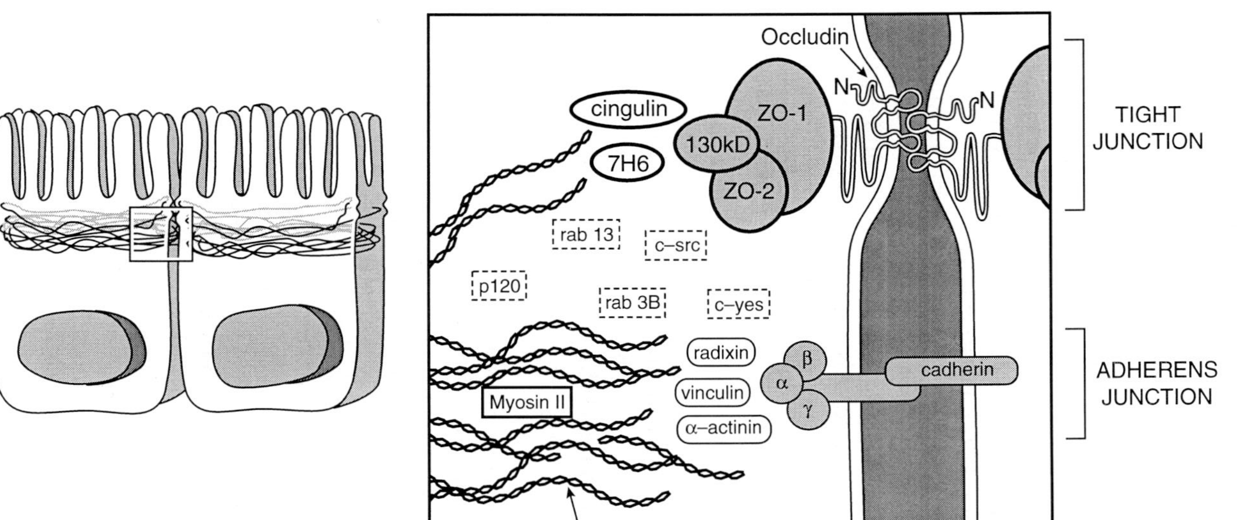

FIGURE 1 (Left) Tight junctions are postioned as continuous contacts at the apical–lateral membrane borders between polarized epithelial cells. Filamentous actin is contentrated under the junction. Boxed region is expanded at right. (Right) Hypothetical model of protein interactions at tight and aderens junction. The intercellular barrier at the tight junction is formed by homotypic contacts of the transmembrane protein occludin, which is bound on the cytoplasmic surface directly to ZO-1. The ZO-1/ZO-2 MAGUK heterodimer binds an uncharacterized 130-kDa protein. Binding interactions for cingulin and the 7H6 antigen are presently undefined. Adherens junctions are formed by homotypic association of cadherin, which associates directly with the cytoplasmic proteins, β- and γ-catenin and indirectly with α-catenin. Several reguatory "signaling" proteins are located with the apical junction complex, incuding c-Src, c-Yes, protein kinase C-zeta, Rab13, and Rab3B. (From Anderson and Van Itallie, 1995.)

normally paired and probably create a functional unit. At variable positions below these are desmosomes, presently thought to serve primarily, or exclusively, an adhesive function.

In transmission electron micrographs, the tight junction is formed by a variable number of very close membrane contacts evocatively referred to as "kisses." Freeze-fracture images of the junction reveal an organized network of long, branching fibrils embedded in the cell membrane; the union of paired longitudinal fibrils from adjacent cells corresponds to a cross-sectional kiss (Madara, 1992). Although there is no specific evidence, the fibrils are believed to be formed by linear polymers of transmembrane proteins circling the cell. Electron microscopic studies using electron-dense tracers demonstrate that the kisses are the actual sites of the paracellular barrier. Consistent with this, there exists an inverse correlation between the number of freeze-fracture fibrils and paracellular permeability when comparing cells among tissues of varying permeability, or "tightness" (Claude and Goodenough, 1973).

While the impression of a static structure that electron microscopic images fostered prevailed for many years, more recent work has shown that this notion is false. The tight junction barrier is dynamic and not only varies among epithelial cell types (Claude and Goodenough, 1973) but is subject to physiological regulation and pathological alteration between individual cells (Balda *et al.,* 1992, Madara *et al.,* 1992). Experimental evidence in whole tissues and cultured cell models suggests that junction assembly and barrier properties are influenced in some fashion by every known cell-signaling pathway. These include pathways involving Ca^{2+}, cAMP, and lipid second messengers. Evidence also supports a role for protein kinase C, tyrosine kinases, heterotrimeric G-proteins, and small GTP-binding proteins (reviewed in Balda *et al.,* 1991, Anderson and Van Itallie, 1995). Lacking at present is any understanding of how these signals regulate molecular interactions within the junction to effect its organization and function.

One unifying theme in controlling the junction appears to be involvement of the perijunctional actin cytoskeleton (Madara, 1992, Anderson and Van Itallie, 1995). Actin is highly concentrated under the adherens junction and to a lesser extent under the tight junction contacts (Fig. 1). Several actin-binding proteins have been identified as components of the adherens junction; in contrast, the cytoskeletal links with the tight junction are not yet defined. Agents that alter the organization of perijunctional actin universally affect the paracellular barrier. This correlation is observed with bacterial toxins that enhance permeability, such as the ZO toxin of *Vibrio cholerae* (Fasano *et al.,* 1991) and toxin a of *Clostridium difficile* (Hecht *et al.,* 1988), actin-disrupting drugs such as phalloidin (Madara, 1992), and hormones such as insulin (McRoberts *et al.,* 1990) and hepatocyte growth factor (Nusrat *et al.,* 1994). One popular hypothesis envisions tension from perijunctional actin–myosin on junction contacts as a mechanism for regulating paracellular permeability. The nature of the physical link between the actin cytoskeleton and transmembrane proteins remains a critical unresolved issue.

The recent identification of several protein components of the tight junction has begun to provide insight into how physiological properties are regulated at a molecular level. A candidate for the barrier-forming protein, termed occludin, has been cloned and characterized by S. Tsukita and his colleagues (Furuse *et al.,* 1993). Immunogold electron microscopy localizes occludin precisely and exclusively at the tight junction membrane kisses. The cDNA, cloned from chicken, predicts a novel 504-amino-acid protein with four hydrophobic membrane-spanning segments. Monoclonal antibodies to a C-terminal epitope label the inner surface of the membrane, supporting the folding topology shown in Fig. 1, with two extracellular loops of 44 and 45 residues. These loops lack charged residues and are unusually high in tyrosine and glycine (25% and 36%, respectively, in the more N-terminal loop). While there is no experimental evidence that occludin is capable of forming an intercellular seal, its unusual chemistry suggests the possibility that the extracellular loops contact loops on an adjacent cell to form the close, molecular seal. Although the sequence is novel, the proposed folding topography is very similar to that of other proteins involved in forming molecular seals, such as connexin of the gap junction and synaptophysin (Furuse *et al.,* 1993). On Western blots of different tissues from the chicken, occludin appears as several bands ranging from 59 to 65 kDa (Furuse *et al.,* 1993). It remains to be determined whether these result from proteolysis or, more interestingly, are products of different genes or alternative RNA splicing. Certainly proteins such as cadherins and integrins are composed of large protein families, which contributes to the variable properties of the junctions they form.

Occludin has been shown to bind directly to the cytoplasmic plaque protein ZO-1 (Furuse *et al.,* 1994). ZO-1 was the first tight junction protein to be cloned and sequenced (Itoh *et al.,* 1993, Willott *et al.,* 1993). This 220-kDa protein localizes precisely under tight junction kisses in polarized epithelial cells but has a wider distribution as well. It is also found in the modified tight junction of glomerular epithelia (Schnabel *et al.,* 1990) and nonepithelial cells such as astrocytes (Howarth *et al.,* 1992). It is capable of associating, through undefined interactions, with cadherin complexes (Itoh *et al.,* 1993) and is a component of the cadherin-rich contacts between cardiac myocytes. The formation of cadherin-based cell contacts often precedes or coincides with the assembly of tight junctions, suggesting that the formation of tight junctions may be intimately linked to the assembly of cadherin-based contacts (Nelson, 1991; Balda *et al.,* 1993).

Eight isoforms of ZO-1 are generated by alternative RNA splicing. These show distinct patterns of tissue expression, suggesting that tight junctions are molecularly quite distinct among cell types (A. Brecher, unpublished observation). One alternatively spliced form of ZO-1 defines an 80-amino-acid region referred to as the α-domain (Fig. 2). Epithelial cell junctions express the α-containing isoform while endothelial cells express the α-lacking isoform, again suggesting that functional differences among cell types are based in molecular diversity of their junction proteins (Balda and Anderson, 1993).

ZO-2 was originally identified through co-immunoprecipitation with ZO-1 (Gumbiner *et al.,* 1991). Although a homologue of ZO-1 (Jesaitis and Goodenough, 1994), this 160-kDa protein appears to be specifically restricted to tight junctions. A 130-kDa polypeptide that co-immunoprecipitates with ZO-1 and ZO-2 has also been observed by several groups (Balda *et al.,* 1993, Stuart and Nigam, 1995). This protein has not been cloned, nor are antibodies available at this time. Two other junction-specific proteins

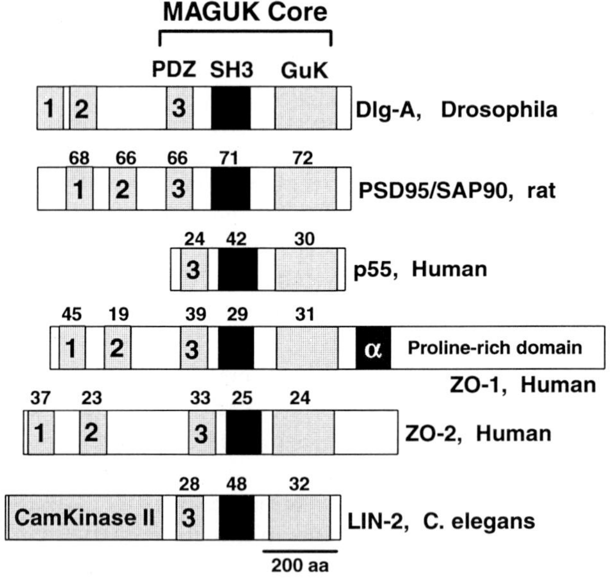

FIGURE 2 Comparative domain organization of several members of the MAGUK protein family. All contain a core of PDZ, SH3, and guanylate kinase homology (GuK) domains. Examples shown contain either three copies or one copy of the PDZ domain. Percent amino acid identities within defined domains are noted in comparison with *Drosophila* DlgA.

have been described, cingulin/140 kDa (Citi *et al.,* 1988) and the 7H6 antigen/160 kDa (Zhong *et al.,* 1993). Their functions are unknown, and the cDNA sequence of 7H6 predicts a novel protein (M. Mori, personal communication). Another group of proteins that are known to have roles in cell signaling have been localized at the apical junction plaque by light microscopy; whether they are actually localized in tight junctions, adherens junctions, or both has not yet been determined. This group includes the small GTP-binding proteins rab13 (Zahraoui *et al.,* 1994) and rab3B (Weber *et al.,* 1994), and the tyrosine kinase proto-oncogenes c-*src* and c-*yes* (Tsukita *et al.,* 1991). Their potential roles in regulating protein interactions in the tight junction remain to be defined.

III. THE MAGUK PROTEIN FAMILY

ZO-1 and ZO-2 are members of the MAGUK protein family (Kim, 1995). A combination of genetic, biochemical, and cell biological studies strongly suggest that these proteins are involved in controlling membrane organization and signaling events ranging from ion transport and receptor signaling to cell differentiation and proliferation. All MAGUK proteins examined to date appear to be located on the cytoplasmic surface of the plasma membrane, often at sites of specialized cell–cell contact. Consistent with the idea that they link junctional membrane events to the cortical cytoskeleton, they have so far been described only in metazoan species. Other MAGUK proteins characterized include DlgA (Lethal(1) Discs large-1) of *Drosophila,* a tumor suppressor molecule (Woods and Bryant, 1989); the gene product of *lin*-2 in the nematode *Caenorhabditis elegans* (Kim, 1995); the human erythrocyte membrane protein p55; the human *discs-large* (hdlg) (Lue *et al.,* 1994); and the rat synaptic junction proteins PSD-95/SAP90 and SAP97 (Cho *et al.* 1992; Kistner *et al.* 1993). Based on its degree of sequence similarity, the synapse-associated protein SAP97 is more likely than hdlg to be the mammalian homologue of *Drosophila* DlgA. All MAGUKs are distinguished by a core of homologous protein domains that include a region homologous to the yeast enzyme guanylate kinase (GuK), a *src*-homology region 3 (SH3) domain, and one or more domains of homology termed PDZ domains (*P*SD-95/*D*lgA/*Z*O-1 homology domains) (Fig. 2), which are discussed at length in Section IV. Presumably this core of domains serves a conserved coordinated binding and signaling function that is presently unknown.

The authentic Guk enzyme catalyzes transfer of the terminal phosphate group from ATP to GMP to produce GDP. It is speculated that MAGUKs regulate focal GMP–GDP levels with control over small GTP-binding proteins, G-protein–coupled receptors, vesicle targeting, or receptor-mediated responses (Woods and Bryant, 1991). The residues required within the active enzyme for binding ATP and GMP have been determined by x-ray crystallography. Comparison of the sequence in the active enzyme to that found in different MAGUKs suggests that the MAGUK family is divided into subgroups based on their presumptive ability to bind GMP and ATP, or function as active kinases. Both p55 (Ruff *et al.,* 1991) and Lin-2 (Hoskins *et al.,* 1995) contain all residues required to bind both nucleotides, although these functions and enzymatic activities have not been reported. A second group, represented by SAP90/PS-D95, SAP97, and Dlg, lacks three residues within the ATP-binding site. Indeed, SAP90 has been shown to bind GMP in the micromolar range; to bind ATP in the millimolar range, which is probably nonphysiological; and to lack Guk activity (Kistner *et al.,* 1995). ZO-1 and ZO-2 represent a third group that has not only deletions in the ATP-binding motif but extensive deletions in the putative GMP-binding region. The tight junction MAGUKs are expected to lack nucleotide-binding and enzymatic activity. Based solely on sequence comparison, we speculate that some MAGUKs are active enzymes while others bind GMP uncoupled from conversion to GDP, and the tight junction MAGUKs lack all these activities. In those MAGUKs lacking enzymatic activity, the GuK domain presumably has a conserved structural rather than functional status.

All MAGUKs contain a single SH3 domain, a motif originally identified in the Src tyrosine kinase and now found in more than 50 other proteins (Musacchio *et al.,* 1992). In other proteins, the SH3 domain controls subcellular localization and mediates protein–protein interactions by recognizing a short proline-rich motif on target proteins (Pawson, 1995). None of the binding targets for MAGUK SH3 domains have been identified. Our unpublished sequence comparison reveals that the SH3 domains of MAGUKs are actually most homologous to that of the adaptor protein Grb2, which links the activity of the epidermal growth factor (EGF) transmembrane tyrosine kinase to the *ras* signaling pathway (Lowenstein and Schlessinger, 1992). SH3 domains have also been shown to mediate interaction with the actin cytoskeleton (Bar-Sagi *et al.,* 1993), suggesting that they may be involved in linking MAGUKs to the cortical cytoskeleton. While the function of MAGUK SH3 domains is unknown, an allele of Dlg giving rise to the neoplastic phenotype encodes a mutation in the SH3 domain (Bryant *et al.,* 1993), indicating that this domain is critical for the normal function of this protein.

Analysis of the mutant phenotypes of two invertebrate MAGUKs strongly suggests they function in membrane signaling. These are the *Drosophila* tumor suppressor gene *lethal(1) Discs large-1,* and the Lin-2A protein of *C. elegans* (Fig. 2). *Lethal(1) Discs large-1* is one of seven tumor suppressor genes in *Drosophila* whose homozygous null mutation results in overgrowth of epithelial cells lining the imaginal discs (Woods and Bryant, 1989). These are embryonic epithelial tissues that develop inside the larva and normally differentiate into structures of the adult fly (legs, wings, antennas) during metamorphosis. The *dlg* gene gives rise to the DlgA protein, the predominant isoform, which is composed of 960 amino acids and contains three PDZ domains, an SH3 domain, and a C-terminal Guk domain (Woods and Bryant, 1991; see Fig. 2). Immunocytochemical analysis shows that DlgA is located on the cytoplasmic face of the septate junction, the invertebrate analogue of the tight junction (Lane, 1991), in various epithelia (Woods and Bryant, 1991). Homozygous null mutations of the *dlg* gene cause neoplastic overgrowth of the imaginal discs in the larva, followed by death in the early pupal stage. The imaginal discs, normally highly polarized single-layered columnar epithelia that contain adherens and septate junctions, become disorganized masses of cuboidal cells that lose their apical–basal polarity and do not form septate junctions (Woods and Bryant, 1991). Alleles including mutations in the GuK or SH3 domains give rise to a phenotype that forms junctions but still demonstrates neoplastic growth (Bryant *et al.*, 1993). The implication is that DlgA serves a dual role; it is required for structural assembly of the septate junction and it plays a role in signal transduction events necessary for controlling cell proliferation in mitotically active tissues.

Study of a MAGUK involved in the formation of the vulva in *C. elegans* also supports a role for MAGUKs in membrane signaling events. The specification of cell fate in vulval precursor cells is initiated by a signal from the gonadal anchor cell that is transduced in vulval precursor cells by a highly conserved receptor tyrosine kinase/Ras pathway (Eisenmann and Kim, 1994). Loss of function mutations in genes of this pathway result in a vulvaless phenotype. Although many of the components of this pathway are found in all cells, a novel subset of genes are specifically required for vulval induction; these are *lin*-2, *lin*-7, and *lin*-10. The *lin*-2 gene encodes a MAGUK that, like other members of this family, contains a PDZ domain, an SH3 domain, and a GuK domain (Hoskins *et al.,* 1995). Surprisingly, the cDNA also predicts a long N-terminal extension with highly significant homology to calcium–calmodulin-dependent protein (CaM) kinase II. Genetic studies indicate that Lin-2 acts on the vulval precursor cell (P6.p) downstream of Let-23, an EGF receptor–like tyrosine kinase, and upstream of let-60 *ras.* The vulvaless phenotype of *lin*-2 mutants is suppressed by a gain-of-function *let*-60 ras allele, indicating that Lin-2 is upstream from *ras* in the inductive signaling pathway. Altered *lin*-2 genes that lack either (CaM) kinase II or guanylate kinase activity are capable of rescuing *lin*-2 mutants, suggesting that Lin-2 may have a structural rather than an enzymatic role in vulval induction (Hoskins *et al.,* 1995).

There is presently no direct evidence that tight junctions are involved in contact inhibition or regulation of cell growth or differentiation. Neither ZO-1 nor ZO-2 maps to the loci of know tumor suppressors in humans or mice (Mohandes *et al.,* 1995, Duclos *et al.,* 1993). Whether all MAGUKs are involved in cell signaling remains to be determined. Resolution of this question will require detailed knowledge of their protein-binding partners.

IV. A ROLE FOR PDZ DOMAINS IN ORGANIZING THE CORTICAL CYTOSKELETON

A. PDZ Domains Mediate Interactions with the Plasma Membrane

The PDZ domain (previously identified as the "DHR" or "GLGF" domain) is an 80- to 90-amino-acid motif present in one or more copies in all MAGUK proteins. PDZ domains are also found in proteins that do not contain the SH3 and GuK domains associated with MAGUKs, such as neuronal nitric oxide synthase (Bredt *et al.,* 1991), the dystrophin-associated protein syntrophin (Adams *et al.,* 1993), the *Drosophila* signal transduction protein dishevelled (Theisen *et al.,* 1994), and several protein tyrosine phosphatases (Gu *et al.,* 1991, Yang and Tonks, 1991; Maekawa *et al.,* 1994). Although sequences can vary considerably among the different PDZ domains, all PDZ domains appear to maintain a core consensus that presumably underlies a common tertiary structural motif (Ponting and Phillips, 1995). It is also apparent, from a diagram of sequence similarity generated by the PILEUP algorithm (Devereux *et al.,* 1984) (Fig. 3, see also Fig. 2), that PDZ domains of different proteins fall into classes in which the PDZ of one protein (e.g., Dlg/hdlg) is more similar to the PDZ domain in a second protein (e.g., PSD-95/SAP90) than it is to other PDZ domains within the same protein. Given the very high identity among some PDZ motifs in different proteins (the second PDZ domains of hdlg and PSD-95/SAP97 are 88% identical), it seems likely that domains of the same class in different proteins have an identical function.

The great majority of proteins that contain PDZ domains appear to be associated with the plasma membrane. This observation has led to the speculation that the PDZ domain is a protein-binding motif that mediates the interaction of proteins with components of the plasma membrane or cortical cytoskeleton. In addition, many proteins with PDZ domains appear to be involved in signal transduction, leading to the further speculation

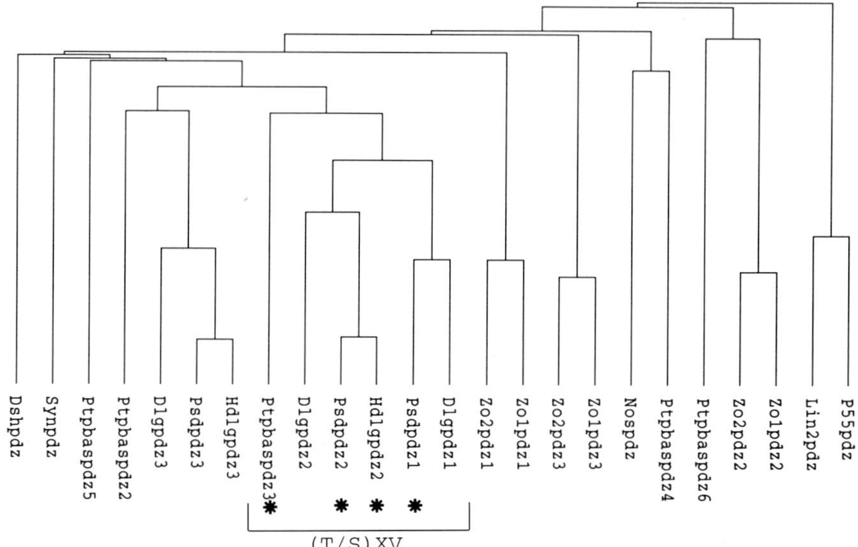

FIGURE 3 Relationships between PDZ domains found in selected MAGUKs. Dendrogram was assembled using the PILEUP program (Devereux *et al.,* 1984). The PDZ domains of several non-MAGUK proteins—PTPbas (also called FAP-1), neuronal nitric oxide synthase (NOS), syntrophin (syn), and dishevelled (dsh)—are included for comparison. Branch length is proportional to percent similiarity between two polypeptides. Domains grouped within each branch are likely to share binding to similar peptide motifs. For example, PDZ1 and -2 of PSD-95, PDZ2 of hdlg, and PDZ3 of the non-MAGUK protein PTPbas have all been shown to bind the (T/S)XV motif.

that this motif may also be an important component of signaling pathways at the plasma membrane.

B. PDZ Domains Are Modular Protein-Binding Motifs

Experimental evidence supports the hypothesis that PDZ domains are indeed protein-binding motifs. This result was provided serendipitously by several groups, all using the yeast two-hybrid system and who were interested in identifying proteins bound to the cytoplasmic tails of various transmembrane proteins. The common result was a sequence motif on the C-terminus of these transmembrane receptors that binds directly to PDZ domains (see Table I). For example, Kornau *et al.* (1995) have identified a 7-amino-acid sequence motif at the C-terminus of the *N*-methyl-D-aspartate (NMDA) receptor NR2 subunit that is both necessary and sufficient for

TABLE I
PDZ Domains and Transmembrane Targets[a]

PDZ-containing protein	PDZ domain	Transmembrane protein	PDZ-binding motif[b]
PSD-95/SAP90	PDZ2	NMDA receptor	
		NR2A subunit	. . . PSIESDV
		NR2B	. . . SSIESDV
		NR2C/D	. . . SSLESEV
		NR1-3/4 subunit	. . . PSVSTVV
PSD-95/SAP90	PDZ2	Shaker-type K^+ channel	
hdlg	PDZ2	Kv1.4	. . . CSNAKAVETDV
KAP5	PDZ?	Kv1.3	. . . KIFTDV
		Kv1.2	. . . KMLTDV
		Kv1.1	. . . KLLTDV
FAP-1 (PTPbas, PTPL1)	PDZ3[c]	Fas receptor	. . . DSENSNFRNEIQSLV
			Consensus (T/S)XV

[a] Data for this table were acquired from Kornau *et al.* (1995) (NMDA receptor); Kim *et al.* (1995) (K^+ channel); and Sato *et al.* (1995) (Fap-1).
[b] The minimum fragment tested determined to be sufficient to mediate PDZ binding.
[c] Corresponds to amino acids 1368–1452 of PTPbas.

binding to the second PDZ domain (PDZ2) of PSD-95/SAP90. A similar sequence in the C-terminus of the NR1 subunit was also found to bind to this PDZ domain. In a separate study, Kim *et al.* (1995) discovered that the C-terminus of the Shaker-type, voltage-gated, K^+ channel protein Kv1.4 binds directly to the PDZ2 domain of PSD-95/SAP90 and hdlg/SAP97, as well as a PDZ domain within KAP5, a previously uncharacterized protein. The same C-terminal motif also binds to a lesser extent to the PDZ1 domain in PSD-95/SAP90. Both studies showed that the distribution of the transmembrane proteins and the MAGUK proteins overlapped in many different tissues and cell types. In addition, the proteins can be co-immunoprecipitated from cultured cells, suggesting that the interaction detected by the yeast two-hybrid experiments was physiologically relevant. In a third study, Sato *et al.* (1995) identified a 15 amino-acid motif at the C-terminus of the Fas receptor that binds directly to one of six PDZ domains within the protein tyrosine phosphatase FAP-1 (also known as PTPbas or PTPL1). Fas is a cell surface receptor that controls a signaling pathway that leads to controlled cell death, or apoptosis (Itoh and Nagata, 1993). Although the C-terminal sequence on these three transmembrane proteins varies, all contain the consensus sequence (T/S)XV, in which T/S is serine or threonine, X is any amino acid, and V is a valine residue at the extreme

C-terminus. Together these three studies establish that the PDZ domains found in diverse proteins are protein-binding motifs.

There is evidence in these experiments that the recognition specificities of different PDZ domains is quite diverse. For example, Fas binds to only one of the six PDZ domains in FAP-1, and the NMDA receptor and K^+ channels bind preferentially to the second PDZ in the postsynaptic density protein PSD-95. Furthermore, the NMDA receptor would not bind to the single PDZ domain in nitric oxide synthase. Specificity is also suggested by the breakdown of PDZ domains into distinct classes based on sequence alignment (Fig. 3). It can be expected that ligands for different classes of PDZ domains will soon be determined, and will reveal complexity analogous to that of SH3 and SH2 binding recognition motifs. Presently the binding interactions of PDZ domains within ZO-1 and ZO-2 are unknown. Their sequences are quite different from those in hdlg and PSD-95, suggesting that the tight junction MAGUKs may not be linked to motifs similar to those found on the transmembrane proteins described here.

C. MAGUK Proteins Organize Components of the Cortical Cytoskeleton

Almost all MAGUKs are associated with actin-rich structures at the plasma membrane, such as the specialized erythrocyte cytoskeleton, cell–cell contacts formed at synapses, tight junctions, or lateral membrane surfaces of polarized epithelial cells. The accumulated evidence suggests that MAGUKs have an important role in the assembly and maintenance of these subcortical domains. For instance, the Dlg protein is a component of the septate junctions of *Drosophila* imaginal disc epithelial cells (Woods and Bryant, 1991). Many mutations of the *dlg* gene result in the loss of septate junctions, suggesting that Dlg is a critical component in the assembly of the septate junction (Woods and Bryant, 1989, 1991). At the tight junction, ZO-1 has been shown to bind directly to the transmembrane protein occludin, and the binding site for ZO-1 on occludin has been partially mapped (Furuse *et al.,* 1994). If the binding site for ZO-1 is deleted in occludin, the occludin protein is not incorporated into cellular tight junctions. This result suggests that ZO-1 is critical for either the assembly or maintenance of the tight junction.

Many MAGUKs may regulate the assembly of membrane complexes by mediating interactions with the cortical actin cytoskeleton. Several MAGUKs, such as the synaptic proteins PSD-95/SAP90 and SAP97, copurify with components of the cortical actin cytoskeleton (Cho *et al.,* 1992; Kistner *et al.,* 1993; Muller *et al.,* 1995). Other MAGUKs have been shown to bind directly to cytoskeletal proteins. For example, the tight junction protein

ZO-1 binds *in vitro* to spectrin (Itoh *et al.,* 1993), and both p55 and hdlg have been shown to bind directly to the cytoskeletal protein band 4.1 (Marfatia *et al.,* 1994; Lue *et al.,* 1994). Protein 4.1 is a member of the ERM family of tyrosine kinase substrates, which includes ezrin, radixin, moesin, and the product of the neurofibromatosis type 2 gene (Trofatter *et al.,* 1993). Protein 4.1 binds fodrin and regulates its affinity for actin. It is required for maintaining erythroid shape and membrane stability by linking spectrin–actin complexes to the plasma membrane via an association with the transmembrane protein glycophorin C. Thus, both p55 and hdlg tether transmembrane proteins to the actin cytoskeleton; p55 tethers to glycophorin C (Marfatia *et al.,* 1994) and hdlg to K^+ channels (Kim *et al.,* 1995). It is possible that all MAGUKs form links between transmembrane proteins and components of the actin cytoskeleton.

There is now evidence that MAGUKs, through interactions mediated by PDZ domains, are directly involved in the formation of physical linkages between components of membrane structures. *In vivo,* proteins such as the NMDA receptor and the K^+ channel are clustered at specific subcellular sites within synapses. Kim *et al.* (1995) found that when the K^+ channel proteins Kv1.4 and PSD-95/SAP90 were individually transfected and expressed in a cell line that expressed neither protein, Kv1.4 was diffusely distributed on the surface of the cell and PSD-95/SAP90 was found throughout the cytosol. However, when both proteins were co-transfected into the same cell, Kv1.4 and PSD-95/SAP90 co-localized at the plasma membrane in discrete irregular patches. These results suggest that PSD-95/SAP90 is directly responsible for the clustering of K^+-channel subunits. PSD-95/SAP90 may also be responsible for clustering of the NMDA receptor. Ehlers *et al.* (1995) have found that splice variants of the NR1 subunit that contain the conserved C-terminal (T/S)XV motif cluster into discrete patches at the plasma membrane, while those that lack this motif are diffusely distributed. The addition of a 15-amino-acid sequence containing the conserved motif to a receptor protein that is normally diffusely distributed was sufficient to mediate the clustering of the chimeric receptor. Both studies suggest that MAGUK proteins (PSD-95/SAP90) are directly involved in the cross-linking and organization of transmembrane proteins in the plasma membrane via interactions mediated by PDZ domains.

Neither the NMDA receptor nor the K^+ channel studies directly addressed the mechanism of cross-linking by MAGUKs. Theoretically, MAGUKs could provide either a direct linkage among the transmembrane molecules themselves or linkage of transmembrane molecules to components of the cortical cytoskeleton. Neither of these models is mutually exclusive. The first model (Fig. 4A) would require that one MAGUK molecule provide two binding sites for a particular type of transmembrane

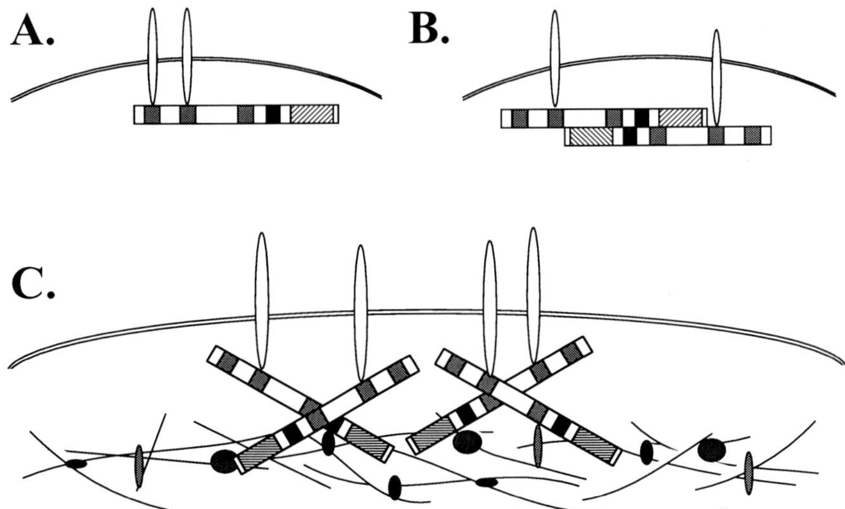

FIGURE 4 Models for how MAGUK proteins are involved in clustering or localizing transmembrane proteins. (A) Multiple functional PDZ domain model; each MAGUK (box) contains multiple binding sites for a transmembrane protein (ovals). The second site could also be provided by the SH3 or another unidentified domain. (B) MAGUK oligomer model; oligomerization of a MAGUK enables cross-linking of a transmembrane protein that binds to only one specific domain within a MAGUK. (C) Cytoskeletal anchor model; transmembrane proteins are cross-linked to the cortical cytoskeleton via MAGUKs. Present evidence suggests all these mechanisms may be used. (Shaded box, PDZ domain; black box, SH3 domain; diagonal hatching, Guk domain.

protein. Presumably one binding site would be provided by a PDZ domain. The second site could be a different PDZ domain or another site on the molecule, such as the SH3 domain. However, the PDZ domains appear to have great specificity; neither the K^+ channel nor the NMDA receptor bound to the other two PDZ domains in PSD-95/SAP90, or to the PDZ domain in nitric oxide synthase (Kornau *et al.,* 1995; Kim *et al.,* 1995). In addition, the cytosolic domains of the NMDA receptor and Kv1.4 did not appear to have any affinity for other domains within PSD-95/SAP90 (Kornau *et al.,* 1995). These observations make it difficult to imagine how a single molecule of PSD-95/SAP90 could cross-link two transmembrane proteins. It is possible that PSD-95/SAP90 forms a dimer (Fig. 4B). Each dimer would have two PDZ domains, which could cross-link two transmembrane proteins. At least two MAGUKs, ZO-1 and ZO-2, are known to exist in a heterodimeric complex *in vivo* (Gumbiner *et al.,* 1991). Thus it is possible that the formation of a dimeric complex by MAGUKs is necessary for clustering of cell surface proteins.

It is just as likely, however, that the clustering of Kv1.4 and the NMDA receptor by PSD-95/SAP90 was a direct result of the cross-linking of Kv1.4 and the NMDA receptor to the cortical cytoskeleton by PSD-95/SAP90 (Fig. 4C). In this third model, a MAGUK would have to provide a binding site for both the transmembrane protein and a cytoskeletal protein. Several MAGUKs, such as p55, hdlg, and ZO-1, have been shown to bind to both transmembrane and cytoskeletal proteins. Presumably binding to the transmembrane protein would be mediated by the PDZ domain. The binding sites for cytoskeletal proteins are still poorly defined, although several MAGUKs have been shown to bind directly to cytoskeletal proteins. The protein 4.1 binding site on p55 and hdlg has been localized to a 34- to 39-amino-acid motif located between the SH3 and GuK domains (Marfatia *et al.*, 1994; Lue *et al.*, 1994). It is also possible that linkages to the cytoskeleton could involve the SH3 domains, which have been shown to mediate interactions with the actin cytoskeleton in non-MAGUK proteins (Bar-Sagi *et al.*, 1993). When more is understood about the binding sites for cytoskeletal proteins in PSD-95/SAP90 (or other MAGUKs), it should be possible to experimentally confirm the third model by repeating the Kv1.4 transfections using a PSD-95/SAP90 molecule in which the cytoskeletal binding domains have been altered or deleted. If linkages to the cortical cytoskeleton are indeed required for receptor clustering, then the alteration of cytoskeletal binding sites in PSD-95/SAP90 should interfere with clustering in co-transfected cells. The reality, however, is that all three models are probably relevant. The formation of a complex cortical structure such as a synapse or a tight junction probably involves linkages between many different transmembrane proteins as well as proteins of the cortical cytoskeleton.

D. *Organization of Cortical Cytoskeletal Structures by MAGUK Proteins May Be Required for Proper Signal Transduction*

Undoubtedly one of the primary roles of the cortical cytoskeleton is to provide a structural scaffold for the plasma membrane. MAGUK proteins appear to be a ubiquitous component of this scaffold, and have been found on almost all membrane domains in fibroblastic, epithelial, and erythroid cells. Thus, some MAGUKs may be generally important in the assembly and maintenance of the cortical cytoskeleton. Many MAGUKs, however, are also specifically localized to specialized structures within the cortical cytoskeleton, such as synapses and cell–cell junctions. Many of these structures are focal points for cell–cell signaling pathways that mediate cell growth and differentiation. It is possible that the transmembrane receptors and channels are organized into many of these structures by MAGUKs.

Furthermore, this organization may be critical for the proper function of the transmembrane signaling protein. The Dlg protein may provide one example. Dlg has been localized to septate junctions of epithelial cells in larval and adult flies (Woods and Bryant, 1991). Among other things, most mutations in Dlg result in a loss of septate junctions as well as neoplastic growth of the disc epithelia (Woods and Bryant, 1989). These observations suggest that Dlg is required for both structural organization and cell signaling at septate junctions.

The *C. elegans lin*-2 gene product provides another example of the importance of MAGUKs in the structural organization of signaling pathways. Lin-2 is a component of the Let-23 receptor tyrosine kinase signaling pathway that regulates vulval development (Hoskins *et al.,* 1995). Although the cellular localization of Lin-2 is unknown, Let-23 is normally distributed on the basolateral surface of vulval precursor cells in close proximity to the belt desmosome (an actin-rich structure that appears to be analogous to the adherens and tight junctions of vertebrate epithelial cells) (Kim, 1995). Unlike *dlg, lin*-2 mutations do not cause changes in cellular morphology, disruption of junctional complexes, or loss of polarity. However, mutations in *lin*-2 that disrupt the Let-23 signaling cascade also result in a mislocalization of a Let-23 reporter protein (Kim, 1995; Hoskins *et al.,* 1995). These observations have led the authors to propose that Lin-2 is required to maintain the polarized distribution of Let-23 on the basolateral surface of vulval precursor cells, and that this polarized distribution is in turn necessary for Let-23 to be activated (Kim, 1995). Thus the Lin-2 MAGUK may function in either receptor localization, cortical organization of downstream signaling proteins, or both. It is tempting to speculate that ZO-1 and ZO-2 are also involved in coordinating transmembrane signals and the protein kinase pathways coupled to the tight junction.

V. REGULATION OF TIGHT JUNCTION PROTEIN INTERACTIONS BY PROTEIN KINASES

A. *Tyrosine Kinases*

The apical junctional complex is an important site of cellular protein tyrosine kinase activity. With the production of anti-phosphotyrosine antibodies, it was early recognized that significant immunofluorescent labeling occurred at cell–substrate and cell–cell junctions of both normal and transformed cells in culture (Maher *et al.,* 1985). In embryonic tissues, cell–cell junctions are heavily labeled with anti-phosphotyrosine antibodies (Takata and Singer, 1988). This relatively high concentration of tyrosine-phosphorylated substrates has been interpreted as reflecting either dynamic junction formation or the importance of this site in signal transduction pathways related to cell proliferation. In this context, it is relevant to note that two proto-oncogene nonreceptor tyrosine kinases, c-*src* and c-*yes* are concentrated under the apical junction complex (Tsukita *et al.,* 1991).

Although some of these tyrosine-phosphorylated substrates at the apical junctional complex are known to be adherens junction components, such as members of the cadherin–catenin complex, some are also likely to be tight junction components. Identification of tyrosine kinase substrates has been impeded by the low level of tyrosine-phosphorylated proteins in normal cells. Inhibition of tyrosine phosphatases has been used to probe effects of increased tyrosine phosphorylation and to identify kinase substrates in a number of different systems. Treatment of MDCK monolayers with pervanadate, a tyrosine phosphatase inhibitor, leads to an increase in tyrosine-phosphorylated proteins at the apical cell junction that can be documented by immunohistochemical staining with antibodies recognizing phosphotyrosine (Volberg *et al.,* 1992). The adherens junction proteins cadherin, β-catenin, and paxillin (Matsuyoshi *et al.,* 1992) are tyrosine phosphorylated in this system, as are the tight junction components ZO-1 and ZO-2 (Staddon *et al.,* 1995). Pervanadate treatment of MDCK cells results in decreases in paracellular resistance that are accompanied by dramatic changes in cellular morphology. In contrast, phenylarsine oxide, a more specific tyrosine phosphatase inhibitor, causes a drop in paracellular resistance with little morphological change at the cell junctions, at least at the resolution level of light microscopy. This agent induces a more focused tyrosine phosphorylation at the junction as visualized by staining with anti-phosphotyrosine antibodies and results in phosphorylation of a restricted set of proteins relative to the action of vanadate (Staddon *et al.,* 1995), including both β-catenin and ZO-1 and ZO-2.

Tyrosine phosphorylation of ZO-1 also occurs under other conditions in which the organization of cell–cell contacts is grossly altered or the paracellular barrier is lost. The filtration slits between the foot processes of renal glomerular epithelial cells are attached to each other by slit diaphragms. These slit diaphragms have been previously identified as modified tight junctions since they arise from tight junctions during development and express ZO-1 (Schnabel *et al.,* 1990). Kurihara *et al.* (1995) have shown that perfusion of rats with the polycation protamine sulfate induces the filtration slits to collapse and take on the appearance of typical occluding junctions. During this reorganization, ZO-1 becomes transiently tyrosine phosphorylated.

EGF and other tyrosine kinase growth factor receptors may also regulate the tight junction MAGUKs ZO-1 and ZO-2. For example, EGF tightens the paracellular barrier between canine gastric mucosal cells in primary culture (A. H. Soll, personal communication), and treatment of cultured human colonic epithelial cell monolayers with insulin (McRoberts *et al.,* 1990) and hepatocyte growth factor (Nusrat *et al.,* 1994) increases paracellular permeability. Although the mechanisms for the changes in paracellular permeability are not known, it was demonstrated that exposure to EGF of the A431 cultured human epidermal cell line results in *de novo* relocalization of actin and ZO-1 into apical junction–like organization. Temporally correlated with this reorganization is new tyrosine phosphorylation of ZO-1 and ZO-2 (Van Itallie *et al.,* 1995).

None of the other MAGUKs have yet been proven to be substrates for tyrosine kinases, although all contain an absolutely conserved tyrosine residue positioned 11 residues N-terminal of the Guk domain. This tyrosine is followed by the consensus sequence EXV in all cases, suggesting that, if phosphorylated, this may bind the SH2 domain of a *src* family member. Given the connection between tyrosine kinase activity and alteration in cell contacts, the participation of MAGUKs in transducing kinase function deserves investigation.

B. Protein Kinase C

A large number of studies have suggested a role for protein phosphorylation by protein kinase C in the assembly and regulation of tight junctions; this information has been reviewed elsewhere (Anderson *et al.,* 1993, Anderson and Van Itallie, 1995). The process of apical junction assembly has been extensively investigated in the "calcium switch" model, in which MDCK cells are maintained in medium without calcium and lack junctions. Addition of calcium to normal levels results in rapid assembly of both adherens and tight junctions, which appears to be dependent on cadherin-mediated cell contacts. The exact mechanism by which calcium promotes junction assembly is unclear. The use of protein kinase C inhibitors in this MDCK cell model both prevents the junction disassembly induced by removal of calcium from the medium (Nigam *et al.,* 1991) and prevents tight junction reassembly after calcium replacement (Balda *et al.* 1991, Stuart and Nigam, 1995). In addition, when the protein kinase C agonist 1,2-dioctanoyl-sn-glycerol (DiC_8) is added to MDCK cells in the absence of extracellular calcium, there is a partial relocalization of ZO-1 and actin to the apical junction complex, and some reestablishment of the paracellular barrier. Cadherin relocalization is unaffected by use of DiC_8 (Balda *et al.,* 1993). Stuart and Nigam (1995) have presented evidence that ZO-1 may be a direct target of protein kinase C, and that inhibitors of this kinase prevent most of the increase in phosphorylation that these authors observed following the formation of cell–cell contacts. These results suggest that protein kinase C may act downstream from cadherin-mediated cell–cell contacts, and may act directly on tight junction proteins.

The calcium switch model for investigation of junction assembly is useful but of limited physiological relevance. More physiological changes in intracellular Ca^{2+} induced by a number of different stimuli, including vasopressin, angiotensin II, and epinephrine, result in increases in paracellular permeability (reviewed in Balda *et al.*, 1992). In a number of cases, this increase in paracellular permeability can be blocked by the use of protein kinase C inhibitors but not by calmodulin inhibitors (Flick *et al.*, 1991). However, results from these experiments are confusing and somewhat contradictory, and a clear hypothesis of how tight junctions are regulated has yet to be formulated. In addition, the role of protein kinase C in regulation of other MAGUKs is presently unknown.

VI. SPECULATION AND FUTURE DIRECTIONS

Further knowledge about the function of other MAGUK proteins seems likely to advance our understanding of the role of ZO-1 and ZO-2 in regulating tight junction organization and function. Yet there are already several reasons to sound a strong cautionary note. Sequence comparison of MAGUK PDZ and GuK domains suggests that the family tree has several branches that have evolved to perform different tasks. The Guk domains of some MAGUKs, such as p55 and Lin-2, are likely to be true kinases, while others, such as ZO-1 and ZO-2, have certainly lost the ability to bind GMP and ATP. The PDZ domains of the tight junction MAGUKs have diverged significantly from those found in other MAGUKs, suggesting that we should not expect their PDZ-binding proteins will be homologues of those associated with other MAGUKs. However, it seems safe to speculate at this time that members of the MAGUK family are involved in the dual functions of structural organization of membrane complexes and permitting membrane signal transduction. It seems likely that the precise function of each MAGUK has been tailored by evolution to fit a particular subdomain of cell cortex, be that a synapse, a septate junction, the lateral membrane, or the tight junction. It will be informative and fascinating to compare and contrast future research on the various members of this protein family.

Acknowledgments

The authors would like to thank the following individuals for thought-provoking discussion about the potential function of MAGUK proteins: Stuart Kim, Peter Bryant, Daniel Woods,

Athar Chishti, Craig Garner, Mary Kennedy, and Morgan Sheng. A. S. Fanning was supported by NRSA award DK09261 from the National Institute of Diabetes and Digestive and Kidney Diseases and the Irwin M. Arias Postdoctoral Research Fellowship from the American Liver Foundation. All authors were supported by National Institutes of Health awards DK45134, CA66263, and DK38979.

References

Adams, M. E., Butler, M. H., Dwyer, T. M., Peters, M. F., Murnane, A. A., and Froehner, S. C. (1993). Two forms of mouse syntrophin, a 58 kd dystrophin-associated protein, differ in primary structure and tissue distribution. *Neuron* **11,** 531–540.

Anderson, J. M., and Van Itallie, C. M. (1995). Tight junctions and the molecular basis for regulation of paracellular permeability. *Am. J. Physiol.* **269,** G467-G475.

Balda, M. S., and Anderson, J. M. (1993). Two classes of tight junctions revealed by ZO-1 isoforms. *Am. J. Physiol.* **264,** C918-C924.

Balda, M. S., Gonzalez-Mariscal, L., Contreras, R. G., Macias-Silva, M., Torres-Marquez, M. E., Garcia-Sainz, J. A., and Cereijido, M. (1991). Assembly and sealing of tight junctions: Possible participation of G-proteins, phospholipase C, protein kinase C and calmodulin. *J. Membr. Biol.* **122,** 193–202.

Balda, M. S., Fallon, M. B., Van Itallie, C. M., and Anderson, J. M. (1992). Structure, function and pathophysiology of tight junctions in the gastrointestinal tract. *Yale J. Biol. Med.* **65,** 725–735.

Balda, M. S., Gonzales-Mariscal, L., Matter, K., Contreras, R. G., Cereijido, M., and Anderson, J. M. (1993). Assembly of the tight junctions: The role of diacylglycerol. *J. Cell Biol.* **123,** 293–302.

Bar-Sagi, D., Rotin, D., Batzer, A., Madiyan, V., and Schlessinger, J. (1993). SH3 domains direct cellular localization of signaling molecules. *Cell* **74,** 83–91.

Bredt, D. S., Hwang, P. M., Glatt, C. E., Lowenstein, C., Reed, R. R., and Snyder, S. H. (1991). Cloned and expressed nitric oxide synthase structurally resembles cytochrome P-450 reductase. *Nature (London)* **351,** 714–751.

Bryant, P. J., Watson, K. L., Justice, R. W., and Woods, D. F. (1993). Tumor suppressor genes encoding protein required for cell interactions and signal transduction in Drosophila. *Dev. Suppl.* 239–249.

Cereijido, M. (1992). Introduction: Evolution of ideas on the tight junction. *In* "Tight Junctions" (M. Cereijido, ed.), pp. 1–14. CRC Press, Boca Raton, FL.

Cho, K.-O., Hunt, C. A., and Kennedy, M. B. (1992). The rat brain postsynaptic density protein contains a region of homology to the drosophila discs-large tumor suppressor protein. *Neuron* **9,** 929–942.

Citi, S., Sabannay, H., Jakes, R., Geiger, B., and Kendrick-Jones, J. (1988). Cingulin, a new peripheral component of tight junctions. *Nature (London)* **333,** 272–276.

Claude, P., and Goodenough, D. A. (1973). Fracture faces of zonula occludentes from "tight" and "leaky" epithelia. *J. Cell Biol.* **58,** 390–400.

Devereux, J., Haeberli, P., and Smithies, O. (1984). A comprehensive set of sequence analysis programs for the VAX. *Nucleic Acids Res.* **12,** 387–395.

Duclos, F., Rodins, F., Wrogemann, K., Mandel, J.-L, and Koenig, M. (1994). The Friedreich ataxia region: Characterization of two novel genes and reduction of the critical region to 300kb. *Hum. Mol. Genet.* **3,** 909–914.

Ehlers, M. D., Tingley, W. G., and Huganir, R. L. (1995). Regulated subcellular distribution of the NR1 subunit of the NMDA receptor. *Nature (London)* **269,** 1734–1737.

Eisenmann, D. M., and Kim, S. (1994). Signal transduction and cell fate specification during *Caenorhabditis elegans* vulval induction. *Curr. Opin. Genet. Dev.* **4,** 508–516.

Fasano, A., Baudry, B., Pumplin, D. W., Wasserman, S. S., Tall, B. D., Ketley, J. M., and Kaper, J. B. (1991). Vibrio cholerae produces a second enterotoxin, which affects intestinal tight junctions. *Proc. Natl. Acad. Sci. U.S.A.* **88,** 5242–5246.

Flick, J. A., Tai, Y. H., Levine, S. L., Watson, A. M., Montrose, J. M., Madara, J. L., and Donowitz, M. (1991). Intracellular Ca^{2+} regulation of tight junction resistance in T84 cell monolayers—a protein kinase C effect [Abstract]. *Gastroenterology* **100,** A686.

Furuse, M., Hirase, T., Itoh, M., Nagafuchi, A., Yonemura, S., Tsukita, S., and Tsukita S. (1993). Occludin: A novel integral membrane protein localizing at tight junctions. *J. Cell Biol.* **123,** 1777–1788.

Furuse, M., Itoh, M., Hirase, T., Nagafuchi, A., Yonemura, S., Tsukita, S., and Tsukita, S. (1994). Direct association between occludin and ZO-1 and its possible involvement in the localization of occludin at tight junctions. *J. Cell Biol.* **127,** 1617–1626.

Gu, M. X., York, J. D., Warshawsky, I., and Majerus, P. W. (1991). Identification, cloning and expression of a cytosolic megakaryocyte protein tyrosine phosphatase with sequence homology to cytoskeletal protein 4.1. *Proc. Natl. Acad. Sci. U.S.A.* **88,** 5867–5871.

Gumbiner, B., Lowenkopf, T., and Apatira, D. (1991). Identification of 160-kDa polypeptide that binds to the tight junction protein ZO-1. *Proc. Natl. Acad. Sci. U.S.A.* **88,** 3460–3464.

Hecht, G., Pothoulakis, G., LaMont, J. T., and Madara, J. L. (1988). Clostridium difficile toxin a perturbs cytoskeletal structure and tight junction permeability of cultured human intestinal epithelial monolayers. *J. Clin. Invest.* **82,** 1516–1524.

Hoskins, R., Hajnal, A. F., Harp., S. A., and Kim, S. K. (1996). The *C. elegans* vulval induction gene *lin*-2 encodes a member of the MAGUK family of cell junction proteins. *Development* **122,** 97–111.

Howarth, A. G., Hughes, M. R., and Stevenson, B. R. (1992). Detection of the tight junction-associated protein ZO-1 in astrocytes and other nonepithelial cell types. *Am. J. Physiol.* **262,** C461-C469.

Itoh, M., Nagafuchi, A., Yonemura, S., Kitani-Yasuda, T., Tsukita, S., and Tsukita S. (1993). The 220-kD protein colocalizing with cadherins in non-epithelial cells is identical to ZO-1, a tight junction-associated protein in epithelial cells: cDNA cloning and immunolocalization. *J. Cell. Biol.* **121,** 491–502.

Itoh, N., and Nagata, S. (1993). A novel protein domain required for apoptosis: Mutational analysis of human Fas antigen. *J. Biol. Chem.* **268,** 10932.

Jesaitis, L. A., and Goodenough, D. A. (1994). Molecular characterization and tissue distribution of ZO-2, a tight junction protein homologous to ZO-1 and the Drosophila discs-large tumor suppressor protein. *J. Cell Biol.* **124,** 949–961.

Kim, E., Niethammer, M., Rothschild, A., Jan, Y. N., and Sheng, M. (1995). Clustering of Shaker-type K+ channels by direct interaction with the PSD-95/SAP90 family of membrane-associated guanylate kinases. *Nature* (*London*) **378,** 85–88.

Kim, S. (1995). Tight junctions, membrane-associated guanylate kinases and cell signaling. *Curr. Opin. Cell Biol.* **7,** 641–649.

Kistner, U., Wenzel, B. M., Veh, R. W., Cases-Langhoff, C., Garner, A. M., Appeltaure, U., Voss, B., Gundelfinger, E. D., and Garner, C. C. (1993). SAP90, a rat presynaptic protein related to the product of the *Drosophila* tumor suppressor gene *dlg-A. J. Biol. Chem.* **268,** 4580–4583.

Kistner, U., Garner, C. C., and Linial, M. (1995). Nucleotide binding by the synapse associated protien SAP90. *FEBS Lett.* **359,** 159–163.

Kornau, H.-C., Schenker, L. T., Kennedy, M. B., and Seeburg, P. H. (1995). Domain interactions between NMDA receptor subunits and the postsyaptic density protein PSD-95. *Science* **269,** 1737–1740.

Kurihara, H., Anderson, J. M., and Farquhar, M. G. (1995). Increased tyr phosphorylation of ZO-1 during modification of tight junctions between glomerular foot processes. *Am. J. Physiol.* **268,** F514-F524.

Lane, N. (1991). Anantomy of the tight junction: Invertebrates. *In* "Tight Junctions" (M. Cereijido, ed.), pp. 23–48. CRC Press, Boca Raton, FL.

Lowenstein, E. J., and Schlessinger, J. (1992). The SH2 and SH3 domain-containing protein GRB2 links receptor tyrosine kinases to ras signaling. *Cell* **70,** 431–442.

Lue, R., Marfatia, S. M., Branton, D., and Chisti, A. H. (1994). Cloning and characterization of hdlg: The human homologue of the Drosophila discs large tumor suppressor binds to protein 4.1. *Proc. Natl. Acad. Sci. U.S.A.* **91,** 9818–9822.

Madara, J. L. (1992). Anatomy of the tight junction: Vertebrates. *In* "Tight Junctions" (M. Cereijido, ed.), pp. 15–22. CRC Press, Bocca Raton, FL.

Madara, J. L., Parkos, C., Colgan, S., Nusrat, A., Atisook, K., and Kaoutzani, P. (1992). The movement of solutes and cells across tight junctions. *Ann. New York Acad. Sci.* **664,** 47–60.

Maekawa, K., Imagawa, N., Nagamatsu, M., and Harada, S. (1994). Molecular cloning of a novel protein-tyrosine phosphatase containing a membrane-binding domain and GLGF repeats. *FEBS Lett.* **337,** 200–206.

Maher, P. A., Pasquale, E. B., Wang, J. Y.J., and Singer, S. J. (1985). Phosphotyrosine-containing proteins are concentrated in focal adhesions and intercellular junctions in normal cells. *Proc. Natl. Acad. Sci. U.S.A.* **82,** 6576–6580.

Marfatia, S. M., Lue, R. A., Branton, D., and Chisti, A. H. (1994). *In vitro* studies suggest a membrane-associated complex between erythroid p55, protein 4.1, and glycophorin C. *J. Biol. Chem.* **269,** 8631–8634.

Matsuyoshi, N., Hamaguchi, M., Taniguchi, S., Nagafuchi, A., Tsukita, S., and Takeichi, M. (1992). Cadherin-mediated cell-cell adhesion is perturbed by v-src tyrosine phosphorylation in metastatic fibroblasts. *J. Cell Biol.* **118,** 703–714.

McRoberts, J. A., Aranda, R., Riley, N., and Kang, H. (1990). Insulin regulates the paracellular permeability of cultured intestinal epithelial cell monolayers. *J. Clin. Invest.* **85,** 1127–1134.

Mohandes, H., Birkenmeier, E., Fanning, A. S., Anderson, J. M., and Korenberg, J. (1995). Localization of the gene for TJP1 to human chromosome 15q13, distal to the Prader-Willi/Angelman critical region (15q11–13) and to mouse chromosome 7. *Genomics* **30,** 594–597.

Muller, B. M., Kistner, U., Veh, R. W., Cases-Langhoff, C., Becker, B., Gundelfinger, E. D., and Garner, C. (1995). Molecular characterization and spatial distribution of SAP97, a novel presynaptic protein homologous to SAP90 and the Drosophila discs-large tumor suppressor protein. *J. Neurosci.* **15,** 2354–2366.

Musacchio, A., Gibson, T., Lehto, V.-P., and Saraste, M. (1992). SH3—an abundant protein domain in search of a function. *FEBS Lett.* **307,** 55–61.

Nelson, W. J. (1991). Generation of plasma membrane domains in polarized epithelial cells: Role of cell-cell contacts and assembly of the membrane cytoskeleton. *Biochem. Soc. Trans.* **19,** 1055–1059.

Nigam, S. K., Deniseko, N., Rodriguez-Boulan, E., and Citi, S. (1991). The role of phosphorylation in development of tight junctions in cultured renal epithelial (MDCK) cells. *Biochem. Biophys. Res. Commun.* **181,** 548–553.

Nusrat, A., Parkos, C. A., Bacarra, A. E., Godowski, P. J., Delp-Archer, C., Rosen, E. M., and Madara, J. L. (1994). Hepatocyte growth factor/scatter factor effects on epithelia. Regulation of intercellular junctions in transformed and nontransformed cell lines, basolateral polarization of c-met receptor in transformed and natural intestinal epithelia and induction of rapid wound repair in a transformed model epthielium. *J. Clin. Invest.* **93,** 2056–2065.

Pawson, T. (1995). Protein modules and signalling networks. *Nature* (*London*) **373,** 573–579.

Ponting, C. F., and Phillips, C. (1995). DHR domains in syntrophins, neuronal NO synthase and other interacellular proteins. *TIBS* **20,** 102–103.
Powell, D. W. (1981). Barrier function of epithelia. *Am. J. Physiol.* **41,** G275-G288.
Ruff, P., Speicher, D. W., and Husain-Chisti, A. (1991). Molecular identification of a major palmitoylated erythrocyte membrane protein containing the src 3 homology motif. *Proc. Natl. Acad. Sci. U.S.A.* **88,** 6595–6599.
Sato, T., Irie, S., Kitada, S., and Reed, J. C. (1995). FAP-1: A protein tyrosine phosphatase that associates with Fas. *Science* **268,** 411–415.
Schnabel, E., Anderson, J. M., and Farquhar, M. G. (1990). The tight junction protein ZO-1 is concentrated along slit diaphragms of the glomerular epithelium. *J. Cell Biol.* **111,** 1255–1263.
Staddon, J. M., Herrenknecht, K., Smales, C., and Rubin, L. L. (1995). Evidence that tyrosine phosphorylation may increase tight junction permeability. *J. Cell Sci.* **108,** 609–619.
Stuart, R. O., and Nigam, S. K. (1995). Regulated assembly of tight junctions by protein kinase C. *Proc. Natl. Acad. Sci. U.S.A.* **92,** 6072–6076.
Takata, K., and Singer, S. J. (1988). Phosphotyrosine-modified proteins are concentrated at the membranes of epithelial and endothelial cells during tissue development in chick embryos. *J. Cell Biol.* **106,** 1757–1764.
Thiesen, H., Purcell, J., Bennett, M., Kansagara, D., Syed, A., and Marsh, J. L. (1994). *dishevelled* is required during *wingless* signaling to establish both cell polarity and identity. *Development* **120,** 347–360.
Trofatter, J. A., MacCollin, M. M., Rutter, J. L., Murrell, J. R., Duyao, M. P., Parry, D. M., Eldredge, R., Kley, N., Menon, A. G., and Pulaski, K. (1993). A novel moesin-, ezrin-, radixin-like gene is a candidate for the neurofibromatosis 2 tumor suppressor. *Cell* **72,** 791–800.
Tsukita, S., Oishi, K., Akiyama, T., Yamanashi, Y., Yamamoto, T., and Tsukita, S. (1991). Specific proto-oncogenic tyrosine kinases of src family are enriched in cell-to-cell adherens junctions where the level of tyrosine phosphorylation is elevated. *J. Cell. Biol.* **113,** 867–879.
Van Itallie, C. M,, Balda, M. S., and Anderson, J. M. (1995). EGF induces ZO-1 translocation and tyrosine phosphorylation in A431 cells. *J. Cell Sci.* **108,** 1735–1742.
Volberg, T., Zick, Y., Dror, R., Sabanay, I., Gilon, C., Levitzki, A., and Geiger, B. (1992). The effect of tyrosine-specific protein phosphorylation on the assembly of adherens-type junctions. *EMBO J.* **11,** 1733–1742.
Warren, S. L., Handel, L. M., and Nelson, W. J. (1988). Elevated expression of pp60$^{c\text{-}src}$ alters a selective morphogenetic property of epithelial cells in vitro without a mitogenic effect. *Mol. Cell. Biol.* **8,** 632–646.
Weber, E., Berta, G., Tousson, A., St. John, P., Gree, M. W., Gopalokrishnam, U., Jilling, T., Sorscher, E. J., Elton, T. S., Abrahamson, D. R., and Kirk, K. L. (1994). Expression and polarized targeting of a Rab3 isoform in epithelial cells. *J. Cell Biol.* **125,** 583–594.
Willott, E., Balda, M. S., Fanning, A. S, Jameson, B., Van Itallie, C., and Anderson, J. M. (1993). The tight junction protein ZO-1 is homologous to the *Drosophila discs-large* tumor suppressor protein of septate junctions. *Proc. Natl. Acad. Sci. U.S.A.* **90,** 7834–7838.
Woods, D. F., and Bryant, P. J. (1989). Molecular cloning of the *Lethal(1) Discs Large-1* oncogene of *Drosophila. Dev. Biol.* **134,** 222–235.
Woods, D. F., and Bryant, P. J. (1991). The discs-large tumor suppressor gene of Drosophila encodes a guanylate kinase homolog localized at septate junctions. *Cell* **66,** 451–464.
Yang, Q., and Tonks, N. K. (1991). Isolation of a cDNA clone encoding a human protein-tyrosine phosphatase with homology to the cytoskeletal associated proteins band 4.1, ezrin, and talin. *Proc. Natl. Acad. Sci. U.S.A.* **88,** 5949–5953.

Zahraoui, A., Joberty, G., Arpin, M., Fontaine, J. J., Hellio, R., Tavitian, A., and Louvard, D. (1994). A small rab GTPase is distributed in cytoplasmic vesicles in non polarized cells but colocalized with the tight junction marker ZO-1 in polarized epithelial cells. *J. Cell Biol.* **124,** 101–115.

Zhong, Y., Saitoh, T., Minase, T., Sawada, N., Enomoto, K., and Mori, M. (1993). Monoclonal antibody 7H6 reacts with a novel tight junction-associated protein distinct from ZO-1, cingulin and ZO-2. *J. Cell Biol.* **120,** 477–483.

CHAPTER 12

Regulation of Membrane Protein Organization at the Neuromuscular Junction

Gregory G. Gayer, James T. Campanelli, and Richard H. Scheller
Howard Hughes Medical Institute, Department of Molecular and Cellular Physiology, Stanford University, Stanford, California 94305

I. INTRODUCTION

During nervous system development, local signaling between neurons and target cells is critical for synapse formation. Cell surface and secreted molecules direct the organization of both the presynaptic axon terminal and postsynaptic cell membrane specializations that are essential for the efficient flow of information between neurons. Identifying these signaling molecules and their effects on pre- and postsynaptic membrane cytoarchitecture is fundamental to our understanding of nervous system development.

Perhaps the most widely studied synapse in developmental neurobiology is the synapse between motor neuron and muscle, the neuromuscular junction (NMJ). This is due to its relative large size, ease of access, and homogeneous synaptic population, compared to central synapses. Like central synapses, the NMJ presynaptic nerve terminal, postsynaptic muscle cell, and intervening synaptic cleft are highly specialized structures compared to nonsynaptic regions (for review, see Hall and Sanes, 1993). Thus, the NMJ provides a model synapse for studying synaptogenesis.

Presynaptic nerve terminal morphological and molecular specializations include the localized accumulation of synaptic vesicles and calcium channels at the active zone, a region of the nerve terminal membrane adjacent to the muscle cell (Robitaille *et al.,* 1990; Hall and Sanes, 1993). Restricting vesicle and ion channel location to the active zone is necessary for the efficient release of acetylcholine (ACh) in response to an action potential and the subsequent intracellular Ca^{2+} rise.

In the muscle, postsynaptic specializations are also apparent. Collectively called the postsynaptic apparatus, this muscle region has morphological and molecular characteristics necessary for the receipt and transduction of the ACh signal, resulting in muscle contraction. Electron micrographs reveal that the muscle membrane opposite the nerve is invaginated, forming structures termed junctional folds (Salpeter, 1987). The fold crests have Ach receptors (AChRs) at concentrations as high as 10, 000 receptors/μm^2 (Fertuck and Salpeter, 1976). The AChR concentration at the NMJ is 1000-fold higher than that in extrajunctional regions. The junctional fold crests also contain rapsyn, syntrophin, and utrophin (see later), while dystrophin, sodium channels, and ankyrin are found localized in the depths of the folds (Flucher and Daniels, 1989). These specializations are thought to maximize the fidelity of synaptic transmission between nerve and muscle by ensuring that each action potential generates a muscle cell contraction.

The synaptic cleft is a 50-nm-wide space between presynaptic and postsynaptic cells filled with an electron-dense form of extracellular matrix called the basal lamina, which is distinct from that in nonsynaptic regions. For example, isoforms of acetylcholinesterase are found concentrated in the synaptic cleft, where they can efficiently terminate synaptic transmission by degrading ACh (Hall, 1973; Katz and Miledi, 1973; McMahan *et al.,* 1978; Salpeter, 1967). Other matrix components, including heparin proteoglycans (Anderson and Fambrough, 1983), agrin (Reist *et al.,* 1987; Rupp *et al.,* 1991), collagens $\alpha 3$(IV), $\alpha 4$(IV), and laminin A (Sanes *et al.,* 1990), S-laminin (Hunter *et al.,* 1989b), and neuregulin growth factors (Jo *et al.,* 1995), are also present specifically in synaptic basal lamina.

II. NEURONAL SIGNALING MOLECULES

How do the developing presynaptic motor neuron and postsynaptic muscle cell coordinate this molecular registry of their components? Molecules released by the nerve that occupy the extracellular matrix have been demonstrated to regulate aspects of this organization (Jennings and Burden, 1993; McMahan, 1990). Neural signals act both to increase local gene expression of AChR subunits at synaptic nuclei and to redistribute preexisting AChRs. For example, prior to innervation muscle membrane AChRs are distributed diffusely, and after contact with the motor neuron they become clustered at the nerve–muscle contact site (Bevan and Steinbach, 1977; Fertuck and Salpeter, 1976). Several neuron-derived signaling molecules have been implicated in this process, including the extracellular matrix protein agrin, AChR-inducing activity (ARIA), and calcitonin gene–related peptide (CGRP). CGRP is present in motor neuron–dense core granules (New and Mudge, 1986) and increases AChR subunit expression through stimulation of the adenylyl cyclase signaling cascade (Fontaine *et al.,* 1987; New and Mudge, 1986). ARIA is a neu differentiation factor family homologue. Members of this family bind to and stimulate the neu receptor tyrosine kinase (Falls *et al.,* 1993). Exposure of myotubes to ARIA increases the rate of synthesis and insertion of AChR subunits into the membrane, as well as receptor cluster size (Usdin and Fischbach, 1986). Furthermore, ARIA is expressed in the ventral horn of the spinal chord, an area that contains motor neuron cell bodies (Falls *et al.,* 1993). Unlike ARIA and CGRP, agrin does not increase AChR subunit synthesis but does redistribute preexisting AChRs (McMahan, 1990). Thus, the mature NMJ architecture is probably a product of these and other factors interacting with specific receptors that direct the local synthesis, accumulation, and anchorage of synaptic components.

Agrin molecules reside in the basal lamina separating the motor neuron and muscle membrane. The importance of the basal lamina in NMJ development is well established. For example, synaptic specializations reform at the site of original innervation in frog muscle after ablation of both the nerve and muscle, even when the nerve, or the muscle, was prevented from regenerating (Burden *et al.,* 1979; Sanes *et al.,* 1978). In these experiments, only the stable basal lamina was present after ablation of pre- and postsynaptic cells. Thus, signaling molecules that regulate synapse formation are released by the nerve and/or muscle and occupy the basal lamina. Several studies indicate that agrin mediates basal lamina–induced formation of synaptic specializations. First, agrin is synthesized in motor neurons and secreted at axon terminals during synapse formation (Magill-Solc and McMahan, 1988, 1990a,b; Rupp *et al.,* 1991). Second, the basal lamina of the

NMJ is labeled with anti-agrin antibodies (Reist *et al.*, 1987), and antibodies specific to agrin applied to muscle–nerve co-cultures prevented nerve-induced AChR aggregation (Reist *et al.*, 1992). Third, agrin expression levels are the highest during development, when motor neurons are contacting muscle cells (Hoch *et al.*, 1993). Finally, agrin application to cultured myotubes, either presented by cells expressing recombinant agrin (Campanelli *et al.*, 1991; see also Fig. 1) or bath applied as a purified soluble molecule (Ferns *et al.*, 1993; Nitkin *et al.*, 1987; Ruegg *et al.*, 1992), results in localized AChR accumulation either at the contact site or as random patches, respectively. The role of agrin in synapse development is the focus of this chapter.

III. AGRIN

A. Domain Structure

Structural analysis of agrin reveals several features that are likely to be involved in synaptogenesis. Agrin is a large, structurally diverse secreted glycoprotein that is conserved in vertebrates. For example, ray, rat, and chick agrin cDNA show 60% amino acid identity over the full length of the molecule, as well as many more highly conserved functional domains (Nastuk and Fallon, 1993; Rupp *et al.*, 1991; Smith *et al.*, 1987; Tsim *et al.*, 1992). Analysis of rat agrin cDNA clones predicts a 1940-amino-acid protein that is highly cysteine rich (Rupp *et al.*, 1991). Western blot analyses of embryonic rat brain, muscle, and spinal cord using anti-agrin antibodies demonstrate that a 200-kDa protein is similar to the protein size predicted from cDNA analysis (Rupp *et al.*, 1991).

Agrin structural domains include a signal sequence, regions that are structurally related to both protease inhibitors and follistatin, epidermal growth factor– (EGF-) like sequences, and domains that resemble portions of laminin (Fig. 2). Beginning at the agrin amino terminus, the hydrophobic amino acid sequence that follows the start methionine is most likely to be the signal sequence that directs agrin to the secretory pathway. Agrin has nine amino-terminal domains that resemble the pancreatic secretory trypsin inhibitor (Kazal-type) family of protease inhibitors (Laskowski and Kato, 1980). Multiple inhibitor domains on a single polypeptide chain, such as in agrin, are common among protease inhibitors and are thought to arise through repeated duplication by unequal crossover events (Laskowski and Kato, 1980). It is unknown if these agrin protease inhibitor–like domains are functional. Presumably, the ability to withstand proteolysis would stabilize the basal lamina. Alternatively, glia-derived nexin, a factor that pro-

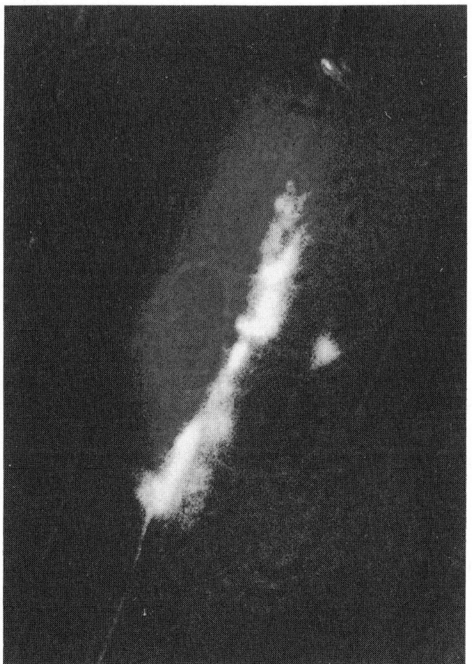

FIGURE 1 Agrin clusters AChRs at cell–cell contact sites. Immunofluorescence microscopy of agrin-expressing Chinese hamster ovary (CHO) cells and C2 myotubes in co-culture. Myotube AChRs are labeled with rhodamine α-bungarotoxin and agrin-expressing CHO cells are labeled with cascade blue.

motes neurite outgrowth, is an example of a very potent serine protease inhibitor; perhaps the agrin amino-terminal region regulates protease activity during neuronal migration (Gloor *et al.,* 1986; Monrad, 1988). The protease inhibitor domains also display sequence homology to follistatin (Patthy and Nikolics, 1993). Follistatin binds to the transforming growth factor-β (TGF-β) family members activin and inhibin, and is thought to play an important role in neural tissue formation (Hemmati-Brivanlou and Melton, 1994; Hemmati-Brivanlou *et al.,* 1994). It has been proposed that agrin may bind members of the TGF-β or platelet-derived growth factor families, thus providing a localized reservoir of growth factors (Patthy and Nikolics, 1993). However, binding studies of activin and inhibin to the agrin amino terminus, which contains only the nine follistatin-like domains, do not thus far support this hypothesis.

The agrin amino terminus also contains a cysteine- and proline-rich region that encodes a domain homologous to domain III of B1 and B2 laminin,

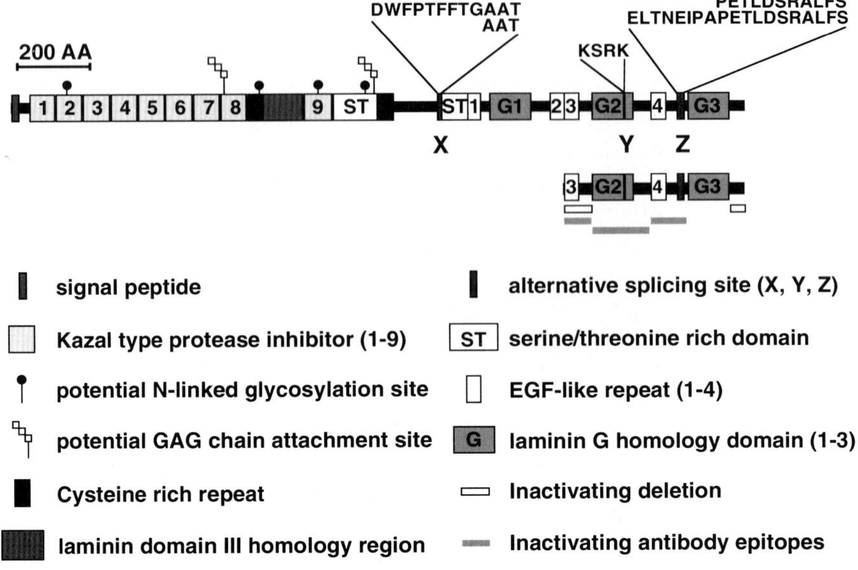

FIGURE 2 Schematic diagram of agrin, a multidomain protein (see text for details).

as well as domain IIIb of laminin A (Rupp *et al.,* 1991; Tsim *et al.,* 1992). Again, the function of these domains in agrin are not known. In laminin, domains III-B1 and -B2 promote cell attachment, chemotaxis, and neuronal attachment but not neurite outgrowth (Graf *et al.,* 1987; Kleinman *et al.,* 1989), while domain IIIb-A can promote process extension of neuroblastoma cells and endothelial cell adhesion (Sephel *et al.,* 1989). Furthermore, a laminin proteolytic fragment that contains the cysteine-rich repeats of domain III has mitogenic activity and binds nidogen (Panayotou *et al.,* 1989). Thus, these laminin domains are important in both cell attachment and cell growth, suggesting that agrin's laminin-like domain may serve a similar function. COS and Chinese hamster ovary cell transfection experiments demonstrate that the agrin amino terminus is associated with the cell membrane or extracellular matrix while the carboxyl terminus is secreted (G. G. Gayer, unpublished data; Ferns *et al.,* 1993). Taken together, these data suggest that agrin's amino terminal region is responsible for extracellular matrix interactions.

Recently, agrin was demonstrated to be a predominant heparan sulfate proteoglycan in embryonic chick brain (Tsen *et al.,* 1995). Two sequences between the last protease inhibitor domain and the first EGF repeat are rich in serine and threonine residues, suggesting that they may serve as *O*-linked glycosylation sites. These agrin regions are not well conserved between species in precise sequence, but the high serine and threonine content is maintained. Sequences between protease inhibitor domain 8 and serine–threonine domain 1 have conserved serine–glycine repeats flanked by acidic residues that in other proteoglycans have been defined as glycosaminoglycan (GAG) chain attachment sites (Jackson *et al.,* 1991).

Another sequence similarity to extracellular matrix proteins is agrin's two Leu-Arg-Glu (LRE) sequences. The LRE tripeptide sequence is not common; however, it is found in proteins located at the NMJ, such as S-laminin, laminin A, B2 laminin, and acetylcholinesterase (Hunter *et al.,* 1989a; Porter *et al.,* 1995; Saski *et al.,* 1988; Saski and Yamada, 1987). In S-laminin, the LRE sequences are involved in neural cell adhesion as well as concentrating and localizing S-laminin at the synapse (Porter *et al.,* 1995).

The carboxyl half of agrin contains four cysteine-rich regions that are 25% identical to a cysteine-rich region of EGF (Rupp *et al.,* 1991; Tsim *et al.,* 1992). Similar cysteine-rich EGF-like domains are found in many extracellular matrix proteins in addition to agrin, including laminin, tenascin, thrombospondin (Davis, 1990), and neurexin (Ushkaryov *et al.,* 1992). EGF repeats in extracellular matrix proteins may function as spacer elements between critical regions or may act like the EGF growth factor itself, which activates intracellular signaling through a tyrosine kinase receptor (Carpenter, 1987). The agrin carboxyl terminus contains three domains that have a high sequence identity to the globular (G) domains of laminin and merosin (Rupp *et al.,* 1991; Tsim *et al.,* 1992). In laminin, the G domains are located at the base of the cruciform structure and are involved in binding heparan-sulfated proteoglycans and heparin (Sakashita *et al.,* 1980), as well as binding to α-dystroglycan (α-DG) (Ervasti and Campbell, 1993b).

Various truncation deletions were used to define the active regions of agrin. AChR clustering activity is localized to the carboxyl terminus, which has many domains that could mediate clustering. The sequence starting with the third EGF-like region through the end of agrin was required for agrin-induced cluster formation (Hoch *et al.,* 1994; Gesemann *et al.,* 1995; see also Fig. 2, truncated fragment). Further deletions of this subfragment produce proteins that are not active in inducing AChR clusters when added to myotubes (Hoch *et al.,* 1994; Fig. 2). In addition, monoclonal antibodies that inhibit agrin-induced cluster formation also bind to this fragment (Hoch *et al.,* 1994; Fig. 2). Since the minimally active region contains many structural domains, agrin's interaction with the cell membrane may be multivalent.

B. Alternative Splicing of Agrin

Agrin undergoes alternative RNA splicing at three positions termed X, Y, and Z (see Fig. 2 for locations and specific sequences). Agrin X position isoforms are expressed with no inserts or 3- or 12-amino-acid inserts. There is no detected difference in biological activity of X position isoforms and the various forms are not regulated in their expression patterns (Ferns *et al.*, 1992; Hoch *et al.*, 1993). The Y splice site, located in the second laminin G-like domain, has two possible structures: no insert (Y0) or with the four basic amino acids, KSRK (Y4) (Ruegg *et al.*, 1992; Rupp *et al.*, 1991). Y4-agrin is approximately 10 times more active in producing AChR clusters on myotubes, compared to Y0-agrin (Ferns *et al.*, 1992). Insertion of KSRK into the G domain increases the ability of Y4-agrin to bind heparin, which may be reflected in the increased clustering activity seen when compared to Y0-agrin (J. T. Campanelli and G. G. Gayer, unpublished data). Z position inserts dramatically increase agrin's ability to form AChR clusters (Ferns *et al.*, 1992, 1993; Gesemann *et al.*, 1995; Ruegg *et al.*, 1992). For example, rat agrin can be found with either no insert (Z0) or an 8- (Z8), 11- (Z11), or 19- (Z19) amino-acid insert at the Z position. Z8-agrin is 1000 times more active in producing AChR clusters when compared to Z0-agrin on cultured myotubes (Ferns *et al.*, 1993). The 19-amino-acid splice variant is a combination of two distinct exons encoding the 8- and 11-amino-acid sequences. The relative activity of the splice inserts at the Z position are $8 \geq 19 > 11 > 0$. Exactly how these inserts are involved in enhancing agrin-induced cluster formation is an area of active investigation. Recently, a clue was provided when it was demonstrated that agrin monoclonal antibody 86 bound specifically to Z8-, Z11-, and Z19-agrin but not Z0-agrin, without binding to the insert sequence directly, suggesting that the inserts produce a conformational change in the agrin protein (Hoch *et al.*, 1994).

C. Expression and Localization

Two important characteristics of agrin are revealed by recent studies of agrin expression and localization. First, the most active isoform of agrin (Y4-Z8) is expressed only in neural tissue, while the least active isoform (Y0-Z0) is expressed in muscle (Hoch *et al.*, 1993; O'Connor *et al.*, 1994; Rupp *et al.*, 1991). Furthermore, agrin located at the adult rat NMJ can be labeled with antisera that are specific to neuronal agrin (Fig. 3i). Presumably, the requirement for neural-derived agrin facilitates synapse-specific cluster formation. Second, agrin expression during embryogenesis is temporally correlated with synapse formation. For example, detectable agrin mRNA and protein levels are found in both chick and rat at the time that motor neurons first contact muscle cells (Hoch *et al.*, 1993; Thomas *et al.*, 1993). Interestingly, brain and spinal cord steady-state mRNA levels stay relatively high until birth, at which time the levels are then dramatically reduced (Hoch *et al.,* 1993).

In the brain, agrin mRNA is found in all regions studied. Furthermore, *in situ* hybridization studies in rat demonstrate that agrin expression is not restricted to cholinergic neurons (O'Connor *et al.,* 1994). Agrin is expressed diffusely throughout the brain and is concentrated in cell bodies located in the olfactory bulb, cortex, granule, pyramidal cell layers, and hilus of the hippocampus, as well as the Purkinje cell layer of the cerebellum (O'Connor *et al.,* 1994). High agrin message levels in embryonic and adult brain suggest that agrin may play a role in central synapse formation. Consistent with the neural expression of agrin, it has been demonstrated that primary hippocampal neurons co-cultured with primary muscle cells can induce AChR clustering (Dutton *et al.,* 1995). Localization of the agrin protein using immunohistochemical methods is less clear. Agrin protein levels at central synapses may be too low to detect, or agrin may be complexed with other molecules and inaccessible to antibody labeling. The highest level of immunohistochemical labeling is of the small capillaries, pia, and choroid plexus of the brain (Magill-Solc and McMahan, 1988, 1990b; Rupp *et al.,* 1991). Interestingly, members of the agrin receptor complex (see later) co-localize with agrin in these same areas (G. G. Gayer, unpublished data; Khurana *et al.,* 1992). Although the function of agrin in nonsynaptic areas is unknown, it is interesting to speculate that agrin may have a role in the formation or maintenance of the blood–brain barrier between endothelial cells and astroglia.

Finally, agrin is expressed in many nonneuronal tissues, including muscle, kidney, and gut (Godfrey, 1991; Godfrey *et al.,* 1988). In the rat, kidney immunohistochemical analysis using agrin-specific monoclonal antibodies indicates that the glomeruli are strongly labeled (G. G. Gayer, unpublished data). While it is not known what function agrin serves in the kidney, the glomerulus is an area of the kidney that is used to filter proteins and has tight junctions similar to the blood–brain barrier. Polymerase chain reaction analysis indicates that the isoforms with weak clustering (Z0-agrin) activity are present in nonneural tissues (Hoch *et al.,* 1993). Agrin's presence in nonneuronal tissue suggests that agrin-induced AChR cluster formation is just one action of this multidomain protein.

IV. MECHANISMS OF AGRIN ACTION

A. Carbohydrate Modifications

Insights into how agrin transduces its signal have been provided by several studies that demonstrate the importance of proteoglycans in agrin's activity.

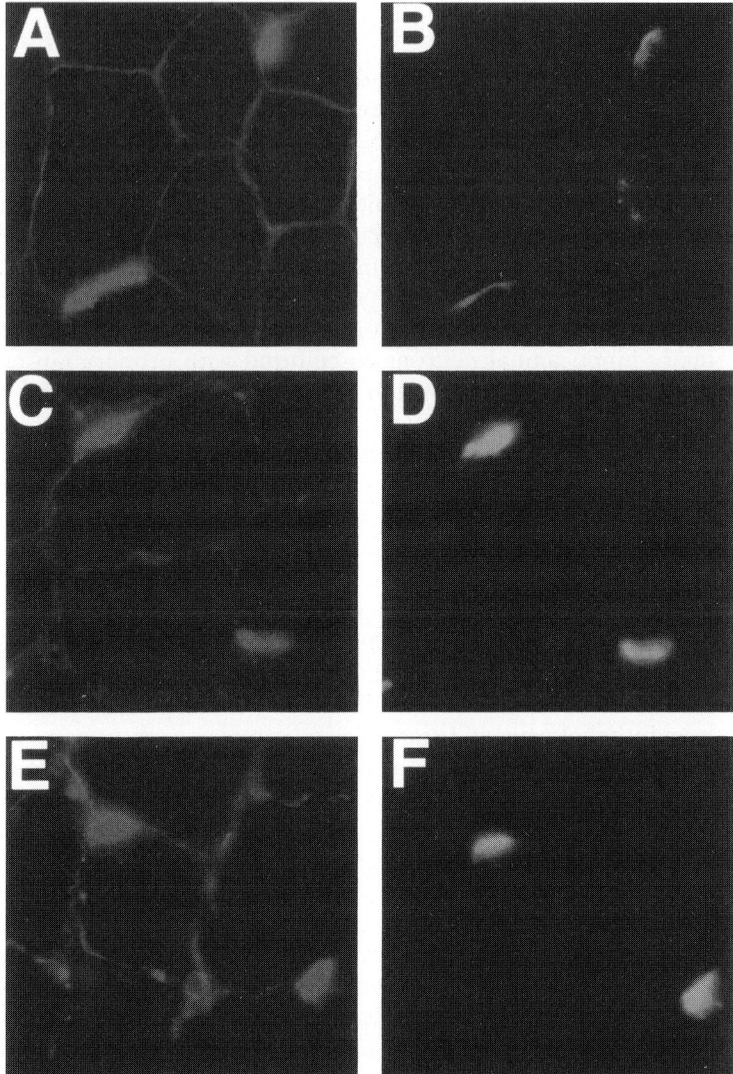

FIGURE 3 Immunofluorescence microscopy of rat NMJ showing extracellular matrix, membrane protein, and cytoskeletal protein localization. Antibody fluorescein labeling for α-DG, dystrophin, α-sarcoglycan, utrophin, and agrin is shown in panels a, c, e, g, and i, respectively. Co-labeling of AChR, using rhodamine α-bungarotoxin, is shown in adjacent panels b, d, f, h, and j, respectively.

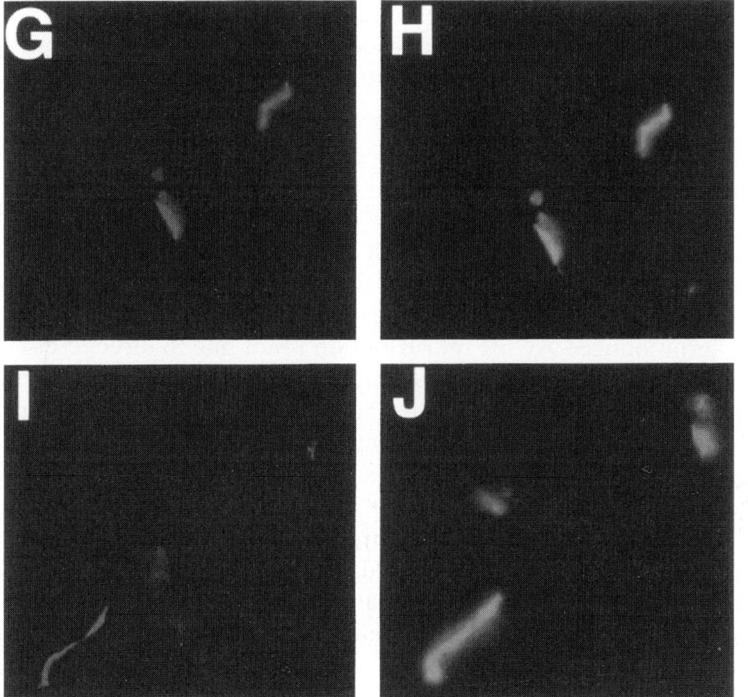

FIGURE 3 *Continued*

For example, addition of heparin and heparan sulfate prevents agrin-induced cluster formation in myotubes (Wallace, 1990). In addition, agrin-induced clustering activity of all agrin forms is decreased when assayed on a genetic variant of the C2 muscle cell line (S27) that is deficient in GAG synthesis (Ferns *et al.,* 1992, 1993). These data taken together suggest three possibilities. First, the agrin receptor is a heparan sulfate proteoglycan, or agrin binding to its receptor is aided by heparin. Second, a proteoglycan may play an auxiliary role in agrin activity. For example, proteoglycans, through a low-affinity interaction, concentrate or present basic fibroblast growth factor to its tyrosine kinase receptor complex (Lopez-Casillas *et al.,* 1991; Rapraeger *et al.,* 1991; Yayon *et al.,* 1991). Finally, cell surface or extracellular matrix proteoglycans might stabilize the localization of agrin-clustered molecules.

In addition to heparin, other carbohydrates are also involved in agrin-induced cluster formation. It has been reported that removal of *N*-acetylgalactosaminyl– (GalNAc-) terminated saccharides by enzymatic digestion prevents agrin-induced cluster formation (Martin and Sanes, 1995). Although the specific role of GalNac saccharides in the molecular mechanism of agrin-induced cluster formation is unclear, they are likely to be involved.

B. Redistribution of AChRs

Agrin interacts with the muscle cell membrane and causes the redistribution of preexisting AChRs without affecting receptor steady-state levels (Anderson and Cohen, 1977; Godfrey *et al.*, 1984; Wallace, 1988). Furthermore, agrin-induced AChR redistribution is not due to direct cross-linking of the AChR oligomers. For example, energy metabolism is required for agrin-induced receptor clustering in chick myotube cultures (Wallace, 1988). Presumably, cross-linking of agrin to AChRs would not require energy. Furthermore, each agrin molecule induces approximately 160 AChRs to cluster (Nitkin *et al.*, 1987), suggesting that agrin acts catalytically. Finally, agrin can induce clustering of other molecules, such as rapsyn, heparan-sulfated proteoglycans, and acetylcholinesterase (Nitkin *et al.*, 1987; Wallace, 1988, 1989). These data taken together support the theory that an agrin receptor–coupled signaling cascade mediates receptor aggregation.

C. Cytoskeleton

Several studies indicate that clustered AChRs at the NMJ are bound to the cytoskeleton. Cytoskeleton-bound proteins are not readily extracted with mild detergent. Thus, resistance to Triton X-100 extraction has been used as a measure of cytoskeletal association. Upon formation of clusters in developing myotubes, AChRs become increasingly resistant to Triton X-100 extraction (Prives *et al.*, 1982; review in Poo, 1985). AChR detergent extractability also correlates with receptor mobility in the plasma membrane of cultured myotubes (Stya and Axelrod, 1983), indicating the receptors are prevented from freely diffusing away from clusters. Finally, agrin-induced AChR clusters are also more resistant to mild detergent extraction (Wallace, 1992). Several cytoskeletal proteins, such as actin, β-spectrin, rapsyn, dystrophin, utrophin, vincullin, talin, paxilin, filamin, α-actinin, and tropomyosin 2, also concentrate at the NMJ (reviewed in Froehner, 1991), which further supports the involvement of cytoskeletal proteins in the formation and stabilization of AChR clusters. While most of these cytoskeletal elements have not been shown to affect clustering, studies do implicate β-

spectrin, actin, rapsyn, syntrophin, and utrophin in AChR aggregation (see later).

In order for a cytoskeletal protein to be involved in receptor clustering, it should be located at the synapse, its expression should correspond to the production of clusters, and removal of the protein should reduce clustering. It is clear that the cytoskeletal protein rapsyn fits all these criteria (reviewed in Apel and Merlie, 1995; Froehner, 1991). Rapsyn, a cytoplasmic 43-kDa protein, co-localizes with AChR clusters on muscle cells both *in vivo* (Froehner *et al.,* 1981) and *in vitro* (Bloch and Froehner, 1987). Furthermore, localization of rapsyn at the NMJ temporally corresponds with AChR aggregation during development (Noakes *et al.,* 1993). The association of rapsyn with AChR clusters increases the ability of AChR to resist Triton X-100 extraction (Phillips *et al.,* 1993). These data suggest that rapsyn binds directly to the cytoplasmic face of AChRs and links them to the cytoskeleton. Transfection studies demonstrate that rapsyn binds to the α, β, γ, or δ AChR subunits (Maimone and Merlie, 1993). However, only the AChR β-subunit is cross-linked to rapsyn in the *Torpedo* electric organ (Burden *et al.,* 1983), and the ratio of rapsyn binding to the AChR oligomer is 1:1 (LaRochelle and Froehner, 1986, 1987), suggesting that rapsyn binds to only one subunit *in vivo*. Conclusive evidence for rapsyn's role in receptor clustering has been demonstrated by adding or removing rapsyn using various methods prior to clustering assays. Reconstitution studies in which rapsyn is transfected into both oocytes and fibroblasts demonstrate that rapsyn can cluster itself, as well as co-transfected AChR subunits, into microclusters (Froehner *et al.,* 1990; Maimone and Merlie, 1993; Phillips *et al.,* 1991). When AChR subunits are transfected alone, they do not form clusters, which suggests that rapsyn aggregation may be an initial event in receptor clustering. Finally, mice genetic knockouts that do not express rapsyn are unable to form AChR clusters and die soon after birth (Gautam *et al.,* 1995). Taken together, these data indicate that rapsyn is critically involved in mediating receptor clustering. Thus the molecular mechanisms that underlie receptor clustering are likely to involve the protein interactions between AChR oligomers, rapsyn, and other associated proteins.

How are receptors tethered to the cytoskeleton? Increasing evidence suggests that a network of actin- and spectrin-like molecules is involved. This network has been shown to consist of a 2-dimensional lattice that provides structural stability to erythrocytes and restricts the mobility of the anion transporter in erythrocytes and the neural cell adhesion molecule in neural cells (reviewed in Bennet, 1990; Gumbiner, 1993; and other chapters in this book). Molecular analysis reveals a family of proteins with structures similar to spectrin, including several spectrin isoforms, α-actinin, dystrophin, utrophin, syntrophin, and the 87-kDa protein. The importance of actin and β-spectrin in the formation and maintenance of muscle cell AChR clusters was found when removal of actin and β-spectrin by mild detergent extraction or enzymatic digestion resulted in AChR redistribution (Bloch, 1986; Bloch and Morrow, 1989).

Syntrophin (formerly 58-kDa) and an 87-kDa protein also co-localize with rapsyn at the NMJ (reviewed in Apel and Merlie, 1995; Sunada and Campbell, 1995; Tinsley *et al.,* 1994). Syntrophin is a member of a multigene family, and the members of the family vary in subcellular distribution (Sunada and Campbell, 1995). The acidic α-syntrophin (mouse syntrophin-1 and rabbit 59-kDa dystrophin-associated protein 1 [DAP-1]) is expressed throughout the sarcolemma in skeletal and cardiac muscle, while the basic β1-syntrophin is expressed in various tissues (Adams *et al.,* 1993; Ahn *et al.,* 1994; Peters *et al.,* 1994). Unlike the diffuse distribution of α- and β1-syntrophin, β2-syntrophin is localized specifically to the NMJ (Peters *et al.,* 1994), suggesting that the β2 form has a synapse-specific function. In support of syntrophin involvement in postsynaptic membrane function, *Torpedo* syntrophin has been found to co-localize at junctional folds where AChRs and rapsyn molecules are concentrated (Froehner *et al.,* 1987). In myotubes, syntrophin and AChRs are present in an area that contains a network of actin-like filaments (Bloch *et al.,* 1991). Thus, syntrophin may provide a link between AChR clusters and the actin cytoskeleton. Further data in support of this hypothesis are provided by studies demonstrating that the long, rodlike molecules utrophin and dystrophin can bind both syntrophin at their carboxyl terminus (Suzuki *et al.,* 1994, 1995; Yang *et al.,* 1994, Kramarcy *et al.,* 1994) and F-actin at their amino terminus (Hemmings *et al.,* 1992; Levine *et al.,* 1992; Way *et al.,* 1992; Winder *et al.,* 1995). The exact molecular mechanisms involved in the cross-linking are unclear, but they are likely to involve the 87-kDa protein as well. The 87-kDa protein is homologous to the carboxyl-terminal domains of dystrophin (Wagner *et al.,* 1993). This protein co-localizes specifically with syntrophin at the NMJ and copurifies with syntrophin and dystrophin, suggesting that these proteins are components of a complex (Butler *et al.,* 1992; Froehner *et al.,* 1987).

Evidence suggests that the 427-kDa spectrin-like protein dystrophin is not directly involved in linking AChRs to the cytoskeleton, but rather has other roles. The gene for dystrophin is mutated in both Duchenne's and Becker's muscular dystrophy (reviewed in Campbell, 1995). Patients with this disease present with severe muscle necrosis that results in death usually by the third decade of life. The molecular mechanisms involved in dystrophin-mediated muscular dystrophy are unclear. Dystrophin may provide a structural link with the extracellular matrix, thereby protecting the myofiber from mechanical stress, or it may act directly in AChR clustering or the organizing of other membrane proteins. The apparent normal NMJs and clustered AChRs in the mouse model (*mdx* mouse) for muscular dystrophy, in which the dystrophin gene is missing, argue against dystrophin's direct involvement in AChR clustering (Matsumura *et al.*, 1992). In addition, although dystrophin is concentrated at the NMJ, it is also found surrounding the sarcolemma (Sealock *et al.*, 1991; see also Fig. 3c), suggesting a function not specific to the postsynaptic apparatus. Dystrophin does co-localize with Na^+ channels and ankyrin at the troughs of junctional folds (Byers *et al.*, 1991; Flucher and Daniels, 1989). It is therefore possible that Na^+ channels are stabilized by a spectrin-like dystrophin cytoskeleton. Finally, dystrophin interacts with syntrophin and four transmembrane proteins that comprise the dystrophin–glycoprotein complex (DGC). The DGC links the extracellular matrix to the actin cytoskeleton (Ervasti and Campbell, 1993a,b). It has been suggested that muscle cells without dystrophin would not have the stabilizing effect of linking the cytoskeleton to the extracellular matrix and, therefore, would become more susceptible to mechanical stress.

Utrophin (or dystrophin-related protein) is highly homologous to dystrophin and shares many structural, as well as functional, characteristics (Tinsley *et al.*, 1994). These features include an amino terminus that binds actin (Winder *et al.*, 1995) and a carboxyl terminus that interacts with syntrophin (Kramarcy *et al.*, 1994; Yang *et al.*, 1995b). In addition, utrophin co-localizes with AChRs at the NMJ (Bewick *et al.*, 1992; Tinsley *et al.*, 1994). However, unlike dystrophin, it is not localized to extrajunctional areas (Khurana *et al.*, 1992; Ohlendieck *et al.*, 1991, see also Fig. 3g) and is found at junctional fold crests where AChRs reside (Bewick *et al.*, 1992; Tinsley *et al.*, 1994). In addition, utrophin co-localizes with AChRs at the earliest stages of muscle development (Phillips *et al.*, 1993). Intriguingly, microclusters of AChRs produced by overexpressing rapsyn in C2 muscle cells are not associated with utrophin, while the larger aggregates do have utrophin immunofluorescence (Phillips *et al.*, 1993). It has been suggested that utrophin might therefore be involved in the growth of AChR clusters (Phillips *et al.*, 1993). This increase in size of AChR clusters is likely to be directed by extracellular matrix–induced changes in members of the DGC, utrophin, and actin cytoskeleton.

V. AGRIN GLYCOPROTEIN COMPLEX

A schematic of the glycoprotein complex (GC) is shown in Fig. 4. The complex consists of the transmembrane proteins, α-DG and β-dystroglycan, and α- (adhalin or 50-dystrophin-associated glycoprotein [DAG]); β-(43-DAG), δ- (25-DAP), and γ- (35-DAG) sarcoglycan (reviewed in Apel and Merlie, 1995; Sunada and Campbell, 1995). Other proteins that associ-

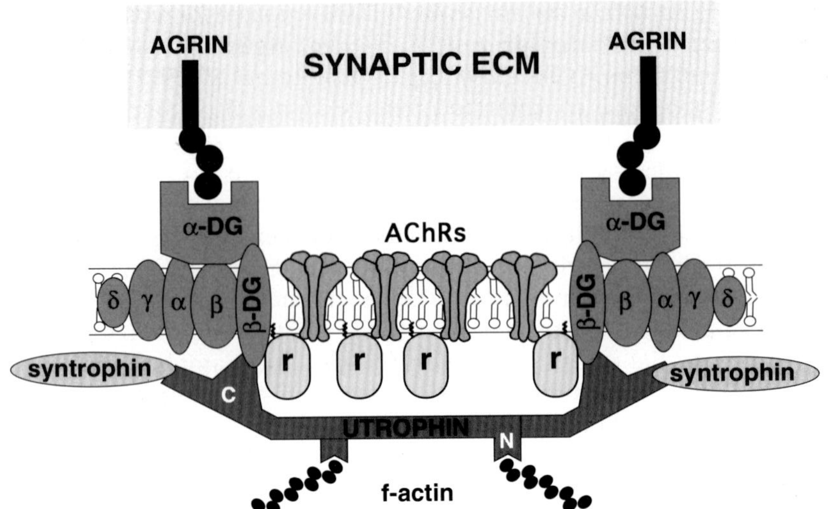

FIGURE 4 Organization of molecules at the NMJ. Rapsyn, α-, β-, δ-, and γ-sarcoglycan, and α- and β-dystroglycan are represented by r, α, β, δ, γ, α-DG, and β-DG, respectively.

ate with this complex are the intracellular proteins syntrophin and either utrophin or dystrophin. The transmembrane proteins form a tight association with each other. Dystrophin binds to members of the GC and F-actin, suggesting that it is a link between the GC and the actin cytoskeleton (Suzuki *et al.,* 1994). Utrophin binds F-actin and, by virtue of its homology to dystrophin, may bind to the GC, suggesting that it may also link the GC to the cytoskeleton.

The syntrophin–utrophin–actin-based cytoskeleton has been implicated in AChR clustering. A hypothesis for the molecular mechanism by which agrin induces receptor clustering arose from the demonstration that agrin binds directly to α-DG, a GC component (Bowe *et al.,* 1994; Campanelli *et al.,* 1994; Gee *et al.,* 1994; Sugiyama *et al.,* 1994; reviewed in Sealock and Froehner, 1994). Agrin and α-DG co-localize along with the other GC components at AChR clusters (Bowe *et al.,* 1994; Campanelli *et al.,* 1994; Gee *et al.,* 1994; Sugiyama *et al.,* 1994; see also Fig. 3). Furthermore, when antibodies specific to α-DG are co-incubated with agrin, they interfere with AChR cluster formation on myotubes, suggesting that agrin binding to α-DG is required for cluster formation (Campanelli *et al.,* 1994; Gee *et al.,* 1994). In one study (Campanelli *et al.,* 1994), the inhibition of agrin-induced clusters by α-DG antibodies resulted in loose aggregates of microclusters. Similar observations have also been made for nerve-induced clusters formed in the presence of anti-DG antibody (Cohen *et al.,* 1995). The accumulated data support a model in which small clusters are formed by the interaction of rapsyn molecules with AChRs, followed by the grouping of these microclusters into larger conglomerates.

If, indeed, the effects of agrin on clustering are mediated though α-DG, this might explain how heparin, proteoglycans, and Ca^{2+} are involved in agrin's activity. The agrin binding to α-DG is Ca^{2+} dependent and inhibited by heparin, similar to inhibition of agrin-induced cluster formation *in vivo* (Campanelli *et al.,* 1994; Gee,*et al.,* 1994; Ma *et al.,* 1993; Sugiyama *et al.,* 1994). In addition, because α-DG is a proteoglycan, it could explain the reduced agrin activity in muscle cells with improper GAG chain synthesis. For example, *in vitro* binding assays of α-DG derived from these mutant cell lines showed a much lower affinity to agrin (Campanelli *et al.,* 1994; Gee *et al.,* 1994; Sugiyama *et al.,* 1994).

The experiments described above support a role for α-DG in mediating agrin's signaling to the cytoskeleton. However, they do not rule out the involvement of other receptors or even other proteoglycans. This later hypothesis is supported by a study that showed that heparin binding to agrin was not affected by splice variants at the Z position (J. T. Campanelli, G. G. Gayer, and R. H. Scheller, unpublished data), suggesting that agrin binding to heparin is distinct from α-DG binding. In addition, the isoform with the least clustering activity (Z0) binds with a higher affinity to α-DG than the most active isoform (Z8) when measured either by overlay assays (Sugiyama *et al.,* 1994) or with agrin affinity chromatography (J. T. Campanelli, G. G. Gayer, and R. H. Scheller, unpublished data). This suggests two possibilities: (1) the affinity of agrin binding to the α-DG does not reflect in the magnitude of its ability to cluster receptors, or (2) agrin binds to an as yet unidentified proteoglycan distinct from α-DG.

A model depicting component organization in the agrin–GC–AChR complex is shown in Fig. 4. In this model, agrin binds to the GC via α-DG. The fact that rapsyn co-localizes with α-DG clusters independently of AChRs in reconstitution assays (Apel *et al.,* 1995) suggests that rapsyn binds to both AChRs and the GC. Utrophin, by virtue of its homology to dystrophin, may bind both β-dystroglycan and F-actin (see earlier). Thus, AChR linkage to the cytoskeleton would be through the agrin–GC–rapsyn complex binding to the actin cytoskeleton through utrophin. However, a direct interaction between agrin and the cytoskeleton via GC, similar to that found when laminin binds to a-DG (Ervasti and Campbell, 1993b), remains to be determined. The model suggests a molecular mechanism that would allow a single agrin molecule to cluster many AChRs. It is possible that agrin bound to the GC forms a submembraneous cytoskeletal coral surrounding the AChRs, thus trapping the receptors and preventing lateral migration in the plane of the membrane. Alternatively, agrin may transduce a signal through the GC or some as yet unidentified protein, which through unknown mechanisms produces a more stable cytoskeleton–AChR interaction.

VI. AGRIN INTRACELLULAR SIGNALING

Several studies have implicated tyrosine phosphorylation in agrin's mechanism of action. Agrin increases the phosphorylation of the AChR subunits β, γ, and δ (Wallace *et al.,* 1991). β-Subunit phosphorylation occurs, as detected by labeling with anti-phosphotyrosine antibodies, shortly after agrin addition to chick myotubes (Wallace *et al.,* 1991). In addition, inhibition of agrin-induced β-subunit phosphorylation prevents AChR clustering (Wallace, 1994). Using a combination of protein kinase and phosphatase inhibitors, it was shown that agrin induces tyrosine phosphorylation via kinase activation rather than phosphatase inhibition (Wallace, 1995). Agrin-induced AChR γ- and δ-subunit phosphorylation is blocked by protein serine kinase inhibitors without preventing AChR clustering (Wallace, 1994; Wallace *et al.,* 1991), indicating that these sites are not critical for the process. These data suggest that agrin-induced protein tyrosine kinase activation results in β-subunit phosphorylation, which, in turn, is involved in agrin-induced cluster formation.

Increases in AChR β-subunit tyrosine phosphorylation also correlate with decreases in AChR detergent extractability (Wallace, 1995). As discussed previously, resistance to detergent extraction indicates cytoskeletal interactions. Thus, agrin-induced tyrosine phosphorylation of AChR subunits may increase the association of AChRs with the cytoskeleton. Rapsyn is an obvious candidate for mediating this effect because it also binds to the AChR β-subunit. However, removal of the tyrosine phosphorylation consensus sequence from the β-subunit did not affect AChR binding to rapsyn (Yu and Hall, 1994), suggesting that other cytoskeletal proteins may be involved. Possible candidates include the 87-kDa protein, syntrophin, utrophin, dystrophin, members of the GC, and β-spectrin. Intriguingly, the 87-kDa protein, syntrophin, and dystrophin are all substrates for protein tyrosine kinases as well (Wagner *et al.,* 1993; Wagner and Huganir, 1994). However, the role of tyrosine phosphorylation of these proteins in receptor aggregation is unknown.

The protein tyrosine kinase that is involved in agrin-induced AChR aggregation is unknown. There are, however, some likely candidates. For example, in the *Torpedo* electric organ two protein tyrosine kinases, Fyn and Fyk, have been shown to specifically associate at high levels with

AChRs at postsynaptic membranes (Swope and Huganir, 1993), suggesting that these kinases are involved in postsynaptic membrane function. A mammalian protein tyrosine kinase, MuSK (muscle-specific kinase), was identified that is expressed specifically in skeletal muscle (Valenzuela *et al.*, 1995). In the adult muscle, MuSK is localized exclusively at the NMJ (Valenzuela *et al.*, 1995). A possible link of this kinase to muscle function is implied by findings that demonstrate that the MuSK gene maps to the same chromosomal region as Fukuyama's muscular dystrophy (Valenzuela *et al.*, 1995). Thus, protein tyrosine kinases are found to co-localize at the postsynaptic area with AChR, suggesting that agrin-induced β-subunit phosphorylation and oligomer receptor clustering may involve these kinases.

It is important to note that multiple signaling events may be involved in mediating agrin's activity. For example, the adapter molecule Grb2 binds to β-dystroglycan through a SH3–proline-rich domain interaction (Yang *et al.*, 1995a). Grb2 is involved in linking the small GTP-binding protein and protein tyrosine kinase signaling pathways (Medema and Bos, 1993). Thus, it is possible that a more complex signaling cascade, including both the protein tyrosine kinases and GTP-binding protein signaling cascades, is required for agrin's activity.

VII. SUMMARY

Agrin is an extracellular matrix protein released by the motor neuron at the time of synapse formation. It resides in the synaptic cleft of the NMJ, where it forms part of the stable basal lamina. In this location, agrin induces the formation of AChR clusters under the nerve–muscle synapse. Furthermore, the underlying molecular mechanisms of agrin's actions are likely to involve tethering AChR to the cytoskeleton. Several lines of evidence support this hypothesis. First, agrin induces cluster formation by redistributing AChR already present on the muscle cell membrane and has no effect on AChR subunit synthesis. Second, AChR clusters are more resistant to detergent extraction than unclustered receptors. Third, many spectrin-like molecules, including syntrophin, utrophin, β-spectrin, the 87-kDa protein, and rapsyn, are specifically co-localized with AChRs at the NMJ. These molecules are likely to serve as a link between AChRs and the actin cytoskeleton. Finally, agrin binds to α-DG, a GC member. The GC is linked to the cytoskeleton by binding dystrophin or utrophin, spectrin-like proteins known to bind F-actin. These data provide a model in which agrin, by binding to α-DG, traps the AChRs as they diffuse into the agrin–receptor–cytoskeleton complex.

This model does not explain the involvement of second messenger cascades in agrin-mediated clusters. Agrin-induced tyrosine phosphorylation of the AChR β-subunit has been suggested to increase the association of AChR with the cytoskeleton. Agrin signaling may be produced by agrin binding to α-DG, which then activates as yet unidentified signaling cascades. Alternatively, agrin may bind to an as yet unidentified protein that is able to increase protein tyrosine kinase activity. These two mechanisms are not mutually exclusive. For example, agrin binding to α-DG may stabilize receptor clusters, while agrin activating a protein tyrosine kinase may increase the ability of AChRs to associate with the cytoskeleton.

Finally, agrin activity may not be restricted to NMJ development and regeneration. In adult organisms, rearrangement and maintenance of synapses are likely to utilize informational molecules similar to those involved in neuronal development. The molecular interplay that results in the generation, maintenance, and alteration of synaptic structures is a form of synaptic communication, not unlike the interchange of information central to classical synaptic transmission. However, this form of communication is likely to be slow and long term, due to structural alterations, rather than the rapid functional changes produced during classical synaptic transmission. Modification of these contacts in the adult nervous system may play a crucial role in higher order processes, such as learning and memory.

References

Adams, M. E., Butler, M. H., Dwyer, T. M., Peters, M. F., Murnane, A. A., and Froehner, S. C. (1993). Two forms of mouse syntrophin, a 58 kd dystrophin-associated protein, differ in primary structure and tissue distribution. *Neuron* **11,** 531–540.

Ahn, A. H., Yoshida, M., Anderson, M. S., Feener, C. A., Selig, S., Hagiwara, Y., Ozawa, E., and Kunkel, L. M. (1994). Cloning of human basic A1, a distinct 59-kDa dystrophin-associated protein encoded on chromosome 8q23–24. *Proc. Natl. Acad. Sci. U.S.A.* **91,** 4446–4450.

Anderson, M. J., and Cohen, M. W. (1977). Nerve-induced and spontaneous redistribution of acetylcholine receptors on cultured muscle cells. *J. Physiol.* **268,** 757–773.

Anderson, M. J., and Fambrough, D. M. (1983). Aggregates of acetylcholine receptors are associated with plaques of a basal lamina heparan sulfate proteoglycan. *J. Cell. Biol.* **97,** 1396–1411.

Apel, E. D., and Merlie, J. P. (1995). Assembly of the postsynaptic apparatus. *Curr. Opin. Cell Biol.* **5,** 62–67.

Apel, E. D., Roberds, S. L., Campbell, K. P., and Merlie, J. P. (1995). Rapsyn may function as a link between the acetylcholine receptor and the agrin-binding dystrophin-associated glycoprotein complex. *Neuron* **15,** 115–126.

Bennet, V. (1990). Spectrin-based membrane skeleton: A multipotential adapter between plasma membrane and cytoplasm. *Physiol. Rev.* **70,** 1029–1065.

Bevan, S., and Steinbach, J. H. (1977). The distribution of α-bungarotoxin binding sites on mammalian skeletal muscle developing *in vivo*. *J. Physiol.* **267,** 195–213.

Bewick, G. S., Nicholson, L. V., Young, C., O'Donnell, E., and Slater, C. R. (1992). Different distributions of dystrophin and related proteins at nerve-muscle junctions. *Neuroreport* **3,** 857–860.

Bloch, R. J. (1986). Actin at receptor-rich domains of isolated acetylcholine receptor clusters. *J. Cell Biol.* **102,** 1447–1458.

Bloch, R. J., and Froehner, S. C. (1987). The relationship of the postsynaptic 43K protein to acetylcholine receptors in receptor clusters isolated from cultured rat myotubes. *J. Cell Biol.* **104,** 645–654.

Bloch, R. J., and Morrow, J. S. (1989). An unusual beta-spectrin associated with clustered acetylcholine receptors. *J. Cell Biol.* **108,** 481–493.

Bloch, R. J., Resneck, W. G., O'Neill, A., Strong, J., and Pumplin, D. W. (1991). Cytoplasmic components of acetylcholine receptor clusters of cultured rat myotubes: The 58-kD protein. *J. Cell Biol.* **115,** 435–446.

Bowe, M. A., Deyst, K. A., Leszyk, J. D., and Fallon, J. R. (1994). Identification and purification of an agrin receptor from Torpedo postsynaptic membranes: A heteromeric complex related to the dystroglycans. *Neuron* **12,** 1173–1180.

Burden, S. J., Sargent, P. B., and McMahan, U. J. (1979). Acetylcholine receptors in regenerating muscle accumulate at original synaptic sites in the absence of the nerve. *J. Cell Biol.* **82,** 412–425.

Burden, S. J., DePalma, R. L., and Gottesman, G. S. (1983). Cross-linking of proteins in acetylcholine receptor-rich membranes: Association between the β-subunit and the 43 kd subsynaptic protein. *Cell* **35,** 687–692.

Butler, M. H., Douvile, K., Murnane, A. A., Kramarcy, N. R., Cohen, J. B., Sealock, R., and Froehner, S. C. (1992). Association of the Mr 58,000 postsynaptic protein of electric tissue with Torpedo dystrophin and the Mr 87,000 postsynaptic protein. *J. Biol. Chem.* **267,** 6213–6218.

Byers, M., Kunkel, L., and Watkins, S. (1991). The subcellular distribution of dystrophin in mouse skeletal, cardiac, and smooth muscle. *J. Cell Biol.* **115,** 411–421.

Campanelli, J. T., Hoch, W., Rupp, F., Kreiner, T., and Scheller, R. H. (1991). Agrin mediates cell contact-induced acetylcholine receptor clustering. *Cell* **67,** 909–916.

Campanelli, J. T., Roberds, S. L., Campbell, K. P., and Scheller, R. H. (1994). A role for dystrophin-associated glycoproteins and utrophin in agrin-induced AChR clustering. *Cell* **77,** 663–674.

Campbell, K. P. (1995). Three muscular dystrophies: Loss of cytoskeleton-extracellular matrix linkage. *Cell* **80,** 675–679.

Carpenter, G. (1987). Receptors for epidermal growth factor and other polypeptide mitogens. *Annu. Rev. Biochem.* **56,** 881–914.

Cohen, M. W., Jacobson, C., Godfrey, E. W., Campbell, K. P., and Carbonetto, S. (1995). Distribution of alpha-dystroglycan during embryonic nerve-muscle synaptogenesis. *J. Cell Biol.* **129,** 1093–1101.

Davis, C. G. (1990). The many faces of epidermal growth factor repeats. *New Biol.*, **2,** 410–419.

De Bleecker, J. L., Engel, A. G., and Winkelmann, J. C. (1993). Localization of dystrophin and beta-spectrin in vacuolar myopathies. *Am. J. Pathol.* **143,** 1200–1208.

Dutton, E. K., Uhm, S. J., Samuelsson, S. J., Schaffner, A. E., Fitzgerald, S. C., and Daniels, M. P. (1995). Acetylcholine receptor aggregation at nerve-muscle contacts in mammalian cultures: Induction by ventral spinal cord neurons is specific to axons. *J. Neurosci.* **15,** 7401–7416.

Ervasti, J. M., and Campbell, K. P. (1993a). Dystrophin and the membrane skeleton. *Curr. Opin. Cell Biol.* **5,** 82–87.

Ervasti, J. M., and Campbell, K. P. (1993b). A role for the dystrophin-glycoprotein complex as a transmembrane linker between laminin and actin. *J. Cell Biol.* **122,** 809–823.

Falls, D. L., Rosen, K. M., Corfas, G., Lane, W. S., and Fischbach, G. D. (1993). ARIA, a protein that stimulates acetylcholine receptor synthesis, is a member of the neu ligand family. *Cell* **72,** 801–815.

Ferns, M., Hoch, W., Campanelli, J. T., Rupp, F., Hall, Z. W., and Scheller, R. H. (1992). RNA splicing regulates agrin-mediated acetylcholine receptor clustering activity on cultured myotubes. *Neuron* **8,** 1079–1086.

Ferns, M. J., Campanelli, J. T., Hoch, W., Scheller, R. H., and Hall, Z. (1993). The ability of agrin to cluster AChRs depends on alternative splicing and on cell surface proteoglycans. *Neuron* **11,** 491–502.

Fertuck, H. C., and Salpeter, M. M. (1976). Quantitation of junctional and extrajunctional acetylcholine receptors by electron microscope autoradiography after ^{125}I-alpha bungarotoxin binding at mouse neuromuscular junctions. *J. Cell Biol.* **69,** 144–158.

Flucher, B. E., and Daniels, M. P. (1989). Distribution of Na^{+} channels and ankyrin in neuromuscular junctions is complementary to that of acetylcholine receptors and the 43 kd protein. *Neuron* **3,** 163–175.

Fontaine, B., Klarsfeld, A., and Changeux, J. P. (1987). Calcitonin gene-related peptide and muscle activity regulate acetylcholine receptor alpha-subunit mRNA levels by distinct intracellular pathways. *J. Cell Biol.* **105,** 1337–1342.

Froehner, S. C. (1991). The submembrane machinery for nicotinic acetylcholine receptor clustering. *J. Cell Biol.* **114,** 1–7.

Froehner, S. C., Gulbrandsen, V., Hyman, C., Jeng, A. Y., Neubig, R. R., and Cohen, J. B. (1981). Immunofluorescence localization at the mammalian neuromuscular junction of the Mr 43, 000 protein of Torpedo postsynaptic membranes. *Proc. Natl. Acad. Sci. U.S.A.* **78,** 5230–5234.

Froehner, S. C., Murnane, A. A., Tobler, M., Peng, B. H., and Sealock, R. (1987). A postsynaptic Mr 58, 000 (58k) protein concentrated at acetylcholine receptor-rich sites in Torpedo electroplaques and skeletal muscle. *J. Cell Biol.* **104,** 1633–1646.

Froehner, S. C., Luetje, C. W., Scotland, P., and Patrick, J. (1990). The postsynaptic 43K protein clusters muscle nicotinic acetylcholine receptors in Xenopus oocytes. *Neuron* **5,** 403–410.

Gautam, M., Noakes, P. G., Mudd, J., Nichol, M., Chu, G. C., Sanes, J. R., and Merlie, J. P. (1995). Failure of postsynaptic specialization to develop at neuromuscular junctions of rapsyn-deficient mice [see comments]. *Nature* (*London*) **377,** 232–236.

Gee, S. H., Montanaro, F., Lindenbaum, M. H., and Carbonetto, S. (1994). Dystroglycan-alpha, a dystrophin-associated glycoprotein, is a functional agrin receptor. *Cell* **77,** 675–686.

Gesemann, M., Denzer, A. J., and Ruegg, M. A. (1995). Acetylcholine receptor-aggregating activity of agrin isoforms and mapping active site. *J. Cell Biol.* **128,** 625–636.

Gloor, S., Odink, K., Geunther, J., Nick, H., and Monrad, D. (1986). A glia-derived neurite promoting factor with protease inhibitory activity belongs to the protease nexins. *Cell* **47,** 687–693.

Godfrey, E. W. (1991). Comparison of agrin-like proteins from the extracellular matrix of chicken kidney and muscle with neural agrin, a synapse organizing protein. *Exp. Cell Res.* **195,** 99–109.

Godfrey, E. W., Nitkin, R. M., Wallace, B. G., Rubin, L. L., and McMahan, U. J. (1984). Components of Torpedo electric organ and muscle that cause aggregation of acetylcholine receptors on cultured muscle cells. *J. Cell Biol.* **99,** 615–627.

Godfrey, E. W., Dietz, M. E., Morstad, A. L., Wallskog, P. A., and Yorde, D. E. (1988). Acetylcholine receptor-aggregating proteins are associated with the extracellular matrix of many tissues in Torpedo. *J. Cell Biol.* **106,** 1263–1272.

Graf, J., Sasaki, M., Martin, G. R., Kleinman, H. K., Robey, F. A., and Yamada, Y. (1987). Identification of an amino sequence in laminin mediating cell attachment, chemotaxis and receptor binding. *Cell* **48,** 989–996.

Gumbiner, B. M. (1993). Proteins associated with the cytoplasmic surface of adhesion molecules. *Neuron* **11,** 551–564.

Hall, Z. (1973). Multiple forms of acetylcholinesterases and their distribution of endplate and non-endplate regions of rat diaphragm muscle. *J. Neurobiol.* **4,** 343–361.

Hall, Z. W., and Sanes, J. R. (1993). Synaptic structure and development: The neuromuscular junction. *Cell* **72,** 99–121.

Hemmati-Brivanlou, A., and Melton, D. A. (1994). Inhibition of activin receptor signaling promotes neuralization in Xenopus. *Cell* **77,** 273–281.

Hemmati-Brivanlou, A., Kelly, O. G., and Melton, D. A. (1994). Follistatin, an antagonist of activin, is expressed in the Spemann organizer and displays direct neuralizing activity. *Cell* **77,** 283–295.

Hemmings, L., Kuhlman, P. A., and Critchley, D. R. (1992). Analysis of the actin-binding domain of alpha-actinin by mutagenesis and demonstration that dystrophin contains a functionally homologous domain. *J. Cell Biol.* **116,** 1369–1380.

Hoch, W., Ferns, M., Campanelli, J. T., Hall, Z. W., and Scheller, R. H. (1993). Developmental regulation of highly active alternatively spliced forms of agrin. *Neuron* **11,** 479–490.

Hoch, W., Campanelli, J. T., Harrison, S., and Scheller, R. H. (1994). Structural domains of agrin required for clustering of nicotinic acetylcholine receptors. *Embo J.* **13,** 2814–2821.

Hunter, D. D., Porter, B. E., Bulock, J. W., Adams, S. P., Merlie, J. P., and Sanes, J. R. (1989a). Primary sequence of a motor neuron-selective adhesive site in the synaptic basal lamina protein S-laminin. *Cell* **59,** 905–913.

Hunter, D. D., Shah, V., Merlie, J. P., and Sanes, J. R. (1989b). A laminin-like adhesive protein concentrated in the synaptic cleft of the neuromuscular junction. *Nature* (*London*) **338,** 229–233.

Jackson, R., Busch, S., and Cardin, A. (1991). Glycosaminoglycans: Molecular properties, protein interactions, and role in physiological processes. *Physiol. Rev.* **71,** 481–539.

Jennings, C. G., and Burden, S. J. (1993). Development of the neuromuscular synapse. *Curr. Opin. Neurobiol.* **3,** 75–81.

Jo, S. A., Zhu, X., Marchionni, M. A., and Burden, S. J. (1995). Neuregulins are concentrated at nerve-muscle synapses and activate ACh-receptor gene expression. *Nature* (*London*) **373,** 158–161.

Katz, B., and Miledi, R. (1973). The binding of acetylcholine to receptors and its removal from the synaptic cleft. *J. Physiol.* **231,** 549–574.

Khurana, T. S., Watkins, S. C., and Kunkel, L. M. (1992). The subcellular distribution of chromosome 6-encoded dystrophin-related protein in the brain. *J. Cell Biol.* **119,** 357–366.

Kleinman, H. K., Graf, J., Iwamoto, Y., Sasaki, M., Schasteen, C. S., Yamada, Y., Martin, G. R., and Robey, F. A. (1989). Identification of a second active site in laminin for promotion of cell adhesion and migration and inhibition of *in vivo* melanoma lung colonization. *Arch. Biochem. Biophys.* **272,** 39–45.

Kramarcy, N. R., Vidal, A., Froehner, S. C., and Sealock, R. (1994). Association of utrophin and multiple dystrophin short forms with the mammalian M(r) 58, 000 dystrophin-associated protein (syntrophin). *J. Biol. Chem.* **269,** 2870–2876.

LaRochelle, W. J., and Froehner, S. C. (1986). Determination of the tissue distributions and relative concentrations of the post-synaptic 43-kDa protein and the acetylcholine receptor in Torpedo. *J. Biol. Chem.* 261, 5270–5274.

LaRochelle, W. J., and Froehner, S. C. (1987). Comparison of the postsynaptic 43-kDa protein from muscle cells that differ in acetylcholine receptor clustering activity. *J. Biol. Chem.* **262,** 8190–8195.

Laskowski, M., and Kato, I. (1980). Protein inhibitors of proteinases. *Annu. Rev. Biochem.* **49,** 593–626.

Levine, B. A., Moir, A. J., Patchell, V. B., and Perry, S. V. (1992). Binding sites involved in the interaction of actin with the N-terminal region of dystrophin. *FEBS Lett.* **298,** 44–48.

Lopez-Casillas, F., Cheifetz, S., Doody, J., Andres, J. L., Lane, W. S., and Massague, J. (1991). Structure and expression of the membrane proteoglycan betaglycan, a component of the TGF-beta receptor system. *Cell* **67,** 785–795.

Ma, J., Nastuk, M. A., McKechnie, B. A., and Fallon, J. R. (1993). The agrin receptor. Localization in the postsynaptic membrane, interaction with agrin, and relationship to the acetylcholine receptor. *J. Biol. Chem.* **268,** 25108–25117.

Magill-Solc, C., and McMahan, U. J. (1988). Motor neurons contain agrin-like molecules. *J. Cell. Biol.* **107,** 1825–1833.

Magill-Solc, C., and McMahan, U. J. (1990a). Agrin-like molecules in motor neurons. *J. Physiol.* 84, 78–81.

Magill-Solc, C., and McMahan, U. J. (1990b). Synthesis and transport of agrin-like molecules in motor neurons. *J. Exp. Biol.* **153,** 1–10.

Maimone, M. M., and Merlie, J. P. (1993). Interaction of the 43 kd postsynaptic protein with all subunits of the muscle nicotinic acetylcholine receptor. *Neuron* **11,** 53–66.

Martin, P. T., and Sanes, J. R. (1995). Role for a synapse-specific carbohydrate in agrin-induced clustering of acetylcholine receptors. *Neuron* **14,** 743–754.

Matsumura, K., Ervasti, J. M., Ohlendieck, K., Kahl, S. D., and Campbell, K. P. (1992). Association of dystrophin-related protein with dystrophin-associated proteins in mdx mouse muscle. *Nature (London)* **360,** 588–591.

McMahan, U. J. (1990). The agrin hypothesis. *Cold Spring Harbor Symp. Quant. Biol.* **55,** 407–518.

McMahan, U. J., Sanes, J. R., and Marshall, L. M. (1978). Cholinesterase is associated with the basal lamina at the neuromuscular junction. *Nature (London)* **271,** 172–174.

Medema, R. H., and Bos, J. L. (1993). The role of p21 ras in receptor tyrosine kinase signaling. *Crit. Rev. Oncog.* **4,** 615–661.

Monrad, D. (1988). Cell-derived proteases and protease inhibitors as regulators of neurite outgrowth. *Trends Neurosci.* **11,** 541–544.

Nastuk, M. A., and Fallon, J. R. (1993). Agrin and the molecular choreography of synapse formation. *Trends Neurosci.* **16,** 72–76.

New, H. V., and Mudge, A. W. (1986). Calcitonin gene-related peptide regulates muscle acetylcholine receptor synthesis. *Nature (London)* **323,** 809–811.

Nitkin, R. M., Smith, M. A., Magill, C., Fallon, J. R., Yao, Y. M., Wallace, B. G., and McMahan, U. J. (1987). Identification of agrin, a synaptic organizing protein from Torpedo electric organ. *J. Cell Biol.* **105,** 2471–2578.

Noakes, P. G., Phillips, W. D., Hanley, T. A., Sanes, J., R., and Merlie, J. P. (1993). 43k protein and acetylcholine receptors co-localize during the initial stages of neuromuscular synapse formation *in vivo*. *Dev. Biol.* **155,** 275–280.

O'Connor, L. T., Lauterborn, J. C., Gall, C. M., and Smith, M. A. (1994). Localization and alternative splicing of agrin mRNA in adult rat brain: transcripts encoding isoforms that aggregate acetylcholine receptors are not restricted to cholinergic regions. *J. Neurosci.* **14,** 1141–1152.

Ohlendieck, K., Ervasti, J. M., Matsumura, K., Kahl, S. D., Leveille, C. J., and Campbell, K. P. (1991). Dystrophin-related protein is localized to neuromuscular junctions of adult skeletal muscle. *Neuron* **7,** 499–508.

Panayotou, G., End, P., Aumailley, M., Timpl, R., and Engel, J. (1989). Domains of laminin with growth-factor activity. *Cell* **56,** 93–101.

Patthy, L., and Nikolics, K. (1993). Functions of agrin and agrin-related proteins. *Trends Neurosci.* **16,** 76–81.

Peters, M. F., Kramarcy, N. R., Sealock, R., and Froehner, S. C. (1994). Beta-2 syntrophin: Localization at the neuromuscular junction in skeletal muscle. *Neuroreport* **5,** 1577–1580.

Phillips, W. D., Kopta, C. P. B., Gardner, P. D., Steinbach, J. H., and Merlie, J. P. (1991). ACh receptor-rich membrane domains organized in fibroblasts by recombinant 43-kilodalton protein. *Science* **251,** 568–570.

Phillips, W. D., Noakes, P. G., Roberds, S. L., Campbell, K. P., and Merlie, J. P. (1993). Clustering and immobilization of acetylcholine receptors by 43-kD protein: A possible role of dystrophin-related protein. *J. Cell Biol.* **123,** 729–740.

Poo, M.-M. (1985). Mobility and localization of proteins in excitable membranes. *Annu. Rev. Neurosci.* **8,** 369–406.

Porter, B. E., Weis, J., and Sanes, J. R. (1995). A motoneuron-selective stop signal in the synaptic protein S-laminin. *Neuron* **14,** 549–559.

Prives, J., Fulton, A. B., Penman, S., Daniels, M. P., and Christian, C. N. (1982). Interaction of the cytoskeletal framework with acetylcholine receptor on the surface of embryonic muscle cells in culture. *J. Cell Biol.* 92, 231–236.

Rapraeger, A. C., Krufka, A., and Olwin, B. B. (1991). Requirement of heparan sulfate for bFGF-mediated fibroblast growth and myoblast differentiation. *Science* **252,** 1705–1708.

Reist, N. E., Magill, C., and McMahan, U. J. (1987). Agrin-like molecules at synaptic sites in normal, denervated, and damaged skeletal muscles. *J. Cell Biol.* **105,** 2457–2469.

Reist, N. E., Werle, M. J., and McMahan, U. J. (1992). Agrin released by motor neurons induces the aggregation of acetylcholine receptors at neuromuscular junctions. *Neuron* **8,** 865–868.

Robitaille, R., Adler, E. M., and Charlton, M. P. (1990). Strategic location of calcium channels at transmitter release sites of frog neuromuscular synapses. *Neuron* **5,** 773–779.

Ruegg, M. A., Tsim, K. W., Horton, S. E., Kroger, S., Escher, G., Gensch, E. M., and McMahan, U. J. (1992). The agrin gene codes for a family of basal lamina proteins that differ in function and distribution. *Neuron* **8,** 691–699.

Rupp, F., Payan, D. G., Magill-Solc, C., Cowan, D. M., and Scheller, R. H. (1991). Structure and expression of a rat agrin. *Neuron* **6,** 811–823.

Sakashita, S., Engvall, E., and Ruoslahti, E. (1980). Basement membrane glycoprotein laminin binds to heparin. *FEBS Lett.* **116,** 243–246.

Salpeter, M. M. (1967). Electron microscope autoradiography of a quantitative tool in enzyme cytochemistry. I. The distribution of acetylcholinesterase at motor endplate of a vertebrate twitch muscle. *J. Cell Biol.* **32,** 379–389.

Salpeter, M. M. (1987). "Vertebrate Neuromuscular Junctions: General Morphology, Molecular Organization, and Functional Consequences." Alan R. Liss, New York.

Sanes, J. R., Marshall, L. M., and McMahan, U. J. (1978). Reinnervation of muscle fiber basal lamina after removal of myofibers. Differentiation of regenerating axons at original synaptic sites. *J. Cell Biol.* **78,** 176–198.

Sanes, J. R., Engvall, E., Butkowski, R., and Hunter, D. D. (1990). Molecular heterogeneity of basal laminae: Isoforms of laminin and collagen IV at the neuromuscular junction and elsewhere. *J. Cell Biol.* **11,** 1685–1699.

Saski, M., and Yamada, Y. (1987). The laminin B2 chain has a multidomain structure homologous to the B1 chain. *J. Biol. Chem.* **262,** 17111–17117.

Saski, M., Kleinman, H. K., Huber, H., Deutzman, R., and Yamada, Y. (1988). Laminin, a multidomain protein. The A chain has unique globular domain and homology with the basement membrane proteoglycan and the laminin B chains. *J. Biol. Chem.* **263,** 16536–16544.

Sealock, R., and Froehner, S. C. (1994). Dystrophin-associated proteins and synapse formation: Is alpha-dystroglycan the agrin receptor? *Cell* **77,** 617–619.

Sealock, R., Butler, M. H., Kramarcy, N. R., Gao, K. X., Murnane, A. A., Douville, K., and Froehner, S. C. (1991). Localization of dystrophin relative to acetylcholine receptor domains in electric tissue and adult and cultured skeletal muscle. *J. Cell Biol.* **113,** 1133–1144.

Sephel, G. C., Tashiro, K. I., Sasaki, M., Greatorex, D., Martin, G. R., Yamada, Y., and Kleinman, H. K. (1989). Laminin A chain synthetic peptide which supports neurite outgrowth. *Biochem. Biophys. Res. Commun.* **162,** 821–829.

Smith, M. A., Yao, Y. M., Reist, N. E., Magill, C., Wallace, B. G., and McMahan, U. J. (1987). Identification of agrin in electric organ extracts and localization of agrin-like molecules in muscle and central nervous system. *J. Exp. Biol.* **132,** 223–230.

Stya, M., and Axelrod, D. (1983). Mobility and detergent extractability of acetylcholine receptors on cultured rat myotubes: A correlation. *J. Cell Biol.* **97,** 48–51.

Sugiyama, J., Bowen, D. C., and Hall, Z. W. (1994). Dystroglycan binds nerve and muscle agrin. *Neuron* **13,** 103–115.

Sunada, Y., and Campbell, K. P. (1995). Dystrophin-glycoprotein complex: Molecular organization and critical roles in skeletal muscle. *Curr. Opin. Neurol.* **8,** 379–384.

Suzuki, A., Yoshida, M., Hayashi, K., Mizuno, Y., Hagiwara, Y., and Ozawa, E. (1994). Molecular organization at the glycoprotein-complex-binding site of dystrophin: Three dystrophin-associated proteins bind directly to the carboxy-terminal portion of dystrophin. *Eur. J. Biochem.* **220,** 283–292.

Suzuki, A., Yoshida, M., and Ozawa, E. (1995). Mammalian α1- and β1-syntrophin bind to the alternative splice-prone region of dystrophin COOH terminus. *J. Cell Biol.* **128,** 363–371.

Swope, S. L., and Huganir, R. L. (1993). Molecular cloning of two abundant protein tyrosine kinases in Torpedo electric organ that associate with the acetylcholine receptor. *J. Biol. Chem.* **268,** 25152–25161.

Thomas, W. S., O'Dowd, D. K., and Smith, M. A. (1993). Developmental expression and alternative splicing of chick agrin RNA. *Dev. Biol.* **158,** 523–535.

Tinsley, J. M., Blake, D. J., Zuellig, R. A., and Davies, K. E. (1994). Increasing complexity of the dystrophin-associated protein complex. *Proc. Natl. Acad. Sci. U.S.A.* **91,** 8307–8313.

Tsen, G., Halfter, W., Kroger, S., and Cole, G. J. (1995). Agrin is a heparan sulfate proteoglycan. *J. Biol. Chem.* **270,** 3392–3399.

Tsim, K. W., Ruegg, M. A., Escher, G., Kroger, S., and McMahan, U. J. (1992). cDNA that encodes active agrin. *Neuron* **8,** 677–689.

Usdin, T. B., and Fischbach G. D. (1986). Purification and characterization of a polypeptide from chick brain that promotes the accumulation of acetylcholine receptors in chick mykotubes. *J. Cell Biol.* **103,** 493–507.

Ushkaryov, Y. A., Petrenko, A. G., Geppert, M., and Sudhof, T. C. (1992). Neurexins: Synaptic cell surface proteins related to the alpha-latrotoxin receptor and laminin. *Science* **257,** 50–56.

Valenzuela, D. M., Stitt, T. N., DiStefano, P. S., Rojas, E., Mattsson, K., Compton, D. L., Nunez, L., Park, J. S., Stark, J. L., Gles, D. R., Thomas, S., Le Beau, M. M., Fernald, A. A., Copeland, N. G., Jenkins, N. A., Burden, S. J., Glass, D. J., and Yancopoulos, G. D. (1995). Receptor tyrosine kinase specific for the skeletal muscle lineage: Expression in embryonic muscle, at the neuromuscular junction, and after injury. *Neuron* **15,** 573–584.

Wagner, K. R., and Huganir, R. L. (1994). Tyrosine and serine phosphorylation of dystrophin and the 58-kDa protein in the postsynaptic membrane of Torpedo electric organ. *J. Neurochem.* **62,** 1947–1952.

Wagner, K. R., Cohen, J. B., and Huganir, R. L. (1993). The 87K postsynaptic membrane protein from Torpedo is a protein-tyrosine kinase substrate homologous to dystrophin. *Neuron* **10,** 511–522.

Wallace, B. G. (1988). Regulation of agrin-induced acetylcholine receptor aggregation by Ca^{++} and phorbol ester. *J. Cell Biol.* **107,** 267–278.

Wallace, B. G. (1989). Agrin-induced specializations contain cytoplasmic, membrane, and extracellular matrix-associated components of the postsynaptic apparatus. *J. Neurosci.* **9,** 1294–1302.

Wallace, B. G. (1990). Inhibition of agrin-induced acetylcholine-receptor aggregation by heparin, heparan sulfate, and other polyanions. *J. Neurosci.* **10,** 3576–3582.

Wallace, B. G. (1992). Mechanism of agrin-induced acetylcholine receptor aggregation. *J. Neurobiol.* **23,** 592–604.

Wallace, B. G. (1994). Staurosporine inhibits agrin-induced acetylcholine receptor phosphorylation and aggregation. *J. Cell Biol.* **125,** 661–668.

Wallace, B. G. (1995). Regulation of the interaction of nicotinic acetylcholine receptors with the cytoskeleton by agrin-activated protein tyrosine kinase. *J. Cell Biol.* **128,** 1121–1129.

Wallace, B. G., Qu, Z., and Huganir, R. L. (1991). Agrin induces phosphorylation of the nicotinic acetylcholine receptor. *Neuron* **6,** 869–878.

Way, M., Pope, B., Cross, R. A., Kendrick-Jones, J., and Weeds, A. G. (1992). Expression of the N-terminal domain of dystrophin in E. coli and demonstration of binding to F-actin. *FEBS Lett.* **301,** 243–245.

Winder, S. J., Hemmings, L., Maciver, S. K., Bolton, S. J., Tinsley, J. M., Davies, K. E., Critchley, D. R., and Kendrick-Jones, J. (1995). Utrophin actin binding domain: Analysis of actin binding and cellular targeting. *J. Cell Sci.* **108,** 63–71.

Yang, B., Ibraghimov-Beskrovnaya, O., Moomaw, C. R., Slaughter, C. A., and Campbell, K. P. (1994). Heterogeneity of the 59-kDa dystrophin-associated protein revealed by cDNA cloning and expression. *J. Biol. Chem.* **269,** 6040–6044.

Yang, B., Jung, D., Motto, D., Meyer, J., Koretzky, G., and Campbell, K. P. (1995a). SH3 domain-mediated interaction of dystroglycan and Grb2. *J. Biol. Chem.* **270,** 11711–11714.

Yang, B., Jung, D., Rafael, J. A., Chamberlain, J. S., and Campbell, K. P. (1995b). Identification of a-syntrophin binding to syntrophin triplet, dystrophin, and utrophin. *J. Biol. Chem.* **270,** 4975–4978.

Yayon, A., Klagsbrun, M., Esko, J. D., Leder, P., and Ornitz, D. M. (1991). Cell surface, heparin-like molecules are required for binding of basic fibroblast growth factor to its high affinity receptor. *Cell* **64,** 841–848.

Yu, X. M., and Hall, Z. W. (1994). The role of the cytoplasmic domains of individual subunits of the acetylcholine receptor in 43 kDa protein-induced clustering in COS cells. *J. Neurosci.* **14,** 785–795.

CHAPTER 13

Integrin Signaling and the Platelet Cytoskeleton

Martin Eigenthaler* and Sanford J. Shattil*,†

Departments of *Vascular Biology and †Molecular and Experimental Medicine, The Scripps Research Institute, La Jolla, California 92037

I. INTRODUCTION

During hemostasis, platelets are converted from nonadhesive, disc-shaped cells into adhesive multicell aggregates at the site of blood vessel injury. While this normal hemostatic response is necessary for survival, platelet adhesion and aggregation at the wrong time or place can lead to thrombotic occlusion of arteries in vascular diseases as diverse in pathogenesis and incidence as atherosclerosis and thrombotic thrombocytopenic purpura (Fig. 1). Thus, the adhesive phenotype of platelets must be closely regulated.

Platelet adhesion and aggregation are controlled through complex interactions between extracellular adhesive ligands, plasma membrane adhesion

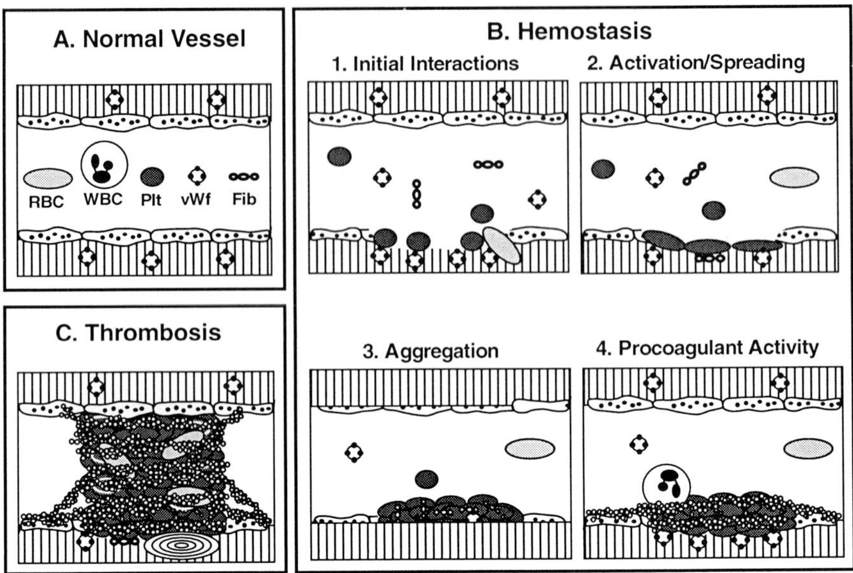

FIGURE 1 Adhesive platelet functions in hemostasis and thrombosis. (A) In a normal blood vessel, platelets (Plt) are quiescent and fail to interact with other blood cells (RBC, WBC), plasma adhesive proteins (von Willebrand factor, vWf; fibrinogen, Fib), or the extracellular matrix, which is shielded from platelets by a nonthrombogenic layer of endothelial cells. (B) When a blood vessel is damaged and hemostasis is triggered, the platelets undergo (1) initial adhesive interactions with components of the exposed extracellular matrix (e.g., vWf, collagen); (2) activation and spreading in response to contact with soluble agonists and extracellular matrix components; (3) platelet–platelet aggregation, which results in the formation of a primary hemostatic platelet plug and the initial arrest of bleeding; and (4) an increase in procoagulant activity stimulated by certain agonists such as thrombin and collagen, which generate a catalytic surface on the platelets for assembly of coagulation enzyme–cofactor–substrate complexes. This leads to formation of a secondary fibrin plug for more durable hemostasis. In addition, P-selectin on activated platelets recruits WBC to the hemostatic plug, thereby facilitating wound repair. (C) In certain pathological conditions, such as atherosclerosis, endothelial cell damage or rupture of an atherosclerotic plaque may expose the platelets to the extracellular matrix at the wrong time and place (e.g., a coronary artery). If the thrombogenic stimulus is great enough, complete thrombotic occlusion of the vessel may result and cause tissue ischemia distally.

receptors, and cytoskeletal proteins. This interplay is coordinated by intracellular signaling molecules that regulate the activation state of the platelet. In the case of one well-characterized adhesion receptor, integrin $\alpha_{IIb}\beta_3$, binding of a soluble, cognate ligand (e.g., fibrinogen) to the receptor is regulated by signaling reactions. Ligand binding in turn stimulates additional reactions that affect the organization and composition of the cytoskeleton and the adhesive, secretory, and procoagulant responses of the cell. This chapter discusses emerging relationships between $\alpha_{IIb}\beta_3$, the cytoskeleton, and signaling pathways in the platelet, a model system that may be relevant to similar relationships in other cells. The reader is also referred to several excellent reviews for additional perspectives and references on this topic (Hynes, 1992; Clark and Brugge, 1995; Sastry and Horwitz, 1993; Williams *et al.,* 1994; Pavalko and Otey, 1994; Fox, 1993; Stossel, 1993; Burridge *et al.,* 1992; Barkalow and Hartwig, 1995; Hemmings *et al.,* 1995; Schwartz *et al.,* 1995).

II. THE PLATELET CYTOSKELETON

Actin filament systems in the platelet have been studied by a variety of microscopic and biochemical techniques. Despite considerable differences among these methods, there is a consensus that the approximately 2000 F-actin filaments in a resting platelet are distributed primarily in two locales: a *membrane skeleton* that coats the lipid bilayer of the plasma membrane and a *cytoplasmic "core" cytoskeleton* (Hartwig and DeSisto, 1991; Fox, 1993; Barkalow and Hartwig, 1995). In biochemical and morphological studies of Triton X-100–insoluble residues from platelets, the core cytoskeleton has been defined operationally as the residue that sediments at low *g* forces (15,600*g*), while the membrane skeleton sediments at high forces (100,000*g*) (Fox *et al.,* 1988; Fox, 1993). In platelets that are truly not activated, even the cytoplasmic actin filaments are unstable in Triton X-100 and can only be sedimented at high *g* forces. However, under the usual conditions of platelet isolation from plasma, where the cells have been minimally activated, the cytoplasmic actin filaments are cross-linked and can be sedimented at low *g* forces. The cytoskeleton is responsible for the disc shape of resting platelets, and changes in the composition and subcellular distribution of cytoskeletal components in activated platelets help to mediate disk-to-sphere transformation of the cell as well as filopodial extension, platelet spreading on extracellular matrices, and fibrin clot retraction (Hartwig, 1992; Fox, 1993; Barkalow and Hartwig, 1995). The shape of resting platelets may also be influenced by a circumferential, submembranous microtuble coil that becomes redistributed upon platelet activation (Escolar *et al.,* 1987).

A. Resting Platelets

The concentration of actin in unstimulated platelets is approximately 500 μmol/liter, and 40% of this is filamentous (Hartwig *et al.,* 1995). The

300-μM pool of actin monomers is prevented from polymerizing by monomer-sequestering proteins, such as thymosin β4 and profilin, and by barbed-end–capping proteins, such as gelsolin and CapG (Hartwig *et al.,* 1995; Stossel, 1993; Nachmias, 1993; Nachmias *et al.,* 1993). The cytoplasmic core actin filament network of resting platelets contains actin-binding protein, α-actinin and tropomyosin (Hartwig and DeSisto, 1991; Fox, 1993).

The membrane skeleton has the appearance on electron microscopy of a tightly woven planar sheet composed of a spectrin-rich network of short actin filaments connected to the central cytoplasmic core network by radial actin filaments (Hartwig and DeSisto, 1991). Spectrin supports this structure by attaching to one end of the radial filaments. The membrane skeleton is physically associated with complexes made up of actin-binding protein (filamin) and a transmembrane receptor for von Willebrand factor (vWf), glycoprotein (GP) Ib-IX. It is also associated with talin, vinculin, dystrophin-related protein, several signaling molecules ($p120^{Ras}$ GTPase-activating protein (GAP), Src, and Yes), and other unidentified proteins (Fox, 1993; Earnest *et al.,* 1995). In addition, about 20% of the total pool of integrin $\alpha_{IIb}\beta_3$ is associated with the membrane skeleton (Suzuki *et al.,* 1991; Kouns *et al.,* 1991; Zhang *et al.,* 1992; Fox *et al.,* 1993a). The specific interactions that hold the platelet membrane skeleton in apposition to the plasma membrane are not well characterized. However, in the case of the GP Ib-IX–actin-binding protein interaction, it has been shown that GP Iba binds to residues near the C-terminus of actin-binding protein (Ezzell *et al.,* 1988; Meyer *et al.,* 1995). The association of a pool of $\alpha_{IIb}\beta_3$ with the membrane skeleton may be facilitated by the presence of talin and vinculin in this location. Talin can interact with the β_3 cytoplasmic tail and with vinculin, and vinculin binds to F-actin (Horwitz *et al.,* 1986; Johnson and Craig, 1995; Hemmings *et al.,* 1995).

B. Activated Platelets

When platelets in suspension are stimulated with thrombin, there is a rapid and marked increase in F-actin to 60–70% of total actin and a reorganization of the actin networks, associated with a redistribution of $\alpha_{IIb}\beta_3$ and many other proteins to the 15,600*g* Triton-insoluble core cytoskeleton (Hartwig, 1992; Barkalow and Hartwig, 1995; Fox, 1993). The increase in F-actin is due primarily to the generation of actin filament free barbed ends, which serve as nuclei to promote actin polymerization. This occurs through a combination of filament severing and release of barbed-end–capping proteins. Filament severing is mediated in a Ca^{2+}-dependent manner by gelsolin and other severing proteins (Stossel, 1993; Barkalow and Hartwig, 1995). It is of interest, therefore, that gelsolin-deficient platelets can still form filopodia but not lamellae (Witke *et al.*, 1995). The release of barbed-end–capping proteins appears to be mediated by phosphatidylinositol 4,5-biphosphate, whose synthesis is stimulated by thrombin in a manner that is dependent on activation of the small GTP-binding protein Rac (Hartwig *et al.*, 1995). Actin polymerization proceeds in large part because the affinity of new barbed ends for actin monomers is greater than that for the major capping protein, thymosin $\beta 4$. In addition, cell activation may increase the concentration of ATP-actin monomers at the expense of ADP–actin monomers, and the former polymerize more readily to filament barbed ends (Goldschmidt-Clermont *et al.*, 1992).

Cytoskeletal reorganization in thrombin-stimulated platelets occurs in two phases, an early one independent of platelet aggregation that is complete within seconds to a minute and a later one dependent on aggregation that is complete within several minutes. The aggregation-independent phase is associated with an immediate change in platelet shape from discoid to spherical, accompanied by the formation of actin filament networks in developing filopodia and at the cell periphery. Actin-binding protein and tropomyosin concentrate within the filopodia, and the bulk of talin and α-actinin now take on a submembrane distribution. In contrast, myosin associates with the cytoplasmic actin filament network, coincident with centralization and secretion of platelet granules (Hartwig, 1992). Biochemical studies confirm that the amount of F-actin increases in the Triton-insoluble cytoplasmic core cytoskeleton, which also becomes enriched in myosin and cortactin (Fox, 1993; Ozawa *et al.*, 1995). The latter 80- to 85-kDa cortical actin-binding protein also becomes phosphorylated on tyrosine residues in response to thrombin (Clark *et al.*, 1994a; Ozawa *et al.*, 1995).

The second phase of cytoskeletal reorganization requires the binding of fibrinogen (or vWf) to $\alpha_{IIb}\beta_3$, close cell–cell contact (such as that obtained experimentally by stirring the platelet suspension), and platelet aggregation. During aggregation, a number of proteins redistribute from the Triton-insoluble membrane skeleton to the cytoplasmic core cytoskeleton (e.g., $\alpha_{IIb}\beta_3$, spectrin, dystrophin-related protein, vinculin, talin, $p120^{Ras}$ GAP, Src, and Yes) (Fox *et al.*, 1993a; Fox, 1993; Earnest *et al.*, 1995). In addition, many proteins originally present in the Triton-soluble fraction of resting platelets are now found in the low-speed pellet, including $\alpha_{IIb}\beta_3$, protein kinases (e.g., $pp60^{Src}$, $pp72^{Syk}$, $pp125^{FAK}$, protein kinase C), a lipid kinase (e.g., phosphatidylinositol-3-kinase (PI-3K)), protein phosphatases (e.g., phosphotyrosine phosphatase 1B and 1C (PTP-1B and PTP-1C), a phospholipase (e.g., phospholipase $C_{\gamma 1}$), GTP-binding proteins (RhoA, Rac, CDC42Hs, Rap1B), and other molecules implicated in various phases of platelet function (e.g., PECAM-1, HSP27, diacylglycerol, calpain, and unidentified tyrosine-phosphorylated proteins) (Fox, 1993; Clark and Brugge, 1993; Clark *et al.*, 1994b; Zhang *et al.*, 1992; Grondin *et al.*, 1991; Li *et al.*, 1994; Guinebault *et al.*, 1994,1995; Ezumi *et al.*, 1995; Dash *et al.*, 1995a,b; Tohyama *et al.*, 1994; Nemoto *et al.*, 1992; Altmüller and Presek, 1995; Zhu *et al.*, 1994; Torti *et al.*, 1994; Fox *et al.*, 1993b; Newman *et al.*, 1992). In most of these cases, protein redistribution has been shown to be dependent on fibrinogen binding to $\alpha_{IIb}\beta_3$ since it can be inhibited by function-blocking antibodies to $\alpha_{IIb}\beta_3$ or Arg-Gly-Asp (RGD) peptides or peptidomimetics. Actin polymerization is also required, since protein redistribution is inhibited by pretreatment of platelets with cytochalasin B, D, or E, which prevents actin monomer deposition onto actin filament barbed ends. Based on studies with tyrosine kinase inhibitors, in some cases tyrosine phosphorylation may also be required for protein redistribution.

A similar process of actin polymerization and cytoskeletal reorganization takes place when individual platelets are allowed to spread on glass or fibrinogen (White, 1988; Hartwig, 1992; Nachmias and Golla, 1991; Bearer, 1995). Under these conditions, the amount of polymerized actin increases by a factor of 2 within 30 s, coinciding with the appearance of filopodial actin bundles and a circumferential band of orthogonally arranged short actin filaments within lamellipodia (Hartwig, 1992). The formation of filopodia may be due primarily to uncapping of actin barbed ends, while the formation of lamellipodia may be due to severing of actin filaments (Witke *et al.*, 1995; Hartwig *et al.*, 1995). Over the course of 15–60 min, the platelets progressively spread. Eventually, actin stress fibers form, connecting the core cytoskeleton with membrane-based structures that contain $\alpha_{IIb}\beta_3$ and vinculin (Nachmias and Golla, 1991). These may be similar to adhesion plaques found in other adherent cells in culture (Burridge *et al.*, 1988).

III. ADHESIVE AND SIGNALING FUNCTIONS OF INTEGRIN $\alpha_{IIb}\beta_3$

Like all integrins, $\alpha_{IIb}\beta_3$ is a heterodimer composed of α and β type I transmembrane subunits (Hynes, 1992). α_{IIb} consists of an 871-amino-acid extracellular heavy chain that is disulfide linked to a 137-residue light chain that spans the plasma membrane and contains a 20-residue cytoplasmic tail (Fitzgerald *et al.*, 1987a). The β_3 subunit contains a 692-amino-acid extracellular domain, a 23-residue transmembrane domain, and a 47-residue cytoplasmic tail (Fitzgerald *et al.*, 1987b). The ligand-binding site in $\alpha_{IIb}\beta_3$ is likely composed of several discontinuous regions in the N-terminal portions of both subunits (Haas and Plow, 1994). One of these regions in β_3 may be analogous to the "I domain" (present in some α subunits but not α_{IIb}), which is known to coordinate with Mg^{2+} and participate in ligand binding (Lee *et al.,* 1995). Extracellular Ca^{2+} and Mg^{2+} are essential for stability of the $\alpha_{IIb}\beta_3$ heterodimer and for ligand binding, likely reflecting different functions for specific divalent cation–binding sites within each subunit (Brass *et al.,* 1985; Shattil *et al.,* 1986). A low-resolution model of $\alpha_{IIb}\beta_3$ based on electron microscopic images of detergent-soluble heterodimers depicts an 8 × 12-nm N-terminal globular head connected to two flexible 18-nm tails extending from one side (Weisel *et al.,* 1992).

There are approximately 80,000 copies of $\alpha_{IIb}\beta_3$ per platelet, and most of these are expressed on the surface of the resting cell. An additional pool of receptors within α-granules and open canalicular membranes becomes surface expressed when platelets are stimulated by strong agonists, such as thrombin (Niiya *et al.,* 1987; Wencel Drake *et al.,* 1986). $\alpha_{IIb}\beta_3$ can engage the RGD-containing ligands fibrinogen, vWf, fibronectin, and vitronectin. In the case of vWf and vitronectin, studies with mutant forms of these ligands indicate that the RGD sequence plays a direct role in binding to $\alpha_{IIb}\beta_3$ (Beacham *et al.,* 1992; Cherny *et al.,* 1993). In contrast, RGD sites in the fibrinogen Aα chain appear dispensable, at least for initial interactions of the ligand with the receptor, while a sequence at the C-terminus of the fibrinogen γ chain is necessary (Farrell and Thiagarajan, 1994).

For all soluble and most immobilized adhesive ligands, high-affinity (or avidity) binding to $\alpha_{IIb}\beta_3$ requires "inside-out" signaling, whereby agonist activation of the platelet converts $\alpha_{IIb}\beta_3$ to a ligand-competent state. Productive interactions of $\alpha_{IIb}\beta_3$ with ligands depend on the activation state of the platelet and the physical state of the ligand (Fig. 2). Fibrinogen binding to $\alpha_{IIb}\beta_3$ triggers "outside-in" signaling, manifested initially by tyrosine phosphorylation of specific platelet proteins (Shattil *et al.,* 1994a). Mechanisms of adhesive signaling in platelets may be relevant to integrin function in nucleated cells. For example, many other integrins appear to be subject to some form of affinity (or avidity) regulation (Diamond and Springer, 1994; van Kooyk and Figdor, 1993). Furthermore, cell adhesion and cytoskeletal remodeling triggered through integrins have been shown to regulate motility, growth, differentiation, and programmed death in several cell types (Juliano and Haskill, 1993; Clark and Brugge, 1995; Schwartz *et al.,* 1995; Roskelley *et al.,* 1994).

IV. INSIDE-OUT SIGNALING AND THE PLATELET CYTOSKELETON

Resting platelets do not bind soluble fibrinogen with measurable affinity or avidity, while platelets stimulated with agonists, such as thrombin, bind an average of 40,000 fibrinogen molecules/platelet. Binding is half-maximal at fibrinogen concentrations of 0.1–0.3 μmol/liter, or about 1/10th the fi-

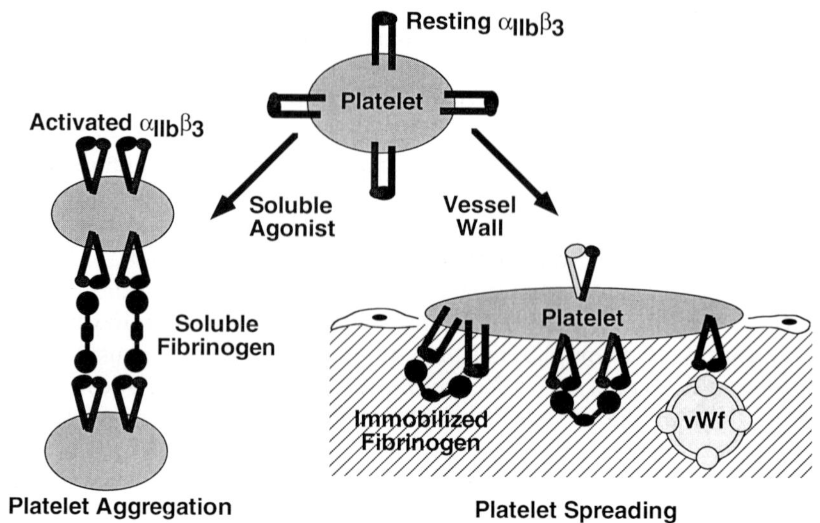

FIGURE 2 The adhesive functions of platelet integrin $\alpha_{IIb}\beta_3$ and its ligands. (Left) Agonist-induced binding of soluble dimeric fibrinogen (or multimeric vWf) to $\alpha_{IIb}\beta_3$ mediates platelet aggregation. Ligand binding requires conversion of $\alpha_{IIb}\beta_3$ to a high-affinity state and aggregation requires the formation of fibrinogen bridges between integrins on adjacent platelets. (Right) Adherence of platelets via $\alpha_{IIb}\beta_3$ to immobilized fibrinogen or vWf results in platelet spreading on the vascular matrix. In this case, platelet interaction with fibrinogen does not require initial platelet activation, presumably because of conformational changes in the immobilized ligand, but platelet interaction with vWf does.

brinogen concentration of plasma (Bennett and Vilaire, 1979). Exposure of the internal pool of platelet $\alpha_{IIb}\beta_3$ is not required for ligand binding. Therefore, platelet activation must trigger a fundamental change in surface-expressed $\alpha_{IIb}\beta_3$. Activation is believed to induce structural changes within the $\alpha_{IIb}\beta_3$ heterodimer that result in improved access of fibrinogen to binding sites in the receptor (Ginsberg *et al.,* 1992). The binding of fibrinogen to $\alpha_{IIb}\beta_3$ is complex. Surface plasmon resonance studies of the purified ligand and receptor show that binding is composed of at least two consecutive processes, each with its own kinetics (Huber *et al.,* 1995).

Platelet activation may also result in receptor oligomerization, a process that could increase ligand binding either by reducing the off-rate of the reaction or by promoting multivalent interactions between fibrinogen and adjacent integrin heterodimers. Since a subpopulation of $\alpha_{IIb}\beta_3$ is already associated with the membrane skeleton in resting platelets (Suzuki *et al.,* 1991; Kouns *et al.,* 1991; Zhang *et al.,* 1992; Fox *et al.,* 1993a), and the β_3 tail can interact directly with α-actinin and talin (Horwitz *et al.,* 1986; Otey *et al.,* 1993; Pavalko and Otey, 1994), it is plausible that receptor clustering may be promoted by the reorganization of the cytoskeleton that takes place during platelet activation. The two proposed mechanisms of $\alpha_{IIb}\beta_3$ regulation, conformational change and receptor clustering, are not mutually exclusive. By analogy with other plasma membrane receptors, conformational changes in $\alpha_{IIb}\beta_3$ brought about by cell activation or ligand binding might even promote receptor oligomerization (Frelinger *et al.,* 1991; Heldin, 1995).

When platelets are stimulated by thrombin or other agonists in the presence of radiolabeled fibrinogen and divalent cations, ligand binding is observed within seconds and reaches steady state within several minutes. During this time, almost all of the binding can be reversed by inhibitors of ligand binding (calcium chelators, RGD peptides) or by compounds that stimulate production of cAMP or cGMP and reverse the platelet activation process (Graber and Hawiger, 1982). However, if allowed to proceed for >10 min, most of the fibrinogen binding becomes irreversible and the ligand can now be recovered in the 15,600*g* Triton-insoluble pellet (Fig. 3) (Peerschke, 1991,1992,1994). Several observations suggest that the confor-

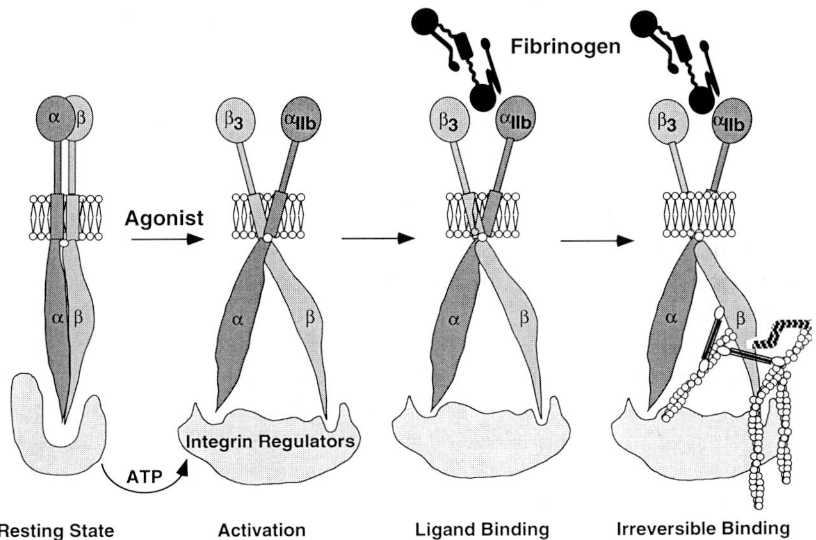

FIGURE 3 Sequence of changes in $\alpha_{IIb}\beta_3$ during platelet activation. Platelet activation induces a modification in $\alpha_{IIb}\beta_3$ that enables the ligand-binding site to become more accessible to fibrinogen. This may be triggered by agonist-induced phosphorylation of one or more proteins that regulate receptor affinity through an interaction with the integrin cytoplasmic tails. Fibrinogen binding is initially reversible but subsequently irreversible. The cytoskeleton may be involved in the process of irreversible fibrinogen binding because pretreatment of platelets with cytochalasin D or E inhibits the irreversible phase of fibrinogen binding.

mational change mechanism is the primary one responsible for the initial reversible phase of fibrinogen binding to $\alpha_{IIb}\beta_3$ and that subsequent cytoskeletal events may contribute to the later, irreversible phase of ligand binding.

First, *binding studies* with a fibrinogen mimetic monoclonal antibody, PAC1, suggest that a conformational change takes place in $\alpha_{IIb}\beta_3$ in the first few seconds after platelet activation to expose binding sites for ligand (Shattil *et al.,* 1992; Frojmovic *et al.,* 1991). Since platelet activation is required even for the binding of recombinant, monovalent PAC1 Fab to $\alpha_{IIb}\beta_3$, ligand valency must not be a necessary determinant of the reaction (Abrams *et al.,* 1994). Conversely, an RYD sequence in the context of a specific hypervariable antibody loop (H-CDR3) of PAC1 is a necessary determinant of binding (Taub *et al.,* 1989; Abrams *et al.,* 1994; Kunicki *et al.,* 1996). Furthermore, fibrinogen and PAC1 binding to purified $\alpha_{IIb}\beta_3$ immobilized onto plastic can be induced by certain anti-α_{IIb} or anti-β_3 "activating" antibodies under experimental conditions in which inducible clustering of the integrin cannot take place (Kouns *et al.,* 1992).

Second, an *inherited point mutation* in the β_3 cytoplasmic tail can abrogate ligand binding to $\alpha_{IIb}\beta_3$ in response to platelet agonists (Chen *et al.,* 1992). This is consistent with a role for integrin cytoplasmic tails in inside-out signaling. Third, certain *monoclonal antibodies* to α_{IIb} or β_3 have been shown to propagate long-range conformational changes in $\alpha_{IIb}\beta_3$ that lead to fibrinogen binding to platelets, to Chinese hamster ovary (CHO) cells transfected with $\alpha_{IIb}\beta_3$, and to purified preparations of the receptor (Du *et al.,* 1993). Finally, *cytochalasin D and E* prevent the later, irreversible phase of fibrinogen binding to platelets, but not the earlier, reversible phase (Peerschke, 1984; Shattil *et al.,* 1994b).

The signaling pathways responsible for modulation of $\alpha_{IIb}\beta_3$ have not been worked out in complete detail. Figure 4 presents a model of thrombin-induced platelet activation based on current information. Thrombin binds to a G-protein–linked receptor, resulting in dissociation of Gα- and $\beta\gamma$-subunits and activation of protein tyrosine kinases (PTKs), among them pp60^{Src} (Brass *et al.,* 1993; Clark *et al.,* 1994a). Then, Gα, $\beta\gamma$, and PTKs initiate multiple signaling reactions. For example, the $\beta\gamma$-subunits may be responsible for activation of pp60^{Src} and other PTKs (Touhara *et al.,* 1995) as well as for activation of an isoform of PI-3K (PI-3K γ) (Zhang *et al.,* 1995). In Swiss 3T3 cells, Gα_{12} and Gα_{13} regulate Rho-dependent responses (Buhl *et al.,* 1995). Intriguingly, these G-proteins are also present in platelets, suggesting that this may be one route to activation of Rho A in these cells (Brass *et al.,* 1993). As discussed previously, thrombin stimulation also leads to activation of Rac in platelets (Hartwig *et al.,* 1995). Rho, Rac, and cdc42Hs are regulated by nucleotide exchange. The functions of specific

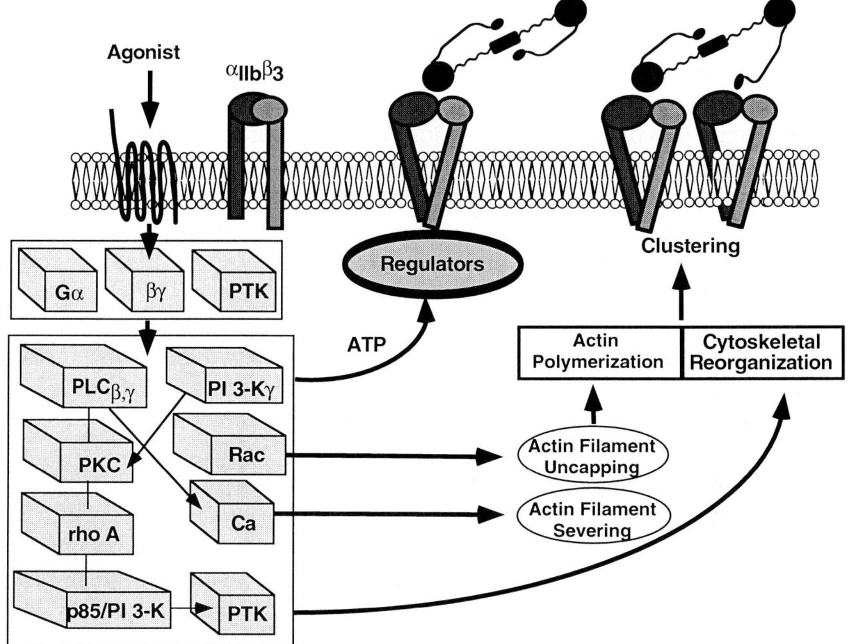

FIGURE 4 A working model of the signaling pathways in platelets that regulate fibrinogen binding to $\alpha_{IIb}\beta_3$. As emphasized in the text, many steps and intermediates in the process remain to be defined.

guanine nucleotide exchange factors, guanine nucleotide dissociation inhibitors, and GAPs in the regulation of these proteins in platelets remain to be determined.

Two other Gα-subunits, Giα and Gqα, can activate phospholipase C_β resulting in inositol 1,4,5-triphosphate–induced Ca^{2+} transients and diacylglycerol-mediated activation of Ca^{2+}-dependent isoforms of protein kinase C (PKC). In addition, specific isoforms of PKC may become activated by PI-3K γ following activation of phospholipase Cγ by unidentified PTKs (Fig. 4) (Tate and Rittenhouse, 1993; Guinebault *et al.,* 1994; Kovacsovics *et al.,* 1995b). PKC may promote activation of Rho A, which in turn is involved in the activation of the p85/p110 isoform of PI-3K in platelets (Zhang *et al.,* 1992,1995). Based on studies with selective PTK inhibitors in fibroblasts and platelets, unidentified PTKs may be interposed between Rho A, PI-3K, and their downstream effects (Romer *et al.,* 1992; Bachelot *et al.,* 1992). These signaling molecules are drawn within a box in Fig. 4, from which there are several outputs, among them (1) regulation of integrin affinity, (2) actin polymerization in response to actin filament uncapping and severing, and (3) reorganization of the cytoskeleton.

How are the reversible and irreversible phases of fibrinogen binding to $\alpha_{IIb}\beta_3$ ultimately regulated? We propose that the initial, reversible phase is set into motion by the phosphorylation of a hypothetical integrin regulator(s) by PKC or some other serine–threonine kinase or PTK. Through an interaction with the cytoplasmic tail of α_{IIb} and/or β_3, the penultimate regulatory element would influence the relative orientation of the subunits and, therefore, the conformation and accessibility of the ligand-binding interface. The reported low stoichiometry of phosphorylation of β_3 in thrombin-treated platelets (<5%) argues against (but does not formally exclude) direct phosphorylation of the β_3 tail as a regulatory event (Hillery *et al.*, 1990). Activated Rho A might regulate integrin conformation and/or promote receptor clustering through an effect on the cytoskeleton (Morii *et al.*, 1992). In fact, agonist-induced platelet aggregation is inhibited by C3 exoenzyme (Morii *et al.*, 1992), a specific inhibitor of Rho A, but it is not known if Rho A functions at a step proximal or distal to ligand binding to $\alpha_{IIb}\beta_3$. Lipid products of PI-3K may play some role in maintaining $\alpha_{IIb}\beta_3$ in a high-affinity state, perhaps by activation of a Ca^{2+}-independent isoform of PKC (Kovacsovics *et al.*, 1995a,b) or through some undefined direct effect on the integrin or the cytoskeleton. The final stabilization of fibrinogen binding to $\alpha_{IIb}\beta_3$ may be a consequence of events resulting from the oligomerization of the receptor, both by the multivalent ligand and by the process of cytoskeletal reorganization.

Studies of recombinant $\alpha_{IIb}\beta_3$ expressed in CHO cells are consistent with the model in Fig. 4. In CHO cells, wild-type $\alpha_{IIb}\beta_3$ is in a default low-affinity state, as assessed by binding of soluble fibrinogen or PAC1. However, substitution of the α_{6A} cytoplasmic tail for the α_{IIb} tail converts the extracellular portion of $\alpha_{IIb}\beta_3$ into a high-affinity state that is dependent on cellular energy metabolism (O'Toole *et al.*, 1994). Expression of isolated β_3 tail chimeras in CHO cells exerts a dominant-negative effect on this activated form of $\alpha_{IIb}\beta_3$, suggesting competition between the chimeric and heterodimeric tails for unidentified intracellular regulatory molecules (Chen *et al.*, 1994).

V. OUTSIDE-IN SIGNALING AND THE PLATELET CYTOSKELETON

Experiments on platelets were among the first to show that ligand occupancy and clustering of integrins trigger intracellular signaling reactions, such as tyrosine phosphorylation. Anchorage-dependent signaling in the platelet may facilitate platelet spreading on extracellular matrices and the conversion of small platelet aggregates into larger ones. In nucleated cells, the phenotypic responses to integrin signaling are more diverse, and include cell migration and induction of gene programs that regulate cell growth, differentiation and programmed death (Juliano and Haskill, 1993; Schwartz *et al.,* 1995; Boudreau *et al.,* 1995).

Three complementary experimental protocols have been useful in defining signaling events downstream of fibrinogen binding to $\alpha_{IIb}\beta_3$. In the first, platelets in suspension are activated by a conventional agonist (e.g., thrombin), and platelet responses that are dependent on fibrinogen binding are defined as those that are inhibited by pretreatment of the platelets with a specific inhibitor of ligand binding to $\alpha_{IIb}\beta_3$ (e.g., an RGD peptide or peptidomimetic or a function-blocking anti-$\alpha_{IIb}\beta_3$ antibody) (Golden *et al.,* 1990; Clark *et al.,* 1994a; Shattil *et al.,* 1994a). In a variant of this strategy, Glanzmann thrombasthenia platelets with an inherited deficiency of $\alpha_{IIb}\beta_3$ are studied in parallel with normal platelets, and those responses missing in the thrombasthenic platelets are ascribed to $\alpha_{IIb}\beta_3$. In the second protocol, fibrinogen binding to unstirred platelets is induced directly by the Fab fragment of an activating anti-β_3 monoclonal antibody (Huang *et al.,* 1993). The advantage of this approach is that the Fab, unlike a conventional agonist such as thrombin, does not trigger generalized activation of the cell, thus allowing only signaling reactions secondary to fibrinogen binding to be monitored. The effects of fibrinogen-mediated platelet aggregation can also be studied in this system if the platelets are stirred. In the third protocol, washed platelets are allowed to adhere to and spread on a solid fibrinogen matrix. Here, full spreading is not only dependent on platelet adhesion via $\alpha_{IIb}\beta_3$, but also requires the action of an agonist—for example, ADP released from the adherent platelets (Haimovich *et al.,* 1993). Similar adhesion studies can be performed in CHO cells transfected with $\alpha_{IIb}\beta_3$ (Leong *et al.,* 1995). Taken together, these experiments have established that ligand binding to $\alpha_{IIb}\beta_3$ triggers an initial wave of protein tyrosine phosphorylation, followed by additional tyrosine phosphorylation coincident with platelet aggregation or spreading (Fig. 5).

Since tyrosine phosphorylation is the earliest biochemical response that has been detected after ligation of $\alpha_{IIb}\beta_3$, attention has focused on PTKs and phosphatases in platelets whose activity may be modified by fibrinogen binding, aggregation, or spreading. Figure 6 depicts the sequence of morphological, cytoskeletal, and biochemical changes that occurs in platelets during adhesion to a fibrinogen matrix.

A. Responses during Integrin Occupancy and Clustering

Fibrinogen binding to $\alpha_{IIb}\beta_3$ on platelets stimulates tyrosine phosphorylation and activation of the PTK $pp72^{Syk}$, as well as tyrosine phosphorylation

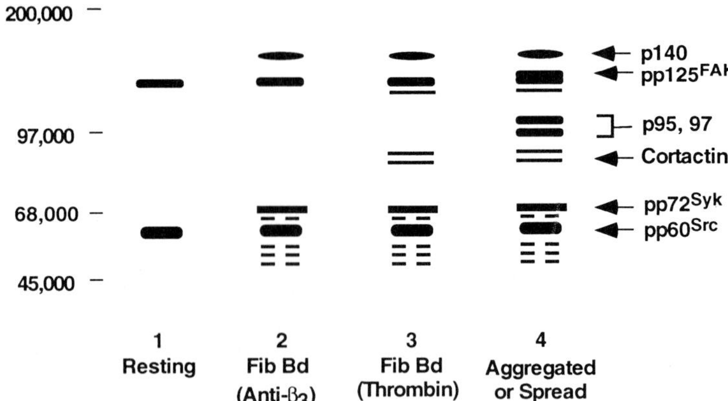

FIGURE 5 Schematic representation of phosphotyrosine-containing proteins in platelets. Platelet lysates were immunoprecipitated with a polyclonal anti-phosphotyrosine antibody and Western blots probed with monoclonal anti-phosphotyrosine antibodies. (Lane 1) Resting platelets show a basal pattern of tyrosine phosphorylation of $pp60^{Src}$. (Lane 2) Effect of fibrinogen binding to unstirred platelets in response to an "activating" anti-β_3 antibody Fab. Bands between 50 and 72 kDa, including $pp72^{Syk}$, and a band at 140 kDa are now phosphorylated. (Lane 3) Effect of fibrinogen binding to unstirred platelets in response to thrombin. Additional bands are apparent, including the actin-binding protein cortactin. (Lane 4) Effect of thrombin-induced platelet aggregation. Additional bands at 95, 97, and 125 kDa ($pp125^{FAK}$) are now present.

of several unidentified substrates of 50–68 kDa and 140 kDa (Huang *et al.*, 1993; Clark *et al.*, 1994b) (Figs. 5 and 6). This response is observed within seconds of the binding of multivalent RGD ligands, such as fibrinogen. Since tyrosine phosphorylation is not observed upon binding of monovalent ligands to $\alpha_{IIb}\beta_3$ and is prevented by pretreatment of the platelets with cytochalasin D or E, this response would appear to require both ligand-induced oligomerization of $\alpha_{IIb}\beta_3$ complexes and actin polymerization (Huang *et al.*, 1993; Clark *et al.*, 1994c; Abrams *et al.*, 1994). Indeed, fibrinogen binding to $\alpha_{IIb}\beta_3$ does induce a small but significant increase in platelet F-actin content (Shattil *et al.*, 1994b).

One unresolved mystery is how ligation of $\alpha_{IIb}\beta_3$ by multivalent ligands triggers protein tyrosine phosphorylation of $pp72^{Syk}$ and other substrates. Unlike receptor tyrosine kinases, the cytoplasmic tails of α_{IIb} and β_3 do not possess catalytic domains. Of note, $pp72^{Syk}$ is involved in the activation of a number of hematopoietic cells following ligation of their immune response receptors (Cambier *et al.*, 1994). In these cases, engagement of the receptor by ligand leads to phosphorylation of tyrosine residues within an "ITAM" motif on a cytoplasmic tail of a receptor subunit by a Src family kinase. This results in the binding of the tandem SH2 domains of

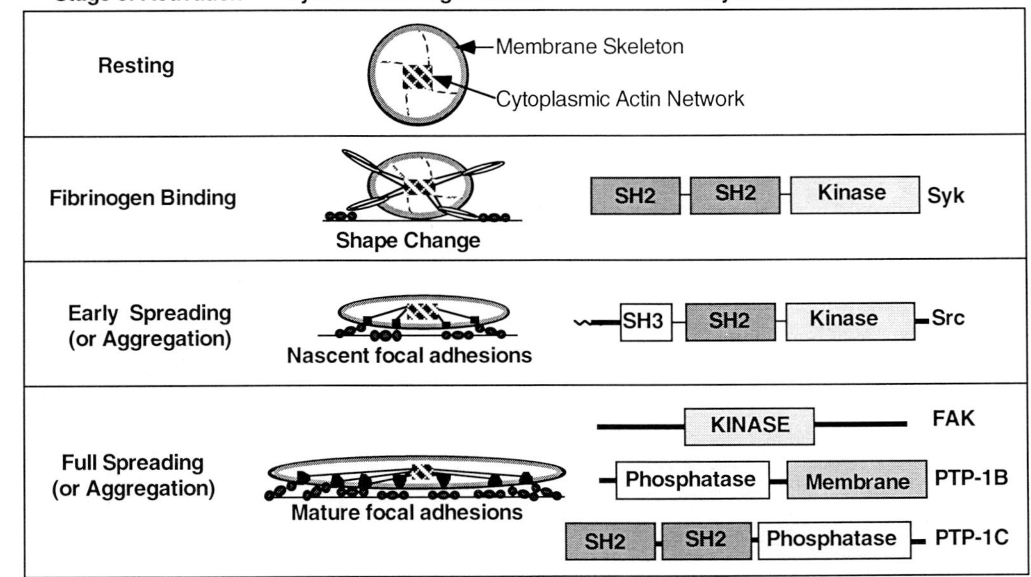

FIGURE 6 Changes in cytoskeletal organization and activation of PTKs and phosphatases following fibrinogen binding to platelet $\alpha_{IIb}\beta_3$. The example shown is for platelets adherent to a fibrinogen matrix. A similar sequence of events accompanies agonist-induced platelet activation and aggregation.

$pp72^{Syk}$ to the receptor subunit and in activation of Syk. Except for the cytoplasmic tail of the β_4 integrin subunit (Mainiero *et al.,* 1995), integrin cytoplasmic tails do not contain anything approaching a classical ITAM motif. Therefore, Syk might become activated in platelets through interactions with unidentified tyrosine-phosphorylated proteins that are recruited to the vicinity of the cytoplasmic tails of $\alpha_{IIb}\beta_3$ during an early phase of cytoskeletal reorganization. However, at present, the mechanism of $\alpha_{IIb}\beta_3$-mediated Syk activation remains obscure. It should be emphasized here that $pp72^{Syk}$ (and $pp60^{Src}$) also becomes activated in agonist-stimulated platelets in a manner independent of $\alpha_{IIb}\beta_3$ (Maeda *et al.,* 1993; Rezaul *et al.,* 1994; Taniguchi *et al.,* 1993; Clark *et al.,* 1994b; Yanaga *et al.,* 1995). A number of other platelet proteins become tyrosine phosphorylated in an $\alpha_{IIb}\beta_3$-independent manner, including $p120^{Ras}$ GAP, p42 mitogen-activated protein kinase, and p80–85 cortactin (Clark and Brugge, 1993; Clark *et al.,* 1994a,b). Thus, the platelet may contain discrete subcellular pools of PTKs, protein tyrosine phosphatases, and substrates that are differentially affected by ligation of G-protein–linked receptors and $\alpha_{IIb}\beta_3$.

B. Early Responses during Platelet Spreading or Aggregation

$pp60^{Src}$ becomes activated early in the process of fibrinogen-dependent spreading or aggregation (Fig. 6). While the mechanism for Src activation in aggregating platelets is not clear, it could be due to dephosphorylation of a C-terminal negative-regulatory tyrosine (Tyr^{527}) by a tyrosine phosphatase or to displacement of phosphorylated Tyr^{527} from its binding site in the Src SH2 domain by a competing tyrosine-phosphorylated protein (Clark and Brugge, 1995; Erpel and Courtneidge, 1995). Either mechanism might be facilitated by formation of nascent adhesion plaques during the early phase of platelet aggregation or spreading and the assembly of a submicroscopic, integrin-based multiprotein signaling complex. Platelets contain four other Src family kinases, Yes, Fyn, Lyn, and Hck. After thrombin stimulation, Fyn and Src associate with PI-3K (Gutkind *et al.,* 1990), while Fyn, Yes, and Lyn associate with $p120^{Ras}$ GAP (Cichowski *et al.,* 1992). The responses of these other Src family kinases to integrin ligation have not been studied.

C. Late Responses during Platelet Spreading or Aggregation

If platelets are permitted to spread fully on fibrinogen or to form very large aggregates, tyrosine phosphorylation and activation of $pp125^{FAK}$ are observed, along with tyrosine phosphorylation of unidentified substrates of 95 and 97 kDa (Figs. 5 and 6) (Lipfert *et al.,* 1992; Haimovich *et al.,* 1993). In contrast to the earlier responses to $\alpha_{IIb}\beta_3$ ligation, these late responses also require co-stimulation of platelets through a G-protein–linked receptor (Shattil *et al.,* 1994b). Mutational analyses of $\alpha_{IIb}\beta_3$ in CHO cells suggest that the β_3 cytoplasmic tail is required for tyrosine phosphorylation of pp125^{FAK} in response to cell adhesion to fibrinogen (Leong *et al.,* 1995). The requirement for the β_3 tail may be explained by a direct interaction of focal adhesion kinase (FAK) with the tail (Schaller *et al.,* 1995). Alternatively, the tail may be involved indirectly in recruiting other PTKs (e.g., pp60^{Src}) that can bind to and activate FAK (Cobb *et al.,* 1994). Indeed, tyrosine phosphorylation and activation of FAK during platelet spreading occur coincident with the formation of mature adhesion plaque–like structures (Nachmias and Golla, 1991). Thus, signals from G-protein–linked receptors and $\alpha_{IIb}\beta_3$ may converge and become integrated within these adhesion complexes.

Many of the proteins that become tyrosine phosphorylated during integrin ligation, platelet spreading, and aggregation eventually undergo dephosphorylation. Pertinent in this regard, platelets contain several cytoplasmic tyrosine kinases, and two of these, PTP-1B and PTP-1C, become activated and redistribute to the Triton-insoluble core cytoskeleton in fully aggregated platelets (Fig. 6) (Frangione *et al.,* 1993; Takayama *et al.,* 1993; Ezumi *et al.,* 1995; Li *et al.,* 1994). The redistribution of PTP-1B and the tyrosine phosphorylation of PTP-1C are dependent on platelet aggregation, while redistribution of PTP-1C is partly dependent on aggregation and partly dependent on some other aspect of the thrombin activation process.

Vasodilator-stimulated phosphoprotein (VASP) is an ~50-kDa serine–threonine phosphoprotein that localizes to the periphery of lamellipodia, to focal adhesions, and along stress fibers in fibroblasts (Reinhard *et al.,* 1992). It contains binding sites for F-actin, profilin, and an 83-kDa zyxin-related protein (Haffner *et al.,* 1995; Reinhard *et al.,* 1995). VASP is present in high concentrations in platelets, where it is a substrate for both cAMP- and cGMP-dependent protein kinases (Eigenthaler *et al.,* 1992; Nolte *et al.,* 1994). In spread platelets, VASP is located predominantly in the distal parts of radial microfilament bundles and in peripheral microfilaments (Reinhard *et al.,* 1992). Interestingly, cyclic nucleotides inhibit agonist-induced activation of $\alpha_{IIb}\beta_3$ and platelet aggregation at the same concentrations that they stimulate serine phosphorylation of VASP (Mellion *et al.,* 1981; Graber and Hawiger, 1982; Horstrup *et al.,* 1994; Eigenthaler and Walter, 1994). Thus, phosphorylation of VASP may play some role in downregulating cytoskeletal assembly and adhesive signaling.

During the adhesion of fibroblasts to extracellular matrices via β_1 integrins, three Rho family members (cdc42Hs, Rac, and Rho) have been implicated in the formation of filopodia, lamellipodia, and adhesion plaques, respectively (Chong *et al.,* 1994; Nobes and Hall, 1995). As already discussed, in platelets Rac has been implicated in actin polymerization and Rho A in platelet aggregation, and all three Rho family members redistribute to the Triton-insoluble cytoplasmic cytoskeleton during platelet aggregation. Moreover, platelets contain p62^{PAK}, a serine–threonine kinase that binds to and is activated by cdc42Hs and Rac and may be a downstream effector of these proteins (Teo *et al.,* 1995). Yet it remains to be determined how these GTP-binding proteins influence the adhesive and signaling functions of $\alpha_{IIb}\beta_3$ and, conversely, how outside-in signaling through $\alpha_{IIb}\beta_3$ influences the functions of Rho family proteins in organizing filopodia, lamellipodia, and adhesion plaques.

Outside-in signaling through integrins has been documented in many types of cells, including monocytes, fibroblasts, epithelial cells, and endothelial cells (Juliano and Haskill, 1993; Clark and Brugge, 1995; Boudreau *et al.,* 1995; Schwartz *et al.,* 1995). As in platelets, integrin ligation and clustering in fibroblasts trigger a hierarchical assemblage of enzymes and their substrates within a scaffold of cytoskeletal proteins (Miyamoto *et al.,* 1995a,b). A central element of current models of adhesive signaling by integrins is that specific intracellular regulatory proteins interact with integrin cytoplasmic tails to initiate the signaling process. While this model is being tested, it should be kept in mind that integrin function may also be influenced by transmembrane proteins and membrane lipids (Zhou and Brown, 1993; Berditchevski *et al.,* 1995; Hermanowski-Vosatka *et al.,* 1992). In this regard, transmembrane proteins such as CD9 and CD47 (integrin-associated protein) have been shown to co-immunoprecipitate with $\alpha_{IIb}\beta_3$ from platelet lysates (Slupsky *et al.,* 1989; Lindberg *et al.,* 1994). However, no integrin-associated transmembrane protein or membrane lipid has yet been shown to directly affect $\alpha_{IIb}\beta_3$ function in platelets (Smyth *et al.,* 1992; Fujimoto *et al.,* 1995). For that matter, relatively few proteins have been shown to interact *directly* with integrin cytoplasmic tails. The α_{IIb} tail has been reported to interact with calreticulin through a membrane-proximal sequence that is highly conserved among all α-subunits (Leung-Hagesteijn *et al.,* 1994). The β_3 tail has been shown to interact with α-actinin (Otey *et al.,* 1993), talin (Horwitz *et al.,* 1986), pp125^{FAK} (Schaller *et al.,* 1995), and β_3-endonexin (Shattil *et al.,* 1995). Some of these interactions may be specific for the β_3 tail (e.g., β_3-endonexin), while others are not (e.g., α-actinin, talin, pp125^{FAK}). In many cases, direct interactions have been demonstrated only *in vitro.* Given the availability of powerful techniques to detect protein–protein interactions (Phizicky and Fields, 1995), the list of integrin-binding proteins is sure to grow (Hannigan *et al.,* 1995). The challenge will be to determine how each of these affects integrin function under physiological circumstances.

VI. SUMMARY

Platelets contain membrane and cytoplasmic cytoskeletons that control the discoid shape of the resting cell. These cytoskeletons are remodeled during platelet activation, adhesion, and aggregation, resulting in the appearance of a variety of F-actin structures, including filopodia, lamellipodia, and adhesion plaques. Remodeling of the cytoskeleton is mediated, in part, through platelet-signaling pathways that include G-protein–linked receptors, heterotrimeric and small GTP-binding proteins, and protein and lipid kinases. It is also facilitated by signaling reactions triggered by fibrinogen binding to integrin $\alpha_{IIb}\beta_3$. The cytoskeleton of activated platelets may serve as a scaffold to localize enzymes and substrates at the right time and place to achieve optimal platelet responses to vascular injury. Further clarification of the complex relationships between $\alpha_{IIb}\beta_3$, cytoskeletal proteins, and the signaling machinery of the platelet will have diagnostic and therapeutic implications for hemostasis and thrombosis. It may also facilitate our understanding of anchorage-dependent cellular responses in nucleated cells, where integrins and the cytoskeleton provide a conduit for the transfer of information from the extracellular matrix to the nucleus.

References

Abrams, C., Deng J., Steiner, B., and Shattil, S. J. (1994). Determinants of specificity of a baculovirus-expressed antibody Fab fragment that binds selectively to the activated form of integrin $\alpha_{IIb}\beta_3$. *J. Biol. Chem.* **269,** 18781–18788.

Altmüller, A., and Presek, P. (1995). Rapid protein tyrosine phosphorylation in the cytoskeleton of stimulated human platelets. *Biochim. Biophys. Acta Mol. Cell Res.* **1265,** 61–66.

Bachelot, C., Cano, E., Grelac, F., Saleun, S., Druker, B. J., Levy-Toledano, S., Fischer, S., and Rendu F. (1992). Functional implications of tyrosine protein phosphorylation in platelets. Simultaneous studies with different agonists and inhibitors. *Biochem. J.* **284,** 923–928.

Barkalow, K., and Hartwig, J. H. (1995). The role of actin filament barbed-end exposure in cytoskeletal dynamic and cell motility. *Biochem. Soc. Trans.* **23,** 451–456.

Beacham, D. A., Wise, R. J., Turci, S. M., and Handin, R. I. (1992). Selective inactivation of the Arg-Gly-Asp-Ser (RGDS) binding site in von Willebrand factor by site-directed mutagenesis. *J. Biol. Chem.* **267,** 3409–3415.

Bearer, E. L. (1995). Cytoskeletal domains in the activated platelet. *Cell Motil. Cytoskeleton* **30,** 50–66.

Bennett, J. S., and Vilaire, G. (1979). Exposure of platelet fibrinogen receptors by ADP and epinephrine. *J. Clin. Invest.* **64,** 1393–1401.

Berditchevski, F., Bazzoni, F., and Hemler, M. E. (1995). Specific association of CD63 with the VLA-3 and VLA-6 integrins. *J. Biol. Chem.* **270,** 17784–17790.

Boudreau, N., Myers, C., and Bissell, M. J. (1995). From laminin to lamin-regulation of tissue-specific gene expression by the ECM. *Trends Cell Biol.* **5,** 1–4.

Brass, L. F., Shattil, S. J., Kunicki, T. J., and Bennett, J. S. (1985). Effect of calcium on the stability of the platelet membrane glycoprotein IIb-IIIa complex. *J. Biol. Chem.* **260,** 7875–7881.

Brass, L. F., Hoxie, J. A., and Manning, D. R. (1993). Signaling through G proteins and G protein-coupled receptors during platelet activation. *Thromb. Haemost.* **70,** 217–223.

Buhl, A. M., Johnson, N. L., Dhanasekaran, N., and Johnson, G. L. (1995). $G\alpha_{12}$ and $G\alpha_{13}$ stimulate rho-dependent stress fiber formation and focal adhesion assembly. *J. Biol. Chem.* **270,** 24631–24634.

Burridge, K., Fath, K., Kelly, T., Nuckolls, G., and Turner, C. (1988). Focal adhesions: Transmembrane junctions between the extracellular matrix and the cytoskeleton. *Annu. Rev. Cell Biol.* **4,** 487–525.

Burridge, K., Petch, L. A., and Romer, L. H. (1992). Signals from focal adhesions. *Curr. Biol.* **2,** 537–539.

Cambier, J. C., Pleiman, C. M., and Clark, M. R. (1994). Signal transduction by the B cell antigen receptor and its coreceptors. *Annu. Rev. Immunol.* **12,** 457–486.

Chen, Y.-P., Djaffar, I., Pidard, D., Steiner, B., Cieutat, A.-M., Caen, J. P., and Rosa, J.-P. (1992). Ser-752 → Pro mutation in the cytoplasmic domain of integrin β_3 subunit and defective activation of platelet integrin $\alpha_{IIb}\beta_3$ (glycoprotein IIb-IIIa) in a variant of Glanzmann thrombasthenia. *Proc. Natl. Acad. Sci. U.S.A.* **89,** 10169–10173.

Chen, Y.-P., O'Toole, T. E., Shipley, T., Forsyth, J., La Flamme, S. E., Yamada, K. M., Shattil, S. J., and Ginsberg, M. H (1994). "Inside-out" signal transduction inhibited by isolated integrin cytoplasmic domains. *J. Biol. Chem.* **269,** 18307–18310.

Cherny, R. C., Honan, M. A., and Thiagarajan, P. (1993). Site-directed mutagenesis of the arginine-glycine-aspartic acid in vitronectin abolishes cell adhesion. *J. Biol. Chem.* **268,** 9725–9729.

Chong, L. D., Traynor-Kaplan, A., Bokoch, G. M., and Schwartz, M. A. (1994). The small GTP-binding protein Rho regulates a phosphatidylinositol 4-phosphate 5-kinase in mammalian cells. *Cell* **79,** 507–513.

Cichowski, K., McCormick, F., and Brugge, J. S. (1992). $p21^{ras}$GAP association with Fyn, Lyn, and Yes in thrombin-activated platelets. *J. Biol. Chem.* **267,** 5025–5028.

Clark, E. A., and Brugge, J. S. (1993). Redistribution of activated $pp60^{c\text{-}src}$ to integrin-dependent cytoskeletal complexes in thrombin-stimulated platelets. *Mol. Cell. Biol.* **13,** 1863–1871.

Clark, E. A., and Brugge, J. S. (1995). Integrins and signal transduction pathways. The road taken. *Science* **268,** 233–239.

Clark, E. A., Shattil, S. J., and Brugge, J. S. (1994a). Regulation of protein tyrosine kinases in platelets. *Trends Biochem. Sci.* **19,** 464–469.

Clark, E. A., Shattil, S. J., Ginsberg, M. H., Bolen, J., and Brugge, J. S. (1994b). Regulation of the protein tyrosine kinase, $pp72^{syk}$, by platelet agonists and the integrin, $\alpha_{IIb}\beta_3$. *J. Biol. Chem.* **46,** 28859–28864.

Clark, E. A., Trikha, M., Markland, F. S., and Brugge, J. S. (1994c). Structurally distinct disintegrins contortrostatin and multisquamatin differentially regulate platelet tyrosine phosphorylation. *J. Biol. Chem.* **269,** 21940–21943.

Cobb, B. S., Schaller, M. D., Leu, T.-H., and Parsons, J. T. (1994). Stable association of $pp60^{src}$ and $pp59^{fyn}$ with the focal adhesion-associated protein tyrosine kinase, $pp125^{FAK}$. *Mol. Cell. Biol.* **14,** 147–155.

Dash, D., Aepfelbacher, M., and Siess, W. (1995a). The association of pp125^{FAK}, pp60^{Src}, CDC42Hs and Rap1B with the cytoskeleton of aggregated platelets is a reversible process regulated by calcium. *FEBS Lett.* **363,** 231–234.

Dash, D., Aepfelbacher, M., and Siess, W. (1995b). Integrin $\alpha_{IIb}\beta_3$-mediated translocation of CDC42Hs to the cytoskeleton in stimulated human platelets. *J. Biol. Chem.* **270,** 17321–17326.

Diamond, M. S., and Springer, T. A. (1994). The dynamic regulation of integrin adhesiveness. *Curr. Biol.* **4,** 506–517.

Du, X., Gu, M., Weisel, J., Nagaswami, C., Bennett, J. S., Bowditch, R., and Ginsberg, M. H. (1993). Long range propagation of conformational changes in integrin $\alpha_{IIb}\beta_3$. *J. Biol. Chem.* **268,** 23087–23092.

Earnest, J. P., Santos, G. F., Zuerbig, S., and Fox, J. E. B. (1995). Dystrophin-related protein in the platelet membrane skeleton. *J. Biol. Chem.* **270,** 27259–27265.

Eigenthaler, M., and Walter, U. (1994). Signal transduction and cyclic nucleotides in human platelets. *Thromb. Haemorrh. Disorders* **8,** 41–46.

Eigenthaler, M., Nolte, C., Halbrugge, M., and Walter, U. (1992). Concentration and regulation of cyclic nucleotides, cyclic-nucleotide-dependent protein kinases and one of their major substrates in human platelets. Estimating the rate of cAMP-regulated and cGMP-regulated protein phosphorylation in intact cells. *Eur. J. Biochem.* **205,** 471–481.

Erpel, T., and Courtneidge, S. A. (1995). Src family protein-tyrosine kinases and cellular signal transduction pathways. *Curr. Opin. Cell Biol.* **7,** 176–182.

Escolar, G., Sauk, J., Bravo, M. L., Krumwiede, M., and White, J. G. (1987). Immunogold staining of microtubules in resting and activated platelets. *Am. J. Hematol.* **24,** 177–188.

Ezumi, Y., Takayama, H., and Okuma, M. (1995). Differential regulation of protein-tyrosine phosphatases by integrin $\alpha_{IIb}\beta_3$ through cytoskeletal reorganization and tyrosine phosphorylation in human platelets. *J. Biol. Chem.* **270,** 11927–11934.

Ezzell, R. M., Kenney, D. M., Egan, S., Stossel, T. P., and Hartwig, J. H. (1988). Localization of the domain of actin-binding protein that binds to membrane glycoprotein Ib and actin in human platelets. *J. Biol. Chem.* **263,** 13303–13309.

Farrell, D. H., and Thiagarajan, P. (1994). Binding of recombinant fibrinogen mutants to platelets. *J. Biol. Chem.* **269,** 226–231.

Fitzgerald, L. A., Poncz, M., Steiner, B., Rall, S. C. Jr., Bennett, J. S., and Phillips, D. R. (1987a). Comparison of cDNA-derived protein sequences of the human fibronectin and vitronectin receptor alpha-subunits and platelet glycoprotein IIb. *Biochemistry* **26,** 8158–8165.

Fitzgerald, L. A., Steiner, B., Rall, S. C. Jr., Lo S. S., and Phillips, D. R. (1987b). Protein sequence of endothelial glycoprotein IIIa derived from a cDNA clone: Identity with platelet glycoprotein IIIa and similarity to "integrin". *J. Biol. Chem.* **262,** 3936–3939.

Fox, J. E. B. (1993). The platelet cytoskeleton. *Thromb. Haemost.* **70,** 884–893.

Fox, J. E., Boyles, J. K., Berndt, M. C., Steffen, P. K., and Anderson, L. K. (1988). Identification of a membrane skeleton in platelets. *J. Cell Biol.* **106,** 1525–1538.

Fox, J. E. B., Lipfert, L., Clark, E. A., Reynolds, C. C., Austin, C. D., and Brugge, J. S. (1993a). On the role of the platelet membrane skeleton in mediating signal transduction: Association of GP IIb-IIIa, pp60$^{c\text{-}src}$, pp62$^{c\text{-}yes}$, and the p21^{ras} GTPase-activating protein with the membrane skeleton. *J. Biol. Chem.* **268,** 25973–25984.

Fox, J. E.B., Taylor, R. G., Taffarel, M., Boyles, J. K., and Goll, D. E. (1993b). Evidence that activation of platelet calpain is induced as a consequence of binding of adhesive ligand to the integrin, glycoprotein IIb-IIIa. *J. Cell Biol.* **120,** 1501–1507.

Frangione, J. V., Oda, A., Smith, M., Salzman, E. W., and Neel, B. G. (1993). Calpain-catalyzed cleavage and subcellular relocation of protein phosphotyrosine phosphatase 1B (PTP-1B) in human platelets. *EMBO J.* **12,** 4843–4856.

Frelinger, A. L. III, Du, X., Plow, E. F., and Ginsberg, M. H. (1991). Monoclonal antibodies to ligand-occupied conformers of integrin $\alpha_{IIb}\beta_3$ (glycoprotein IIb-IIIa) alter receptor affinity, specificity, and function. *J. Biol. Chem.* **266,** 17106–17111.

Frojmovic, M., Wong, T., and Van de Ven, T. (1991). Dynamic measurements of the platelet membrane glycoprotein II_b-III_a receptor for fibrinogen by flow cytometry. I. Methodology, theory and results for two distinct activators. *Biophys. J.* **59,** 815–827.

Fujimoto, T., Fujimura, K., Noda, M., Takafuta, T., Shimomura, T., and Kuramoto, A. (1995). 50-kD integrin-associated protein does not detectably influence several functions of glycoprotein IIb-IIIa complex in human platelets. *Blood* **86,** 2174–2182.

Ginsberg, M. H., Du, X., and Plow, E. F. (1992). Inside-out integrin signalling. *Curr. Opin. Cell Biol.* **4,** 766–771.

Golden, A., Brugge, J. S., and Shattil, S. J. (1990). Role of platelet membrane glycoprotein IIb-IIIa in agonist-induced tyrosine phosphorylation of platelet proteins. *J. Cell Biol.* **111,** 3117–3127.

Goldschmidt-Clermont, P., Furman, M. I., Wachsstock, D., Safer, D., Nachmias, V. T., and Pollard, T. D. (1992). The control of actin nucleotide exchange by thymosin β4 and profilin—a potential regulatory mechanism for actin polymerization in cells. *Mol. Biol. Cell* **3,** 1015–1024.

Graber, S. E., and Hawiger, J. (1982). Evidence that changes in platelet cyclic AMP levels regulate the fibrinogen receptor on human platelets. *J. Biol. Chem.* **257,** 14606–14609.

Grondin, P., Plantavid, M., Sultan, C., Breton, M., Mauco, G., and Chap, H. (1991). Interaction of $pp60^{c\text{-}src}$, phospholipase C, inositol-lipid, and diacylglycerol kinases with the cytoskeletons of thrombin-stimulated platelets. *J. Biol. Chem.* **266,** 15705–15709.

Guinebault, C., Payrastre, B., Mauco, G., Breton, M., Plantavid, M., and Chap, H. (1994). Rapid and transient translocation of PLC-gamma1 to the cytoskeleton of thrombin-stimulated platelets. Evidence for a role of tyrosine kinases. *Cell. Mol. Biol.* **40,** 687–693.

Guinebault, C., Payrastre, B., Racaud-Sultan, C., Mazarguil, H., Breton, M., Mauco, G., Plantavid, M., and Chap, H. (1995). Integrin-dependent translocation of phosphoinositide 3-kinase to the cytoskeleton of thrombin-activated platelets involves specific interactions of p85α with actin filaments and focal adhesion kinase. *J. Cell Biol.* **129,** 831–842.

Gutkind, J. S., Lacal, P. M., and Robbins, K. C. (1990). Thrombin-dependent association of phosphatidylinositol-3 kinase with $p60^{c\text{-}src}$ and $p59^{fyn}$ in human platelets. *Mol. Cell. Biol.* **10,** 3806–3809.

Haas, T. A., and Plow, E. F. (1994). Integrin ligand interactions—a year in review. *Curr. Opin. Cell Biol.* **6,** 656–662.

Haffner, C., Jarchau, T., Reinhard, M., Hoppe, J., Lohmann, S. M., and Walter, U. (1995). Molecular cloning, structural analysis and functional expression of the proline-rich focal adhesion and microfilament-associated protein VASP. *EMBO J.* **14,** 19–27.

Haimovich, B., Lipfert, L., Brugge, J. S., and Shattil, S. J. (1993). Tyrosine phosphorylation and cytoskeletal reorganization in platelets are triggered by interaction of integrin receptors with their immobilized ligands. *J. Biol. Chem.* **268,** 15868–15877.

Hannigan, G. E., Leung-Hagesteijn, C., Fitz-Gibbon, L., Coppolono, M. G., Radeva, G., Filmus, J., Bell, J. C., and Dedhar, S. (1996). Regulation of cell adhesion and anchorage-dependent cell growth by a novel β_1 integrin-linked protein kinase. *Nature* (*London*). **379,** 91–96.

Hartwig, J. H. (1992). Mechanisms of actin rearrangements mediating platelet activation. *J. Cell Biol.* **118,** 1421–1442.

Hartwig, J. H., and DeSisto, M. (1991). The cytoskeleton of the resting human blood platelet: Structure of the membrane skeleton and its attachment to actin filaments. *J. Cell Biol.* **112,** 407–425.

Hartwig, J. H., Bokoch, G. M., Carpenter, C. L., Janmey, P. A., Taylor, L. A., Toker, A., and Stossel, T. P. (1995). Thrombin receptor ligation and activated rac uncap actin filament barbed ends through phosphoinositide synthesis in permeabilized human platelets. *Cell* **82,** 643–653.

Heldin, C.-H. (1995). Dimerization of cell surface receptors in signal transduction. *Cell* **80,** 213–223.

Hemmings, L., Barry, S. T., and Critchley, D. R. (1995). Cell-matrix adhesion: Structure and regulation. *Biochem. Soc. Trans.* **23,** 619–626.

Hermanowski-Vosatka, A., Strijp, J. A.G., Swiggard, W. J., and Wright, S. D. (1992). Integrin modulating factor-1: A lipid that alters the function of leukocyte integrins. *Cell* **68,** 341–352.

Hillery, C. A., Smyth, S. S., and Parise, L. V. (1990). Stoichiometry of glycoprotein IIIa phosphorylation in whole platelets and in vitro. *Blood* **76,** 459a.

Horstrup, K., Jablonka, B., Honig-Liedl, P., Just, M., Kochsiek, K., and Walter, U. (1994). Phosphorylation of focal adhesion vasodilator-stimulated phosphoprotein at Ser157 in intact human platelets correlates with fibrinogen receptor inhibition. *Eur. J. Biochem.* **225,** 21–27.

Horwitz, A., Duggan, K., Buck, C., Beckerle, M. C., and Burridge, K. (1986). Interaction of plasma membrane fibronectin receptor with talin-a transmembrane linkage. *Nature (London)* **320,** 531–533.

Huang, M.-M., Lipfert, L., Cunningham, M., Brugge, J. S., Ginsberg, M. H., and Shattil, S. J. (1993). Adhesive ligand binding to integrin $\alpha_{IIb}\beta_3$ stimulates tyrosine phosphorylation of novel protein substrates before phosphorylation of pp125^{FAK}. *J. Cell Biol.* **122,** 473–483.

Huber, W., Hurst, J., Schlatter, D., Barner, R., Hubscher, J., Kouns, W. C., and Steiner, B. (1995). Determination of kinetic constants for the interaction between the platelet glycoprotein IIb-IIIa and fibrinogen by means of surface plasmon plasmon resonance. *Eur. J. Biochem.* **227,** 647–656.

Hynes, R. O. (1992). Integrins: Versatility, modulation, and signaling in cell adhesion. *Cell* **69,** 11–25.

Johnson, R. P., and Craig, S. W. (1995). F-actin binding site masked by the intramolecular association of vinculin head and tail domains. *Nature (London)* **373,** 261–264.

Juliano, R. L., and Haskill, S. (1993). Signal transduction from the extracellular matrix. *J. Cell Biol.* **120,** 577–585.

Kouns, W. C., Fox, C. F., Lamoreaux, W. J., Coons, L. B., and Jennings, L. K. (1991). The effect of glycoprotein IIb-IIIa receptor occupancy on the cytoskeleton of resting and activated platelets. *J. Biol. Chem.* **266,** 13891–13900.

Kouns, W. C., Kirchhofer, D., Hadváry, P., Edenhofer, A., Weller, T., Pfenninger, G., Baumgartner, H. R., Jennings, L. K., and Steiner, B. (1992). Reversible conformational changes induced in glycoprotein IIb-IIIa by a potent and selective peptidomimetic inhibitor. *Blood* **80,** 2539–2547.

Kovacsovics, T. J., Bachelot, C., Toker, A., Vlahos, C. J., Duckworth, B., Cantley, L. C., and Hartwig, J. H. (1995a). Phosphoinositide 3-kinase inhibition spares actin assembly in activating platelets but reverses platelet aggregation. *J. Biol. Chem.* **270,** 11358–11366.

Kovacsovics, T. J., Hartwig, J. H., Cantley, L. C., and Toker, A. (1995b). Irreversible platelet aggregation and prolonged pleckstrin phosphorylation are mediated by the lipid products of phosphphatidylinositol 3-kinase. [Abstract]. *Blood* **86,** 454a.

Kunicki, T. J., Annis, D., and Shattil, S. J. (1996). A molecular basis for affinity modulation of Fab ligand binding to integrin $\alpha_{IIb}\beta_3$. *J. Biol. Chem.* In press.

Lee, J. O., Riev, P., Aranout, M. A., and Liddington, R. (1995). Crystal structure of the A-domain from the A-subunit of integrin CR3 (CD11B/CD18). *Cell* **80,** 631–638.

Leong, L., Hughes, P., Ginsberg, M., and Shattil, S. J. (1995). Integrin signaling: Roles for the cytoplasmic tails of $\alpha_{IIb}\beta_3$ in the tyrosine phosphorylation of pp125^{FAK}. *J. Cell Sci.* **108,** 3817–3825.

Leung-Hagesteijn, C. Y., Milankov, K., Michalak, M., Wilkins, J., and Dedhar, S. (1994). Cell attachment to extracellular matrix substrates is inhibited upon downregulation of expression of calreticulin, an intracellular integrin α-subunit-binding protein. *J. Cell Sci.* **107,** 589–600.

Li, R. Y., Ragab, A., Gaits, F., Ragab-Thomas, J. M. F., and Chap, H. (1994). Thrombin-induced redistribution of protein-tyrosine-phosphatases to the cytoskeletal complexes in human platelets. *Cell. Mol. Biol.* **40,** 665–675.

Lindberg, F. P., Lublin, D. M., Telen, M. J., Veile, R. A., Miller, Y. E., Donis-Keller, H., and Brown, E. J. (1994). Rh-related antigen CD47 is the signal-transducer integrin-associated protein. *J. Biol. Chem.* **269,** 1567–1570.

Lipfert, L., Haimovich, B., Schaller, M. D., Cobb, B. S., Parsons, J. T., and Brugge, J. S. (1992). Integrin-dependent phosphorylation and activation of the protein tyrosine kinase pp125^{FAK} in platelets. *J. Cell Biol.* **119,** 905–912.

Maeda, H., Taniguchi, T., Inazu, T., Yang, C., Nakagawara, G., and Yamamura, H. (1993). Protein-tyrosine kinase p72^{syk} is activated by thromboxane A_2 mimetic U44069 in platelets. *Biochem. Biophys. Res. Commun.* **197,** 62–67.

Mainiero, F., Pepe, A., Wary, K. K., Spinardi, L., Mohammadi, M., Schlessinger, J., and Giancotti, F. G. (1995). Signal transduction by the $\alpha_6\beta_4$ integrin: Distinct β_4 subunit sites mediate recruitment of Shc/Grb2 and association with the cytoskeleton of hemidesmosomes. *EMBO J.* **14,** 4470–4481.

Mellion, B. T., Ignarro, L. J., Ohlstein, E. H., Pontecorvo, E. G., Hyman, A. L., and Kadowitz, P. J. (1981). Evidence for the inhibitory role of guanosine 3′,5′-monophosphate in ADP-induced human platelet aggregation in the presence of nitric oxide and related vasodilators. *Blood* **57,** 946.

Meyer, S., Zuerbig, S., Bissel, T., Cunningham, C. C., Hartwig, J. H., and Fox, J. E. B. (1995). Characterization of the interaction between platelet glycoprotein Ib-IX and actin-binding protein [Abstract]. *Thromb. Haemost.* **73,** 1201.

Miyamoto, S., Akiyama, S. K., and Yamada, K. M. (1995a). Synergistic roles for receptor occupancy and aggregation in integrin transmembrane function. *Science* **267,** 883–885.

Miyamoto, S., Teramoto, H., Coso, O. A., Gutkind, J. S., Burbelo, P. D., Akiyama, S. K., and Yamada, K. M. (1995b). Integrin function: Molecular hierarchies of cytoskeletal and signaling molecules. *J. Cell Biol.* **131,** 791–805.

Morii, N., Teru-uchi, T., Tominaga, T., Kumagai, N., Kozaki, S., Ushikubi, F., and Narumiya, S. (1992). A *rho* gene product in human blood platelets. II. Effects of the ADP-ribosylation by botulinum C3 ADP-ribosyltransferase on platelet aggregation. *J. Biol. Chem.* **267,** 20921–20926.

Nachmias, V. T. (1993). Small actin-binding proteins: The β-thymosin family. *Curr. Opin. Cell Biol.* **5,** 56–62.

Nachmias, V. T., and Golla, R. (1991). Vinculin in relation to stress fibers in spread platelets. *Cell Motil. Cytoskeleton* **20,** 190–202.

Nachmias, V. T., Cassimeris, L., Golla, R., and Safer, D. (1993). Thymosin beta 4 ($T\beta_4$) in activated platelets. *Eur. J. Cell Biol.* **61,** 314–320.

Nemoto, Y., Namba, T., Teru-uchi, T., Ushikubi, F., Morii, N., and Narumiya, S. (1992). A *rho* gene product in human blood platelets. I. Identification of the platelet substrate for botulinum C3 ADP-ribosyltransferase as *rhoA* protein. *J. Biol. Chem.* **267,** 20916–20920.

Newman, P. J., Hillery, C. A., Albrecht, R., Parise, L. V., Berndt, M. C., Mazurov, A. V., Dunlop, L. C., Zhang, J., and Rittenhouse, S. E. (1992). Activation-dependent changes in human platelet PECAM-1: Phosphorylation, cytoskeletal association, and surface membrane redistribution. *J. Cell Biol.* **119,** 239–246.

Niiya, K., Hodson, E., Bader, R., Byers, Ward V., Koziol, J. A., Plow, E. F., and Ruggeri, Z. M. (1987). Increased surface expression of the membrane glycoprotein IIb/IIIa complex induced by platelet activation. Relationship to the binding of fibrinogen and platelet aggregation. *Blood.* **70,** 475–483.

Nobes, C. D., and Hall, A. (1995). Rho, Rac, and Cdc42 GTPases regulate the assembly of multimolecular focal complexes associated with actin stress fibers, lamellipodia, and filopodia. *Cell* **81,** 53–62.

Nolte, C., Eigenthaler, M., Horstrup, K., Hönig-Liedl, P., and Walter, U. (1994). Synergistic phosphorylation of the focal adhesion-associated vasodilator-stimulated phosphoprotein in intact human platelets in response to cGMP- and cAMP-elevating platelet inhibitors. *Biochem. Pharmacol.* **48,** 1569–1575.

Otey, C. A., Vasquez, G. B., Burridge, K., and Erickson, B. W. (1993). Mapping of the α-actinin binding site within the β_1 integrin cytoplasmic domain. *J. Biol. Chem.* **268,** 21193–21197.

O'Toole, T. E., Katagiri, Y., Faull, R. J., Peter, K., Tamura, R., Quaranta, V., Loftus, J. C., Shattil, S. J., and Ginsberg, M. H. (1994). Integrin cytoplasmic domains mediate inside-out signaling. *J. Cell Biol.* **124,** 1047–1059.

Ozawa, K., Kashiwada, K., Takahashi, M., and Sobue, K. (1995). Translocation of cortactin (p80/85) to the actin-based cytoskeleton during thrombin receptor-mediated platelet activation. *Exp. Cell Res.* **221,** 197–204.

Pavalko, F. M., and Otey, C. A. (1994). Role of adhesion molecule cytoplasmic domains in mediating interactions with the cytoskeleton. *Proc. Soc. Exp. Biol. Med.* **205,** 282–293.

Peerschke, E. I. B. (1984). Observations on the effects of cytochalasin B and cytochalasin D on ADP- and chymotrypsin-treated platelets. *Proc. Soc. Exp. Biol. Med.* **175,** 109–115.

Peerschke, E. I.B. (1991). Time-dependent association between platelet-bound fibrinogen and the Triton X-100 insoluble cytoskeleton. *Blood* **77,** 508–514.

Peerschke, E. I.B. (1992). Events occurring after thrombin-induced fibrinogen binding to platelets. *Semin. Thromb. Hemost.* **18,** 34–43.

Peerschke, E. I. B. (1994). Stabilization of platelet-fibrinogen interactions is an integral property of the glycoprotein IIb-IIIa complex. *J. Lab. Clin. Med.* **124,** 439–446.

Phizicky, E. M., & Fields, S. (1995). Protein-protein interactions—methods for detection and analysis. *Microbiol. Rev.* **59,** 94–123.

Reinhard, M., Halbrügge, M., Scheer, U., Wiegand, C., Jockusch, B. M., and Walter, U. (1992). The 46/50 kDa phosphoprotein VASP purified from human platelets is a novel protein associated with actin filaments and focal contacts. *EMBO J.* **11,** 2063–2070.

Reinhard, M., Giehl, K., Abel, K., Haffner, C., Jarchau, T., Hoppe, V., Jockusch, B. M., and Walter, U. (1995). The proline-rich focal adhesion and microfilament protein VASP is a ligand for profilins. *EMBO J.* **14,** 1583–1589.

Rezaul, K., Yanagi, S., Sada, K., Taniguchi, T., and Yamamura H. (1994). Protein-tyrosine kinase $p72^{syk}$ is activated by platelet activating factor in platelets. *Thromb. Haemost.* **72,** 937–941.

Romer, L. H., Burridge, K., and Turner C. E. (1992). Signaling between the extracellular matrix and the cytoskeleton: Tyrosine phosphorylation and focal adhesion assembly. *Cold Spring Harbor Symp. Quant. Biol.* **57,** 193–202.

Roskelley, C. D., Desprez, P. Y., and Bissell M. J. (1994). Extracellular matrix-dependent tissue-specific gene expression in mammary epithelial cells requires both physical and biochemical signal transduction. *Proc. Natl. Acad. Sci. U.S.A.* **91,** 12378–12382.

Sastry, S. K., and Horwitz, A. F. (1993). Integrin cytoplasmic domains: Mediators of cytoskeletal linkages and extra- and intracellular initiated transmembrane signaling. *Curr. Opin. Cell Biol.* **5,** 819–831.

Schaller, M. D., Otey, C. A., Hildebrand, J. D., and Parsons, J. T. (1995). Focal adhesion kinase and paxillin bind to peptides mimicking β integrin cytoplasmic domains. *J. Cell Biol.* **130,** 1181–1187.

Schwartz, M. A., Schaller, M. D., and Ginsberg, M. H. (1995). Integrins: Emerging paradigms of signal transduction. *Annu. Rev. Cell Biol.* **11,** 549–599.

Shattil, S. J., Motulsky, H. J., Insel, P. A., Flaherty, L., and Brass, L. F. (1986). Expression of fibrinogen receptors during activation and subsequent desensitization of human platelets by epinephrine. *Blood* **68,** 1224–1231.

Shattil, S. J., Weisel, J. W., and Kieber-Emmons, T. (1992). Use of monoclonal antibodies to study the interaction between an integrin adhesion receptor, GP IIb-IIIa, and its physiological ligand, fibrinogen. *Immunomethods* **1,** 53–63.

Shattil, S. J., Ginsberg, M. H., and Brugge, J. S. (1994a). Adhesive signaling in platelets. *Curr. Opin. Cell Biol.* **6,** 695–704.

Shattil, S. J., Haimovich, B., Cunningham, M., Lipfert, L., Parsons, J. T., Ginsberg, M. H., and Brugge, J. S. (1994b). Tyrosine phosphorylation of $pp125^{FAK}$ in platelets requires coordinated signaling through integrin and agonist receptors. *J. Biol. Chem.* **269,** 14738–14745.

Shattil, S. J., O'Toole, T., Eigenthaler, M., Thon, V., Williams, M., Babior, B. M., and Ginsberg, M. H. (1995). β_3-Endonexin, a novel polypeptide that interacts specifically with the cytoplasmic tail of the integrin β_3 subunit. *J. Cell Biol.* **131,** 807–816.

Slupsky, J. R., Seehafer, J. G., Tang, S. C., Masellis Smith, A., and Shaw, A. R. (1989). Evidence that monoclonal antibodies against CD9 antigen induce specific association between CD9 and the platelet glycoprotein IIb-IIIa complex. *J. Biol. Chem.* **264,** 12289–12293.

Smyth, S. S., Hillery, C. A., and Parise, L. V. (1992). Fibrinogen binding to purified platelet glycoprotein IIb-IIIa (integrin $\alpha_{IIb}\beta_3$) is modulated by lipids. *J. Biol. Chem.* **267,** 15568–15577.

Stossel, T. P. (1993). On the crawling of animal cells. *Science* **260,** 1086–1094.

Suzuki, H., Tanoue, K., and Yamazaki, H. (1991). Morphological evidence for the association of plasma membrane glycoprotein IIb/IIIa with the membrane skeleton in human platelets. *Histochemistry* **96,** 31–39.

Takayama, H., Ezumi, Y., Ichinohe, T., and Okuma, M. (1993). Involvement of GPIIb-IIIa on human platelets in phosphotyrosine-specific dephosphorylation. *Biochem. Biophys. Res. Commun.* **194,** 472–477.

Taniguchi, T., Kitagawa, H., Yasue, S., Yanagi, S., Sakai, K., Asahi, M., Ohta, S., Takeuchi, F., Nakamura, S., and Yamamura, H. (1993). Protein-tyrosine kinase $p72^{syk}$ is activated by thrombin and is negatively regulated through Ca^{2+} mobilization in platelets. *J. Biol. Chem.* **268,** 2277–2279.

Tate, B. F., and Rittenhouse, S. E. (1993). Thrombin activation of human platelets causes tyrosine phosphorylation of PLC-gamma$_2$. *Biochim. Biophys. Acta Mol. Cell Res.* **1178,** 281–285.

Taub, R., Gould, R. J., Garsky, V. M., Ciccarone, T. M., Hoxie, J., Friedman, P. A., and Shattil, S. J. (1989). A monoclonal antibody against the platelet fibrinogen receptor contains a sequence that mimics a receptor recognition domain in fibrinogen. *J. Biol. Chem.* **264,** 259–265.

Teo, M., Manser, D., and Lim, L. (1995). Identification and molecular cloning of a $p21^{cdc42/rac1}$-activated serine/threonine kinase that is rapidly activated by thrombin in platelets. *J. Biol. Chem.* **270,** 26690–26697.

Tohyama, Y., Yanagi, S., Sada, K., and Yamamura, H. (1994). Translocation of $p72^{syk}$ to the cytoskeleton in thrombin-stimulated platelets. *J. Biol. Chem.* **269,** 32767.

Torti, M., Ramaschi, G., Sinigaglia, F., Lapetina, E. G., and Balduini, C. (1994). Glycoprotein IIb-IIIa and the translocation of Rap2B to the platelet cytoskeleton. *Proc. Natl. Acad. Sci. U.S.A* **91,** 4239–4243.

Touhara, K., Hawes, B., van Bieson, T., and Lefkowitz, R. J. (1995). G protein $\beta\gamma$ subunits stimulate phosphorylation of Shc adaptor protein. *Proc. Natl. Acad. Sci. U.S.A.* **92,** 9284–9287.

van Kooyk, Y., and Figdor, C. G. (1993). Activation and inactivation of adhesion molecules. *Curr. Top. Microbiol. Immunol.* **184,** 235–248.

Weisel, J. W., Nagaswami, C., Vilaire, G., and Bennett, J. S. (1992). Examination of the platelet membrane glycoprotein IIb-IIIa complex and its interaction with fibrinogen and other ligands by electron microscopy. *J. Biol. Chem.* **267,** 16637–16643.

Wencel Drake, J. D., Plow, E. F., Kunicki, T. J., Woods, V. L., Keller, D. M., and Ginsberg, M. H. (1986). Localization of internal pools of membrane glycoproteins involved in platelet adhesive responses. *Am. J. Pathol.* **124,** 324–334.

White, J. G. (1988). Platelet membrane ultrastructure and its changes during platelet activation. *Prog. Clin. Biol. Res.* **283,** 1–32.

Williams, M. J., Hughes, P. E., O'Toole, T. E., and Ginsberg, M. H. (1994). The inner world of cell adhesion: Integrin cytoplasmic domains. *Trends Cell Biol.* **4,** 109–112.

Witke, W., Sharpe, A. H., Hartwig, J. H., Azuma, T., Stossel, T. P., and Kwiatkowski, D. J. (1995). Hemostatic, inflammatory and fibroblast responses are blunted in mice lacking gelsolin. *Cell* **81,** 41–51.

Yanaga, F., Poole, A., Asselin, J., Blake, R., Schieven, G. L., Clark, E. A., and Watson, S. P. (1995). Syk interacts with tyrosine-phosphorylated proteins in human platelets activated by collagen and cross-linking of the Fcgamma-IIA receptor. *Biochem. J.* **311,** 471–478.

Zhang, J., Fry, M. J., Waterfield, M. D., Jaken, S., Liao, L., Fox, J. E. B., and Rittenhouse, S. E. (1992). Activated phosphoinositide 3-kinase associates with membrane skeleton in thrombin-exposed platelets. *J. Biol. Chem.* **267,** 4686–4692.

Zhang, J., Benovic, J. L., Sugai, M., Wetzker, R., Gout, I., and Rittenhouse, S. E. (1995). Sequestration of a G-protein βgamma subunit or ADP-ribosylation of Rho can inhibit thrombin-induced activation of platelet phosphoinositide 3-kinases. *J. Biol. Chem.* **270,** 6589–6594.

Zhou, M., and Brown, E. J. (1993). Leukocyte response integrin and integrin-associated protein act as a signal transduction unit in generation of a phagocyte respiratory burst. *J. Exp. Med.* **178,** 1165–1174.

Zhu, Y., O'Neill, S., Saklatvala, J., Tassi, L., and Mendelsohn, M. E. (1994). Phosphorylated HSP27 associates with the activation-dependent cytoskeleton in human platelets. *Blood* **84,** 3715–3723.

CHAPTER 14

Interactions between the Membrane–Cytoskeleton and CD44 during Lymphocyte Signal Transduction and Cell Adhesion

Lilly Y. W. Bourguignon
Department of Cell Biology and Anatomy, School of Medicine, University of Miami, Miami, Florida 33101

I. INTRODUCTION

CD44 (also called $gp90^{Hermes}$, GP85/Pgp-1, ECMRIII, or H-CAM) is a family of cell adhesion molecules that are expressed in a variety of cells and tissues (see Section II). Nucleotide sequence analysis reveals that CD44 exists in different forms that are variants of the common or standard hemopoietic form, CD44s (see Section II). Certain CD44 isoforms (or variants) are preferentially expressed on the surface of tumor cells during the progression of lymphoid carcinomas and other solid tumor cancers. Therefore, CD44 variant expression has recently been used as an indicator of tumor metastasis (see Section II).

As transmembrane glycoproteins, CD44 isoforms have been shown to bind extracellular matrix (ECM) components [e.g., hyaluronic acid (HA) at the N-terminus of the extracellular domain] (see Section III) and to contain an ankyrin-binding site within the 70-amino-acid (aa) C-terminus of the cytoplasmic domain (see Section IV). The binding of HA to CD44's external region and the binding of ankyrin to CD44's internal tail region are tightly regulated and closely coupled (see Sections III, IV, and V). Several mechanisms for the regulation of CD44-mediated function have been suggested. These include modifications by an additional exon-coded structure (via an alternative splicing process) (see review by Herrlich *et al.,* 1993), variable *N-/O*-linked glycosylation on the CD44's extracellular domain (Lesley *et al.,* 1995), and modulations of the CD44's cytoplasmic domain by the cytoskeletal proteins, such as ankyrin. This chapter focuses on the interactions between the membrane–cytoskeleton and CD44 during HA-mediated signal transduction (an outside-in signaling event) and the expression of CD44's adhesion function (an inside-out signaling event) during lymphocyte activation.

II. EXPRESSION OF CD44 IN LYMPHOCYTES

The interaction of lymphocytes with endothelial cells in high endothelial venules (HEV) is an important heterotypic cell–cell interaction required for the circulation of lymphocytes between the cardiovascular and lymphatic systems (Stamper and Woodruff, 1976). In humans, a lymphocyte surface glycoprotein of 85–95 kDa, $gp90^{Hermes}$ [also called CD44 or P-glycoprotein 1 (Pgp-1)], has been identified as the lymphocyte HEV homing receptor (Jalkanen *et al.,* 1987; Goldstein *et al.,* 1989). CD44 is also involved in the adhesion of T cells to erythrocytes (Haynes *et al.,* 1989), stromal cells to B cells (Miyake *et al.,* 1990a), T-cell proliferation (Haynes *et al.,* 1989), and homotypic cell aggregation (Belitsos *et al.,* 1990; Droll *et al.,* 1995). Due to its widespread occurrence and its role in cell adhesion, CD44 is also called the homing cellular adhesion molecule (H-CAM). The murine homologue of CD44 is an 85-kDa transmembrane glycoprotein, GP85 (also called Pgp-1) (Trowbridge *et al.,* 1982; Lesley and Trowbridge, 1982; Lesley *et al.,* 1985a,b; Kalomiris and Bourguignon, 1988, 1989; Bourguignon *et al.,* 1991, 1993b; Lokeshwar and Bourguignon, 1991, 1992; Lokeshwar *et al.,* 1994).

CD44 is also considered to be a T-cell differentiation antigen (Lesley *et al.,* 1985a,b) and has been detected on many different cell types, including B cells, macrophages, granulocytes, fibroblasts, endothelial cells, and epithelial cells (Letarte *et al.,* 1985; Isacke *et al.,* 1986; Stamenkovic *et al.,* 1991; Bourguignon *et al.,* 1992; Iida and Bourguignon, 1995a; Welsh *et al.,* 1995).

Studies indicate that the expression of CD44 changes profoundly during tumor metastasis, particularly during the progression of lymphoid carcinomas (Horst *et al.,* 1990; Quackenbush *et al.,* 1990; Jalkanen *et al.,* 1991). In fact, CD44 expression has been used as an indicator of metastasis in non-Hodgkin's lymphoma (NHL). NHL with a low level of CD44 expression is less malignant, disseminates less frequently, and has a favorable prognosis. However, NHL with a high level of CD44 expression is more malignant and metastatic (Horst *et al.,* 1990; Jalkanen *et al.,* 1991). Furthermore, cells with a high level of CD44 expression show enhanced HA binding, which increases their migration to various lymphoid tissues (Horst *et al.,* 1990; Jalkanen *et al.,* 1991). Therefore, it has been suggested that cells with a high level of CD44 expression phenotype play an important role in the dissemination of NHL and, via this mechanism, exert an unfavorable prognostic influence (Horst *et al.,* 1990). In addition to NHL, many malignant leukemias, such as null acute lymphocytic leukemia (ALL), common ALL, and chronic lymphocytic leukemia, as well as other types of carcinomas of epithelial and glial origins, also show a high level of CD44 expression phenotype (Quackenbush *et al.,* 1990; Gunthert *et al.,* 1991). Therefore, the presence of a high level of CD44 expression is emerging as an important metastatic tumor marker in a number of carcinomas, and is also implicated in the unfavorable prognosis of these diseases.

Nucleotide sequence data indicate that the CD44 gene contains 19 exons, of which 10 exons can potentially be alternatively spliced (Fig. 1A) (Arch *et al.,* 1992; Screaton *et al.,* 1992; Tolg *et al.,* 1993). The most common hemopoietic form, CD44s (CD44 standard form), contains exons 1–4 (N-terminal 150 aa), exons 5, 15, and 16 (membrane proximal 85 aa), exon 17 (transmembrane domain), and a portion of exons 17 and 19 (cytoplasmic tail, 70 aa) (designated as the common form of CD44s) (Fig. 1C) (Screaton *et al.,* 1992). In addition, there is another rare form of CD44s in hemopoietic cells. In this CD44s, exon 18 is included (instead of exon 19), resulting in a CD44 isoform that contains a cytoplasmic domain with only 3 aa (designated as the rare form of CD44s). (Fig. 1B) (Screaton *et al.,* 1992). Certain other exons (e.g., 6–14) are spliced and expressed in a variety of cells. For example, epithelial cells contain additional exons 12–14, which are inserted into the CD44s transcripts (Fig. 1D). This isoform has been designated as CD44E (Screaton *et al.,* 1992).

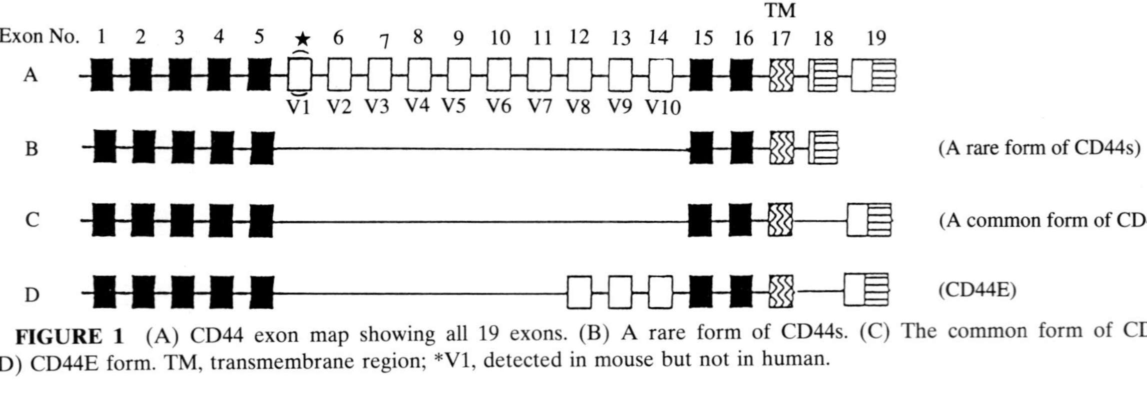

FIGURE 1 (A) CD44 exon map showing all 19 exons. (B) A rare form of CD44s. (C) The common form of CD44s. (D) CD44E form. TM, transmembrane region; *V1, detected in mouse but not in human.

A number of tumor cells and tissues express different CD44 variant (CD44v) isoforms in addition to CD44s and CD44E (Gunthert *et al.,* 1991; Arch *et al.,* 1992; Matsumura and Tarin, 1992; Screaton *et al.,* 1992; Heider *et al.,* 1993; Cooper and Dougherty, 1995). These CD44v isoforms have the same aa sequences at the two ends of the molecule but differ in a middle region that contains additional exon insertions (exons 6–14) within the CD44 membrane-proximal region located at the external side of the membrane (Arch *et al.,* 1992; Screaton *et al.,* 1992). The variable primary aa sequence of different CD44v isoforms is further modified by extensive *N*- and *O*-glycosylations and glycosaminoglycan addition (Jackson *et al.,* 1995; Jalkanen *et al.,* 1988; Lokeshwar and Bourguignon, 1991; Bennet *et al.,* 1995; Cooper and Dougherty, 1995). In particular, a CD44v is transiently expressed in certain macrophages and T and B cells in response to antigen stimulation (Arch *et al.,* 1992), suggesting that CD44v isoforms play an essential role in the maturation of lymphocytes to form immune cells in the lymph nodes. The expression of certain CD44v isoforms, such as CD44v6, is also shown to be associated with the development of NHL (Ristamaki *et al.,* 1995). Most importantly, a number of research groups discovered that, during metastasis, tumor cells express CD44v structures that, in effect, mimic developing lymphocytes and allow them to escape immune surveillance (Arch *et al.,* 1992). Therefore, CD44s and CD44v isoforms are thought to be critically important for both immune responses and disease development.

III. EXTRACELLULAR DOMAIN OF CD44 FUNCTIONS AS A HA RECEPTOR

HA is one of the major components of the ECM. Primary locations of HA accumulation are in connective tissues such as the dermis of the skin, the lamina propria of mucous membranes, and the adventitia surrounding blood vessels (Laurent and Fraser, 1992). HA is known to cause cell aggregation with a number of different cell types (Green *et al.,* 1988), and has been implicated in the stimulation of cell proliferation (West and Kumar, 1989), cell migration (Turley *et al.,* 1991), cell adhesion (Miyake and Kincade, 1990; Miyake *et al.,* 1990b; Bourguignon *et al.,* 1993b), and angiogenesis (Rooney *et al.,* 1995). CD44 is considered to be one of the major HA receptors. It shares a large amount of structural similarity with the HA receptor purified from mouse fibroblasts (Underhill *et al.,* 1985) and BHK cells (Underhill *et al.,* 1987). In addition, N-terminal regions in the CD44 extracellular domain are homologous to cartilage proteoglycan core and link proteins that specifically bind HA (Goldstein *et al.,* 1989). A number of laboratories, including our own, have shown that mouse T-lymphoma

CD44 binds to specific ECM molecules such as HA and collagen (types I and VI) (Carter and Wayner, 1988; Lesley *et al.,* 1990; Lokeshwar and Bourguignon, 1991; Bourguignon *et al.,* 1993b). Specifically, using [^{3}H]HA to measure the binding of HA to mouse T-lymphoma cells *in vivo,* we have identified a single high-affinity HA receptor that has a dissociation constant of 0.3 nmol/liter (Bourguignon *et al.,* 1993b). The binding of [^{3}H]HA to these cells is effectively inhibited by anti-CD44 antibody, establishing that CD44 is an HA receptor. Furthermore, purified mouse CD44 also binds specifically to HA but not to glycosaminoglycans in general (Bourguignon *et al.,* 1993b). Other studies also support the observation that HA binding is inhibited by CD44-specific antibodies and unconjugated HA, but not by antibodies directed against other cell surface molecules or glycosaminoglycans (Lesley *et al.,* 1990). In addition, it has been shown that the binding of a B-cell hybridoma to a cloned stromal cell line is inhibited by hyaluronidase and certain CD44-specific antibodies (Miyake *et al.,* 1990a). Also, the binding of a CD44–immunoglobulin fusion protein to primary cultures of lymph node high endothelial cells can be blocked by hyaluronate or hyaluronidase treatment of cultured cells or tissues (Lesley *et al.,* 1992). In addition, it has been shown that CD44 modified by chondroitin sulfate is capable of interacting with other extracellular matrix molecules such as fibronectin (Jalkanen and Jalkanen, 1992). A study by Sorimachi *et al.* (1995) shows that CD44 also binds a hematopoietic cell lineage–specific proteoglycan, serglycin. All of these findings suggest that CD44 is an important cell adhesion receptor that is needed for both cell–cell and cell–ECM interactions.

IV. INTERACTION OF THE CYTOPLASMIC DOMAIN OF CD44 WITH THE CYTOSKELETAL PROTEIN ANKYRIN

Ankyrin is known to bind to a number of plasma membrane–associated proteins, including band 3, two other members of the anion exchange gene family (Bennett, 1992; Davis *et al.,* 1989), Na^+,K^+-ATPase (Nelson and Veshnock, 1987; Morrow *et al.,* 1989; Shull *et al.,* 1989), the amiloride-sensitive Na^+ channel (Smith *et al.,* 1991), and the voltage-dependent Na^+ channel (Kordeli and Bennett, 1991; Kordeli *et al.,* 1995). The cytoplasmic domain of CD44 [a portion of the C-terminus including exons 17 and 19 (~70 aa long)] is highly conserved (≥90%) in most of the CD44 isoforms that bind to cytoskeletal proteins such as ankyrin (Bourguignon *et al.,* 1986,1991,1992; Kalomiris and Bourguignon, 1988,1989; Lokeshwar and Bourguignon, 1991, 1992; Lokeshwar *et al.,* 1994; Welsh *et al.,* 1995). Specifically, our laboratory has determined that in lymphocytes the CD44 cytoplasmic domain is closely associated with an ankyrin-like protein *in vivo* in a large 16S complex that accumulates underneath CD44 cap structures (Bourguignon *et al.,* 1986). Both CD44 precursors [p42 (nonglycosylated) and p52 (*N*-linked glycosylated)] and mature CD44 (*N* linked and *O* linked) contain ankyrin-binding site(s). Therefore, the expression of ankyrin-binding site(s) occurs at a very early stage of CD44 biosynthesis (Lokeshwar and Bourguignon, 1991). Furthermore, results of a Scatchard plot analysis indicate that purified lymphocyte CD44 binds directly and specifically to ankyrin ($K_d \sim 1.94$ nmol/liter) in a saturable manner *in vitro* (Bourguignon *et al.,* 1993b).

It has been reported that certain CD44 isoforms, such as the CD44E protein, interact with the ERM (ezrin, radixin, and moesin) family of cytoskeletal proteins. However, it should be noted that these observations are solely based on co-immunoprecipitation experiments (Tsukita *et al.,* 1994). Direct binding between purified CD44E and the ERM family of cytoskeletal proteins *in vitro* has not yet been demonstrated (Tsukita *et al.,* 1994). Using a similar co-immunoprecipitation approach designed to preserve the membrane–cytoskeleton complexes, our laboratory has failed to observe any other CD44-immunoprecipitated materials, such as the ERM cytoskeletal proteins. Thus, it appears that CD44s directly associates only with ankyrin and with no other cytoskeletal proteins in lymphocytes.

Furthermore, the interaction between CD44s and ankyrin appears to be regulated by fatty acylation (Bourguignon *et al.,* 1991), protein kinase C (PKC) (Kalomiris and Bourguignon, 1989), and GTP binding (Lokeshwar and Bourguignon, 1992) as described in this section.

A. Regulation of CD44s–Ankyrin Interaction by Fatty Acylation

Ester–thioester-linked acylation (e.g., palmitylation and/or myristylation) is considered to be critical for the targeting of certain proteins to the membrane (Schmidt, 1989). Previously, we have demonstrated that palmitic acid is incorporated into CD44s *in vivo,* and that the amount of palmitic acid incorporation is greatly increased during lymphoma receptor cap formation (Bourguignon *et al.,* 1991). The majority of the incorporated palmitic acid appears to be covalently linked to CD44s. According to the predicted sequence obtained from cDNA cloning of the mouse CD44s gene (Zhou *et al.,* 1989), one can tentatively identify the possible cysteine residues in the transmembrane domain (^{288}Cys and ^{297}Cys) that may be modified by the addition of palmitic acid or some other fatty acid. We have also established that deacylation of CD44s *in vitro* (removal of the palmitic acid moiety from CD44s by 1-*M* hydroxylamine treatment) significantly reduces the binding affinity between CD44s and ankyrin, and reacylation of CD44s restores the binding affinity. These findings strongly suggest that fatty acid acylation of CD44s by palmitic acid is required for the stable attachment of the cytoskeleton to the lymphoma plasma membrane (Bourguignon *et al.,* 1991).

B. Regulation of CD44s–Ankyrin Interaction by PKC

PKC is known to play a pivotal role in signal transduction (Nishizuka, 1992). In a previous study we determined that CD44s can be phosphorylated by purified PKC. Phosphoamino acid analysis indicates that this phosphorylation occurs primarily at serine residues and to a lesser extent at threonine residues, and does not occur at all at tyrosine residues. These data suggest that CD44s is one of the cellular substrates for PKC (Kalomiris and Bourguignon, 1989). Based on the predicted sequence obtained from cDNA cloning data of mouse CD44s (Zhou *et al.,* 1989), we have tentatively identified several potential PKC phosphorylation sites. These include ^{341}T, ^{347}T, ^{351}T, ^{318}S, ^{325}S, ^{327}S, and ^{339}S (Fig. 2). These aa (e.g., T and S) are also present in CD44s derived from human and other animal species (Fig. 2). Most importantly, we have established that phosphorylation of CD44s by PKC significantly enhances the binding between CD44s and ankyrin (Kalomiris and Bourguignon, 1989). Therefore, PKC-mediated phosphorylation of CD44s may play an important role in plasma membrane–cytoskeleton interactions.

C. Regulation of CD44s–Ankyrin Interaction by GTP Binding

Previously, we determined that CD44s is a GTP-binding protein and displays GTPase activity (Lokeshwar and Bourguignon, 1992). Further analysis indicates that the overall sequence of CD44s does not display significant homology with any known G-proteins. However, if the aa sequences of the CD44s cytoplasmic domain are compared with critical regions of known G-proteins, we have found that certain similarities exist between CD44s and several classes of G-proteins (e.g., the Ras class) within particular consensus regions (e.g., G-1, G-2, G-3, and G-4) (Bourne *et al.,* 1991). Specifically, the G-4 region of Ras, a guanidine-binding sequence of common G-proteins (i.e., ^{330}MVHLVNK^{337}E) (Fig. 2) is found in mouse CD44s (Fig. 2). This sequence is also present in CD44s derived from human and other animal species (Fig. 2). Since GTP binding significantly enhances the interaction of CD44s with ankyrin, these data suggest that GTP plays an important role in promoting the interaction between CD44s and the

Region I (Ankyrin Binding Domain)

Mouse CD44	^{306}N G G N G T V E D R K P S E ^{320}L
Human CD44	^{304}N **S** G N G **A** V E D R K P S **G** ^{318}L
Rat CD44	^{292}N **S** G N G T V E D R K P S E ^{306}L
Hamster CD44	^{305}N **S** G N G **K** V E D R K P S E ^{319}L
Bovine CD44	^{309}N **N** G N G T **M** E **E** R K P S **G** ^{323}L
Horse CD44	^{302}N **N** G N G **A** V **D** D R K **A** S **G** ^{316}L

Region II (Regulatory Domain)

^{321}N G E A S K S Q E M V H L V N K E P S E T P D Q C . T A D E T R N L ^{355}Q

^{319}N G E A S K S Q E M V H L V N K E **S** S E T P D Q **F I** T A D E T R N L ^{353}Q

^{307}N G E A S K S Q E M V H L V N K E P T E T P D Q **F I** T A D E T R N L ^{341}Q

^{320}N G E A S K S Q E M V H L V N K E P S E T P D Q **F I** T A D E T R N L ^{354}Q

^{324}N G E A S K S Q E M V H L V N K **G** S S E T P D Q **F M** T A D E T R N L ^{358}Q

^{317}N G E A S **R** S Q E M V H L V N K E S S E T Q D Q **F M** T A D E T R N L ^{351}Q

FIGURE 2 Sequence comparison of CD44 cytoplasmic domain (with region I and region II) from various species. Bold letters indicate a conserved substitution; shaded letters (e.g., S and T) represent the putative PKC phosphorylation site(s); underlined letters represent the putative guanidine-binding region.

membrane–cytoskeleton via ankyrin. A model based on the current structural and functional information regarding CD44s is shown in Fig. 3.

V. CD44–ANKYRIN INTERACTION IS REQUIRED FOR HA-INDUCED CELLULAR FUNCTIONS

It has been suggested that several regions in the extracellular domain of CD44 containing clusters of conserved basic residues (Peach *et al.,* 1993; Yang *et al.,* 1994; Liao *et al.,* 1995) play an important role in CD44 binding to HA. However, not all CD44 isoforms constitutively bind HA. In some cell types, activation of cells by certain anti-CD44 antibodies is required for HA binding to CD44 (Hyman *et al.,* 1991; Lesley *et al.,* 1992; Liao *et al.,* 1993). Since HA binding plays an important role in CD44-mediated cell–cell and cell–ECM interactions, it becomes important to determine

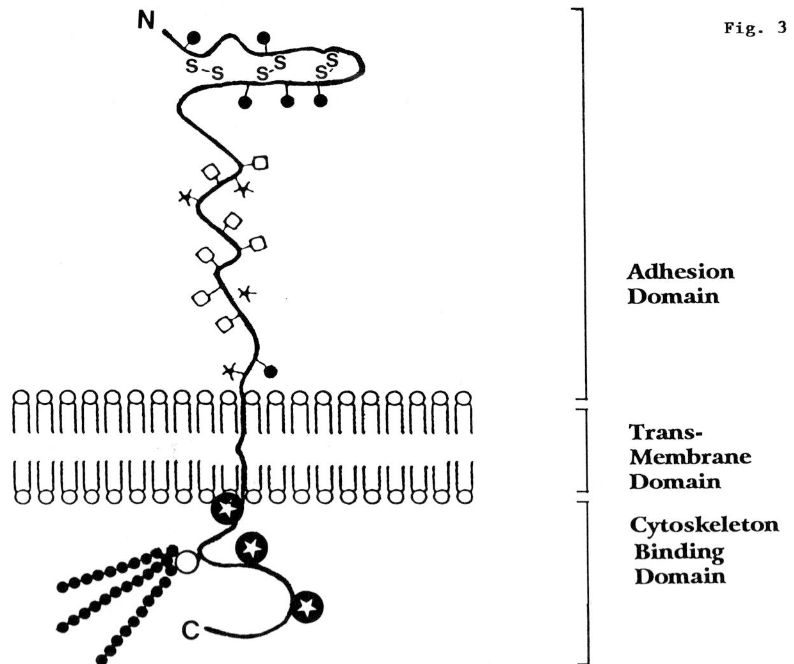

FIGURE 3 A schematic illustration of CD44 illustrating an adhesion domain with heavy glycosylation (●, *N*-glycosylation, □, *O*-glycosylation; and ★, glycosaminoglycan) addition, a transmembrane domain, and the cytoskeleton-binding domain [containing an ankyrin-binding site (○), associated fodrin–actin microfilaments (●●●), and regulatory regions (✪) for fatty acylation, PKC phosphorylation, and GTP binding].

those factors involved in regulating the expression of the HA-binding domain in CD44.

The interaction between HA and its receptor(s) on the cell surface does not appear to require divalent cations (Underhill, 1982). However, a number of hypotheses are possible concerning the mechanism involved in regulating the HA binding to cells. First, it is possible that the membrane-proximal region of the CD44 extracellular domain is important for binding HA, although it is not directly involved in the binding. This region is the least conserved in all CD44 isoforms that have been characterized to date (Hofmann *et al.,* 1991). However, a report by He *et al.* (1992) suggests that only the N-terminus of CD44 is sufficient for HA binding, and that changes in the membrane proximal region do not abolish HA binding. Therefore, the role of the membrane proximal region in determining HA binding remains uncertain. A second hypothesis suggests that the carbohydrate chains on CD44 are important for HA binding. For example, the *N*-linked and *O*-linked oligosaccharides on CD44 have been found to be important for HA binding (Lokeshwar and Bourguignon, 1991; Lesley *et al.,* 1995; Katoh *et al.,* 1995). It is possible that certain carbohydrate chains modulate the overall conformation of CD44, resulting in efficient HA binding to the N-terminal region of the CD44 molecule. A third hypothesis proposes that HA binding is regulated by an additional exon-coded structure (via an alternative splicing mechanism) in cells expressing CD44v isoforms (Herrlich *et al.,* 1993). Although all CD44v isoforms contain similar HA-binding motifs, certain CD44v isoforms display significantly less HA binding than CD44s (Stamenkovic *et al.,* 1991; Iida and Bourguignon, 1995b). It is possible that the CD44v-encoded structure induces surface rearrangement of HA-binding site(s), resulting in a loss of HA binding. Finally, a fourth hypothesis proposes that HA binding is regulated by the interaction between CD44 and the cytoskeleton. The cytoplasmic domain (C-terminus) of CD44 binds specifically to ankyrin, one of the important linker molecules connecting certain membrane proteins to cytoskeletal proteins such as fodrin (a spectrin-like molecule). Fodrin, in turn, binds specifically to actomyosin-containing microfilaments (Bourguignon *et al.,* 1986,1991,1992,1993b; Kalomiris and Bourguignon, 1988,1989; Lokeshwar and Bourguignon, 1991,1992; Bourguignon, 1992; Lokeshwar *et al.,* 1994; Welsh *et al.,* 1995; Isacke, 1994). We believe that the transmembrane interaction between CD44 and ankyrin may play an important role in HA-mediated signal transduction.

A. HA Induces CD44–Ankyrin Interaction during Signal Transduction (an Outside-In Signaling Event)

The hyaluronate polysaccharide is known to induce a variety of cellular activities in various cell types depending on the size and the concentration of HA (West and Kumar, 1989). Our data indicate that elevation of the intracellular Ca^{2+} concentration (measured by Fura-2 fluorescence) occurs within seconds after the addition of HA to the lymphocytes (Bourguignon *et al.*, 1993b). This suggests that Ca^{2+} is one of the earliest signals to occur following HA binding to mouse T-lymphoma cells. Following Ca^{2+} mobilization, HA also induces its lymphocyte receptors (CD44) to form capped structures on the cell surface and to adhere to HA-coated plates (Bourguignon *et al.*, 1993b). Although the coincidence of increased intracellular Ca^{2+} activity and HA-induced capping and adhesion does not establish a cause–effect relationship between these three events, the fact that the rise in intracellular Ca^{2+} (at ~5 min after HA addition) precedes the time required to reach the maximal level for HA-induced capping (~20 min after HA addition) and adhesion (~1 h after HA addition) strongly suggests that Ca^{2+} may be one of the second messenger systems required for the initial events leading to HA-induced capping and subsequent adhesion (Bourguignon *et al.*, 1993b).

Double immunofluorescence staining results indicate that intracellular ankyrin is preferentially accumulated underneath HA-induced receptor-capped structures (Bourguignon *et al.*, 1993b). We proposed that the transmembrane linkage between CD44 and the cytoskeleton may be important for the adhesion function of CD44. This hypothesis is consistent with previous findings showing that a portion of CD44 is constitutively associated with the cytoskeleton in lymphocytes (Bourguignon *et al.*, 1986; Bourguignon, 1992). In addition, the fact that ionomycin (a Ca^{2+} ionophore) stimulates, and ethyleneglycol-bis-(B-aminoethylether) N, N′-tetraacetic acid (a Ca^{2+} chelator) and nefedipine/bepridil (Ca^{2+} channel blockers) strongly inhibit, HA-mediated receptor capping and adhesion suggests that Ca^{2+} is needed for the onset of HA receptor redistribution and adhesion. Furthermore, microfilament inhibitors (but not microtubule inhibitors) block HA-induced receptor capping, cell aggregation, and adhesion to HA (Bourguignon *et al.*, 1993a). Together, these results suggest that the binding of CD44 to the cytoskeletal network is required for HA-induced signal transduction (i.e., outside-in signaling events).

B. *CD44–Ankyrin Interaction Regulates the Expression of CD44's HA-Mediated Binding and Adhesion Function (an Inside-Out Signaling Event)*

We have mapped the ankyrin-binding domain of CD44 by deleting various portions of the cytoplasmic region, followed by expression of these truncated cDNAs in COS cells (Lokeshwar *et al.*, 1994). Our deletion mutation analysis indicates that the ankyrin-binding domain of CD44 resides between aa 305 and 355. However, at least two subregions within this domain contribute to ankyrin binding: region I, containing 15 aa between 305 and 320; and region II, containing 35 aa between 320 and 355 (Fig. 2). Region II appears to be required for high-affinity ankyrin binding since its deletion results in a 2- to 2.5-fold decrease in the dissociation constant for ankyrin binding. This region is highly conserved in various CD44 proteins from different species (Fig. 2). Nevertheless, no sequence homology has been detected between CD44 region II and other ankyrin-binding proteins (e.g., band 3 and Na^+,K^+-ATPase) using a Best fit program (Genetics Computer Group Inc.) (Lokeshwar *et al.,* 1994). However, studies indicate that point mutations at certain PKC phosphorylation sites (but not every such putative site) in region II result in an alteration of HA-binding properties (Pure *et al.,* 1995; Uff *et al.,* 1995). Therefore, it is possible that region II represents *a regulatory domain* (e.g., PKC-mediated phosphorylation or GTP-binding–GTPase activity) (Fig. 2) responsible for the upregulation of CD44–ankyrin interaction (Kalomiris and Bourguignon, 1989; Lokeshwar and Bourguignon, 1992) required for HA binding. Deletion or mutation of some aa in this regulatory domain would result in the modification of high-affinity binding between CD44 and ankyrin, which could lead to changes in HA binding on the cell surface.

The deletion of both regions I and II of CD44 leads to a complete loss of ankyrin binding. Consequently, region I (^{306}NGGNGTVEDRKPSE^{320}L) is required for ankyrin binding. Biochemical analyses using competition binding assays and a synthetic peptide sharing sequence homology with the specific 15-aa sequence (^{306}NGGNGTVEDRKPSE^{320}L) between aa 305 and 320 further support the notion that this region contains an ankyrin-binding site (Lokeshwar *et al.,* 1994). Most importantly, deletion of the above ankyrin-binding sequence (i.e., region I and region II) from CD44 results in a drastic reduction ($\geq$90%) of the HA-binding ability by the mutant CD44 molecules expressed in COS cells (Lokeshwar *et al.,* 1994). A similar reduction of HA binding was also observed in studies described by Perschl *et al.* (1995) on lymphocytes transfected with CD44 mutant cDNAs lacking most of the cytoplasmic tail (including both region I and region II domains). However, deletion of region I alone or region II alone is not sufficient to block HA binding (Perschl *et al.,* 1995). These findings strongly suggest that the cytoplasmic domain of CD44—both the ankyrin-binding region (region I) and the adjacent regulatory region (region II) (Fig. 3)—plays a pivotal role in modulating the surface expression of CD44 required for HA binding and/or HA-mediated cell adhesion. It is possible that ankyrin binding to the CD44 cytoplasmic domain may promote the dimerization (clustering–patching) of CD44 molecules required for high-affinity HA binding.

We have also found that certain cytoskeletal proteins, such as ankyrin, interact with intracellular Ca^{2+} channels such as the inositol 1, 4, 5-triphosphate (IP_3) receptor (Bourguignon *et al.,* 1993a; Bourguignon and Jin, 1995) and the ryanodine receptor (Bourguignon *et al.,* 1995) in lymphocytes. The CD44 ankyrin-binding domain shares a large amount of sequence homology with the ankyrin-binding domain located in the IP_3 receptor (Bourguignon and Jin, 1995). Most importantly, the binding of ankyrin to the IP_3 receptors blocks IP_3 binding and IP_3-induced Ca^{2+} release activity (Bourguignon *et al.,* 1993a; Bourguignon and Jin, 1995). The ankyrin-binding domain of these Ca^{2+} channels appears to be located at the putative "pore region" of the channels, required for the regulation of Ca^{2+} activities (Yamamo-Hino *et al.,* 1994).

The fact that ankyrin not only binds to the surface molecule, CD44, but also binds to intracellular proteins, such as the IP_3 receptor, suggests that ankyrin may be involved in multiple functions during cellular regulation. In lymphocytes, for example, ankyrin appears to play a pivotal role in linking surface CD44 adhesion molecules and intracellular Ca^{2+} storage organelle membrane proteins (e.g., IP_3 receptor and ryanodine receptor) (Fig. 4) to the cytoskeleton. This ankyrin-based linkage between plasma membrane proteins and organelle Ca^{2+} channel molecules may be critically important for receptor-mediated signal transduction during lymphocyte activation.

VI. SUMMARY

CD44 denotes a family of glycoproteins that are expressed in a variety of cells and tissues derived from hemopoietic, epithelial, endothelial, and mesodermal origins. At the present time, the best characterized function of CD44 is its ability to mediate both cell adhesion to ECM components such as HA, and homotypic cell aggregation. Data indicate that HA induces CD44-mediated signal transduction events (e.g., Ca^{2+} mobilization and CD44–ankyrin interaction) followed by cell adhesion and proliferation, suggesting that CD44 is involved in an "outside-in" signaling event during lymphocyte activation.

The evidence that posttranslational modification of the CD44 cytoplasmic domain by PKC, acylation, or GTP binding promotes the binding between CD44 and ankyrin suggests that the interaction between these two molecules is tightly regulated.

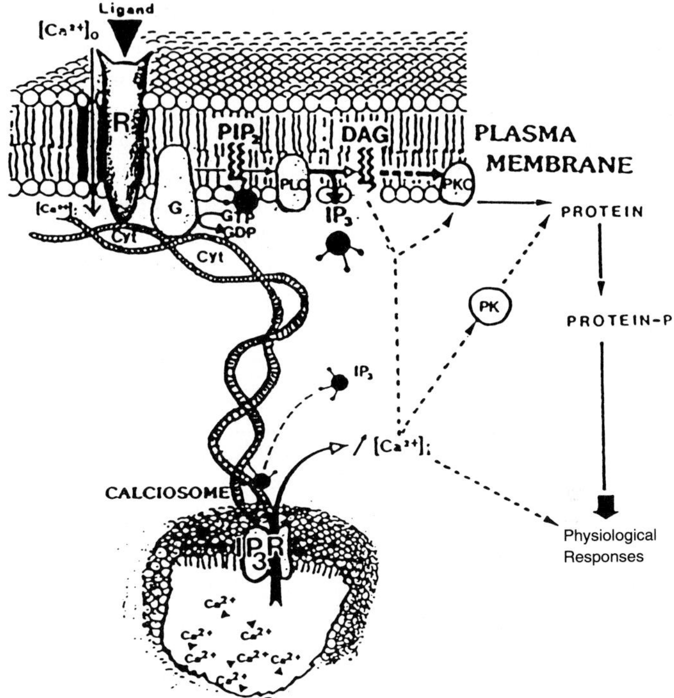

FIGURE 4 A model of CD44–ankyrin interactions in ligand-induced signal transduction. Upon binding of the ligand (e.g., HA) to its specific receptors (e.g., CD44), a cascade of biochemical events takes place. These events include an influx of Ca^{2+} ($[Ca^{2+}]_o \rightarrow [Ca^{2+}]_i$) and activation of G-protein- (G-) coupled phospholipase C (PLC), which hydrolyzes phosphoinositides (PIP_2) into diacylglycerol (DAG) and inositol triphosphate (IP_3). DAG is known to activate protein kinase C (PKC). IP_3 binds to IP_3 receptors (IP_3R) causing internal Ca^{2+} release. Ankyrin may play a pivotal role in linking surface CD44 adhesion molecules (R) and intracellular Ca^{2+} storage organelles [also called calciosomes; known to contain IP_3 receptor, ryanodine receptor, and Ca^{2+}-binding proteins (e.g., calreticulin and calsequestrin)] (Bourguignon *et al.*, 1993a, 1995; Bourguignon and Jin, 1995) to the cytoskeleton. This linkage may be very important for signal transduction and the regulation of Ca^{2+} activities needed for the activation of Ca^{2+}-dependent enzymes or protein kinases (PK) during the onset of physiological responses.

Deletion mutation analysis indicates that at least two subregions within the CD44 cytoplasmic domain contribute to ankyrin binding—region I (i.e., the high-affinity ankyrin-binding region) and region II (i.e., the regulatory region). These ankyrin-binding regions appear to play a pivotal role in regulating the expression of CD44's HA-mediated adhesion functions (so-called inside-out signaling events).

Most importantly, the CD44 ankyrin-binding domain shares a large amount of sequence homology with the ankyrin-binding domain located in at least two Ca^{2+} channels (IP_3 receptor and ryanodine receptor). Therefore, in lymphocytes ankyrin may play a pivotal role in linking surface CD44 adhesion molecules and intracellular Ca^{2+} storage organelle membrane proteins (e.g., IP_3 receptor and ryanodine receptor) to the cytoskeleton. In conclusion, we believe that the CD44–ankyrin interaction is not only very important for presenting CD44 properly for HA binding, but is also required for interactions with other organelles (e.g., Ca^{2+} channels) needed for signal transduction during lymphocyte adhesion, homing, hemopoiesis, migration, and tumor metastasis.

Acknowledgments

I gratefully acknowledge Dr. Gerard J. Bourguignon's assistance in preparing this manuscript. I would also like to thank Dr. N. Iida and Ms. D. Zhu for their suggestions and reviewing the manuscript. This work was supported by National Institutes of Health grants (CA 66163 and GM36353) and a Department of Defense grant.

References

Arch, R., Wirth, K., Hofmann, M., Ponta, H., Matzku, S., Herrlich, P., and Zoller, M. (1992). Participation in normal immune responses of a metastasis-inducing splicing variant of CD44. *Science* **257,** 682–685.

Belitsos, P. C., Hilderth, J. E. K., and August, J. T. (1990). Homotypic cell aggregation induced by anti-CD44(Pgp-1) monoclonal antibodies and related to CD44(Pgp-1) expression. *J. Immunol.* **144,** 1661–1670.

Bennet, K., Jackson, D. G., Simon, J. C., Tanczos, E., Peach, R., Modrell, B., Stamenkovic, I., Plowman, G., and Aruffo, A. (1995). CD44 isoforms containing exon v3 are responsible for the presentation of heparin binding growth factors. *J. Cell Biol.* **128,** 687–698.

Bennett, V. (1992). Ankyrins: Adapters between diverse plasma membrane proteins and the cytoplasm. *J. Biol. Chem.* **267,** 8703–8706.

Bourguignon, L. Y. W. (1992). Membrane skeleton linker proteins in immune cells. *In* "Encyclopedia of Immunology" (I. M. Roitt and P. J. Delves, eds.) pp. 1044–1046. Academic Press, New York.

Bourguignon, L. Y. W., and Jin, H. (1995). Identification of the ankyrin-binding domain of the mouse T-lymphoma cell inositol 1,4,5-triphosphate (IP_3) receptor and its role in the regulation of IP_3-mediated internal Ca^{2+} release. *J. Biol. Chem.* **270,** 7257–7260.

Bourguignon, L. Y. W., Walker, G., Suchard, S., and Balazovich, K. (1986). A lymphoma plasma membrane-associated protein with ankyrin-like properties. *J. Cell Biol.* **102,** 2115–2124.

Bourguignon, L. Y. W., Kalomiris, E., and Lokeshwar, V. B. (1991). Acylation of the lymphoma transmembrane glycoprotein, GP85, may be required for GP85-ankyrin interaction. *J. Biol. Chem.* **266,** 11761–11765.

Bourguignon, L. Y. W., Lokeshwar, V. B., He, J., Chen, X., and Bourguignon, G. J. (1992). A CD44-like endothelial transmembrane glycoprotein (GP116) interacts with extracellular matrix and ankyrin. *Mol. Cell. Biol.* **12,** 4464–4471.

Bourguignon, L. Y. W., Jin, H., Iida, N., Brandt, N., and Zhang, S. H. (1993a). The involvement of ankyrin in the regulation of inositol 1, 4, 5-triphosphate receptor-mediated internal

Ca^{2+} release from Ca^{2+} storage vesicles in mouse T-lymphoma cells. *J. Biol. Chem.* **268,** 7290–7297.

Bourguignon, L. Y. W., Lokeshwar, V. B., Chen, X., and Kerrick, W. G. L. (1993b). Hyaluronic acid-induced lymphocyte signal transduction and HA receptor (GP85/CD44)-cytoskeleton interaction. *J. Immunol.* **151,** 6634–6644.

Bourguignon, L. Y. W., Chu, A., Jin, H., and Brandt, N. R. (1995). Ryanodine receptor-ankyrin interaction regulates internal Ca^{2+} release in mouse T-lymphoma cells. *J. Biol. Chem.* **270,** 17917–17922.

Bourne, H. R., Sanders, D. A., and McCormik, F. (1991). The ATPase superfamily: Conserved structural and molecular mechanism. *Nature (London)* **349,** 117–127.

Carter, W. G., and Wayner, E. A. (1988). Characterization of the class III collagen receptor, a phosphorylated transmembrane glycoprotein expressed in nucleated human cells. *J. Biol. Chem.* **263,** 4193–4201.

Cooper, D. L., and Dougherty, G. J. (1995). To metastasize or not? Selection of CD44 splice sites. *Nature Med.* **1,** 635–637.

Davis, J., Lux, S. E., and Bennett, V. (1989). Mapping the ankyrin binding site of human erythrocyte anion exchanger. *J. Biol. Chem.* **264,** 9665–9672.

Droll, A., Dougherty, S. T., Chiu, R. K., Dirks, J. F., McBride, W. H., Cooper, D. L., and Dougherty, G. J. (1995). Adhesive interactions between alternatively spliced CD44 isoforms. *J. Biol. Chem.* **270,** 11567–11573.

Goldstein, L. A., Zhou, D. F. H., Picker, L. J., Minty, C. N., Bargatze, R. F., Ding, J. F., and Butcher, E. C. (1989). A human lymphocyte homing receptor, the hermes antigen, is related to cartilage proteoglycan core and link proteins. *Cell* **56,** 1063–1072.

Green, S. J., Tarone, G., and Underhill, C. B. (1988). Aggregation of macrophages and fibroblasts is inhibited by a monoclonal antibody to the hyaluronate receptor. *Exp. Cell Res.* 178, 224–232.

Gunthert, U., Hoffman, M., Rudy, W., Reber, S., Zoller, M., Haussmann, I., Matzku, S., Wenzel, A., Ponta, H., and Herrlich, P. (1991). A new variant of glycoprotein CD44 confers metastatic potential to rat carcinoma cells. *Cell* **65,** 13–24.

Haynes, B. F., Telen, M. J., Hale, L. P., and Denning, S. M. (1989). CD44—a molecule involved in leukocyte adherence and T-cell activation. *Immunol. Today* **10,** 423–428.

He, Q., Lesley, J., Hyman, R., Ishihara, K., and Kincade, P. W. (1992). Molecular isoforms of murine CD44 and evidence that the membrane proximal domain is not critical for hyaluronate recognition. *J. Cell Biol.* **119,** 1711–1719.

Heider, K. H., Hofmann, M., Hors, E., Berg, F. V. D., Ponta, H., Herrlich, P., and Pal, S. T. (1993). A human homologue of the rat metastasis associated variant of CD44 is expressed in colorectal carcinoma and adenomatous polyps. *J. Cell Biol.* **120,** 227–233.

Herrlich, P., Zoller, M., Pals, S. T., and Ponta, H. (1993). CD44 splice variants: Metastases meet lymphocytes. *Immunol. Today* **14,** 395–399.

Hofmann, M., Rudy, W., Zoller, M., Tolg, C., Ponta, H., Herrlich, P., and Gunthert, U. (1991). CD44 splice variants confer metastatic behavior in rats: Homologous sequences are expressed in human tumor cells. *Cancer Res.* **51,** 5292–5297.

Horst, E., Meijer, C. J., Radaszkiewicz, T., Ossekoppele, G. J., Van Krieken, J. H., and Pals, S. T. (1990). Adhesion molecules in the prognosis of diffuse large-cell lymphoma: Expression of a lymphocyte homing receptor (CD44), LFA (CD11a/18), and ICAM-1 (CD54). *Leukemia* **4,** 595–599.

Hyman, R., Lesley, J., and Schulte, R. (1991). Somatic cell mutants distinguish CD44 expression and hyaluronic acid binding. *Immunogenetics* **33,** 392–395.

Iida, N., and Bourguignon, L. Y. W. (1995a). New CD44 splice variants associated with human breast cancers. *J. Cell Physiol.* **162,** 127–133.

Iida, N., and Bourguignon, L. Y. W. (1995b). Alteration of cell adhesion and breast tumor cell behavior by CD44 variant isoforms. *Proc. Am. Assoc. Cancer Res.* **36,** 62.

Isacke, C. M. (1994). The role of the cytoplasmic domain in regulating CD44 function. *J. Cell Sci.* **107,** 2353–2359.

Isacke, C. M., Sauvage, C. A., Hyman, R., Lesley, J., Schulte, R., and Trowbridge, I. S. (1986). Identification and characterization of the human Pgp-1 glycoprotein. *Immunogenetics* **23,** 326–332.

Jackson, D. G., Bell, J. I., Dickinson, R., Timans, J., Shields, J., and Whittle, N. (1995). Proteoglycan forms of the lymphocyte homing receptor CD44 are alternatively spliced variants containing the v3 exon. *J. Cell Biol.* **128,** 673–685.

Jalkanen, S., and Jalkanen, M. (1992). CD44 binds the COOH-terminal heparin-binding domain of fibronectin. *J. Cell Biol.* **116,** 817–825.

Jalkanen, S., Bargatze, R. F., de los Toyos, J., and Butcher, E. C. (1987). Lymphocyte recognition of high endothelium antibodies to distinct epitopes of an 85–95kD glycoprotein antigen differentially inhibit lymphocyte binding to lymph node, mucosal, or synovial endothelial cells. *J. Cell Biol.* **105,** 983–990.

Jalkanen, S., Jalkanen, M., Bargatze, R., Tammi, M., and Butcher, E. C. (1988). Biochemical properties of glycoproteins involved in lymphocyte recognition of high endothelial venules in man. *J. Immunol.* **141,** 1615–1623.

Jalkanen, S., Joensuu, H., Soderstrom, K. O., and Klemi, P. (1991). Lymphocyte homing and clinical behavior of non-Hodgkin's lymphoma. *J. Clin. Invest.* **87,** 1835–1840.

Kalomiris, E. L., and Bourguignon, L. Y. W. (1988). Mouse T-lymphoma cells contain a transmembrane glycoprotein (GP85) which binds ankyrin. *J. Cell Biol.* **106,** 319–327.

Kalomiris, E. L., and Bourguignon, L. Y. W. (1989). Lymphoma protein kinase C is associated with the transmembrane glycoprotein, GP85, and may function in GP85-ankyrin binding. *J. Biol. Chem.* **264,** 8113–8119.

Katoh, S., Zheng, Z., Oritani, K., Shimozato, T., and Kincade, P. W. (1995). Glycosylation of CD44 negatively regulates its recognition of hyaluronan. *J. Exp. Med.* **182,** 419–429.

Kordeli, E., and Bennett, V. (1991). Distinct ankyrin isoforms at neuron cell bodies and nodes of Ranvier resolved using erythrocyte ankyrin-deficient mice. *J. Cell Biol.* **114,** 1243–1259.

Kordeli, E., Lambert, S., and Bennett, V. (1995). AnkyrinG, a new ankyrin gene with neural-specific isoforms localized at the axonal initial segment and node of Ranvier. *J. Biol. Chem.* **270,** 2352–2359.

Laurent, T. C., and Fraser, J. R. (1992). Hyaluronan. *FASEB J.* **6,** 2397–2404.

Lesley, J., and Trowbridge, I. S. (1982). Genetic characterization of a murine polymorphic cell surface glycoprotein. *Immunogenetics* **15,** 313–320.

Lesley, J., Hyman, R., and Schulte, R. (1985a). Evidence that the Pgp-1 glycoprotein is expressed on thymus-homing progenitor cells of the thymus. *Cell. Immunol.* **91,** 397–403.

Lesley, J., Trotter, J., and Hyman, R. (1985b). The Pgp-1 antigen is expressed on early fetal thymocytes. *Immunogenetics* **22,** 149–157.

Lesley, J. R., Schulte, R., and Hyman, R. (1990). Binding of hyaluronic acid to lymphoid cell lines is inhibited by monoclonal antibodies against Pgp-1. *Exp. Cell Res.* **187,** 224–233.

Lesley, J., He, Q., Miyake, K., Hamann, A., Hyman, R., and Kincade, P. (1992). Requirement for hyaluronic acid binding by CD44: A role for the cytoplasmic domain and activation by antibody. *J. Exp. Med.* **175,** 257–266.

Lesley, J., English, N., Perschl, A., Gregoroff, J., and Hyamn, R. (1995). Variant cell lines selected for alterations in the function of the hyaluronan receptor CD44 show differences in glycosylation. *J. Exp. Med.* **182,** 431–437.

Letarte, M., Iturbe, S., and Quackenbush, E. J. (1985). A glycoprotein of molecular weight 85, 000 on human cells of B-lineage: Detection with a family of monoclonal antibodies. *Mol. Immunol.* **22,** 113–124.

Liao, H. X., Levesque, M. C., Patton, K., Bergamo, B., Jones, D., Moody, A., Telen, M. J., and Haynes, B. F. (1993). Regulation of human CD44H and CD44E isoform binding to hyaluronan by PMA and anti-CD44 monoclonal and polyclonal antibodies. *J. Immunol.* **151**(Suppl. 11), 6490.

Liao, H. X., Lee, D. M., Levesque, M. C., and Haynes, B. F. (1995). N-terminal and central regions of the human CD44 extracellular domain participate in cell surface hyaluronan binding. *J. Immunol.* **155,** 3938–3945.

Lokeshwar, V. B., and Bourguignon, L. Y. W. (1991). Post-translational protein modification and expression of ankyrin-binding site(s) in GP85 (Pgp-1.CD44) and its biosynthesis precursors during T-lymphoma membrane biosynthesis. *J. Biol. Chem.* **266,** 17983–17989.

Lokeshwar, V. B., and Bourguignon, L. Y. W. (1992). The lymphoma transmembrane glycoprotein GP85 (CD44) is a novel guanine nucleotide-binding protein which regulates GP85 (CD44)-ankyrin interaction. *J. Biol. Chem.* **267,** 22073–22078.

Lokeshwar, V. B., Fregien, N., and Bourguignon, L. Y. W. (1994). Ankyrin-binding domain of CD44(GP85) is required for the expression of hyaluronic acid-mediated adhesion function. *J. Cell Biol.* **126,** 1099–1109.

Matsumura, Y., and Tarin, D. (1992). Significance of CD44 gene products for cancer diagnosis and disease evaluation. *Lancet* **340,** 1053–1058.

Miyake, K., and Kincade, P. W. (1990). A new cell adhesion mechanism involving hyaluronate and CD44. *Curr. Top. Microbiol. Immunol.* **166,** 87–90.

Miyake, K., Medina, K. L., Hayashi, S. I., Ono, S., Hamaoka, T., and Kincade, P. W. (1990). Monoclonal antibodies to Pgp-1/CD44 block lympho-hemopoiesis in long-term bone marrow culture. *J. Exp. Med.* **171,** 477–488.

Miyake, K., Underhill, C. B., Lesley, J., and Kincade, P. W. (1990b). Hyaluronate can function as a cell adhesion molecule and CD44 participate in hyaluronate recognition. *J. Exp. Med.* **172,** 69–75.

Morrow, J. S., Cianci, C. D., Ardito, A. S., and Kashgarian, M. (1989). Ankyrin links fodrin to the alpha subunit of Na^+/K^+-ATPase in Madin-Darby canine kidney cells and in intact renal tubule cells. *J. Cell Biol.* **108,** 455–465.

Nelson, W. J., and Veshnock, P. J. (1987). Ankyrin binding to Na^+/K^+-ATPase and implications for the organizations of membrane domains in polarized cells. *Nature (London)* **328,** 533–536.

Nishizuka, Y. (1992). Intracellular signaling by hydrolysis of phospholipids and activation of protein kinase C. *Science* 258, 607–614.

Peach, R. J., Hollenbaugh, D., Stamenkovic, I., and Aruffo, A. (1993). Identification of hyaluronic acid binding sites in the extracellular domain of CD44. *J. Cell Biol.* **122,** 257–264.

Perschl, A., Lesley, J., English, N., Trowbridge, I., and Hyman, R. (1995). Role of CD44 cytoplasmic domain in hyaluronan binding. *Eur. J. Immunol.* **25,** 495–501.

Pure, E., Camp, R. L., Peritt, D., Panettieri, P. A. Jr., Lazaar, A. L., and Nayak, S. (1995). Defective phosphorylation and hyaluronate binding of CD44 with point mutations in the cytoplasmic domain. *J. Exp. Med.* **181,** 55–62.

Quackenbush, E. J., Vera, S., Greaves, A., and Letarte, M. (1990). Confirmation by peptide sequence and co-expression on various cell types of the identity of CD44 and p85 glycoprotein. *Mol. Immunol.* **27,** 947–955.

Ristamaki, R., Joensuu, H., Soderstrom, K. O., and Jalkanen, S. (1995). CD44v6 expression in non-Hodgkin's lymphoma: An association with low histological grade and poor prognosis. *J. Pathol.* **176,** 259–267.

Rooney, P., Kumar, S., and Wang, M. (1995). The role of hyaluronan in tumor neovascularization. *Int. J. Cancer* **60,** 632–636.

Schmidt, M. F. G. (1989). Fatty acylation of proteins. *Biochim. Biophys. Acta* **988,** 411–426.

Screaton, G. R., Bell, M. V., Jackson, D. G., Cornelis, F. B., Gerth, U., and Bell, J. I. (1992). Genomic structure of DNA coding the lymphocyte homing receptor CD44 reveals 12 alternatively spliced exons. *Proc. Natl. Acad. Sci. U.S.A.* **89,** 12160–12164.

Shull, M. M., Pugh, D. G., and Lingrel, J. B. (1989). Characterization of the human Na^+/K^+-ATPase $\alpha 2$ gene and identification of the intragenic restriction fragment length polymorphism. *J. Biol. Chem.* **264,** 17532–17543.

Smith, P. R., Saccomani, G., Joe, E. H., Angelides, K. J., and Benos, D. J. (1991). Amiloride sensitive sodium channel is linked to the cytoskeleton in renal epithelial cells. *Proc. Natl. Acad. Sci. U.S.A.* **88,** 6971–6975.

Sorimachi, N. T., Sorimachi, H., Tobita, Y., Kitamura, F., Yagita, H., Suzuki, K., and Miyasaka, M. (1995). A novel ligand for CD44 is serglycin, a hematopoietic cell lineage-specific proteoglycan. *J. Biol. Chem.* **270,** 7437–7444.

Stamenkovic, I., Amiot, M., Pesando, J. M., and Seed, B. (1991). The hemopoietic and epithelial forms of CD44 are distinct polypeptides with different adhesion potentials for hyaluronon-bearing cells. *EMBO J.* **10,** 343–347.

Stamper, H. B. Jr., and Woodruff, J. J. (1976). Lymphocyte homing into lymph nodes: In vitro demonstration of the selective affinity of recirculating lymphocytes for high endothelial venules. *J. Exp. Med.* **144,** 828–833.

Tolg, C., Hofmann, M., Herrlich, P., and Ponta, H. (1993). Splicing choice from ten variant exons establishes CD44 variability. *Nucleic Acids Res.* **21,** 1225–1229.

Trowbridge, I. S., Lesley, J., Sculte, R., Hyman, R., and Trotte, J. (1982). Biochemical characterization and cellular distribution of a polymorphic murine cell-surface glycoprotein expressed on lymphoid tissues. *Immunogenetics* **15,** 299–312.

Tsukita, T., Oishi, K., Sato, N., Sagara, J., Kawai, A., and Tsukita, S. (1994). ERM family members as molecular linkers between CD44 and actin-based cytoskeleton. *J. Cell Biol.* **126,** 391–421.

Turley, E. A., Austen, L., Vandeligt, K., and Clary, C. (1991). Hyaluronan and a cell-associated hyaluronan binding protein regulate the locomotion of ras-transformed cells. *J. Cell Biol.* **112,** 1041–1047.

Uff, C. R., Neame, S. J., and Isacke, C. M. (1995). Hyaluronic binding by CD44 is regulated by a phosphorylation-independent mechanism. *Eur. J. Immunol.* **25,** 1883–1887.

Underhill, C. B. (1982). Interaction of hyaluronate with the surface of simian virus 40-transformed 2T2 cells: Aggregation and binding studies. *J. Cell Sci.* **56,** 177–189.

Underhill, C. B., Thurn, A. L., and Lacy, B. E. (1985). Characterization and identification of the hyaluronate-binding site from membranes of SV-3T3 cells. *J. Biol. Chem.* **260,** 8128–8133.

Underhill, C. B., Green, S. J., Comoglio, P. M., and Tarone, G. (1987). The hyaluronate receptor is identical to a glycoprotein of 85, 000 Mr. (gp85) as shown by a monoclonal antibody that interferes with binding activity. *J. Biol. Chem.* **262,** 13142–13146.

Welsh, C. F., Zhu, D., and Bourguignon, L. Y. W. (1995). Interaction of CD44 variant isoforms with hyaluronic acid and the cytoskeleton in human prostate cancer cells. *J. Cell Physiol.* **164,** 605–612.

West, D. C., and Kumar, S. (1989). The effect of hyaluronate and its oligosaccharides on endothelial cell proliferation and monolayer integrity. *Exp. Cell Res.* **183,** 179–196.

Yamamo-Hino, M., Sugiyama, T., Hikichi, K., Mattei, M. G., Hasegawa, K., Sekine, S., Sakurada, K., Miyawaki, A., Furuichi, T., Hasegawa, M., and Mikoshiba, K. (1994). Cloning and characterization of human type 2 and type 3 inositol 1,4,5-triphosphate receptors. *Receptors Channels* **2,** 9–22.

Yang, B., Yang, B. L., Savani, R. C., and Turley, E. A. (1994). Identification of a common hyaluronan binding motif in the hyaluronan binding proteins RHAMM, CD44 and link protein. *EMBO J.* **13,** 286–296.

Zhou, D. F. H., Ding, J. F., Picker, L. J., Bargatze, R. F., Butcher, E. C., and Goeddel, D. V. (1989). Molecular cloning and expression of Pgp-1: the mouse homolog of the human H-CAM (Hermes) lymphocyte homing receptor. *J. Immunol.* **143,** 3390–3395.

CHAPTER 15

Dynamic Properties of the Lymphocyte Membrane–Cytoskeleton: Relationship to Lymphocyte Activation Status, Signal Transduction, and Protein Kinase C

Elizabeth A. Repasky and Jennifer D. Black*

Departments of Molecular Immunology and *Experimental Therapeutics, Roswell Park Cancer Institute, Buffalo, New York 14263

I. INTRODUCTION

Lymphocytes are required to undergo rapid alterations in functional state in response to recognition of foreign antigen. Binding of antigen to specialized receptors on the plasma membrane of T and B cells results in activation, a process associated with a variety of cellular responses, including altered cell morphology and development of polarity, entry into the cell cycle, cell differentiation, and changes in immune effector activity. The changes in cell morphology that accompany lymphocyte activation have long been of interest to cell biologists investigating the function of cytoskeletal proteins. Research into the role of the cytoskeleton in lymphocyte membrane-related events began in the early 1970s with the observation that "patching" and "capping" of B-cell surface receptors (Taylor *et al.,* 1971; de Petris and Raff, 1973) is accompanied by a coincident rearrangement of underlying cytoskeletal elements (reviewed by Raff, 1976; Schreiner and Unanue, 1978; Braun and Unanue, 1983; Bourguignon and Bourguignon, 1984). Evidence for coincident movement of cytoskeletal elements and antigen receptors was one of the first indications of a link between the cell surface and intracellular events, and was thought to be an important component of activation-related signaling pathways controlling lymphocyte proliferation and/or differentiation. While knowledge of the molecular events associated with lymphocyte activation has advanced significantly in recent years (e.g., Cambier, 1992; Cambier *et al.,* 1994; Szamel and Resch, 1995), understanding of the role of the cytoskeleton in this process and in subsequent effector function remains limited.

Studies by our group have focused on characterization of the spectrin-based membrane–cytoskeleton in T and B lymphocytes and its role in antigen-induced signaling. These studies revealed several properties of the lymphocyte membrane–cytoskeleton that were unexpected based on the well-characterized erythrocyte model of membrane–cytoskeleton interaction. Of particular interest are the following observations:

1. The subcellular organization of spectrin and ankyrin exhibits marked variability among lymphocytes *in vivo* and among cultured lymphocyte cell lines. While many T and B lymphocytes express a membrane-associated cytoskeleton, many others appear to lack a spectrin-based cortical cytoskeletal network and, instead, express spectrin and ankyrin in a single large or several small aggregates at various locations in the cytoplasm. Still others express these proteins in discrete patches or caps at the plasma membrane.
2. Alterations in lymphocyte membrane-skeletal organization are observed in various disorders of the immune system and can be induced *in vivo* by exogenous treatments that modulate lymphocyte function.
3. The subcellular organization of the lymphocyte membrane skeleton is linked to that of several other proteins, including the β isoform of protein kinase C (PKC β) and the 70-kDa heat shock protein. Together, these proteins exhibit dynamic behavior in response to lymphocyte activation

signals, and may play an important role in signaling cascades that regulate proliferation, differentiation, and adhesive potential of T and B cells *in vivo*.

These findings suggest that there are inherent differences between erythrocytes and lymphocytes, and between lymphocyte subsets themselves, with regard to the function of membrane-associated cytoskeletal proteins. An understanding of the structural properties of the lymphocyte spectrin-based cytoskeleton, and of its regulation within the lymphocyte cytoplasm, may provide important clues regarding its role in activation and subsequent effector activities in normal and neoplastic T and B cells. This chapter summarizes current understanding of spectrin–ankyrin regulation in lymphocytes and concludes with a discussion of the role of the membrane skeleton in PKC function, and the importance of this interaction in T and B lymphocyte activation.

II. ORGANIZATION OF THE SPECTRIN-BASED CYTOSKELETON IN LYMPHOCYTES

A. Heterogeneity in the Subcellular Organization of the Spectrin-Based Cytoskeleton

1. Immunofluorescence Analysis of Spectrin Ankyrin in Tissue Lymphocytes

Until the early 1980s, the spectrin-based cytoskeleton was studied primarily in mature erythrocytes, where it forms a meshwork underlying the plasma membrane (see reviews by Branton *et al.,* 1981; Bennett, 1989; Bennett and Gilligan, 1993). Erythrocyte spectrin, which accounts for 75% of the cytoskeletal protein mass in these cells, mediates the linkage of short actin filaments to the plasma membrane via the cytoskeletal protein ankyrin, which binds to the cytoplasmic domain of the erythrocyte anion exchanger (band 3). The spectrin-based meshwork is a strong, flexible, and elastic structure that is responsible for many of the unique properties of the erythrocyte membrane. It is involved in maintaining the biconcave shape of the mature erythrocyte, conferring reversible deformability and membrane structural integrity, restricting the lateral mobility of integral membrane proteins or receptors, stabilizing the compositional and phase asymmetries of the lipid bilayer, and preventing membrane vesiculation (see Davies and Lux, 1989; Bennett and Gilligan, 1993). The erythrocyte cytoskeleton is a remarkable, highly specialized structure that stabilizes the membranes of otherwise fragile cells. Spectrin, a long "floppy" molecule that assumes no rigidly defined morphology, is ideally suited as the major structural component of a cell whose membrane must be capable of undergoing extensive deformation. Unlike other cells, the mammalian erythrocyte contains no transcellular cytoskeleton and has no microtubules or intermediate filaments and no internal organelles; thus, it relies on the plasma membrane and the spectrin network to serve as cytoskeleton and to generate shape. In view of the variability exhibited by lymphocytes regarding the subcellular distribution of spectrin and ankyrin (see later), it is important to note that little or no heterogeneity is seen among mature mammalian erythrocytes or nucleated avian erythrocytes (Repasky *et al.,* 1982) with respect to the organization of the spectrin-based cytoskeleton; absence or decreased levels of membrane-associated spectrin or ankyrin are observed *only* in cases of hereditary defects in erythrocyte membrane-skeletal structure (Bodine *et al.,* 1984; Greenquist *et al.,* 1978; Davies and Lux, 1989). Thus, in erythrocytes, the function of the spectrin–ankyrin meshwork is apparently best served by a static association with the plasma membrane.

Components of the erythrocyte membrane skeleton such as spectrin and ankyrin are now known to be expressed in most nonerythroid cells (Goodman *et al.,* 1981; Glenney *et al.,* 1982; Repasky *et al.,* 1982; Bennett *et al.,* 1982; Burridge *et al.,* 1982; Palfrey *et al.,* 1982). Their widespread distribution suggests that they perform important functions in different cell types, and their specific role(s) in cells where more extensive cytoskeletal networks determine shape, internal spatial organization, and motility are currently under investigation in several laboratories (see other chapters in this volume). Studies on the subcellular localization of spectrin and ankyrin in many nonerythroid cell types, including studies from our group on peripheral blood lymphocytes (Evans *et al.,* 1993) and most cultured lymphocyte cell lines (Pauly *et al.,* 1986), revealed a pattern consistent with that observed in the mammalian or avian erythrocyte. Thus, a tight cortical band of staining delineating the cell periphery was seen in the majority of cell types examined. It was therefore surprising to find that spectrin and ankyrin exhibit striking heterogeneity in their subcellular organization in T and B lymphocytes *within lymphoid organs* (Repasky *et al.,* 1984; Lee and Repasky, 1987; Black *et al.,* 1988; Gregorio *et al.,* 1994). These unexpected cytoskeletal patterns were found to occur naturally, in the absence of exogenously added cross-linking agents or modulators of lymphocyte function. Patterns of lymphocyte spectrin–ankyrin distribution detected in spleen, thymus, Peyer's patches, and lymph node include the "classical" tight cortical band delineating the cell periphery, membrane-associated accumulations resembling the "patches" and "caps" induced by membrane receptor cross-linking, small aggregates scattered throughout the cytoplasm or near the nuclear membrane, and a large single focal accumulation at one pole of the cell (see Fig. 1a,b). Perhaps the most unexpected finding was that lymphocyte spectrin and ankyrin *often appear to be entirely absent from the cell periphery and can occur as a single large or several smaller aggregates at some distance from the plasma membrane.* Ultrastructural studies revealed that the single polar accumulation of spectrin and ankyrin is frequently located near the Golgi apparatus or at the nuclear membrane (Black *et al.*, 1988; Gregorio *et al.*, 1994).

Each of the patterns of membrane–cytoskeletal organization described here is represented in all mammalian lymphoid organs thus far examined, and can be detected in both T and B cells. Moreover, patterns of lymphocyte spectrin distribution are developmentally regulated and anatomically segregated (Repasky *et al.*, 1984). These properties are clearly evident in the thymus (Fig. 1c), a lymphoid organ that represents a useful model system in which to examine spectrin–ankyrin organization in relation to lymphocyte maturation stage. The thymus is the site where T-cell precursors attain a state of functional competence. Each of its two lobes is divided into multiple lobules consisting of an outer cortex and an inner medulla. Precursors that are committed to the T-cell lineage enter the thymic cortex via the circulation and are thought to migrate subsequently to the medulla, where they acquire a surface phenotype characteristic of mature T lymphocytes. Thus, immature and mature cells are segregated into two histologically distinct regions, with the medulla consisting mainly of mature T cells. Cells with polar aggregates of spectrin are detectable in murine thymus by day 19 of gestation (Repasky *et al.*, 1984), at a time that coincides with acquisition of the T-cell receptor by many thymocytes (von Boehmer, 1990). Immunofluorescence analysis reveals a distinct segregation of cells expressing a polar accumulation of cytoskeletal elements to the medullary region, with most cells of the cortex expressing lower levels of spectrin and ankyrin in a diffuse pattern at the cell periphery. These findings strongly suggest that factors regulating T-lymphocyte maturation modulate the expression and organization of components of the T-lymphocyte membrane cytoskeleton.

In summary, morphological analysis revealed an unexpected heterogeneity in the organization of the mammalian lymphocyte spectrin-based cytoskeleton that appears to be anatomically and developmentally regulated. It is intriguing that lymphocytes from chicken lymphoid tissues, including spleen, bursa, and thymus, express only two patterns of membrane-skeletal organization: tight cortical rings delineating the cell periphery or "caps" associated with a single domain of the plasma membrane. Cells with a single large, or several smaller, aggregates of spectrin–ankyrin are not detected in avian lymphoid tissues (P. Schmitz and E. A. Repasky, unpublished data). The basis for the species differences in lymphocyte cytoskeletal organization is unknown.

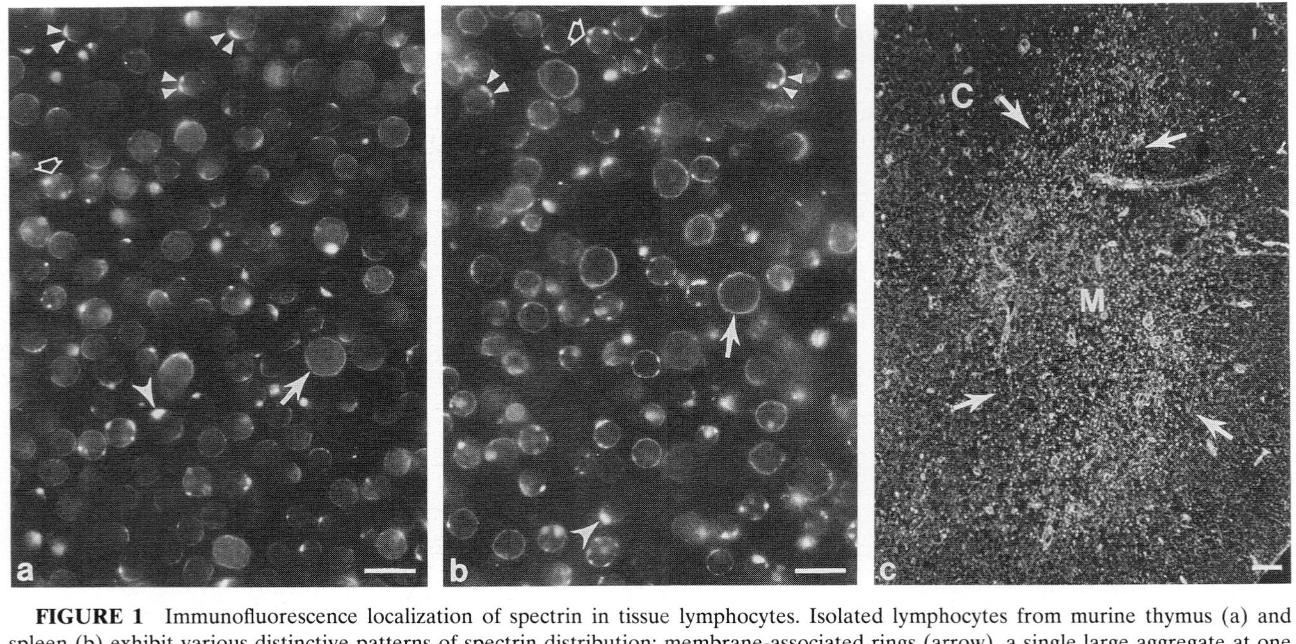

FIGURE 1 Immunofluorescence localization of spectrin in tissue lymphocytes. Isolated lymphocytes from murine thymus (a) and spleen (b) exhibit various distinctive patterns of spectrin distribution: membrane-associated rings (arrow), a single large aggregate at one pole of the cell (arrowhead), membrane-associated caps (double arrowhead), or several patches of variable size and distribution (open arrow). Bars, 10 μm. A cryosection of murine thymus (c) immunostained for spectrin reveals that cells that contain large single cytoskeletal aggregates are concentrated in the medulla (M) (demarcated by arrows), whereas cells of the cortex (C) exhibit a diffuse staining pattern for spectrin. Bar, 40 μm. Ankyrin exhibits identical distribution patterns in murine thymus (not shown).

2. Membrane-Skeletal Organization in Lymphocyte Cell Lines and Hybridomas

Long-term *in vitro* cultures of murine and human T and B lymphocytes can be distinguished based on the stable expression of a particular pattern of spectrin–ankyrin distribution (Pauly *et al.,* 1986; Black *et al.,* 1988; Lee *et al.,* 1988). In the majority of T-cell lines and in *all* of the B-cell lines examined, spectrin and ankyrin are uniformly expressed at the cell periphery. In contrast, several murine and human T-cell lines and all of the more mature, antigen-specific, major histocompatibility complex– (MHC-) restricted T-cell hybridomas examined were found to constitutively express spectrin and ankyrin in a single large aggregate, with little or no expression at the plasma membrane (Black *et al.,* 1988; Lee *et al.,* 1988; see Fig. 2). These findings suggest that the pattern of spectrin–ankyrin organization can be an inherent characteristic of transformed lymphocyte cell lines, and likely reflects a component of the phenotype associated with a particular lymphocyte maturation stage. Possible explanations for the absence of B-cell lines or hybridomas that constitutively express cytoskeletal aggregates include (i) expression of the aggregate is not associated with the maturation

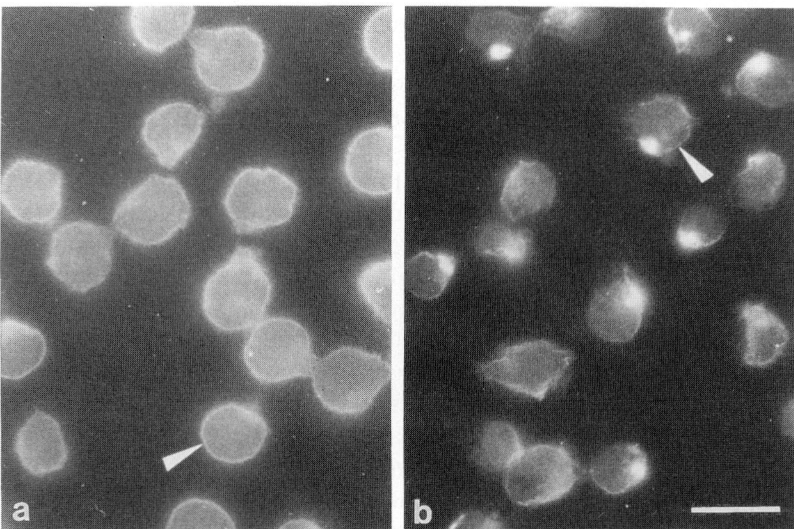

FIGURE 2 Lymphocyte cell lines exhibit different patterns of cytoskeletal subcellular organization. As described in the text, lymphocyte cell lines can be distinguished based on constitutive expression of a particular pattern of spectrin–ankyrin organization. Shown here are two human T-cell lines: (a) Molt-4 cells, which exhibit diffuse staining for spectrin with a higher concentration of the protein at the plasma membrane, and (b) JM cells (a derivative of Jurkat cells), which exhibit distinct aggregates of spectrin. Bar, 20 μm.

state(s) of B cells that survive in culture or, (ii) B-cell lines at the appropriate stage of maturation have yet to be examined for cytoskeletal organization. However, as discussed later, cultured B cells can be induced by specific treatments to express a single large cytoskeletal aggregate (K. Lin, R. Ward, J. D. Black, and E. A. Repasky, manuscript in preparation).

B. *Analysis of Naturally Occurring Accumulations of Spectrin-Ankyrin in Lymphocytes*

1. Ultrastructural Characterization of the Polar Cytoskeletal Aggregate

To characterize further the various patterns of cytoskeletal distribution that occur naturally in lymphocytes, and to establish the relationship of lymphocyte spectrin–ankyrin with the plasma membrane, a preembedding immunocytochemical labeling technique was used to localize these proteins at the ultrastructural level in various lymphoid tissues *in situ* (Black *et al.*, 1988; Gregorio *et al.*, 1994). These studies clearly demonstrated that, in many tissue lymphocytes, spectrin and ankyrin are confined to a single discrete polar aggregate at some distance from the plasma membrane, often in close association with the nuclear membrane, the Golgi apparatus, or the centrioles. Lymphocyte spectrin and ankyrin were also found to occur as several smaller aggregates scattered throughout the cytoplasm, in close association with the plasma membrane, or near the nucleus. Few cells were found to exhibit uniform staining for these proteins at the plasma membrane.

The availability of lymphocyte cell lines that homogeneously express a single cytoskeletal phenotype enabled a detailed ultrastructural characterization of the spectrin–ankyrin aggregate detected in tissue lymphocytes (Black *et al.*, 1988; Gregorio *et al.*, 1994). Ultrastructural immunoperoxidase staining in the DO.11.10 functionally mature T-cell hybridoma revealed that spectrin and ankyrin are accumulated in a discrete region (1–2 μm in diameter) at one pole of the cell (Fig. 3), several microns from the cell surface and, usually, in close association with the microtubule organizing center–centriolar complex and *trans*-region of the Golgi apparatus. The stained area was frequently surrounded by many small membranous vesicles and cisternae. The location of spectrin–ankyrin reaction product was found to correspond to that of a cytoplasmic structure composed primarily of a fine meshwork of densely packed filaments. Conventional thin-section microscopy revealed that this aggregate of filaments, which appears more electron dense than the surrounding cytoplasm, is not membrane bound but excludes most other cellular organelles. In addition to membranous vesicles, it contains many electron-dense ribosome-like granules. Moreover,

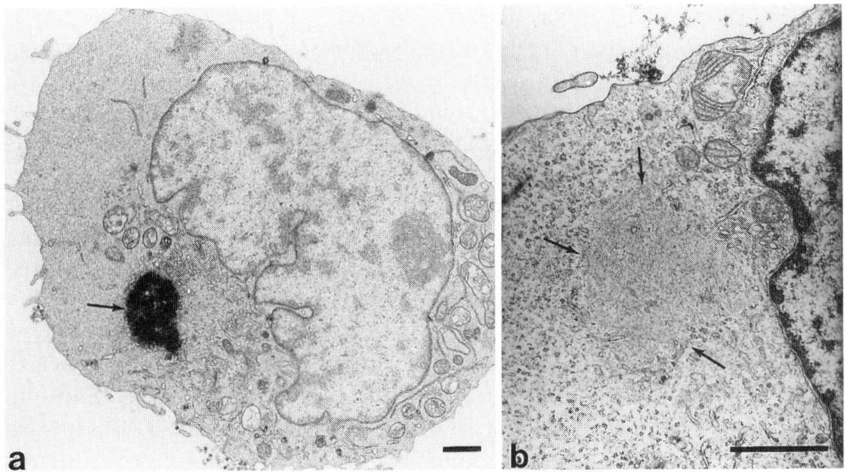

FIGURE 3 Ultrastructural analysis of the cytoskeletal aggregate in lymphocytes. (a) Ultrastructural immunoperoxidase localization of spectrin in a DO.11.10 T cell. The arrow indicates a region of dense immunoprecipitate several microns from the plasma membrane. (b) Using thin-section electron microscopy, this region is shown to correspond to a non-membrane-bound, filament-rich structure. The region often contains ribosome-like particles and membranous vesicles. Bar, 1 μm.

numerous intermediate filaments often surround and course through the structure. It is noteworthy that a similar region of specialized, spectrin-rich cytoplasm termed the fusome–spectrosome was recently identified in *Drosophila* germline cysts (Lin *et al.,* 1994; Lin and Spradling, 1995). Ultrastructural analysis revealed that this spectrin-rich region is associated with centrioles and that it is present at only one pole of the mitotic spindles in dividing cysts. It is thought to play an important role in cyst formation and oocyte differentiation. The relationship between the lymphocyte spectrin-rich aggregate and the *Drosophila* fusome–spectrosome in germline cysts remains to be determined.

2. Morphological Characterization of Tissue B Lymphocytes Expressing Naturally Occurring "Caps" of Spectrin and Ankyrin and a Coincident Cap of Cell Surface Immunoglobulin

A membrane-associated "cap" of spectrin–ankyrin is another of the naturally occurring patterns of cytoskeletal organization expressed by lymphocytes *in situ* (Repasky *et al.,* 1984; Black *et al.,* 1988; see Fig. 1). Cytoskeletal caps are distinguished from the polar cytoplasmic aggregates described earlier by their close apposition to the plasma membrane. In a study of this cytoskeletal phenotype (P. Masso-Welsh, J. Erikson, J. D. Black, and

E. A. Repasky, 1996), it was determined that "capped" cells are largely B lymphocytes. The percentage of B cells expressing cytoskeletal caps was found to vary among mouse strains and to increase with age. Furthermore, membrane-associated caps were found to be the predominant pattern of lymphocyte spectrin–ankyrin organization in the MRL-*lpr/lpr* mouse model for progressive autoimmunity and in mice transgenic for anti–single-stranded DNA immunoglobulin (Erikson *et al.,* 1991). Since both of these models are characterized by the expression of high levels of self-antigen, these data suggest that the accumulation of spectrin–ankyrin in a caplike zone at the plasma membrane may be related to the level of exposure of B lymphocytes to antigen or autoantigen.

Research into the process of antigen-induced B-cell activation provides evidence for a possible link between the presence of antigen and the formation of cytoskeletal caps *in situ.* Binding of antigen to the B-cell antigen receptor, membrane immunoglobulin (Ig), results in receptor cross-linking, a process that can be mimicked by anti-Ig antibodies. Such cross-linking leads to formation of patches of Ig on the cell surface, which coalesce to form a polar cap in an energy- and cytoskeleton-dependent manner (reviewed by Braun and Unanue, 1983; Bourguignon and Bourguignon, 1984). Ig cap formation *in vitro* is accompanied by coincident recruitment of a number of cytoskeletal proteins, including spectrin (Levine and Willard, 1983; Nelson *et al.,* 1983), ankyrin (Bourguignon and Bourguignon, 1984; Lokeshwar and Bourguignon, 1992a,b), myosin (Braun *et al.,* 1978), actin (Gabbiani *et al.,* 1977; Flanagan and Koch, 1978), α-actinin (Gupta and Woda, 1988), tubulin (Gabbiani *et al.,* 1977; Yahara and Kakimoto-Sameshima, 1978), and vimentin (Dellagi and Brouet, 1982; Lee and Repasky, 1987). Our recent immunofluoresence studies revealed that naturally occurring cytoskeletal caps in tissue B cells are morphologically similiar to those resulting from antibody- or antigen-induced cross-linking of cell surface Ig *in vitro,* and are almost invariably associated with an overlying cap of surface Ig. Therefore, it is tempting to speculate that "capped" tissue B cells reflect the occurrence *in vivo* of antigen-induced Ig cross-linking and capping, a process that presumably takes place with high frequency in autoimmune and transgenic animals expressing abundant self-antigen. The clustered Ig–spectrin–ankyrin phenotype may denote B cells at a stage after antigen receptor cross-linking in which Ig is stabilized by cytoskeletal proteins or antigen. It is noteworthy that, although a role for Ig cross-linking in B-cell activation has been well established, the physiological significance of subsequent Ig cap formation remains unknown. It is possible that development of this cellular polarity may be required for the directional interaction of B cells with T cells, in a manner analogous to the interaction of T cells with target or antigen-presenting cells (Kupfer and Singer, 1989; Kupfer *et al.,* 1994). Future work will determine whether particular patterns of spectrin–ankyrin and Ig clustering are indicative of differential activation potential and stage of differentiation of B lymphocytes *in vivo.*

C. *Role of Differential Expression of Spectrin–Ankyrin in Differences in Cytoskeletal-Dependent Functions of the Lymphocyte Plasma Membrane*

Morphological and biochemical analysis clearly demonstrates that lymphocytes differ with respect to the amount and distribution of spectrin and ankyrin at the plasma membrane. In the erythrocyte, the plasma membrane–associated spectrin-based cytoskeleton has been implicated in regulation of a variety of membrane properties (see earlier), including membrane fluidity, maintenance of the asymmetrical distribution of phospholipids across the bilayer, and anchorage of integral membrane proteins or receptors. Therefore, the finding that lymphocyte subpopulations exhibit variability with respect to the level of expression of spectrin and ankyrin at the plasma membrane raises important questions regarding control of membrane organization in these cells.

The plasma membrane of the erythrocyte has been intensively studied as a model for transverse membrane lipid asymmetry (see reviews by Op den Kamp, 1979; Williamson and Schlegel, 1994). In this cell, sphingomyelin and glycolipids are largely confined to the outer leaflet of the lipid bilayer, whereas certain aminophospholipids, particularly phosphatidylserine, are virtually absent from this layer and are concentrated in the inner one (Tanaka and Ohnishi, 1976; Morrot *et al.,* 1986). As a consequence of this phospholipid asymmetry, the inner lipid layer is less ordered and more fluid than the outer one. Although the mechanism by which this asymmetry is maintained is still unclear, it has been proposed to be the result of direct interactions between phosphatidylserine and the membrane cytoskeleton, particularly spectrin (Haest and Deutike, 1976; Haest *et al.,* 1978, 1982, Williamson *et al.,* 1982). More recently, alternate hypotheses regarding maintenance of phospholipid asymmetry that do not implicate the cytoskeleton have been proposed, including the existence of aminophospholipid translocases such as the ATP-dependent "flippase" (Seigneuret and Devaux, 1984; reviewed in Williamson and Schlegel, 1994). Evidence has also been presented to support a role for both the cytoskeleton and translocases in maintenance of transverse lipid asymmetry in the red blood cell (Connor and Schroit, 1990).

Despite the uncertainty regarding the role of the cytoskeleton in maintaining transverse membrane lipid asymmetry and fluidity, there is correlative evidence to suggest that lymphocytes that differ with respect to the amount and distribution of spectrin and ankyrin at the plasma membrane also differ with regard to membrane fluidity. Two studies have examined the relationship between cytoskeletal distribution and membrane lipid organization in lymphocytes (Del Buono *et al.,* 1988; Langner *et al.,* 1992). Del Buono *et al.* (1988) stained isolated populations of cortical and medullary thymocytes expressing different levels of membrane-associated spectrin (Repasky *et al.,* 1984) with the fluorescent lipophilic probe merocyanin 540 (MC540), which binds preferentially to loosely packed, disorganized lipid bilayers. These studies revealed that thymocytes that stained brightly for MC540 exhibited uniform immunofluorescence staining for spectrin at the plasma membrane, while cells that stained dimly for the probe expressed an aggregate of spectrin, with little or no spectrin at the cell periphery. Thus, expression of a spectrin-based membrane-associated skeleton correlates with the presence of a more loosely packed lipid bilayer in lymphocytes.

In the second study (Langner *et al.,* 1992), two related *in vitro* T-cell systems expressing different levels of spectrin at the cell periphery, the DO.11.10 T-cell hybridoma and a spontaneously arising variant (DO.11.10V), were used to determine whether the various patterns of lymphocyte spectrin organization are predictive of differences in plasma membrane lipid properties. Eximer formation in a fluorescent pyrene-labeled phospholipid was used as a probe for the lateral mobility of plasma membrane lipids in these cell lines. Consistent with the findings of del Buono *et al.* (1988), these studies revealed that the plasma membrane lipids of cells that appear to lack membrane-associated spectrin (i.e., the DO.11.10 cell line) are considerably less mobile than those of cells that express a spectrin-based membrane skeleton (the variant cell line). It is noteworthy that fatty acid perturbation of the organization of the lymphocyte lipid bilayer can, in turn, significantly affect the subcellular organization of spectrin (Stephen *et al.,* 1990). Together, these findings indicate that the various patterns of spectrin organization in lymphocytes can be correlated with differences in plasma membrane lipid organization. The significance of these differences in relation to lymphocyte function remains to be determined.

Based on the erythrocyte model of membrane–cytoskeleton function, anchorage and mobility of cell surface macromolecules would also be expected to differ among lymphocyte subsets expressing varying levels of spectrin and ankyrin at the plasma membrane. Involvement of cytoskeletal interactions in regulation of membrane receptor distribution at nonrandom locations has been reported in several nonerythroid systems (e.g., Nelson and Hammerton, 1989; Venkatakrishnan *et al.,* 1991; Scher and Bloch, 1993). In an extensive series of studies using mouse T-lymphoma cells as a model system, spectrin or ankyrin have been shown to associate either directly or indirectly with various plasma membrane components, including CD45 and CD44 (e.g., see Kalomiris and Bourguignon, 1988, 1989; Lokeshwar and Bourguignon, 1992a,b), the inositol 1, 4, 5-trisphosphate (IP_3) receptor (Bourguignon and Jin, 1995), GP116 (indicated to be a CD44-like molecule; Bourguignon *et al.,* 1992), and a membrane-associated GTP–binding protein (Bourguignon *et al.,* 1990). Whether these linkages also occur in lymphocytes that constitutively express spectrin and ankyrin at other cytoplasmic locations, or how these associations are affected by formation of the cytoskeletal aggregate, is currently unclear. A comprehensive understanding of the role of the spectrin-based membrane–cytoskeleton in lymphocytes will require an in-depth comparison of the organization of these membrane proteins in cells expressing different levels of membrane-associated spectrin and ankyrin.

III. PHYSIOLOGICAL SIGNIFICANCE OF THE HETEROGENEITY IN CYTOSKELETAL ORGANIZATION IN TISSUE LYMPHOCYTES

A. Dynamic Properties of Spectrin and Ankyrin in T and B Cells: Modulation of Cytoskeletal Organization by Lymphocyte Activation Signals

Several recent studies in our laboratory have focused on gaining insight into the physiological basis for the heterogeneity in cytoskeletal organization exhibited by mammalian tissue lymphocytes. Efforts have been directed toward testing the hypothesis that individual patterns of cytoskeletal organization reflect particular stages of activation and/or differentiation of T and B lymphocytes induced by naturally occurring stimuli. Data described here provide strong experimental evidence to support this notion.

Lymphocyte activation occurs in response to the binding of antigen to specialized antigen receptors on the cell surface. While B lymphocytes capture native antigen via surface-bound Ig molecules, T cells require antigen to be processed by antigen-presenting cells, such as macrophages, and presented in the context of MHC molecules; processed antigen is recognized by a specific T-cell receptor. For both T and B cells, cross-linking of antigen receptors by either correctly presented antigen or agonistic antibodies triggers the cell via protein tyrosine phosphorylation events and the phosphoinositol lipid turnover pathway (Weiss, 1994; DeFranco, 1994). Using several T- and B-cell model systems, our studies have revealed that activation-related signaling results in dramatic alterations in the organization of the spectrin-based membrane–cytoskeleton

(Gregorio *et al.*, 1992; 1994; K. Lin, J. D. Black, and E. A. Repasky, 1996; see Fig. 4a–c). For example, stimulation of murine lymph node–derived T cells by α/β antigen receptor cross-linking using the monoclonal antibody H57–597 (Kubo *et al.*, 1989) results in rapid, protein synthesis–independent formation of a single large cytoskeletal aggregate in the majority of cells (Gregorio *et al.*, 1992, 1994). Similar formation of a cytoskeletal aggregate was found to occur in splenic B cells in response to exposure to anti-IgM antibody, lipopolysaccharide, or murine recombinant interleukin-4 (Fig. 4c). Recombinant interferon-α treatment of normal and malignant human peripheral blood B cells also results in aggregate formation, although in this case cytoskeletal rearrangement occurs in a protein synthesis–dependent manner over a 24-h period (Evans *et al.*,

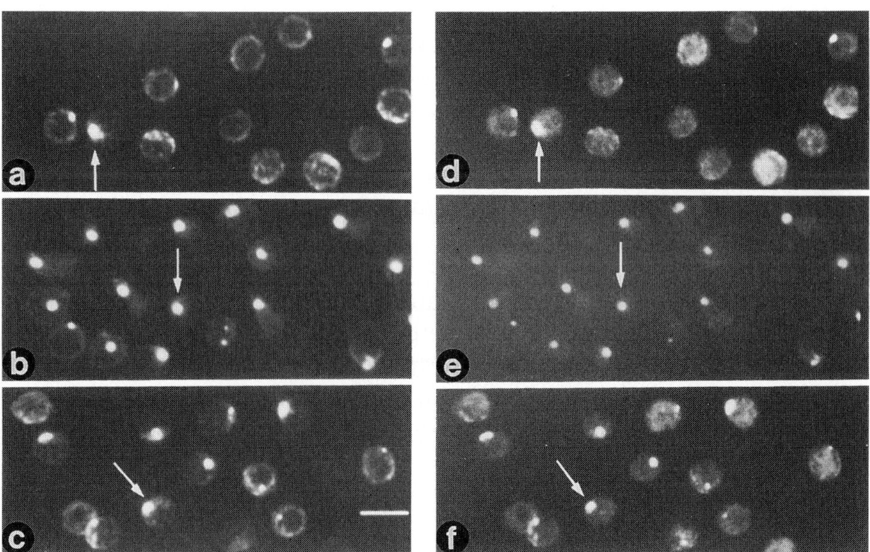

FIGURE 4 Activation-related signals induce a coincident movement of spectrin and PKC β in the B lymphocyte cytoplasm. Double immunofluorescence staining for spectrin (a, b, and c) and PKC β (d, e, and f) in the same isolated splenic B lymphocytes. Control (untreated) cells (a and d) exhibit various distribution patterns for spectrin and PKC β. Cells that contain a focal accumulation of spectrin (arrows) exhibit a coincident accumulation of the enzyme. Direct activation of PKC by PMA (b and e) induces the translocation of both proteins to a cytoplasmic aggregate in nearly all cells; similarly, activation of B cells through the antigen receptor using anti-IgM antibody (c and f) results in an increase in the percentage of cells expressing co-localized aggregates of spectrin and PKC β (bar, 1 μm). Activation-induced cytoskeletal reorganization was also observed in T lymphocytes (Gregorio *et al.*, 1992, 1994). Moreover, ankyrin co-localizes with spectrin under all condition examined (see text for details).

1993). Aggregate formation can be rapidly induced in both T and B cells by treatment with the potent PKC activator phorbol 12-myristate 13-acetate (PMA) or with the diacylglycerol (DAG) analogue 1, 2-dioctanoyl-sn-glycerol (DiC_8) (Gregorio *et al.,* 1992, 1994; Fig. 4b). Thus, translocation of spectrin and ankyrin to a focal center in the lymphocyte cytoplasm appears to be an early response to activation-related signals in T and B cells.

Other patterns of cytoskeletal organization that occur naturally in tissue lymphocytes *in situ* can also be induced *in vitro* in response to specific activation-related signals. For example, stimulation of B cells via MHC class II or cAMP-generating signaling pathways results in formation of multiple small cytoskeletal aggregates near the nucleus. Moreover, in T-cell lines and hybridomas (e.g., the DO.11.10 T-cell hybridoma) that *constitutively express a single cytoskeletal aggregate,* treatment with the PKC activator PMA, with mitogens (e.g., concanavalin A), or with the calcium ionophore A23187 results in rapid *dissipation* of the cytoskeletal aggregate and an increase in the level of spectrin associated with the plasma membrane (Lee *et al.,* 1988). The molecular basis for the differential effects of activation-related signals on the spectrin-based cytoskeleton in the various T- and B-lymphocyte models examined remains to be determined. However, the data suggest that patterns of membrane–cytoskeleton reorganization resulting from lymphocyte activation signals may be dependent upon specific stage of lymphocyte activation and/or differentiation, a notion that is currently under investigation.

Collectively, these data demonstrate the dynamic nature of components of the membrane cytoskeleton in lymphocytes and provide insight into the regulation of cytoskeletal aggregate formation in these cells. Furthermore, they suggest that the naturally occurring heterogeneity in the subcellular distribution of spectrin and ankyrin in tissue T and B cells may arise as a result of engagement of specific signaling pathways that regulate important aspects of lymphocyte function.

B. Regulation of the Integrity of the Cytoskeletal Aggregate in Lymphocytes

The integrity of the cytoskeletal aggregate in lymphocytes can be modulated by cytoskeleton-disrupting drugs (Lee and Repasky, 1987), by calcium (Gregorio *et al.*, 1993), and by ATP (Di *et al.,* 1995). The microfilament-disrupting agent cytochalasin D causes a rapid dispersal of the spectrin–ankyrin aggregate, suggesting that integrity of lymphocyte actin filaments is essential for maintenance of these cytoskeletal proteins in an aggregated state. In contrast, agents that affect microtubule structure, such as colchicine and nocodozole, do not appear to cause disruption of the aggregate itself; however, they have a dramatic effect on the positioning of the aggregate in the lymphocyte cytoplasm (J. K. Lee, J. D. Black, and E. A. Repasky, unpublished data). Treatment of the DO.11.10 T-cell hybridoma with these agents appears to release the aggregate from the *trans*-Golgi region, suggesting that microtubules play an important role in anchorage of the structure in this cytoplasmic location.

The cytoskeletal aggregate has also been shown to be sensitive to the levels of calcium in the lymphocyte cytoplasm (Gregorio *et al.*, 1993). Exposure of DO.11.10 cells to calcium-free medium, or to medium containing the calcium chelator ethyleneglycol-bis-(β-aminoethyl ether) *N,N'*-tetraacetic acid (EGTA) or the calcium channel blocker verapamil, significantly blocks concanavalin A– or calcium ionophore–induced dissipation of the aggregate. Furthermore, DO.11.10 cell activation in the presence of EGTA or verapamil results in abnormalities in aggregate shape; immunofluorescence and ultrastructural analyses revealed that, following activation under these conditions, the aggregate shows a distinctly elongated form. Together, these findings indicate that calcium plays a role in the activation-induced dissipation of the cytoskeletal aggregate in cells that constitutively express this structure. In addition, they indicate that, in the absence of calcium, aggregate dissipation is arrested at an early stage, a phenomenon manifested by abnormalities in aggregate shape.

Introduction of exogenous ATP into lymphocytes using a transient permeabilization procedure also leads to dissipation of the cytoskeletal aggregate (Di *et al.*, 1995). It is interesting in this regard that addition of ATP has been shown to abolish the interaction between spectrin and a 72-kDa heat shock protein, hsp70, another component of the lymphocyte cytoskeletal aggregate. This protein also exhibits dynamic properties in response to lymphocyte activation signals and can be co-precipitated with spectrin in reciprocal immunoprecipitation assays (Di *et al.*, 1995). Given the reported function of hsp70 as a molecular chaperone facilitating the assembly, maturation, and interactions of other proteins (see Welch, 1992, for review), it is tempting to speculate that this molecule plays a role in maintenance of spectrin, ankyrin, and other components of the cytoskeletal aggregate in a nonprecipitated, biologically active state within the lymphocyte cytoplasm. Since hsp70 is also capable of transient relocation to various sites in the cytoplasm and nucleus (Welch and Feramisco, 1984), it may also participate in the redistribution of components of the aggregate in response to lymphocyte activation-related signals.

IV. ROLE OF THE SPECTRIN-BASED CYTOSKELETON IN PKC-MEDIATED, ACTIVATION-INDUCED SIGNAL TRANSDUCTION IN LYMPHOCYTES

A. Coordinate Regulation of the Subcellular Organization of Spectrin, Ankyrin, and PKC in Lymphocyte Subpopulations

Antigen receptor stimulation in T and B lymphocytes leads to protein tyrosine phosphorylation and subsequent activation of the phosphoinositol lipid turnover pathway (Weiss, 1994; DeFranco, 1994). Hydrolysis of phosphatidylinositol 4, 5-bisphosphate results in generation of the second messengers IP_3, which causes an increase in the levels of cytoplasmic free calcium, and DAG, which activates the signal-transducing molecule PKC. Activation of PKC is a critical event in activation-related signaling in T and B cells that has been implicated in control of a variety of downstream cellular responses.

PKC is a multigene family of lipid-regulated serine–threonine kinases (i.e., PKC α, βI, βII, γ, δ, ε, ζ, η, θ, ι and μ; Nishizuka, 1992; Dekker and Parker, 1994) that differ with respect to enzymological properties, substrate specificity, tissue expression, and subcellular localization. It is a well-characterized example of a group of protein kinases that undergo subcellular redistribution in response to exogenous activating stimuli (Kraft and Anderson, 1983). Also included in this group are cAMP-dependent protein kinase, the β-adrenergic receptor kinase, and Raf-1 (see Mochly-Rosen, 1995, for review). It is now well established that, before stimulation, PKC isozymes are present in the cytosol in an inactive conformation. Following activation, which is dependent on phospholipid, DAG, and in some cases calcium, PKC isozymes translocate to the particulate fraction, as determined by cell subfractionation techniques or by morphological analysis. In cells stimulated by growth factors, hormones, phorbol esters, or other PKC agonists, PKC isozymes have been demonstrated to associate with the plasma membrane (Kraft and Anderson, 1983; Ito *et al.*, 1988; Ganesan *et al.*, 1992), cytoskeletal elements (e.g., Mochly-Rosen *et al.*, 1990; Kiley and Jaken, 1990; Spudich *et al.*, 1991; Gregorio *et al.*, 1992, 1994; Goodnight *et al.*, 1995), the nucleus (Hocevar and Fields, 1991; see Malviya and Block, 1993, and Buchner, 1995, for reviews), and various cellular organelles (e.g., endoplasmic reticulum; Goodnight *et al.*, 1995). Individual isozymes have been shown to localize to different subcellular compartments within the same cell (Hocevar and Fields, 1991; Disatnik *et al.*, 1994; Saxon *et al.*, 1994; Goodnight *et al.*, 1995), suggesting that they mediate distinct signaling functions. Although the significance of the differential association of PKC isozymes with different cellular structures is still not well understood, it is likely that translocation targets the activated isozymes to the site of their respective substrates.

Sustained activation of PKC has been determined to be a prerequisite for a variety of long-term cellular responses, including lymphocyte activation (Berry *et al.,* 1990; Berry and Nishizuka, 1990; Szamel *et al.,* 1989, 1990). While T and B lymphocytes have been shown to express all of the members of the PKC family with the exception of PKC γ (Mischak *et al.,* 1991, 1993a,b), increasing evidence supports a link between *sustained* activation of the abundantly expressed β isozyme and lymphocyte activation-related functions *in vivo.* In T cells, for example, a bimodal activation of PKC has been observed in response to stimulation via the antigen receptor, with rapid and short-lived activation of PKC α preceding a sustained activation of PKC β (Szamel *et al.,* 1993). Direct support for the requirement for PKC β in T-lymphocyte activation-related functions comes from the demonstration that introduction of PKC β–specific antibodies into T cells results in inhibition of interleukin-2 (IL-2) synthesis in stimulated cells (Szamel *et al.,* 1993). Moreover, preincubation of T lymphocytes with the immunosuppressive agent cyclosporin A (Szamel *et al.,* 1993) specifically and completely abolishes the sustained activation of PKC β in response to engagement of the T cell receptor, and completely inhibits IL-2 synthesis and cell proliferation. Together, these studies indicate that PKC β in particular appears to play a critical role(s) in lymphocyte activation-related signaling.

Based on the finding that the subcellular distribution of spectrin and ankyrin can be modulated by signals that lead to activation of PKC isozymes (e.g., lymphocyte activation or PMA), combined with the well-documented observation that PKC undergoes rapid translocation to various subcellular compartments upon activation, it was of interest to compare the localization patterns of these molecules in resting and activated lymphocytes. These studies revealed a direct relationship between the organization of the spectrin-based cytoskeleton and the distribution of PKC βII in both T and B cells. Double immunofluorescence analysis demonstrated a striking colocalization of spectrin, ankyrin, and the βII isozyme of PKC in the majority of murine tissue–derived lymphocytes (Gregorio *et al.,* 1992, 1994; Fig. 4d–f). PKC βII was detected in all of the identified spectrin–ankyrin distribution patterns, including plasma membrane–associated rings, membrane-associated caps, a large cytoplasmic aggregate at some distance from the plasma membrane, or small patches scattered throughout the cytoplasm or near the nucleus. In addition, PKC agonists, including PMA and the DAG analogue DiC_8, as well as lymphocyte activation signals, were found to induce a rapid coincident accumulation of these proteins in a focal aggregate in the cytoplasm (Gregorio *et al.,* 1992, 1994; Fig. 4). Both formation and maintenance of the cytoskeletal aggregate were found to require activation of PKC: Redistribution of the molecules was blocked by calphostin C, a PKC inhibitor that interacts with the regulatory domain of the enzyme and inhibits the binding of phorbol esters and DAG analogues (Kobayashi *et al.,* 1989), and incubation with calphostin C resulted in loss of previously formed aggregates (Gregorio *et al.,* 1992, 1994). Examination of the solubility properties of spectrin, ankyrin, and PKC βII in lymphocytes demonstrated that aggregate formation as a result of PKC activating signals is accompanied by a marked increase in the Triton X-100 insolubility of all three proteins (Gregorio *et al.,* 1992,1994). Of particular significance was the finding that lymphocyte spectrin and PKC βII appear to interact physically, either directly or indirectly via another protein(s), as revealed by co-precipitation of the two molecules in reciprocal immunoprecipitation assays (Gregorio *et al.,* 1992). Additional evidence for a link between these two proteins is provided by the demonstration that overproduction of PKC β in a B-cell line containing membrane-associated spectrin resulted in stable expression of spectrin–PKC β aggregates in a large proportion of the cells (K. Lin, J. D. Black, and E. A. Repasky, unpublished results). Moreover, *in vivo* administration in mice of cyclosporin A, an agent that has been shown to block the sustained activation of PKC β in response to ligation of the T-cell receptor (Szamel *et al.,* 1993; see earlier), results in a marked decrease in the percentage of tissue lymphocytes expressing aggregates of spectrin, and prevents aggregate formation in response to *in vivo* administration of PMA (J.-X. Tang, P. Lein, and E. A. Repasky, unpublished results). Together, these data provide strong support for a link between activation of PKC β, formation and maintenance of the cytoskeletal aggregate, and lymphocyte activation-related functions *in vivo.* While an in-depth comparison of the distribution of membrane-skeletal proteins and other PKC isozymes expressed in lymphocytes has yet to be performed, initial studies in our and other laboratories indicate that, in addition to PKC βII, several members of the PKC family can occur as an aggregate or cap in lymphocyte subpopulations (Keenan *et al.,* 1995; C. C. Gregorio, J. D. Black, and E. A. Repasky, unpublished observations). It remains to be determined if these distribution patterns also reflect an association with lymphocyte membrane-skeletal elements.

Evidence for co-localization and/or association of PKC isozymes with cytoskeletal elements has been obtained in several other systems (Wolf and Sahyoun, 1986; Zalewski *et al.,* 1988; Papadopoulos and Hall, 1989; Mochly-Rosen *et al.,* 1990; Jaken *et al.,* 1989; Kiley and Jaken, 1990; Omary *et al.,* 1992; Spudich *et al.,* 1991; Saxon *et al.,* 1994; Disatnik *et al.,* 1994; Goodnight *et al.,* 1995). For example, immunofluorescence analysis of the distribution of PKC α in rat embryo fibroblasts revealed a co-localization of this isozyme with vinculin and talin, cytoskeletal proteins found in focal contacts (Jaken *et al.,* 1989). This organization of PKC α was disrupted by cytochalasin B, indicating an association with the actin-based cytoskeleton.

Mochly-Rosen *et al.* (1990) demonstrated that activation induced the translocation of an unidentified PKC isozyme (probably PKC ε; Disatnik *et al.*, 1994) from the cytosol to myofibrils in cardiac cells. A PKC ε–related kinase has been found to associate with and phosphorylate cytokeratins 8 and 18 (Omary *et al.*, 1992), intermediate filament proteins found in epithelial cells (Moll *et al.*, 1982). In light of our findings in lymphocytes, it is noteworthy that an association of PKC β with cytoskeletal elements has been reported in several systems: Activation-induced association of PKC β with vimentin filaments has been observed in rat basophilic leukemia cells (Spudich *et al.*, 1991) and PKC βII has been shown to associate with cytoskeletal elements in rat brain synaptosomes (Tanaka *et al.*, 1991). Moreover, Goodnight *et al.* (1995) demonstrated that PKC βII overexpressed in NIH 3T3 cells associates with actin filaments following PMA-induced activation. It is also noteworthy that spectrin and PKC βII appear to be linked in other systems. For example, they co-localize in the terminal web of intestinal epithelial cells (Saxon *et al.*, 1994), where they exhibit similar solubility properties, and both molecules can be co-immunoprecipitated from extracts of rat brain synaptosomes in reciprocal immunoprecipitation assays (L. Herceg and J. D. Black, unpublished results). Components of the spectrin-based cytoskeleton and PKC βII may therefore have interrelated functions in a variety of cell types.

B. *Role of the Spectrin Skeleton in Anchorage of Activated PKC βII in the Lymphocyte Cytoplasm*

Elegant studies by Mochly-Rosen and colleagues (Mochly-Rosen *et al.*, 1991a,b; Ron *et al.*, 1994; Ron and Mochly-Rosen, 1995; Mochly-Rosen, 1995) have provided important insight regarding the potential basis for the association of PKC isozymes with cytoskeletal elements. Based on evidence that activated PKC isozymes can be localized on cytoskeletal structures rather than on membranes and that pretreatment of membranes with proteases can abolish phorbol ester–induced, chelator-resistant binding of PKC (Gopalakrishna *et al.*, 1986), these investigators proposed the existence of proteins whose function is to anchor activated PKC at the site of translocation. They have identified a group of these anchoring proteins, collectively termed *r*eceptors for *a*ctivated *C*-*k*inase (RACKs). One RACK protein, RACK1, has been cloned and determined to be a homologue of the β-subunit of G-proteins (Ron *et al.*, 1994). RACKs bind PKC only in the presence of PKC activators; they bind through a site distinct from the substrate binding site and are thus not necessarily substrates of the enzyme. Binding is specific and saturable and appears to stabilize the active conformation of PKC. Evidence indicates that RACKs are critically required for the translocation and subsequent functions of PKC (Ron and Mochly-Rosen, 1995), and thus play an important role in PKC-mediated signal transduction. In addition to RACK proteins and the cytoskeletal proteins listed previously, several other proteins that bind PKC have been described, including annexins and the nuclear protein PICK1 (Mochly-Rosen *et al.*, 1991a,b; Hyatt *et al.*, 1994; Liao *et al.*, 1994; Chapline *et al.*, 1993; Staudinger *et al.*, 1995). It is possible that these proteins also play a role in modulating the subcellular distribution of activated PKC isozymes in different cell types.

The finding that regulation of the subcellular organization of the lymphocyte spectrin-based cytoskeleton is closely linked to that of PKC βII (Gregorio *et al.*, 1992, 1994) suggests that a component(s) of the spectrin-based cytoskeleton may act as a RACK in lymphocytes. Therefore, a member(s) of the spectrin network may serve as an intracellular receptor or chaperone to direct this PKC isozyme to the subcellular site(s) in which its substrates are localized. Lymphocyte spectrin (or an associated molecule) appears to fulfill some of the criteria previously established for RACK proteins (Mochly-Rosen *et al.*, 1991a,b; Ron *et al.*, 1994): (i) the spectrin-based cytoskeleton is present in the Triton-insoluble material of the particulate subcellular fraction; (ii) co-translocation of components of the spectrin-based cytoskeleton and PKC βII involves PKC activation, as demonstrated by the fact that the PKC inhibitor calphostin C blocks formation of the *trans*-Golgi–associated aggregate in response to PMA treatment (Gregorio *et al.*, 1992, 1994); (iii) anti-spectrin antibodies co-precipitate active PKC (Gregorio *et al.*, 1992); and (iv) to date, neither spectrin nor ankyrin has been identified as a PKC substrate in any system, a characteristic of the RACK proteins identified by the Mochly-Rosen group (Ron *et al.*, 1994). Thus, the spectrin-based cytoskeleton may function directly, or indirectly via other associated proteins (e.g., other cytoskeletal elements, a RACK protein, or hsp70), in targeting activated PKC to various subcellular locations. Based on data from our laboratory and from studies in other systems, the following sequence of events can be envisioned. Following exposure of lymphocytes to PKC agonists or to lymphocyte activation signals, activated PKC βII (and/or other PKC isozymes) undergoes a conformational change that increases its binding affinity for its anchoring protein spectrin, an associated component of the spectrin network, or perhaps a previously identified RACK protein (Mochly-Rosen *et al.*, 1991a,b) that can function as an adaptor molecule and associate with cytoskeletal elements. The proteins interact and the assembly becomes anchored as an aggregate in the *trans*-Golgi region or at other subcellular location(s) through a mechanism(s) currently under investigation. Spectrin and/or another member of the protein assembly may help to maintain PKC in an active conformation and may even enhance its activity (Ron *et al.,* 1994). It is tempting to speculate that the *sustained* activation of PKC β required for lymphocyte activation-related signaling is made possible by the formation of such protein assemblies at specific subcellular locations. Active PKC then proceeds to phosphorylate its substrates as part of the activation signaling cascade. Downstream events could include entry into the cell cycle and/or cell differentiation, altered adhesion properties, and synthesis and secretion of cytokines.

The significance of the specific localization of PKC βII in lymphocytes in relation to lymphocyte activation signaling cascades remains to be determined. In many cases, the activation-induced PKC βII–cytoskeleton aggregate has been localized to the *trans*-Golgi region of the lymphocyte cytoplasm. The positioning of PKC in this region as a result of lymphocyte activation signals may reflect a requirement for the phosphorylation of newly synthesized proteins. Alternatively, the association of membrane-skeletal components, PKC, and other proteins with the Golgi region may be coupled to Golgi organization and function (e.g., regulation of cell polarity, cell motility, and intracellular trafficking). It is interesting that spectrin, ankyrin, and PKC βII have been detected in the Golgi region in other cell types: PKC βII has been found to be accumulated near the Golgi complex of pyramidal cells in rat brain hippocampus (Saito *et al.,* 1989), and an erythroid β-spectrin homologue (Beck *et al.,* 1994a) and ankyrin (Beck *et al.,* 1994b) have been found to co-localize with markers of the Golgi complex in a variety of cell types. Thus, the presence of these proteins in the region of the Golgi complex may not be unique to lymphocytes and may reflect a common function for these molecules in some nonerythroid cells. It should be noted, however, that ultrastructural analysis of the localization of spectrin in lymphoid organs revealed that the cytoskeletal aggregate can also be found in close association with the nuclear membrane and at various other cytoplasmic locations (Black *et al.,* 1988). The significance of the variability in aggregate localization in the lymphocyte cytoplasm, as well as of other patterns of lymphocyte membrane-skeletal organization, is an important question that remains to be addressed in future studies.

Another critical question currently under investigation concerns the precise molecular details of the interaction of PKC βII with the spectrin-based cytoskeleton. Whether activated PKC βII binds spectrin (or an associated molecule) directly in a saturable and specific manner remains to be determined. It is intriguing in this regard that several components of the spectrin network have been shown to contain motifs implicated in mediating protein–protein interactions. For example, ankyrin contains 33-residue motifs, known as ankyrin repeats, that appear to be involved in molecular recognition (see Bennett, 1992, for review), and spectrin contains a *src* homology region 3 (SH3) domain, a motif thought to bind to proline-rich regions in several proteins (Tu *et al.,* 1994). Of particular relevance is the presence in β-spectrin of a pleckstrin homology (PH) domain (Haslam *et al.,* 1993; Hu *et al.,* 1992; Macias *et al.,* 1994), a 100-residue protein module originally detected as an internal repeat in pleckstrin, the major substrate of PKC in platelets. The PH domain seems to be particularly abundant in proteins that participate in signal transduction pathways and is thought to be involved in molecular recognition in a manner similar to SH2 and SH3 domains. Based on research into the identification of ligand(s) for the PH domain (Yao *et al.,* 1994; Musacchio *et al.,* 1993; Wang *et al.,* 1994), it is tempting to propose the following two possible mechanisms for the interaction of the spectrin-based cytoskeleton with activated PKC βII in response to lymphocyte activation-related signals (see Fig. 5):

1. Spectrin might interact directly with PKC βII via its PH domain. Support for this mechanism comes from the finding that the PH domain of Bruton tyrosine kinase interacts directly with PKC βI in mast cells (Yao *et al.,* 1994).
2. Spectrin might associate via its PH domain with RACK1 or a RACK1-like protein that, in turn, binds to activated PKC βII. Evidence for this mechanism comes from the demonstration that the PH domain binds to WD40 repeats within the β-subunit of trimeric G-proteins (Musacchio *et al.,* 1993; Wang *et al.,* 1994) and that RACK1 itself is a homologue of the G-protein β-subunit and contains seven WD40 repeat elements (Ron *et al.,* 1994).

It is noteworthy in regard to the potential involvement of RACK1 in the interaction of spectrin with activated PKC βII that (i) the protein is present in lymphocytes and (ii) the levels of cytoskeleton-associated RACK1 are markedly higher in cells expressing PKC–cytoskeletal aggregates (S.-Y. Park, E. A. Repasky, and J. D. Black, unpublished results). Studies in our laboratory are currently examining the proposed mechanisms at the molecular level.

In summary, accumulated evidence supports the intriguing notion that the spectrin-based cytoskeleton plays an important role in PKC βII localization and function in lymphocytes. The heterogeneity in spectrin–ankyrin subcellular distribution in lymphocytes and their dynamic properties in this cell type might reflect, at least in part, their involvement in regulation of PKC movement and activity. In accordance with their structural role in erythrocytes and with their reported role in subcellular targeting events in other cell types, spectrin and ankyrin are ideally suited to provide the scaffolding necessary for the appropriate subcellular placement of signal-transducing molecules. The presence of molecular recognition motifs in these cytoskeletal proteins provides a potential mechanism for formation

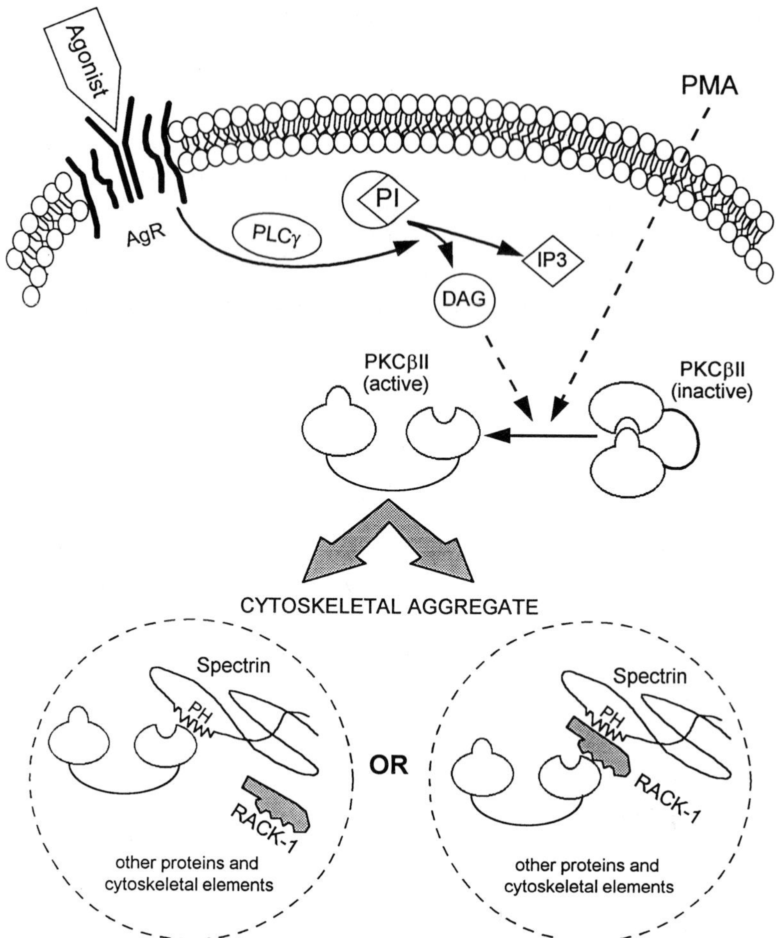

FIGURE 5 Model of the potential mechanisms underlying the interaction of the spectrin-based cytoskeleton with activated PKC βII in response to activation-related signals. Following exposure of lymphocytes to agonist-induced activation signals or to the phorbol ester PMA, PKC βII becomes activated and undergoes a conformational change that allows it to bind (a) directly to the PH domain of spectrin or, alternatively, (b) to a RACK-like protein that interacts with the PH domain of spectrin via its WD40 repeats. The assembly forms an aggregate that contains other cytoskeletal proteins (e.g., ankyrin) and hsp70 and becomes anchored at particular locations in the lymphocyte cytoplasm (see text for further details).

of protein assemblies required to perform specific functions at restricted subcellular sites. Thus, the lymphocyte spectrin-based cytoskeleton may function to position PKC βII (and other PKC isozymes) for phosphorylation of substrates involved in T- and B-cell activation and other lymphocyte functions.

Acknowledgments

The authors thank Dr. Sharon Evans, Dr. Carol Gregorio, and Mr. Shin-Young Park for their helpful comments on this manuscript, and Jeanne Prendergast, Margaret Frey, Cheryl Zuber, and Sue Sabadasz for their help in preparing the text and figures. We would also like to thank the individuals whose ideas and research contributed to the work described here: Drs. John K. Lee, Carol Gregorio, Keng-Mean Lin, Patricia Masso-Welch, and Yuan-Pu Di and Mr. Shin-Young Park. This work was supported by National Institutes of Health grants AI30131, AI26612, and CA16056, NSF grant DCB 8917424, and ACS grant CH-421A.

References

Beck, K. A., Buchanan, J. A., Malhotra, U., and Nelson, W. J. (1994a). Golgi spectrin: Identification of an erythroid beta-spectrin homolog associated with the Golgi complex. *J. Cell Biol.* **127,** 707–723.

Beck, K. A., Buchanan, J. A., and Nelson, W. J. (1994b). Golgi ankyrin: Association of an ankyrin isoform with the Golgi complex. *Mol. Biol. Cell* **5,** 240a.

Bennett, V. (1989). The spectrin-actin junction of erythrocyte membrane skeletons. *Biochim. Biophys. Acta* **988,** 107–121.

Bennett, V. (1992). Ankyrins: Adaptors between diverse plasma membrane proteins and the cytoplasm. *J. Biol. Chem.* **267,** 8703–8706.

Bennett, V., and Gilligan, D. M. (1993). The spectrin-based membrane structure and micron-scale organization of the plasma membrane. *Annu. Rev. Cell Biol.* **9,** 27–66.

Bennett, V., Davis, J., and Fowler, W. E. (1982). Brain spectrin, a membrane-associated protein related in structure and function to erythrocyte spectrin. *Nature* (*London*) **299,** 126–131.

Berry N., and Nishizuka, Y. (1990). Protein kinase C and T cell activation. *Eur. J. Biochem.* **189,** 205–214.

Berry, N., Ase, K., Kishimoto, K., and Nishizuka, Y. (1990). Activation of resting human T cells requires prolonged stimulation of protein kinase C. *Proc. Natl. Acad. Sci. U.S.A.* **87,** 2294–2298.

Black, J. D., Koury, S. T., Bankert, R. B., and Repasky, E. A. (1988). Heterogeneity in lymphocyte spectrin distribution: Ultrastructural identification of a new spectrin-rich cytoplasmic structure. *J. Cell Biol.* **106,** 97–109.

Bodine, D. M., Birkenmeir, C. S., and Barker, J. E. (1984). Spectrin deficient inherited hemolytic anemias in the mouse: Characterization by spectrin synthesis and mRNA activity in reticulocytes. *Cell* **37,** 721–729.

Bourguignon, L. Y. W., and Bourguignon, G. J. (1984). Capping and the cytoskeleton. *Int. Rev. Cytol.* **87,** 195–224.

Bourguignon, L. Y., and Jin, H. (1995). Identification of the ankyrin-binding domain of the mouse T-lymphoma cell inositol 1, 4, 5-trisphosphate (IP_3) receptor and its role in the regulation of IP_3-mediated internal Ca^{2+} release. *J. Biol. Chem.* **270,** 7257–7260.

Bourguignon, L. Y., Walker, G., and Huang, H. S. (1990). Interactions between a lymphoma membrane-associated guanosine 5′-triphosphate-binding protein and the cytoskeleton during receptor patching and capping. *J. Immunol.* **144,** 2242–2252.

Bourguignon, L. Y., Lokeshwar, V. B., He, J., Chen, X., and Bourguignon, G. J. (1992). A CD44-like endothelial cell transmembrane glycoprotein (GP116) interacts with extracellular matrix and ankyrin. *Mol. Cell. Biol.* **12,** 4464–4471.

Branton, D., Cohen, C. M., and Tyler, J. (1981). Interaction of cytoskeletal proteins on the human erythrocyte membrane. *Cell* **24,** 24–32.

Braun, J., and Unanue, E. R. (1983). The lymphocyte cytoskeleton and its control of surface receptor functions. *Semin. Hematol.* **20,** 322–333.

Braun, J., Fujiwara, K., Pollard, T. D., and Unanue, E. R. (1978). Two mechanisms for the redistribution of lymphocyte surface macromolecules. I. Relationship to cytoplasmic myosin. *J. Cell Biol.* **79,** 408–416.

Buchner, K. (1995). Protein kinase C in the transduction of signals toward and within the cell nucleus. *Eur. J. Biochem.* **228,** 211–221.

Burridge, D., Kelly, T., and Mangeat, P. (1982). Nonerythrocyte spectrins: Actin-membrane attachment proteins occurring in many cell types. *J. Cell Biol.* **95,** 478–486.

Cambier, J. C. (1992). Signal transduction by T- and B-cell antigen receptors: Converging structures and concepts. *Curr. Opin. Immunol.* **4,** 257–264.

Cambier, J. C., Pleiman, C. M., and Clark, M. R. (1994). Signal transduction by the B cell antigen receptor and its coreceptors. *Annu. Rev. Immunol.* **12,** 457–486.

Chapline, C., Ramsay, K., Klauck, T., and Jaken, S. (1993). Interaction cloning of protein kinase C substrates. *J. Biol. Chem.* **268,** 6858–6861.

Connor, J., and Schroit, A. J. (1990). Aminophospholipid translocation in erythrocytes: Evidence for the involvement of a specific transporter and an endofacial protein. *Biochemistry* **29,** 37–43.

Davies, K. A., and Lux, S. E. (1989). Hereditary disorders of the red cell membrane skeleton. *Trends Genet.* **5,** 222–227.

DeFranco, A. L. (1994). B lymphocyte activation. *In* "Fundamental Immunology," 3rd ed (W. Paul, ed.), pp. 267–504. Raven Press, New York.

Dekker, L. V., and Parker, P. J. (1994). Protein kinase C—a question of specificity. *Trends Biochem. Sci.* **19,** 73–77.

Del Buono, B. J., Williamson, P. J., and Schlegel, R. A. (1988). Relation between the organization of spectrin and of membrane lipids in lymphocytes. *J. Cell Biol.* **106,** 697–703.

de Petris, S., and Raff, M. C. (1973). Normal distribution, patching and capping of lymphocyte surface immunoglobulin studied by electron microscopy. *Nature New Biol.* **241,** 257–259.

Di, Y.-P., Repasky, E. A., Laszlo, A., Calderwood, S., and Subjeck, J. (1995). HSP70 translocates into a cytoplasmic aggregate during lymphocyte activation. *J. Cell. Physiol.* ***165,*** 228–238.

Disatnik, M.-H., Buraggi, G., and Mochly-Rosen, D. (1994). Localization of protein kinase C isozymes in cardiac myocytes. *Exp. Cell Res.* **210,** 287–297.

Erikson, J., Radic, M. Z., Camper, S. A., Hardy, R. R., Carmack, C., and Wiegert, M. (1991). Expression of anti-DNA immunoglobulin transgenes in non-autoimmune mice. *Nature (London)* **349,** 331–335.

Evans, S. S., Wang, W.-C., Gregorio, C. C., Han, T., and Repasky, E. A. (1993). Interferon-α alters spectrin organization in normal and leukemic human B lymphocytes. *Blood* **81,** 759–766.

Flanagan, J., and Koch, G. L. E. (1978). Cross-linked surface Ig attaches to actin. *Nature (London)* **273,** 278–281.

Gabbiani, G., Chaponnier, C., Zumbe, A., and Vassalli, P. (1977). Actin and tubulin co-cap with surface immunoglobulin in mouse B lymphocytes. *Nature (London)* **269,** 696–698.

Ganesan, S., Calle, R., Zawalich, K., Greenawalt, K., Zawalich, W., Shulman, G. I., and Rasmussen, H. (1992). Immunocytochemical localization of α-protein kinase C in rat pancreatic β-cells during glucose-induced insulin secretion. *J. Cell Biol.* **119,** 313–324.

Glenney, J. R. Jr., Glenney, P., Osborn, M., and Weber, K. (1982). An F-actin and calmodulin-binding protein from isolated intestinal brush borders has a morphology related to spectrin. *Cell* **28,** 843–854.

Goodman, S. R., Zagon, I. S., and Kulikowski, R. R. (1981). Identification of a spectrin like protein in non-erythroid cells. *Proc. Natl. Acad. Sci. U.S.A.* **78,** 7570–7574.

Goodnight, J., Mischak, H., Kolch, W., and Mushinski, J. F. (1995). Immunocytochemical localization of eight protein kinase C isozymes overexpressed in NIH 3T3 fibroblasts. *J. Biol. Chem.* **270,** 9991–10001.

Gopalakrishna, R., Barsky, S. H., Thomas, T. P., and Anderson, W. B. (1986). Factors influencing cholator-stable, detergent-extractable, phorbol diester-induced membrane association of protein kinase C. Differences between Ca^{++} induced and phorbol ester-stabilized membrane binding(s) of protein kinase C. *J. Biol. Chem.* **261,** 16438–16445.

Greenquist, A. C., Shohet, S. B., and Bernstein, S. E. (1978). Marked reduction of spectrin in hereditary spherocytosis in the common house mouse. *Blood* **51,** 1149–1155.

Gregorio, C. C., Kubo, R. T., Bankert, R. B., and Repasky, E. A. (1992). Translocation of spectrin and protein kinase C to a cytoplasmic aggregate upon lymphocyte activation. *Proc. Natl. Acad. Sci. U.S.A.* **89,** 4947–4951.

Gregorio, C. C., Black, J. D., and Repasky, E. A. (1993). Dynamic aspects of cytoskeletal protein distribution in T lymphocytes: Involvement of calcium in spectrin reorganization. *Blood Cells* **19,** 361–373.

Gregorio, C. C., Repasky, E. A., and Black, J. D. (1994). Dynamic properties of ankyrin in T lymphocytes: Colocalization with spectrin and protein kinase Cβ. *J. Cell Biol.* **125,** 345–358.

Gupta, S. K., and Woda, B. A. (1988). Ligand-induced association of surface immunoglobulin with the detergent-insoluble cytoskeleton may involve alpha-actinin. *J. Immunol.* **140,** 176–182.

Haslam, R. J., Koide, H. B., and Hemmings, B. A. (1993). Pleckstrin domain homology. *Nature* (*London*) **363,** 309–310.

Haest, C. W. M. (1982). Interactions between membrane skeleton proteins and the intrinsic domain of the erythrocyte membrane. *Biochim. Biophys. Acta* **694,** 331–352.

Haest, C. W. M., and Deuticke, B. (1976). Possible relationship between membrane proteins and phospholipid asymmetry in the human erythrocyte membrane. *Biochim. Biophys. Acta* **436,** 353–365.

Haest, C. W. M., Plasa, G., Kamp, D., and Deuticke, B. (1978). Spectrin as a stabilizer of phospholipid asymmetry in the human erythrocyte membrane. *Biochim. Biophys. Acta* **509,** 21–32.

Hocevar, B. A., and Fields, A. P. (1991). Selective translocation of βII-protein kinase C to the nucleus of human promyelocytic (HL60) leukemia cells. *J. Biol. Chem.* **266,** 28–33.

Hu, R. J., Watanabe, M., and Bennett, V. (1992). Characterization of human brain cDNA encoding the general isoform of beta spectrin. *J. Biol. Chem.* **267,** 18715–18722.

Hyatt, S. L., Liao, L., Aderem, A., Nairn, A. C., and Jaken, S. (1994). Correlation between protein kinase C binding proteins and substrates in REF52 cells. *Cell Growth Differ.* **5,** 495-502.

Ito, T., Tanaka, T., Yoshida, T., Onoda, K., Ohta, H., Hagiwara, M., Itoh, Y., Ogura, M., Saito, H., and Hidaka, H. (1988). Immunocytochemical evidence for translocation of protein kinase C in human megakaryoblastic leukemic cells: Synergistic effects of Ca^{2+} and activators of protein kinase C on the plasma membrane association. *J. Cell Biol.* **107,** 929–937.

Jaken, S., Leach, K., and Klauck, T. (1989). Association of type 3 protein kinase C with focal contacts in rat embryo fibroblasts. *J. Cell Biol.* **109,** 697–704.

Kalomiris, E. L., and Bourguignon, L. Y. (1988). Mouse T lymphoma cells contain a transmembrane glycoprotein (GP85) that binds ankyrin. *J. Cell Biol.* **106,** 319–327.

Kalomiris, E. L., and Bourguignon, L. Y. (1989). Lymphoma protein kinase C is associated with the transmembrane glycoprotein, GP85, and may function in GP85-ankyrin binding. *J. Biol. Chem.* **264,** 8113–8119.

Keenan, C., Kelleher, D., and Long, A. (1995). Regulation of non-classical protein kinase C isoenzymes in a human T cell line. *Eur. J. Immunol.* **25,** 13–17.

Kiley, S. C., and Jaken, S. (1990). Activation of α-protein kinase C leads to association with detergent-insoluble components of GH_4C_1 cells. *Mol. Endocrinol.* **4,** 59–68.

Kobayashi, E., Nakano, H., Morimoto, M., and Tamaoki, T. (1989). Calphostin C (UCN-1028C), a novel microbial compound, is a highly potent and specific inhibitor of protein kinase C. *Biochem. Biophys. Res. Commun.* **159,** 548–553.

Kraft, A., and Anderson, W. B. (1983). Phorbol esters increase the amount of Ca^{2+} phospholipid-dependent kinase associated with the plasma membrane. *Nature* (*London*) **301,** 621–623.

Kubo, R. T., Born, W., Kappler, J. W., Marrack, P., and Pigeon, M. (1989). Characterization of a monoclonal antibody which detects all murine $\alpha\beta$ T cell receptors. *J. Immunol.* **142,** 2736–2742.

Kupfer, A., and Singer, S. J. (1989). The specific interaction of helper T cells and antigen presenting cells. IV. Membrane and cytoskeletal reorganizations in the bound T cell as a function of antigen dose. *J. Exp. Med.* **170,** 1697–1707.

Kupfer, H., Monks, C. R. F., and Kupfer, A. (1994). Small splenic B cells that bind to antigen-specific T helper cells and face the site of cytokine production in the Th cells selectively proliferate: Immunofluorescence microscopic studies of Th-B antigen-presenting cell interactions. *J. Exp. Med.* **179,** 1507–1515.

Langner, M., Repasky, E. A., and Hui, S.-W. (1992). Relationship between membrane lipid mobility and spectrin distribution in lymphocytes. *FEBS Lett.* **305,** 197–202.

Lee, J. K., and Repasky, E. A. (1987). Cytoskeletal polarity in mammalian lymphocytes *in situ*. *Cell Tissue Res.* **247,** 195–202.

Lee, J. K., Black, J. D., Repasky, E. A., Kubo, R. T., and Bankert, R. B. (1988). Activation induces a rapid reorganization of spectrin in lymphocytes. *Cell* **55,** 807–816.

Levine, J., and Willard, M. (1983). Redistribution of fodrin (a component of the cortical cytoplasm) accompanying capping of cell surface molecules. *Proc. Natl. Acad. Sci. U.S.A* **80,** 191–195.

Liao, L., Hyatt, S. L., Chapline, C., and Jaken, S. (1994). Protein kinase C domains involved in interaction with other proteins. *Biochemistry* **33,** 1229–1233.

Lin, H., and Spradling, A. C. (1995). Fusome asymmetry and oocyte determination in *Drosophila*. *Dev. Genet.* **16,** 6–12.

Lin, H., Yue, L., and Spradling, A. C. (1994). The *Drosophila* fusome, a germline-specific organelle, contains skeletal proteins and functions in cyst formation. *Development* **120,** 947–956.

Lin, K., Black, J. D., and Repasky, E. A. (1996). Subcellular reorganization of protein kinase Cβ and the spectrin-based cytoskeleton during B lymphocyte activation. Submitted.

Lokeshwar, V. B., and Bourguignon, L. Y. (1992a). The lymphoma transmembrane glycoprotein GP85 (CD44) is a novel guanine nucleotide-binding protein which regulates GP85 (CD44)-ankyrin interaction. *J. Biol. Chem.* **267,** 22073–22078.

Lokeshwar, V. B., and Bourguignon, L. Y. (1992b). Tyrosine phosphatase activity of lymphoma CD45 (GP180) is regulated by direct interaction with the cytoskeleton. *J. Biol. Chem.* **267,** 21551–21557.

Macias, M. J., Musacchio, A., Ponstingl, H., Nilges, M., Saraste, M., and Oschkinat, H. (1994). Structure of pleckstrin homology domain from β-spectrin. *Nature* (*London*) **369,** 675–679.

Malviya, A. N., and Block, C. (1993). Nuclear protein kinase C and signal transduction. *Receptor* **3,** 257–275.

Masso-Welch, P., Black, J., Erikson, J., and Repasky, E. A. (1996). B cells from unimmunized mice possess polar domains of immunoglobulin, spectrin and the βII isoform of protein kinase C. Submitted.

Mischak, H., Kolch, W., Goodnight, J., Davidson, W., Rapp, U., Rose-John, S., and Mushinski, J. F. (1991). Expression of protein kinase C genes in hemopoietic cells is cell-type and B cell-differentiation stage specific. *J. Immunol.* **147,** 3981–3987.

Mischak, H., Goodnight, J., Henderson, D. W., Osada, S.-I., Ohno, S., and Mushinski, J. F. (1993a). Unique expression pattern of protein kinase C-θ: High mRNA levels in normal mouse testes and in T-lymphocytic cells and neoplasms. *FEBS Lett.* **326,** 51–55.

Mischak, H., Pierce, J. H., Goodnight, J., Kazanietz, M. G., Blumberg, P. M., and Mushinski, J. F. (1993b). Phorbol ester-induced myeloid differentiation is mediated by protein kinase C-α and -δ and not by protein kinase C-βII, -ε, -ζ, and -η. *J. Biol. Chem.* **268,** 20110–20115.

Mochly-Rosen, D. (1995). Localization of protein kinases by anchoring proteins: A theme in signal transduction. *Science* **368,** 247–251.

Mochly-Rosen, D., Henrich, C. J., Cheever, L., Khaner, H., and Simpson, P. C. (1990). A protein kinase C isozyme is translocated to cytoskeletal elements on activation. *Cell Regul.* **1,** 693–706.

Mochly-Rosen, D., Khanner, H., and Lopez, J. (1991a). Identification of intracellular receptor proteins for activated protein kinase C. *Proc. Natl. Acad. Sci. U.S.A.* **88,** 3997–4000.

Mochly-Rosen, D., Khanner, H., Lopez, J., and Smith, B. L. (1991b). Intracellular receptors for activated protein kinase C. *J. Biol. Chem.* **266,** 14866–14868.

Moll, R., Franke, W. W., and Schiller, D. L. (1982). The catalog of human cytokeratins: Patterns of expression in normal epithelia, tumors and cultured cells. *Cell* **31,** 11–24.

Morrot, G., Cribier, S., Devaux, P. F., Geldwerth, D., Davoust, J., Bureau, J. F., Fellmann, P., Herve, P., and Frilley, B. (1986). Asymmetric lateral mobility of phospholipids in the human erythrocyte membrane. *Proc. Natl. Acad. Sci. U.S.A.* **83,** 6863–6867.

Musacchio, A., Gibson, T., Rice, P., Thompson, J., and Saraste, M. (1993). The PH domain: A common piece of the structural patchwork of signalling proteins. *Trends Biochem. Sci.* **18,** 343–348.

Nelson, W. J., and Hammerton, R. W. (1989). A membrane-cytoskeletal complex containing Na^+,K^+-ATPase, ankyrin, and fodrin in Madin-Darby canine kidney (MDCK) cells: Implications for the biogenesis of epithelial cell polarity. *J. Cell Biol.* **108,** 893–902.

Nelson, W. J., Colaco, C. A. L. S., and Lazarides, E. (1983). Involvement of spectrin in cell-surface receptor capping in lymphocytes. *Proc. Natl. Acad. Sci. U.S.A.* **80,** 1626–1629.

Nishizuka, Y. (1992). Intracellular signaling by hydrolysis of phospholipids and activation of protein kinase C. *Science* **258,** 607–614.

Omary, M. B., Baxter, G. T., Chou, C.-F., Riopel, C. L., Lin, W. Y., and Strulovici, B. (1992). PKCε-related kinase associates with and phosphorylates cytokeratin 8 and 18. *J. Cell Biol.* **117,** 583–593.

Op den Kamp, J. A. F. (1979). Lipid asymmetry in membranes. *Annu. Rev. Biochem.* **48,** 47–71.

Palfrey, H. C., Schiebler, W., and Greengard, P. (1982). A major calmodulin-binding protein common to various vertebrate tissues. *Proc. Natl. Acad. Sci. U.S.A.* **79,** 3780–3784.

Papadopoulos, V., and Hall, P. F. (1989). Isolation and characterization of protein kinase C from Y-1 adrenal cell cytoskeleton. *J. Cell Biol.* **108,** 553–567.

Pauly, J. L., Bankert, R. B., and Repasky, E. A., (1986). Immunofluorescent patterns of spectrin in lymphocyte cell lines. *J. Immunol.* **136,** 246–253.

Raff, M. C. (1976). Cell-surface immunology. *Sci. Am.* **234,** 30–39.

Repasky, E. A., Granger, B. L., and Lazarides, E. (1982). Widespread occurrence of avian spectrin in non-erythroid cells. *Cell* **29,** 821–833.
Repasky, E. A., Symer, D. E., and Bankert, R. B. (1984). Spectrin immunofluorescence distinguishes a population of naturally capped lymphocytes *in situ*. *J. Cell Biol.* **99,** 350–355.
Ron, D., and Mochly-Rosen, D. (1995). An autoregulatory region in protein kinase C: The pseudoanchoring site. *Proc. Natl. Acad. Sci. U.S.A.* **92,** 492–496.
Ron, D., Chen, C.-H., Caldwell, J., Jamieson, L., Orr, E., and Mochly-Rosen, D. (1994). Cloning of an intracellular receptor for protein kinase C: A homolog of the β subunit of G proteins. *Proc. Natl. Acad. Sci. U.S.A.* **91,** 839–843.
Saito, N., Kose, A., Ito, A., Hosoda, K., Mori, M., Hirata, M., Ogita, K., Kikkawa, U., Ono, Y., Igarashi, K., Nishizuka, Y., and Tanaka, C. (1989). Immunocytochemical localization of βII subspecies of protein kinase C in rat brain. *Proc. Natl. Acad. Sci. U.S.A.* **86,** 3409–3413.
Saxon, M. L., Zhao, X., and Black, J. D. (1994). Activation of protein kinase C isozymes is associated with post-mitotic events in intestinal epithelial cells in situ. *J. Cell Biol.* **126,** 747–763.
Scher, M. G., and Bloch, R. J. (1993). Phospholipid asymmetry in acetylcholine receptor clusters. *Exp. Cell Res.* **208,** 485–491.
Schreiner, G. F., and Unanue, E. R. (1977). Capping and the lymphocyte: Models for membrane reorganization. *J. Immunol.* **119,** 1549–1551.
Seigneuret, M., and Devaux, P. F. (1984). ATP-dependent asymmetric distribution of spin-labelled phospholipids in the erythrocyte membrane: Relation to shape change. *Proc. Natl. Acad. Sci. U.S.A.* **81,** 3751–3755.
Spudich, A., Meyer, T., and Stryer, L. (1991). Association of the β isoform of protein kinase C with vimentin filaments. *Cell Motil. Cytoskeleton* **22,** 250–256.
Stephen, F. D., Yokota, S. J., and Repasky, E. A. (1990). The effect of free fatty acids on spectrin organization in lymphocytes. *Cell Biophys.* **17,** 269–282.
Staudinger, J., Zhou, J., Burgess, R., Elledge, S. J., and Olson, E. N. (1995). PICK1: A perinuclear binding protein and substrate for protein kinase C isolated by the yeast two-hybrid system. *J. Cell Biol.* **128,** 263–271.
Szamel, M., and Resch, K. (1995). T-cell antigen receptor-induced signal-transduction pathways: Activation and function of protein kinase C in T lymphocytes. *Eur. J. Biochem.* **228,** 1–15.
Szamel, M., Rehermann, B., Krebs, B., Kurrle, R., and Resch, K. (1989). Activation signals in human lymphocytes. Incorporation of polyunsaturated fatty acids into plasma membrane phospholipids regulates IL-2 synthesis via sustained activation of protein kinase C. *J. Immunol.* **143,** 2806–2813.
Szamel, M., Kracht, M., Krebs, B., Hubner, U., and Resch, K. (1990). Activation signals in human lymphocytes: Interleukin 2 synthesis and expression of high affinity interleukin 2 receptors require differential signalling for the activation of protein kinase C. *Cell Immunol.* **126,** 117–128.
Szamel, M., Bartels, F., and Resch, K. (1993). Cyclosporin A inhibits T cell receptor-induced interleukin-2 synthesis of human T lymphocytes by selectively preventing a transmembrane signal transduction pathway leading to sustained activation of a protein kinase C isoenzyme, protein kinase C-beta. *Eur. J. Immunol.* **23,** 3072–3081.
Tanaka, K. I., and Ohnishi, S. I. (1976). Heterogeneity in the fluidity of intact erythrocyte membranes and its homogenization upon hemolysis. *Biochim. Biophys. Acta* **426,** 218–231.
Tanaka, S., Tominaga, M., Yasuda, I., Kishimoto, A., and Nishizuka, Y. (1991). Protein kinase C in rat brain synaptosomes. *FEBS Lett.* **294,** 267–270.
Taylor, R. B., Duffus, P. H., Raff, M. C., and dePetris, S. (1971). Redistribution and pinocytosis of lymphocyte surface immunoglobulin molecules induced by anti-immunoglobulin antibodies. *Nature New Biol.* **233,** 225–229.

Tu, H., Chen, J. K., Feng, S., Dalgarno, D. C., Brauer, A. W., and Schreiber, S. L. (1994). Structural basis for the binding of proline-rich peptides to SH3 domains. *Cell* **76,** 933–945.

Venkatakrishnan, G., McKinnon, C. A., Pilapil, C. G., Wolf, D. E., and Ross, A. H. (1991). Nerve growth factor receptors are preaggregated and immobile on responsive cells. *Biochemistry* **30,** 2749–2756.

Von Boehmer, H. (1990). Developmental biology of T cells in T cell-receptor transgenic mice. *Annu. Rev. Immunol.* **8,** 531–556.

Wang, D.-S., Shaw, R., Winkelmann, J. C., and Shaw, G. (1994). Binding of PH domains of β-adrenergic receptor kinase and β-spectrin to WD40/β-transducin repeat containing regions of the β-subunit of trimeric G-proteins. *Biochem. Biophys. Res. Commun.* **203,** 29–35.

Weiss, A. (1994). T lymphocyte activation. *In* "Fundamental Immunology," 3rd ed. (W. Paul, ed.), pp. 467–504. Raven Press, New York.

Welch, W. J. (1992). Mammalian stress response: Cell physiology, structure/function of stress proteins, and implications for medicine and disease. *Physiol. Rev.* **72,** 1063–1081.

Welch, W. J., and Feramisco, J. R. (1984). Nuclear and nucleolar localization of the 72,000-dalton heat shock protein in heat-shocked mammalian cells. *J. Biol. Chem.* **259,** 4501–4513.

Williamson, P., and Schlegel, R. A. (1994). Back and forth: The regulation and function of transbilayer phospholipid movement in eukaryotic cells [Review]. *Mol. Membrane Biol.* **11,** 199–216.

Williamson, P., Bateman, J., Kozarsky, K., Mattocks, K., Hermanowicz, N., Choe, H. R., and Schlegel, R. A. (1982). Involvement of spectrin in the maintenance of phase-state asymmetry in the erythrocyte membrane. *Cell* **30,** 725–733.

Wolf, M., and Sahyoun, N. (1986). Protein kinase C and phosphatidylserine bind to M_r 110,000/115,000 polypeptides enriched in cytoskeletal and post-synaptic density preparations. *J. Biol. Chem.* **261,** 13327–13332.

Yahara, I., and Kakimoto-Sameshima, F. (1978). Microtubule organization of lymphocytes and its modulation by patch and cap formation. *Cell* **15,** 251–259.

Yao, L., Kawakami, Y., and Kawakami, T. (1994). The pleckstrin homology domain of Bruton tyrosine kinase interacts with protein kinase C. *Proc. Natl. Acad. Sci. U.S.A.* **91,** 9175–9179.

Zalewski, P. D., Forbes, I. J., Valente, L., Apostolou, S., and Hurst, N. P. (1988). Translocation of protein kinase C to a Triton-insoluble sub-cellular compartment induced by the lipophilic gold compound auranofin. *Biochem. Pharmacol.* **37,** 1415–1417.

CHAPTER 16

Regulation of Epithelial Ion Channel Activity by the Membrane–Cytoskeleton

Peter R. Smith* and Dale J. Benos†

*Department of Physiology, Medical College of Pennsylvania and Hahnemann University, Philadelphia, Pennsylvania, 19129 and †Department of Physiology and Biophysics, University of Alabama at Birmingham, Birmingham, Alabama, 35294

I. INTRODUCTION

The function of epithelial cells as selective permeability barriers between the transcellular and interstitial fluid compartments depends upon the establishment and maintenance of the polarized distribution of membrane transport proteins and ion channels to specific membrane domains. For example, Na^+ reabsorption by the kidney requires the localization of specific transporters, such as the Na^+–H^+ exchanger, Na^+-dependent glucose, amino acid and phosphate co-transporters, and Na^+ channels, to the apical membrane domain and Na^+,K^+-ATPase to the basolateral membrane domain. If these proteins are not segregated into different membrane domains, as can result from neoplastic growth or can occur in renal diseases such as ischemia or adult polycystic kidney disease, renal Na^+ transport would be significantly impaired, and such a compromise in polarity may ultimately lead to renal failure (see Leiser and Molitoris, 1993, for review). Thus establishment and maintenance of transporters and ion channels to specific membrane domains is vital for the normal function of epithelia.

The mechanisms whereby transporters and ion channels are retained within specific membrane domains have been the focus of current research interest. Unique isoforms of spectrin and its associated proteins are differentially distributed to specific membrane domains within cells (Bennett and Gilligan, 1993). These observations, together with the seminal works of Nelson and Veshnock (1987) and Morrow and colleagues (1989) demonstrating that the distribution of Na^+,K^+-ATPase within the basolateral membrane domain of renal epithelial cells is maintained through an interaction with the spectrin-based cytoskeleton, have stimulated researchers to search for interactions between membrane transport proteins and the membrane–cytoskeleton, not only in epithelial cells but also in neurons, muscle cells, and lymphocytes (see Hitt and Luna, 1994, and Mays *et al.,* 1994, for review). These studies have revealed that not only is the spectrin-based membrane–cytoskeleton important in restricting the distribution of membrane transporters and ion channels to specific membrane domains, but it also serves to regulate the activity of these proteins.

In this chapter we present an overview of our current understanding of the interactions between the spectrin-based membrane–cytoskeleton and ion channels in transporting epithelial cells and the role they play in regulating channel activity. Interactions of the membrane–cytoskeleton with epithelial Na^+ channels have received more attention than its interactions with other epithelial ion channels, and these interactions are the major focus of our discussion. We first summarize our current understanding of the biochemical and molecular characteristics of epithelial Na^+ channels and their interactions with the membrane–cytoskeleton. We then turn to a discussion of the role of the membrane–cytoskeleton in the regulation of epithelial Na^+ channels. We next discuss what is known about interactions between the membrane–cytoskeleton and epithelial Cl^- and K^+ channels and the involvement of the cytoskeleton in the regulation of these ion channels. We conclude with perspectives for further research in this emerging field.

II. STRUCTURE OF EPITHELIAL Na^+ CHANNELS AND THEIR ASSOCIATION WITH THE MEMBRANE–CYTOSKELETON

Na^+ reabsorbing epithelia, such as the distal tubules and collecting ducts of the kidney, distal colon, lungs, trachea, and sweat ducts, contain Na^+ channels situated within their apical membranes that are inhibited by the diuretic amiloride and its analogues (Benos *et al.,* 1992, 1995). Epithelial Na^+

channels are rate limiting for net Na^+ reabsorption because they mediate the entry of Na^+ from the luminal fluid into the cells during the first stage of electrogenic transepithelial Na^+ transport (Benos *et al.,*, 1992,1995). Although these channels are commonly referred to as epithelial Na^+ channels, studies have demonstrated that they are expressed in nonepithelial cells such as lymphocytes (Bubien and Warnock, 1993), endothelial cells (Vigne *et al.,* 1989), vascular smooth muscle cells (Van Renterghem and Lazdunski, 1991), and oocytes (Kupitz and Atlas, 1993). Densities of epithelial Na^+ channels range from hundreds to several thousand per cell (Benos *et al.,* 1992,1995). The relative rarity of these proteins has hampered both the biochemical and molecular characterization of these channels, and only recently has the structure of epithelial Na^+ channels been elucidated at the biochemical and molecular levels.

Our laboratory has biochemically characterized a renal epithelial Na^+ channel complex purified from bovine papillary collecting ducts and A6 renal epithelial cells, a cell line derived from the distal nephron of *Xenopus laevis* (Benos *et al.,* 1986, 1987). This Na^+ channel is a heteroligomeric complex consisting of at least six nonidentical polypeptides with molecular masses of 300–315, 130–180, 90–110, 70–85, 55–65, and 41 kDa (Benos *et al.,* 1987). The purified bovine channel complex forms sodium-selective channels when incorporated into planar lipid bilayers (Oh and Benos, 1993) and the 130- to 180-kDa polypeptide isolated from the A6 cell Na^+ channel complex gives a sodium-selective channel in lipid bilayers with a conductance of 5 pS (Sariban-Sohraby *et al.,* 1992). Kleyman and collaborators (1991) have purified an essentially identical heteroligomeric Na^+ channel complex from A6 cells using a monoclonal antibody directed against the amiloride-binding component of the sodium channel. Staub and co-workers (1992) have cloned a 160-kDa apical protein, termed Apx, from A6 cells that may represent either a subunit or an associated regulatory protein of the biochemically purified Na^+ channel because antibodies generated against Apx cross-react with a 130- to 150-kDa polypeptide on immunoblots of purified A6 cell Na^+ channel. Although this protein does not reconstitute amiloride-sensitive Na^+ currents when expressed in *Xenopus* oocytes, coinjection of Apx antisense mRNA inhibits the amiloride-sensitive Na^+ current induced by injection of total mRNA isolated from A6 cells, suggesting Apx has a regulatory function (Staub *et al.,* 1992).

Canessa and co-workers (1993, 1994) and Lingueglia and co-workers (1993, 1994) have independently cloned an epithelial Na^+ channel protein (ENaC) from corticosteroid-stimulated rat distal colon using the *Xenopus* oocyte expression system. This channel is a heteromultimeric complex consisting of three homologous subunits (α, β, and γ) that have predicted molecular masses of 79, 72, and 75 kDa, respectively. The subunits of this channel share a high degree of identity with each other and with the stretch-activated channels of *Caenorhabditis elegans.* It has been postulated that epithelial Na^+ channels and stretch-activated cation channels belong to the same gene family (Canessa *et al.,* 1993; Palmer, 1995). In addition to the colon, this channel has been identified in kidney medulla and cortex, lung, trachea, sweat ducts, parotid glands, and tongue (Duc *et al.,* 1994; Li *et al.,* 1994). This channel has a single-channel conductance of 4 pS when expressed in *Xenopus* oocytes and exhibits ion selectivity, gating kinetics, and an amiloride pharmacological profile similar to that of the 4-pS, highly selective Na^+ channel expressed in native Na^+ reabsorbing epithelia (Canessa *et al.,* 1994; Eaton *et al.,* 1995). Although the relationship of the cloned ENaC to the biochemically purified Na^+ channel complex studied in our laboratory is unclear, antibodies generated against an αENaC fusion protein cross-react with 70- and 90-kDa polypeptides of the purified bovine Na^+ channel (Fuller *et al.,* 1995). Waldman and co-workers (1995) have recently cloned a novel human ENaC (δ) that is expressed in brain, pancreas, testis, and ovary. When expressed in *Xenopus* oocytes with β- and γ-subunits, δENaC induces a channel that exhibits a single-channel conductance of ~12 pS and has a lower sensitivity to amiloride than the 4-pS channel.

The elucidation of the structure of epithelial Na^+ channels at the biochemical and molecular levels has facilitated the characterization of proteins associated with the channels, such as regulatory and cytoskeletal proteins. Using antibodies generated against the purified bovine renal epithelial Na^+ channel (Sorscher *et al.,* 1988), we observed that epithelial Na^+ channels co-localize to the apical membrane with actin and apically associated isoforms of ankyrin and spectrin in Na^+-reabsorbing renal epithelial cells (Smith *et al.,* 1991). This suggested to us that epithelial Na^+ channels interact with the spectrin-based membrane–cytoskeleton and that this interaction may maintain the channels within the apical membrane domain. We examined if spectrin and ankyrin copurify with the renal epithelial Na^+ channel by immunoblotting both bovine and A6-cell partially purified Na^+ channel complexes with antibodies specific for ankyrin and spectrin. α-Spectrin and ankyrin were routinely identified in these preparations, suggesting a direct interaction with the spectrin-based membrane–cytoskeleton (Smith *et al.,* 1991). If this interaction is similar to the linkage of the anion exchanger in red blood cells and Na^+,K^+-ATPase of renal epithelial cells to the spectrin-based cytoskeleton (Nelson and Veshnock, 1987; Morrow *et al.,* 1989; Davis and Bennett, 1990), we predicted it would require ankyrin to bind to a polypeptide of the Na^+ channel complex. We therefore examined if ^{125}I-labeled ankyrin would directly associate with the Na^+ channel in solid-phase assays. Ankyrin was found to bind to the 150-kDa polypeptide of the purified epithelial Na^+ channel complex (Smith *et al.*, 1991). These data indicate that the renal epithelial Na^+ channel is associated with the spectrin-based membrane–cytoskeleton and that this association is involved in maintaining epithelial Na^+ channels within the apical membrane domain.

To further corroborate an association of the epithelial Na^+ channels with the spectrin-based membrane–cytoskeleton, we have examined the lateral mobility of epithelial Na^+ channels on the cell surface of confluent monolayers of filter-grown A6 renal epithelial cells using the technique of fluorescence photobleaching recovery (FPR). The FPR technique has been widely used to measure the lateral mobility of membrane constituents (Zhang *et al.*, 1993), including a variety of membrane transport proteins (Angelides *et al.*, 1988; Chang *et al.*, 1981; Madreperla *et al.*, 1989; Paller, 1994). In the FPR technique, the signal from fluorescently labeled membrane proteins is depleted within a defined spot on the cell surface by photobleaching with a brief, intense laser pulse (Zhang *et al.*, 1993). Typically the membrane proteins are labeled with fluorescently conjugated specific antibodies or Fab′ fragments of these antibodies (Zhang *et al.*, 1993). Analysis of the kinetics of fluorescence recovery in the bleached area gives a diffusion coefficient for the labeled protein and a mobile fraction, which represents the percentage of unbleached molecules that diffuse into the bleached area during the time course of the experiment (Zhang *et al.*, 1993). Measurement of the lateral mobility of Na^+ channels labeled with rhodamine-conjugated Fab′ fragments of the anti-Na^+ channel antibody revealed that greater than 80% of the Na^+ channels were immobile. There was a small mobile fraction (f) of Na^+ channels ($12 \pm 5\%$) that had a diffusion coefficient (D_L) of 4×10^{-11} $cm^2 \cdot s$ (Fig. 1). We interpret this small mobile fraction to be due to either a reversible equilibrium between cytoskeleton-associated and free channels or channels that are confined within microdomains formed by the spectrin-based cytoskeleton. Taken together, the biochemical and FPR data indicate that the renal epithelial Na^+ channel is linked to the spectrin-based cytoskeleton and these cytoskeletal elements serve to restrict the lateral mobility of the Na^+ channels.

Data of Rotin and colleagues (1994) have established a direct interaction between the cloned ENaC and α-spectrin. The C-terminal domain of the αENaC contains two proline-rich sequences (amino acids 666–674 and 681–691) that resemble the *src* homology region 3 (SH3) binding motifs of signal transduction proteins (Ren *et al.*, 1993). SH3 domains are conserved sequences found in several signal transduction and cytoskeletal proteins, including nonerythroid α-spectrin (see Hitt and Luna, 1994), that mediate protein–protein interactions through their binding to proline-rich motifs. In light of our data demonstrating that nonerythroid α-spectrin copurifies with the renal epithelial Na^+ channel (Smith *et al.*, 1991), Rotin and co-

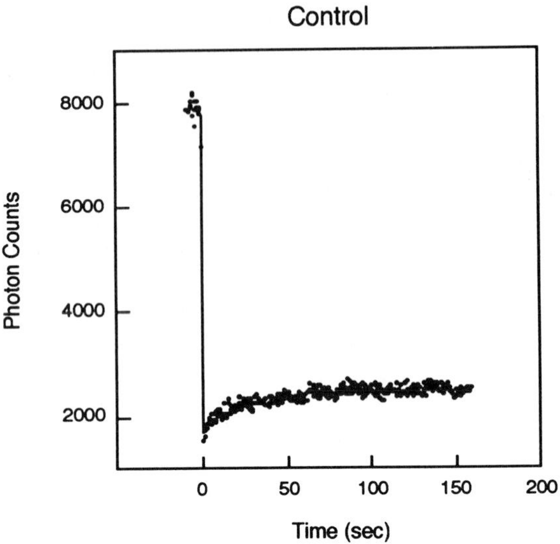

FIGURE 1 Representative fluorescence photobleaching recovery (FPR) curve of fluorescently labeled epithelial Na^+ channels on the apical surface of a filter-grown A6 renal epithelial cell monolayer. This curve illustrates the slow mobility and low mobile fraction. Analysis of this curve gives a diffusion coefficient (D_L) of 4.82×10^{-11} $cm^2 \cdot s$ and a mobile fraction (f) of 14%. (From Smith *et al.*, 1995.)

workers (1994) examined if the proline-rich motif of αENaC cloned from rat colon binds to the SH3 domain of α-spectrin. In an elegant series of experiments, these authors presented four lines of evidence demonstrating that the proline-rich motif mediates binding of αENaC to the SH3 domain of α-spectrin: (1) endogenous α-spectrin co-immunoprecipitated with αENaC from transfected Madin–Darby canine kidney (MDCK) cells overexpressing αENaC; (2) a fusion protein of the C-terminus of αENaC bound to both native α-spectrin and a fusion protein of the SH3 domain of α-spectrin in *in vitro* assays, whereas an N-terminal αENaC fusion protein did not; (3) a fusion protein of the second proline-rich region (amino acids 681–691) bound to the SH3 domain of α-spectrin and binding could be inhibited by mutagenesis of the proline sequence, whereas a fusion protein of the first proline-rich region (amino acid 666–671) did not bind at all; and (4) microinjection of a C-terminal fusion protein into rat alveolar epithelial cells, which express apically restricted α-spectrin, resulted in the apical localization of the fusion protein, whereas a microinjected N-terminal fusion protein remained diffuse within the cytoplasm. Based upon these experiments, Rotin and co-workers (1994) concluded that the interaction of αENaC with α-spectrin is involved in maintaining the polarized distribution of the channel to the apical membrane. Interestingly, a proline-rich C-terminal motif is also present in both the β- and γ-subunits of ENaC (Canessa *et al.*, 1994); however, it is presently unknown if these subunits also associate with α-spectrin. If the β-subunit interacts with spectrin, this may have important implications in Liddle's syndrome, a heritable form of salt-sensitive hypertension resulting from constitutive activation of ENaC (Shimkets *et al.*, 1994). Liddle's syndrome has recently been established to be the result of a mutation in the β-subunit of ENaC that induces a premature stop codon in the β-subunit, resulting in the truncation of the C-terminus and deletion of the proline-rich motif (Shimkets *et al.*, 1994). It has been proposed that such a truncation may prevent an interaction of the β-subunit with α-spectrin, thereby altering the targeting and/or regulation of this subunit by the cytoskeleton and resulting in the constitutive activation of the ENaC (Shimkets *et al.*, 1994).

Because the relationship between the cloned ENaCs and the biochemically purified Na^+ channel is unclear, we have begun to examine whether α-spectrin and ankyrin copurify with αENaC from A6 epithelial cells. In preliminary experiments, we have been able to co-immunoprecipitate α-spectrin and ankyrin with αENaC from sucrose-gradient fractions of A6 cell extracts enriched in epithelial Na^+ channels using an antibody against the *Xenopus* homologue of αENaC, thereby further establishing that ENaC interacts with the spectrin-based membrane–cytoskeleton *in vitro*. A polypeptide that has an M_r of 160 and cross-reacts with anti-Apx antibodies also co-immunoprecipitates with αENaC. The M_r of Apx corresponds to the M_r of the polypeptide that we found to bind ankyrin in the biochemically purified Na^+ channel. However, whether Apx represents the ankyrin-binding component of the Na^+ channel awaits further clarification.

III. REGULATION OF EPITHELIAL Na^+ CHANNELS BY THE MEMBRANE–CYTOSKELETON

The physiological regulation of epithelial Na^+ channels is complex. Activity of these channels has been demonstrated to be regulated by cAMP-dependent protein kinase– (PKA-) mediated phosphorylation, protein kinase C– (PKC-) mediated phosphorylation, G-proteins, methylation, arachidonic acid metabolites, and the membrane–cytoskeleton (see Benos *et al.* 1995, for discussion). The pioneering work of Cantiello and colleagues (1991) has provided compelling evidence for the cytoskeleton being a physiologically important regulator of epithelial Na^+ channels, and it has stimulated researchers to examine the role of the cytoskeleton in regulating other ion channels, both in epithelial and in nonepithelial cells.

Because of the co-localization of epithelial Na^+ channels with ankyrin, spectrin, and actin (Smith *et al.,* 1991), Cantiello and co-workers (1991) hypothesized that the actin component of the membrane–cytoskeleton was involved in the regulation of epithelial Na^+ channels. A6 epithelial cells grown on glass coverslips were used as the model system to test this hypothesis. When grown under these conditions, A6 cells express a poorly selective epithelial Na^+ channel that has a single-channel conductance of 9 pS (Cantiello *et al.,* 1991; Cantiello, 1995). At present it is unclear as to whether the 9-pS channel and the 4-pS highly Na^+-selective ENaC channel described earlier are distinct entities or if they represent different manifestations of the same channel under different growth conditions. A6 cells were incubated in cytochalasin D, an agent that depolymerizes actin filaments, and the effects on Na^+ channel activity were assessed using the patch clamp technique in the cell-attached configuration. Cytochalasin D was found to activate Na^+ channel activity within 5 min of application (Cantiello *et al.,* 1991; Cantiello, 1995). To demonstrate that the effect cytochalasin D had on channel activity was due to changes in actin filament organization, purified actin was added to excised inside-out patches. Addition of actin either induced or increased the activity of Na^+ channels without altering their mean open time. In contrast, application of either actin that had been polymerized to form long filaments or G-actin did not alter channel activity, thereby suggesting that short actin filaments were involved in channel activation (Cantiello *et al.,* 1991; Cantiello, 1995).

Gelsolin is a Ca^{2+}-sensitive actin-severing protein that regulates the length of endogenous actin filaments (Weeds and Maciver, 1993). It has been demonstrated that the extent to which gelsolin will shorten actin filaments is dependent on the gelsolin–actin molar ratio (Yin *et al.,* 1980). Therefore, to test the hypothesis that short actin filaments are involved in Na^+ channel activation, Cantiello and co-workers (1991) polymerized actin in the presence of Ca^{2+} and varying concentrations of gelsolin known to form actin filaments of specific lengths. Actin–gelsolin complexes at molar ratios that give short actin oligomers and tetramers ($<8:1$) were found to induce channel activity, whereas ratios yielding long filaments did not (Cantiello *et al.,* 1991; Cantiello, 1995). It was further hypothesized, based upon these data, that the availability of actin filaments near the channel should be an important factor in regulating the actin-mediated effects on channel activity and channel activity should be modulated by actin-binding proteins. To directly test this hypothesis, filamin, an actin-binding protein (Hitt and Luna, 1994) expressed in renal epithelial cells, was added to excised inside-out patches. Addition of filamin inhibited channel activity, thereby further supporting the thesis of Cantiello and co-workers that, under basal conditions, stabilized actin filaments may contribute to maintaining Na^+ channels in the closed state and that actin depolymerization results in channel activation, either by affecting the membrane environment or by interacting with other membrane–cytoskeleton proteins associated with the channel (Cantiello *et al.,* 1991; Cantiello, 1995).

Further work by Cantiello and co-workers (Prat *et al.,* 1993a,b) examined the role of actin in the vasopressin-mediated upregulation of the 9-pS epithelial Na^+ channel in A6 cells. The hormone vasopressin induces a two- to fourfold increase in transepithelial Na^+ transport across Na^+-reabsorbing epithelial cells such as A6 cells (see Benos *et al.,* 1992, 1995). Vasopressin stimulates an increase in intracellular adenylate cyclase activity and intracellular cAMP levels and the activation of PKA. Vasopressin has also been shown to induce a transient depolymerization of an apical submembrane actin pool in both toad urinary bladder and mammalian inner medullary collecting duct that may permit the fusion of vesicles containing water channels with the apical membrane, thereby facilitating transepithelial water movement (Ding *et al.,* 1991; Simon *et al.,* 1993). This actin pool is also sensitive to cytochalasin D (Franki *et al.,* 1992). In light of vasopressin-induced depolymerization of actin in renal epithelial cells, Cantiello and colleagues investigated the role of changes in actin filament length on the vasopressin- and PKA-mediated activation of the 9-pS Na^+ channel using the patch clamp technique in the excised inside-out configuration (Prat *et al.,* 1993a,b). Pretreatment of cells with cytochalasin D inhibited channel activation by PKA; however, subsequent addition of short actin filaments to patches induced channel activity. These data were interpreted to mean that actin filaments are required for the PKA-mediated activation of the channels, and that actin may be a target for PKA (Prat *et al.,* 1993b). To test if actin is a substrate for PKA-mediated phosphorylation, actin was incubated with PKA and ATP in varying molar ratios of actin to ATP. Both G- and F-actin were phosphorylated by PKA under physiologically relevant conditions, and phosphorylation could be inhibited by a specific inhibitor of PKA, 5- to 24-amide (Prat *et al.,* 1993b; Cantiello, 1995). When the effects of phosphorylated G- and F-actin on channel activity in excised patches were examined, only phosphorylated F-actin filaments were capable of activating Na^+ channels. It was therefore proposed, based upon these data, that PKA-dependent effects of actin on Na^+ channel activity are mediated either directly through a molecular interaction between actin and the channel protein or through actin-binding proteins associated with the channel (Prat *et al.* 1993b; Cantiello, 1995). Because it remains to be established if actin binds directly to the channel, at present the most parsimonious explanation for the observed effects of actin on Na^+ channel activity is they are mediated through actin-binding proteins such as spectrin.

Although the relationship of the 9-pS Na^+ channel studied by Cantiello and co-workers and the cloned ENaCs is not understood at the molecular level, cloned ENaCs share significant homology with mechanosensitive channels found in *C. elegans* (Canessa *et al.,* 1993,1994). We have tested the hypothesis that ENaCs are mechanosensitive channels (Awayda *et al.,* 1995). Incorporation of the α-subunit of the bovine renal ENaC (bENaC) (Fuller *et al.,* 1995) into planar lipid bilayers demonstrated that it can be activated by a hydrostatic pressure gradient across the bilayer, whereas the biochemically purified renal bovine Na^+ channel complex was not activated under these conditions. Stretch activation of αENaC was further confirmed by electrophysiological recordings of αbENaC expressed in *Xenopus* oocytes, which were induced to swell by the injection of a 100-m*M* KCl solution. Based on these data, we concluded that the α-subunit of ENaC encodes a stretch-activated channel and that interactions between the α-subunit and associated polypeptides stabilize it in the membrane and render it mechanically insensitive (Awayda *et al.,* 1995). Furthermore, these data corroborate the hypothesis that the actin-mediated activation of epithelial Na^+ channels observed by Cantiello and co-workers occurs through a direct interaction of actin with the channel, or a channel-associated actin-binding protein, rather than channel activation being induced by membrane stretch or membrane deformation resulting from the experimental conditions.

In light of the data of Cantiello and co-workers that actin modulates activity of the 9-pS Na^+ channel in response to vasopressin (Prat *et al.,* 1993b; Cantiello, 1995), we have examined whether the membrane–cytoskeleton proteins directly associated with the channels, ankyrin and spectrin, are also involved in the vasopressin- and aldosterone-induced upregulation of Na^+ transport. (Smith *et al.,* 1995). Although the mechanism by which aldosterone activates Na^+ channels is unclear, recent biochemical and electrophysiological data indicate that methylation of the channel protein may be one such mechanism (Ismailov *et al.,* 1994; Kemendy *et al.,* 1992; Sariban-Sohraby *et al.,* 1984, 1993; Sariban-Sohraby and Fisher, 1995; Weisman *et al.,* 1985). The methylation inhibitor 3-deazaadenosine has been shown to prevent aldosterone-induced upregulation of Na^+ channel activity (Kemendy *et al.,* 1992). We used the technique of FPR to examine whether the spectrin-based membrane–cytoskeleton is involved in the vasopressin- and aldosterone-induced upregulation of epithelial Na^+ channel activity in filter-grown A6 cell monolayers (Smith *et al.,* 1995). FPR has previously been used to demonstrate an interaction of the spectrin-based membrane–cytoskeleton with the voltage-dependent Na^+ channel of neurons (Angelides *et al.,* 1988), Na^+,K^+-ATPase of photoreceptors (Madreperla *et al.,* 1989) and renal epithelial cells (Paller , 1994), and the anion exchanger AE1 of erythrocytes (Chang *et al.,* 1981; Golan and Veatch, 1980). We hypothesized that, if the spectrin-based membrane–cytoskeleton is involved in the upregulation of transepithelial Na^+ transport by vasopressin and aldosterone, then hormone mediated changes in the membrane–cytoskeleton should be reflected in changes in the lateral mobility (D_L) and/or mobile fraction (f) of Na^+ channels on the cell surface.

Under control conditions (no hormonal treatment), greater than 80% of the Na^+ channels were observed to be immobile ($f = 14.2 \pm 0.77\%$), with those that are mobile having a D_L of $9.9 \pm 1.2 \times 10^{-11}$ $cm^2 \cdot s$. To determine if vasopressin alters f and D_L of the Na^+ channels, A6 cell monolayers were stimulated with vasopressin for 20 min prior to FPR analysis. Vasopressin did not significantly alter either f ($12.7 \pm 2.4\%$) or D_L ($5.9 \pm 1.56 \times 10^{-11}$ $cm^2 \cdot s$) of the Na^+ channels. Because vasopressin and cytochalasin D depolymerize the same pool of actin (Ding *et al.,* 1991; Franki *et al.,* 1992; Simon *et al.,* 1993), we examined the effect of 20-min incubation in cytochalasin D on f and D_L to corroborate our findings for vasopressin. Like vasopressin, cytochalasin D did not significantly affect either f ($7.33 \pm 2.18\%$) or D_L ($8.2 \pm 2.4 \times 10^{-11}$ $cm^2 \cdot s$) of the Na^+ channels. We interpret these data to mean that the interaction between the spectrin-based membrane–cytoskeleton and Na^+ channels is not directly involved in the naturiferic response of vasopressin. Our findings are supported by the observations of Gao and co-workers (1993,1994), who reported that vasopressin, at least at the electron microscopic level, does not alter the distribution of spectrin and ankyrin within the microvilli of Na^+-reabsorbing renal epithelia. Cytochalasins have been shown to have no effect on the lateral mobility or mobile fraction of transport proteins linked directly to ankyrin and spectrin, such as band 3 (Golan and Veatch, 1980) and the voltage-dependent Na^+ channel (Angelides *et al.,* 1988).

Because actin is not directly involved in restricting channel mobility, FPR did not allow us to determine if actin is directly involved in vasopressin-mediated upregulation of Na^+ channel activity in filter-grown A6 cells. Therefore, we used patch clamp analysis to examine if cytochalasin D would activate the 4-pS highly selective Na^+ channel, which is the predominant Na^+ channel expressed in A6 cells grown on filter supports (Stoner *et al.,* 1995; Smith *et al.,* 1995). Application of cytochalasin D was found to inactivate the 4-pS channel in both cell-attached and excised inside-out patches containing Na^+ channels (Fig. 2). We did not observe activation of Na^+ channel activity in quiescent patches following cytochalasin D treatment (Smith *et al.,* 1995). We have observed a similar effect of cytochalasin D on the activity of the 4-pS Na^+ channel expressed in the collecting tubule of the salamander *Ambystoma* (L. C. Stoner, S. C. Viggiano, and P. R. Smith, unpublished observations). These data are in contrast to the findings of Cantiello *et al.* (1991) for the 9-pS, moderately selective Na^+ channel

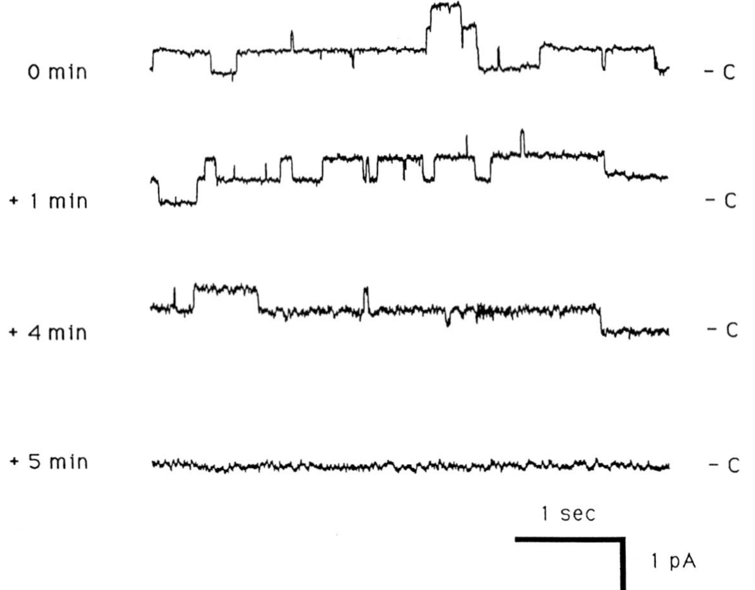

FIGURE 2 Effect of 1-μM cytochalasin D on 4-pS epithelial Na^+ channels expressed in filter-grown A6 renal epithelial cells. Representative channel recording from a cell-attached patch containing three active channels (0 min). Channel activity within the patch was completely abolished after 5-min exposure to 1-μM cytochalasin D. Pipette holding potential was 0 mV. (From Smith *et al.* 1995.)

expressed in coverslip-grown A6 cells, and they suggest that the role of actin in channel regulation differs between the 9-pS and 4-pS Na^+ channels. It is thus conceivable that there may be proteins other than actin that confer vasopressin sensitivity to certain epithelial sodium channels.

To determine if modulation of the membrane–cytoskeleton associated with the Na^+ channels is involved in the aldosterone-induced upregulation of Na^+ transport, we examined the effects of aldosterone on D_L and f of the Na^+ channels. A6 cell monolayers were stimulated with aldosterone for 16 h to allow for the maximal change in hormone-induced Na^+ transport. Although aldosterone did not affect D_L ($6.5 \pm 1.64 \times 10^{-11}$ $cm^2 \cdot s$), it increased f approximately twofold ($f = 32.3 \pm 5.42\%$) compared to control ($f = 14.2 \pm 0.77\%$) and vasopressin-treated ($f = 12.7 \pm 2.4\%$) monolayers (Fig. 3). This twofold increase in mobile fraction of Na^+ channels corresponds to the average increase in Na^+ transport in response to aldosterone in A6 cell monolayers (Wills *et al.*, 1993), indicating that the increase in mobile fraction is correlated with the increase in Na^+ transport. We inter-

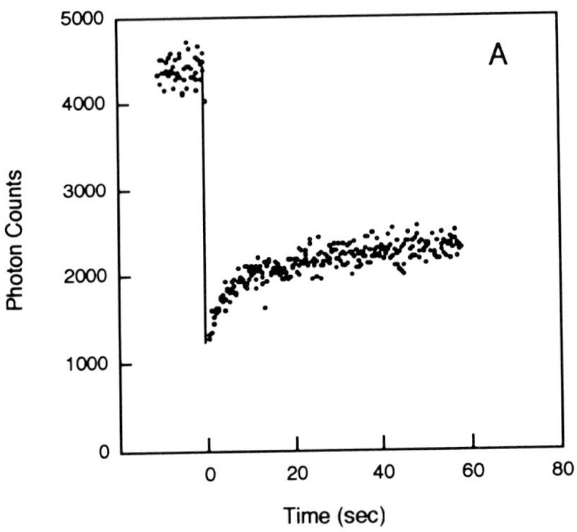

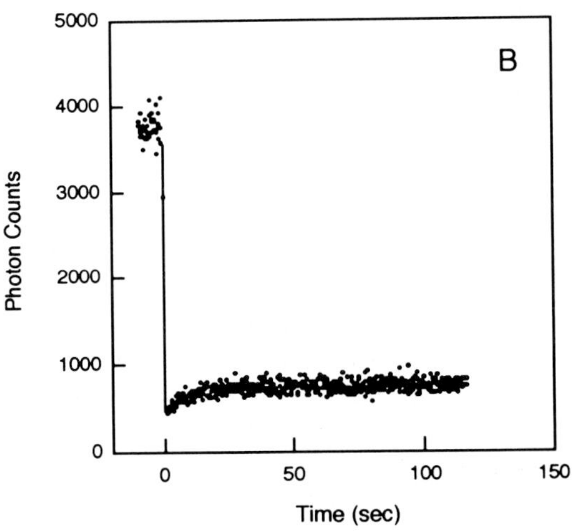

FIGURE 3 Effect of aldosterone on the lateral mobility and mobile fraction of fluorescently labeled epithelial Na^+ channels and sConA receptors on the cell surface of A6 renal epithelial cells. (A) Representative FPR curve for Na^+ channels on the cell surface of an A6 cell monolayer stimulated with aldosterone for 16 h. This curve has a diffusion coefficient (D_L) of 7.7×10^{-11} $cm^2 \cdot s$ and a mobile fraction (f) of 44%. (B) Representative FPR curve for sConA receptors on a A6 cell monolayer stimulated with aldosterone for 16 h. Analysis of this curve yields a diffusion coefficient (D_L) of 9.81×10^{-11} $cm^2 \cdot s$ and a mobile fraction (f) of 10%. D_L and f of the sConA receptors are comparable to untreated monolayers. (From Smith *et al.*, 1995.)

pret the increased mobile fraction of Na^+ channels to be the result of aldosterone-mediated changes either in the affinity of the Na^+ channel for the membrane–cytoskeleton or in the spectrin-based membrane–cytoskeleton itself. It is possible that release of Na^+ channels from the lateral constraints imposed by the spectrin-based cytoskeleton activates the channels, or that recruitment to specific membrane domains on the cell surface promotes channel activity. Although epithelial Na^+ channels in A6 cells are linked directly to the spectrin-based membrane–cytoskeleton, it is possible that the membrane–cytoskeleton also forms a corral-like barrier that restricts the mobility of the channels. This corral-like barrier may be modulated by aldosterone. As pointed out by Edidin and Stroynowski (1991), bleaching of a membrane protein in a spot greater than the dimension of the barrier formed by the spectrin-based membrane–cytoskeleton would make the protein appear immobile. Although these experiments cannot unequivocally allow us to determine whether the aldosterone-associated increase in mobile fraction of Na^+ channels is due to an altered affinity of the Na^+ channels for the membrane–cytoskeleton, or if aldosterone alters a barrier to channel mobility imposed by the membrane–cytoskeleton, they do indicate that the spectrin-based membrane–cytoskeleton is involved in the aldosterone-mediated upregulation of Na^+ channels.

Because aldosterone induces protein synthesis, it is also conceivable that the increase in mobile fraction of Na^+ channels is due to the insertion of newly synthesized channels into the apical membrane that are not linked to the membrane–cytoskeleton. However, three lines of evidence argue against this interpretation: (1) we were unable to detect a significant difference in the binding of fluorescently labeled anti-Na^+ channel antibodies, as determined by photon counting, between control and aldosterone-treated monolayers; (2) Kleyman and co-workers (1992) have demonstrated biochemically that aldosterone does not alter the apical expression of epithelial Na^+ channels in A6 cells following 16 h of stimulation with aldosterone, and (3) Renard and co-workers (1995) have revealed that neither aldosterone nor dexamethasone treatment significantly alters the expression of α-, β-, or γENaC mRNA or protein expression in the rat kidney.

An alternative explanation for the increase in mobile fraction of Na^+ channels is that it is the result of an aldosterone-mediated modification of the apical membrane lipid composition. It has been demonstrated in the toad urinary bladder that aldosterone stimulates fatty acid synthesis and phospholipid deacylation, and that these modifications in the membrane phospholipids are involved in the aldosterone-mediated increase in Na^+ transport (Goodman *et al.*, 1975; Lien *et al.*, 1975). If the increase in mobile fraction of the labeled Na^+ channels was due to a modification of the membrane lipid environment, then we would predict that other transmembrane proteins should exhibit similar increases in their mobile fraction. To test this prediction, we examined the effect of aldosterone on the mobility of succinyl-concanavalin A (sConA) receptors. ConA receptors have restricted rates of lateral mobility and small mobile fractions (10%) on the apical surfaces of A6 cell monolayers (Dragsten *et al.,* 1981). Following aldosterone treatment, we did not observe an increase in the mobile fraction of fluorescently labeled sConA receptors (Fig. 3). This observation suggests that the increase in mobile fraction is specific to the Na^+ channels and that it is related to aldosterone-induced changes in the cytoskeleton and/or affinity of the Na^+ channels for the membrane–cytoskeleton, rather than changes in the membrane lipid composition.

In light of increasing evidence that suggests that methylation of the Na^+ channel is the mechanism whereby aldosterone activates Na^+ channels (Ismailov *et al.,* 1994; Kemendy *et al.,* 1992; Sariban-Sohraby *et al.* 1984,1993; Sariban-Sohraby and Fisher, 1995), we also examined whether the methylation blocker 3-deazaadenosine would inhibit the increased mobile fraction of Na^+ channels associated with aldosterone. Incubation of A6 cell monolayers in aldosterone plus 3-deazaadenosine for 16 h was found to prevent the increase in mobile fraction associated with aldosterone. Although not statistically significant, the mobile fraction of Na^+ channels was slightly higher in the 3-deazaadenosine plus aldosterone monolayers (f = 15 ± 1.4%) when compared to either control (f = 14.2 ± 0.77%) or vasopressin-treated (f = 12.7 ± 2.4%) monolayers (Smith *et al.,* 1995). Although we interpret these data to mean that a methylation event is involved in the aldosterone-associated increase in mobile fraction of Na^+ channels in A6 cells, one potential caveat of these experiments is that 3-deazaadenosine may also have altered DNA transcription, because methylation of histone proteins is required for transcription (Kemendy *et al.,* 1992). We propose that the membrane–cytoskeleton is involved in the aldosterone-induced upregulation of Na^+ channel activity. However, further experiments are clearly necessary to elucidate the biochemical mechanisms underlying the aldosterone-associated increase in mobile fraction of Na^+ channels.

The membrane–cytoskeleton has also been implicated in upregulating the cell surface expression of Na^+ channels in response to vasopressin. Two hypotheses have been proposed to explain the mechanism whereby vasopressin upregulates transepithelial Na^+ transport. One hypothesis proposes that vasopressin activates quiescent channels in the apical membrane through a PKA-mediated phosphorylation of the channel or associated proteins. The alternative hypothesis proposes that vasopressin induces the insertion of new channels into the apical membrane from a subapical pool of Na^+ channel–containing vesicles. Data suggest that both mechanisms may be involved in the vasopressin upregulation of Na^+ transport (Kleyman *et al.,* 1994; Oh *et al.,* 1993; Prat *et al.,* 1993b). Verrey and co-workers (1995) have presented data implicating actin filaments in the vasopressin-mediated insertion of Na^+ channels into the apical membrane, as disruption of actin filaments by cytochalasin D decreased both vasopressin-induced stimulation of Na^+ transport and exocytotic movement. Els and Chou (1993) have reported that cytochalasin B inhibits the hypo-osmolarity–induced increase in apical Na^+ channel density in the frog skin, which has been postulated to be regulated by the recruitment of channels from an intracellular pool. Although these data implicate actin filaments in the stimulus-induced targeting of Na^+ channels to the apical membrane, it is possible that the effects of cytochalasins on Na^+ transport observed in these studies are through actin filaments directly regulating channel activity (Verrey *et al.,* 1995; Els and Chou, 1993). Conversely, Morris *et al.* (1995) found that disruption of either microtubules with colchicine or nocodazole or microfilaments with cytochalasin E did not affect the activation of Na^+ channels in A6 cells by vasotocin.

IV. REGULATION OF EPITHELIAL Cl^- AND K^+ CHANNELS BY THE MEMBRANE–CYTOSKELETON

In epithelial cells, Cl^- and K^+ channels are involved both in the absorption and secretion of Cl^- and K^+ ions, respectively, and in regulatory volume decrease (RVD), a process whereby a swollen cell loses salt and water to restore its original volume. In this section we present evidence that indicates the membrane–cytoskeleton is also involved in regulating the activity of epithelial secretory and absorptive Cl^- and K^+ channels. With the exception of the cystic fibrosis transmembrane regulator (CFTR), the biochemical and molecular identities of the Cl^- and K^+ channels that have been shown to be regulated by the membrane–cytoskeleton by electrophysiological techniques are unknown. For a discussion of the role of the cytoskeleton in regulating ion channels involved in RVD, see Chapter 17 in this volume.

CFTR is a low-conductance Cl^- channel that is regulated by cAMP and ATP and is predominantly expressed in the apical membranes of epithelia (Fuller and Benos, 1992; Riordan, 1993; Cantiello *et al.,* 1994; Stutts *et al.,* 1995b). Mutations in the channel protein have been associated with the lethal autosomal recessive disease cystic fibrosis (Riordan, 1993). Recently Prat and co-workers (1995) have presented electrophysiological data that indicate that the cytoskeleton is also involved in the regulation of CFTR. Addition of cytochalasin D to a mouse mammary adenocarcinoma cell stably transfected with human CFTR was found to activate whole-cell

Cl^- currents that were comparable to Cl^- currents activated by cAMP, suggesting a role of the cytoskeleton in CFTR activation (Prat *et al.,* 1995). Application of cytochalasin D had no effect on the whole-cell currents of mock-transfected cells. Interestingly, activation by PKA in cells pretreated with cytochalasin D had no additive effect on whole-cell chloride currents, suggesting that the two stimulatory mechanisms were interlinked. To examine the molecular mechanism underlying the cytochalasin D–induced activation of CFTR, monomeric (G-) actin was added to excised inside-out patches and the effects on CFTR activation were assessed by patch clamp technique. Actin, either alone or in the presence of ATP, was found to activate CFTR, whereas actin in the presence of filamin or DNase 1, which forms a complex with G-actin and prevents filament formation, inhibited the actin-induced activation of the channel (Prat *et al.,* 1995). In an attempt to determine whether actin binds to CFTR, Prat and co-workers (1995) compared the amino acid sequence of CFTR with the sequences of the known actin-binding domains from a variety of actin-binding proteins. A putative actin-binding domain was identified within each of the two nucleotide-binding domains of CFTR, however, the nature of the molecular interactions between actin and CFTR remains to be elucidated.

The data of Prat *et al.* (1995) indicate that the actin component of the cytoskeleton is involved in the activation and regulation of CFTR and that the regulatory role of actin can be modulated by changing the organization of actin. These authors proposed that long, filamentous actin maintains the channels in the closed state and "severing" of actin into short filaments activates the channels (Prat *et al.,* 1995). Because Prat *et al.* (1993b) had previously shown that phosphorylation of actin by PKA decreases its ability to polymerize into filaments, it was envisioned that actin may be involved in the cAMP-dependent activation of CFTR. This could explain why the effects of cytochalasin D and cAMP on channel activation were not additive.

Although it remains to be determined whether actin interacts directly with CFTR, or if the effects observed by Prat *et al.* (1995) are mediated through actin-binding or regulatory proteins associated with CFTR, their data may have important implications for the role CFTR plays in regulating epithelial Na^+ channels. In cystic fibrosis airway epithelia, which express the highly Na^+-selective 5-pS epithelial Na^+ channels, there is an abnormally high rate of Na^+ absorption and no Cl^- secretion. The abnormally high Na^+ absorption is due to the increased open probability of the Na^+ channels. Stutts and co-workers (1995a) have hypothesized that CFTR acts as a negative regulator of epithelial Na^+ channels. To test this hypothesis, rat colonic ENaC (α-, β-, and γ-subunits) was expressed in MDCK cells either alone or with human CFTR. Cells expressing only ENaC generated large amiloride-sensitive currents that were stimulated by cAMP, whereas coexpression of CFTR with ENaC resulted in smaller basal currents that were inactivated by cAMP. Based upon these data, it has been proposed that CFTR and epithelial Na^+ channels may directly interact through regulatory proteins, soluble extracellular mediators released by CFTR, or cytoskeletal elements (Stutts *et al.,* 1995a; Higgins, 1995). Although there is no direct evidence to support cross-talk between the two channels via cytoskeletal elements, it is intriguing that treatment of cells with cytochalasin D results in the activation of CFTR (Prat *et al.,* 1995) and the downregulation of the 4-pS epithelial Na^+ channel (Smith *et al.,* 1995).

Additional studies have presented evidence that disruption of actin filaments is also involved in the inhibition of epithelial Cl^- channels. Suzuki and co-workers (1993) used the patch clamp technique to examine the role of actin in the regulation of a 33-pS Cl^- channel expressed in the apical membrane of a SV40LT immortalized rabbit proximal tubule cell line. This outwardly rectifying channel has been shown to be activated by parathyroid hormone, PKA, PKC, and membrane depolarization (Suzuki *et al.,* 1993). Application of cytochalasin D was shown to significantly reduce both the mean channel number and the open probability of Cl^- channels that had been activated by depolarization in both cell-attached and inside-out patches. Application of hydrostatic pressure to the patches did not effect channel activity, thereby ruling out that the channel is stretch activated. DNase 1 was not found to alter channel activity in excised inside-out patches. To examine if actin could reactivate Cl^- channels that were inactivated by cytochalasin D, G-actin, short actin filaments, or long actin filaments were added to excised patches pretreated with cytochalasin D. Short actin filaments partially reversed and long actin filaments completely reversed the effects of cytochalasin D, whereas monomeric G-actin did not (Suzuki *et al.,* 1993). A similar inhibitory effect of cytochalasin D has been observed on Cl^- channel activity in the human bronchial epithelial cell line 16HBE14o− (Hug *et al.,* 1995). Incubation of cells in cytochalasin greatly reduced whole-cell currents (Cl^- and K^+ currents) and $^{36}Cl^-$ effluxes and perfusion of the cells with either cytochalasin D or phalloidin largely suppressed the whole current, whereas G-actin had no effect. In excised inside-out patches, neither G-actin, cytochalasin D, nor phalloidin activated channel activity (Hug *et al.,* 1995). Halm and co-workers (1995) reported that cytochalasin B reduced by 40% the prostaglandin E_2–stimulated active chloride secretion in rabbit distal colon. Taken together, the data from these studies indicate that assembled actin filaments are a physiologically important component of the regulatory pathway of Cl^- channels.

In contrast, Fuller *et al.* (1994) found that two inhibitors of microtubule polymerization, nocodazole and colchicine, significantly inhibited the cAMP-, but not the Ca^{2+}-, dependent stimulation of Cl^- secretion in the human colonic T84 cell line as well as in the rat distal colon. These findings were corroborated in immunofluorescence studies (Tousson *et al.,* 1996). Moreover, cytochalasin D was without effect on forskolin-related redistribution of CFTR. In double-label experiments, CFTR was found to colocalize with the microtubule network but not with actin filaments. These results indicate that microtubules but not actin are essential for the recruitment of CFTR to the cell surface in response to cAMP agonists. Grotmol and van Dyke (1992) have observed a similar inhibition of prostaglandin- and theophylline-induced Cl^- secretion in rat distal colon by the microtubule inhibitors colchicine and taxol.

Clapham and associates have characterized a swelling-activated chloride channel expressed in MDCK cells using biochemical, molecular, and electrophysiological techniques (Paulmichl *et al.,* 1992; Krapivinsky *et al.,* 1994). Although this is a stretch-activated channel involved in RVD, it merits discussion here as it is to our knowledge the first biochemical evidence of actin directly binding to an ion channel or a channel-specific regulatory protein. Clapham and co-workers used the *Xenopus* oocyte expression system to clone a novel cDNA from MDCK cells, which encoded a 27-kDa protein (pI_{Cln}) and gave rise to an outwardly rectifying Cl^- channel that displayed the characteristics of a swelling-induced chloride channel when expressed in oocytes (Paulmichl *et al.,* 1992). Although analysis of the primary structure of this protein did not reveal any transmembrane domains, it was proposed that the protein dimerized to make a channel, with a β-barrel forming the pore (Paulmichl *et al.,* 1992). Subsequent work has suggested that pI_{Cln} is not a chloride channel, but rather a predominately cytosolic protein (Krapivinsky *et al.,* 1994). Although only 5% of pI_{Cln} is associated with the membrane fractions, following a hypotonic challenge there is a shift of the protein from the cytosol to the membrane (Krapivinsky *et al.,* 1994; Gschwenter *et al.,* 1995). Monoclonal antibodies generated against pI_{Cln} and antisense oligodeoxynucleotides complementary to pI_{Cln} inhibited activation of the swelling-induced chloride channel (Krapivinsky *et al.,* 1994; Gschwenter *et al.,* 1995). pI_{Cln} was shown to be linked to the cytoskeleton as actin co-immunoprecipitated with pI_{Cln} from MDCK cells overexpressing pI_{Cln} and actin bound to a pI_{Cln} affinity column (Krapivinsky *et al.,* 1994). Based upon these data, it was proposed that pI_{Cln} is a cytoskeleton-associated channel-regulatory protein. However, this remains a point of contention, as Gschwenter and co-workers (1995) have recently argued that the bulk of the data favors pI_{Cln} being the chloride channel itself. Whether pI_{Cln} encodes a regulatory protein or a chloride channel awaits further clarification. This work does, however, provide direct biochemical evidence that actin interacts with ion channels.

Wang and co-workers (1994) have demonstrated that the membrane–cytoskeleton is involved in the inhibition of a low-conductance, secretory K^+ channel expressed in apical membrane of rat cortical collecting duct principal cells. Under physiological conditions, activity of this channel is enhanced by PKA and downregulated by Ca^{2+}-dependent signal transduction pathways, such as PKC and arachidonic acid (Wang *et al.,* 1994; Wang, 1995). To examine if this channel is modulated through the cytoskeleton, the effects of cytochalasins B and D on channel activity were examined using the patch clamp technique (Wang *et al.,* 1994). Cytochalasins B and D were found to inhibit channel activity in cell-attached patches, and the effects could be reversed by washout of the drugs. In contrast, cheatoglobosin C, an analogue of cytochalasins that cannot depolymerize actin, had no effect on actin. In excised inside-out patches, cytochalasins B and D inhibited channel activity; however, the effects were almost irreversible, with restoration of channel activity being observed in only 1 of 15 experiments following washout. Because PKA-induced channel phosphorylation has previously been shown to be essential for maintaining the normal function of this channel (Wang, 1995), the role of PKA in mediating the inhibitory effect of cytochalasins was also examined. Application of PKA prior to or following exposure to cytochalasin B or D did not block the inhibitory effects, suggesting that they did not result from dephosphorylation of the channels. To exclude the possibility that the inhibitory effects of cytochalasins were the result of actin monomers or short filaments directly obstructing the channel pore, actin monomers, which are known to spontaneously form actin filaments of different sizes in the presence of high salt concentrations and ATP, were added to excised inside-out patches. Application of actin monomers and actin filaments were without effect on K^+ channel activity, indicating that channel inhibition was not due to direct obstruction of the channel pore (Wang *et al.,* 1994; Wang, 1995). Because this channel has not been characterized at the biochemical or molecular levels, it is unknown if the channel is directly linked to the actin component of the membrane–cytoskeleton or if interactions are mediated through channel-associated regulatory and/or actin-binding proteins. However, Wang and co-workers (1994) have proposed that the actin is involved in the interaction between the channel complex and the lipid phase of the membrane, and this interaction may be mediated through an actin-binding protein, such as myristoylated alanine-rich C kinase substrate (MARCKS). Phosphorylation of MARCKS by PKC or binding of Ca^{2+}–calmodulin to MARCKS inhibits its cross-linking of actin filaments (Aderem, 1992), and both PKC and Ca^{2+}–calmodulin have been shown to inactivate the K^+ channel (Wang, 1995).

V. PERSPECTIVES

It is evident from the previous discussion that our understanding of the role of the membrane–cytoskeleton in the regulation of epithelial ion channels is just beginning. Work on epithelial Na^+ channels has revealed that epithelial ion channels can be directly associated with the spectrin-based membrane–cytoskeleton. Patch analysis has shown that the application of cytochalasins either activates or inactivates epithelial Na^+, Cl^-, and K^+ channels, suggesting a role for the actin-based cytoskeleton in the regulation of channel activity. A further role for actin in the regulation of epithelial ion channels is supported by the fact that the addition of either monomeric or filamentous actin to excised patches has an effect on channel activity comparable to that of cytochalasin (Table I). However, it is unclear as to whether the effects of actin on channel activity are directly mediated through actin binding either to a channel subunit or a channel-associated actin-binding protein, or if the observed effects on channel activity are mediated through either the activation or inactivation of cytoskeleton-associated second messengers.

Critical to further understanding the membrane–cytoskeleton's role in the regulation of epithelial ion channels will be the identification of membrane–cytoskeleton proteins that directly interact with specific ion channels. The molecular cloning of ion channels will allow sequence homology comparisons with both membrane–cytoskeleton proteins and membrane proteins known to interact with the cytoskeleton, thereby allowing motifs involved in cytoskeletal interactions to be identified and characterized. For example, in addition to the work of Rotin and co-workers (1994) that identified the proline-rich region of ENaC as the α-spectrin–binding domain, Bourguignon and Jin (1995) have identified the ankyrin-binding domain of the T-lymphoma cell inositol 1,4,5-triphosphate (IP_3) receptor based upon shared sequence homology between the C-terminus of brain IP_3 receptor and the ankyrin-binding domain of lymphocyte CD44. This ankyrin-binding motif is involved in regulating activity of the IP_3 receptor as ankyrin binding to the lymphoma cell IP_3 receptor inhibits IP_3 binding and Ca^{2+} release through the IP_3 receptor (Bourguigon *et al.,* 1993), and a synthetic peptide corresponding to the C-terminal ankyrin-binding motif blocks the ankyrin-induced inhibitory effects on IP_3 binding and Ca^{2+} release. Wes and colleagues (1995) have identified ankyrin repeats in a human homologue, TRPC1, of the *Drosophilia* IP_3-activated cation channel. The ankyrin repeats may either bind to the IP_3 receptor and conformational changes in the IP_3 receptor result in the activation of TRPC1 or, alternatively, the channel may be activated through signals that are transduced through spectrin bound to the ankyrin repeats.

TABLE I

Effects of Cytochalasin D and Exogenous Actin on Epithelial Ion Channel Activity

Channel	Cell type	Cytochalasin D	Actin	References
9-pS Na^+ channel	Coverslip-grown A6 cells	Activation	Activation by short filaments; no activation by G-actin or long filaments	Cantiello *et al.* (1991), Cantiello (1995)
4-pS Na^+ channel	Filter-grown A6 cells	Inactivation/ no activation	ND[a]	Smith *et al.* (1995)
CFTR	Transfected mouse mammary adenocarcinoma cell line	Activation	Activation by short filaments	Prat *et al.* (1995)
33-pS Cl^- channel	SV40LT immortalized rabbit proximal tubule cell line	Inactivation	Activation by long filaments, partial activation by short filaments; no activation by G-action	Suzuki *et al.* (1993)
30-pS K^+ channel	Rat cortical collecting duct principal cells	Inactivation/ no activation	No inactivation or activation by G-actin or actin filaments	Wang *et al.* (1994)

[a] ND, not determined.

The cloning of epithelial ion channels will also facilitate use of the yeast two-hybrid system for identifying specific membrane–cytoskeleton proteins or cytoskeleton-associated proteins interacting with ion channels. Kim and co-workers (1995) have recently used the two-hybrid system to identify an association between a human neuronal Shaker K^+ channel and postsynaptic density protein-95 (PSD-95), a membrane-associated putative guanylate kinase. This association is involved in the clustering of Shaker K^+ channels in specific neuronal membrane domains. The identification of cytoskeleton or cytoskeleton-associated proteins interacting with a specific ion channel will allow the role of these proteins in mediating channel activity to be examined using a combination of molecular, biochemical, and biophysical techniques. It will be important to determine whether physiologically relevant hormones or second messengers that up- or downregulate channel activity alter the cytoskeleton–channel interactions through posttranslational modifications such as phosphorylation, dephosphorylation, methylation, acetylation, or myristolation. Ultimately, the role that the membrane–cytoskeleton plays in the coordinate regulation between various ion channels and membrane transport proteins both within and between specific membrane domains will need to be explored.

In this chapter we have attempted to present an overview of our current understanding of the interactions between the membrane–cytoskeleton and ion channels in transporting epithelial cells and the role they play in regulating channel activity. Although it is clear that interactions between the membrane–cytoskeleton and epithelial ion channels are involved in the regulation of channel activity, the mechanisms whereby these interactions regulate channel activity and the physiological roles of these mechanisms remain to be elucidated.

Acknowledgments

This work was supported by funds from the National Institutes of Health to P. R. Smith (DK-46705) and D. J. Benos (DK-37206).

References

Aderem, A. (1992). The MARCKS brothers: A family of protein kinase C substrates. *Cell* **71,** 713–716.

Angelides, K. J., Elmer, L. W., Loftus, D., and Elson, E. (1988). Distribution and lateral mobility of voltage dependent sodium channels in neurons. *J. Cell Biol.* **106,** 1911–1925

Awayda, M. S., Ismailov, I. I., Berdiev, B. K., and Benos, D. J. (1995). A cloned renal epithelial Na^+ channel proteins displays stretch activation in planar lipid bilayers. *Am. J. Physiol.* **268,** C1450–C1459.

Bennett, V., and Gilligan, D. M. (1993). The spectrin-based membrane structure and micron-scale organization of the plasma membrane. *Annu. Rev. Cell Biol.* **9,** 27–66

Benos, D. J., Saccomani, G., Brenner, B. M., and Sariban-Sohraby, S. (1986). Purification and characterization of the amiloride-sensitive sodium channel from A6 cultured cells and bovine renal papilla. *Proc. Natl. Acad. Sci. U.S.A.* **83,** 8525–8529.

Benos, D. J., Saccomani, G., and Sariban-Sohraby, S. (1987). The epithelial sodium channel: Subunit number and location of amiloride-binding site. *J. Biol. Chem.* **262,** 10613–10618.

Benos, D. J., Cunningham, S., Baker, R. R., Beason, K. B., Oh, Y., and Smith, P. R. (1992). Molecular characterization of amiloride-sensitive sodium channels. *Rev. Physiol. Biochem. Pharmacol.* **120,** 31–113.

Benos, D. J., Awayada, M. S., Ismailov, I. I., and Johnson, J. P. (1995). Structure and function of amiloride-sensitive Na^+ channels. *J. Membrane Biol.* **143,** 1–18.

Bourguignon, L. Y. W., and Jin, H. (1995). Identification of the ankyrin-binding domain of the mouse T-lymphoma cell inositol 1,4,5,-triphosphate (IP_3) receptor and its role in the regulation of IP_3-mediated internal Ca^{2+} release. *J. Biol. Chem.* **270,** 7257–7260.

Bourguignon, L. Y. W., Jin, H., Iida, N., Brandt, N. R., and Zhang, S. H. (1993). The involvement of ankyrin in the regulation of inositol 1, 4, 5-triphosphate receptor-mediated internal Ca^{2+} release from Ca^{2+} storage vesicles in mouse T-lymphoma cells. *J. Biol. Chem.* **268,** 7290–7297.

Bubien, J. K., and Warnock, D. G. (1993). Amiloride-sensitive sodium conductance in human B lymphoid cells. *Am. J. Physiol.* **265,** C1175–C1183.

Canessa, C. M., Horisberger, J.-D., and Rossier, B. C. (1993). Epithelial sodium channel related to proteins involved in neurodegeneration. *Nature (London)* **361,** 467–470.

Canessa, C. M., Schild, L., Buell, G., Gautschi, I., Horisberger, J.-D., and Rossier, B. C. (1994). Amiloride-sensitive epithelial Na^+ channel is made up of three homologous subunits. *Nature (London)* **367,** 463–467.

Cantiello, H. F. (1995). Role of the actin cytoskeleton on epithelial Na^+ channel regulation. *Kidney Int.* **48,** 970–984.

Cantiello, H. F., Stow, J. L., Prat, A. G., and Ausiello, D. A. (1991). Actin filaments regulate epithelial Na^+ channel activity. *Am. J. Physiol.* **261,** C882–C888.

Cantiello, H. F., Prat, A. G., Reisin, I. L., Abraham, E. H., Ercole, L. B., Amarara, J. F., Gregory, R. J., and Ausiello, D. A. (1994). External ATP activates the cystic fibrosis transmembrane conductance regulator. *J. Biol. Chem.* **269,** 11224–11232.

Chang, C. H., Takeuchi, H., Ito, T., Machida, K., and Ohnishi, S. I. (1981). Lateral mobility of erythrocyte proteins studied by fluorescence photobleaching recovery technique. *J. Biochem.* **90,** 997–1004.

Davis, J., and Bennett, V. (1990). The anion exchanger and Na^+,K^+ATPase interact with distinct sites on ankyrin in *in vitro* assays. *J. Biol. Chem.* **265,** 17252–17256.

Ding, G., Franki, N., Condeelis, L., and Hays, R. M. (1991). Vasopressin depolymerizes F-actin in toad urinary bladder epithelial cells. *Am. J. Physiol.* **260,** C9–C16.

Dragsten, P. R., Blumenthal, R., and Handler, J. S. (1981). Membrane asymmetry in epithelia: Is the tight junction a barrier to diffusion in the plasma membrane? *Nature (London)* **294,** 718–722.

Duc, C., Farman, N., Canessa, C. M., Bonvalet, J.-P., and Rossier, B. C. (1994). Cell-specific expression of epithelial sodium channel α, β and γ subunits in aldosterone-responsive epithelia from rat: Localization by *in situ* hybridization and immunocytochemistry. *J. Cell Biol.* **127,** 1907–1921.

Eaton, D. C., Becchetti, A., Ma, H., and Ling, B. N. (1995). Renal sodium channels: Regulation and single channel properties. *Kidney Int.* **48,** 941–949.

Edidin, M., and Stroynowski, I. (1991). Differences between the lateral organization of conventional and inositol phospholipid-anchored membrane proteins. A further definition of micrometer scale membrane domains. *J. Cell Biol.* **112,** 1143–1150.

Els, W. J., and Chou, K.-Y. (1993). Sodium-dependent regulation of epithelial sodium channel densities in frog skin: A role for the cytoskeleton. *J. Physiol.* **462,** 447–464.

Franki, N., Ding, G., Gao, Y., and Hays, R. M. (1992). Effect of cytochalasin D on the actin cytoskeleton of the toad bladder epithelial cell. *Am. J. Physiol.* **263,** C995–C1000.

Fuller, C. M., and Benos, D. J. (1992). CFTR! *Am. J. Physiol.* **263,** C267–C283.

Fuller, C. M., Bridges, R. J., and Benos, D. J. (1994). Forskolin- but not ionomycin-evoked Cl^- secretion in colonic epithelia depends upon intact microtubules. *Am. J. Physiol.* **266,** C661–C668.

Fuller, C. M., Awayda, M. S., Arrate, M. P., Bradford, A. L., Morris, R. G., Canessa, C. M., Rossier, B. C., and Benos, D. J. (1995). Cloning of a bovine epithelial Na^+ channel subunit. *Am. J. Physiol.* **269,** C641–C654.

Gao, Y., Franki, N., and Hays, R. M. (1993). Vasopressin-induced redistribution of ankyrin in rat inner medullary collecting duct [Abstract]. *Mol. Biol. Cell* **4,** 57a.

Gao, Y., Franki, N., and Hays., R. M. (1994). The relationship of fodrin to the vasopressin (AVP)-responsive actin cytoskeleton [Abstract]. *J. Am. Soc. Nephrol.* **4,** 853.

Golan, D., and Veatch, W. (1980). Lateral mobility of band 3 in the human erythrocyte membrane studied by fluorescence photobleaching recovery: Evidence for control by cytoskeletal interactions. *Proc. Natl. Acad. Sci. U.S.A.* **77,** 2537–2541.

Goodman, D. B. P., Wong, M., and Rasmussen, H. (1975). Aldosterone induced membrane phospholipid fatty acid metabolism in toad urinary bladder. *Biochemistry* **14,** 2803–2809.

Grotmol, T., and van Dyke, R. W. (1992). Prostaglandin- and theophylline-induced Cl secretion in rat distal colon is inhibited by microtubule inhibitors. *Dig. Dis. Sci.* **37,** 1709–1717.

Gschwenter, M., Nagl, U. O., Wöll, E., Schmarda, A., Ritter, M., and Paulmichl, M. (1995). Antisense oligonucleotides suppress cell-volume-induced activation of chloride channels. *Pflügers Arch.* **439,** 464–470.

Halm, D. R., Halm, S. T., DiBona, D. R., Frizzell, R. A., and Johnson, R. D. (1995). Selective stimulation of epithelial cells in colonic crypts: Relation to active chloride secretion. *Am. J. Physiol.* **269,** C929–C942.

Higgins, C. F. (1995). The ABC of channel regulation. *Cell* **96,** 693–696.

Hitt, A. L., and Luna, E. J. (1994). Membrane interactions with the actin cytoskeleton. *Curr. Opin. Cell Biol.* **6,** 120–130.

Hug, T., Koslowsky, T., Ecke, D., Greger, R., and Kunzelmann, K. (1995). Actin-dependent activation of ion conductances in bronchial epithelial cells. *Pflügers Arch.* **429,** 682–690.

Ismailov, I. I., McDuffie, J. H., Sariban-Sohraby, S., Johnson, J. P., and Benos, D. J. (1994). Carboxyl methylation activates purified renal amiloride-sensitive Na^+ channels in planar lipid bilayers. *J. Biol. Chem.* **269,** 22193–22197.

Kemendy, A. E., Kleyman, T. R., and Eaton, D. C. (1992). Aldosterone alters the open probability of amiloride-blockable sodium channels in A6 epithelia. *Am. J. Physiol.* **263,** C825–C837.

Kim, E., Neithammer, M., Rothschild, A., Jan, Y. N., and Sheng, M. (1995). Clustering of Shaker-type K^+ channels by interaction with a family of membrane-associated guanylate kinases. *Nature* (*London*) **378,** 85–88.

Kleyman, T. R., Kraehenbuhl, J.-P., and Ernst, S. A. (1991). Characterization and cellular localization of the epithelial Na^+ channel. Studies using an anti-Na^+ channel antibody raised by an anti-idiotypic route. *J. Biol. Chem.* **266,** 3907–3915.

Kleyman, T. R., Coupaye-Gerard, B., and Ernst, S. A. (1992). Aldosterone does not alter apical cell surface expression of epithelial sodium channels in the amphibian cell line, A6. *J. Biol. Chem.* **267,** 9622–9628.

Kleyman, T. R., Ernst, S. A., and Coupaye-Gerard, B. (1994). Arginine vasopressin and forskolin regulate apical cell surface expression of epithelial Na^+ channels in A6 cells. *Am. J. Physiol.* **266,** F506–F511.

Krapivinsky, G. B., Ackerman, M. J., Gordon, E. A., Krapivinsky, L. D., and Clapham, D. E. (1994). Molecular characterization of a swelling-induced chloride conductance regulatory protein, pI_{Cln}. *Cell* **76,** 439–448.

Kupitz, Y., and Atlas, D. (1993). A putative ATP-activated Na^+ channel involved in sperm-induced fertilization. *Science* **261,** 484–486.

Leiser, J., and Molitoris, B. A. (1993). Disease processes in epithelia: The role of the actin cytoskeleton and altered surface membrane polarity. *Biochim. Biophys. Acta* **1225,** 1–13.

Li, X.-J., Blackshaw, S., and Snyder, S. H. (1994). Expression and localization of amiloride-sensitive Na^+ channel indicate a role for non-taste cells in taste perception. *Proc. Natl. Acad. Sci. U.S.A.* **91,** 1814–1818.

Lien, E. L., Goodman, D. B. P., and Rasmussen, H. (1975). Effects of an acetyl-coenzyme A carboxylase inhibitor and a sodium-sparing diuretic on aldosterone-stimulated sodium transport, lipid synthesis, and phospholipid fatty acid composition in the toad urinary bladder. *Biochemistry* **14,** 2749–2754.

Lingueglia, E., Voilley, N., Waldmann, R., Lazdunski, M., and Barbry, P. (1993). Expression cloning of an epithelial amiloride-sensitive Na^+ channel. *FEBS Lett.* **318,** 95–99.

Lingueglia, E., Renard, S., Waldman, R., Voilley, N., Champigney, G., Plass, H., Lazdunski, M., and Barbry, P. (1994). Different homologous subunits of the amiloride sensitive Na^+ channel are differently regulated by aldosterone. *J. Biol. Chem.* **269,** 13736–13739.

Madreperla, S. A., Edidin, M., and Adler, R. (1989). Na^+,K^+-adenosine triphosphatase polarity in retinal photoreceptors: A role in cytoskeletal attachments. *J. Cell Biol.* **109,** 1483–1493.

Mays, R. W., Beck, K. A., and Nelson, W. J. (1994). Organization and function of the cytoskeleton in polarized epithelial cells: A component of the protein sorting machinery. *Curr. Opin. Cell Biol.* **6,** 16–24.

Morris, R., Tousson, A., Benos, D. J., and Shafer, J. A. (1995). Microtubule disruption does not prevent vasotocin (AVT) activation of Na^+ channels [Abstract]. *J. Am. Soc. Nephrol.* **6,** 347a.

Morrow, J. S., Cianci, C. D., Ardito, T., Mann, A. S., and Kashgarian, M. (1989). Ankyrin links fodrin to the alpha subunit of Na, K-ATPase in Madin-Darby canine kidney cells and in intact renal tubule cells. *J. Cell Biol.* **108,** 455–465.

Nelson, W. J., and Veshnock, P. J. (1987). Ankyrin binding to (Na^+ + K^+)ATPase and implications for the organization of membrane domains in polarized cells. *Nature* (*London*) **328,** 533–536.

Oh, Y. S., and Benos, D. J. (1993). Single channel characteristics of a purified bovine renal amiloride-sensitive Na^+ channel in planar lipid bilayers. *Am. J. Physiol.* **264,** C1489–C1499.

Oh, Y. S., Smith, P. R., Bradford, A. L., Keeton, D., and Benos, D. J. (1993). Regulation by phosphorylation of purified epithelial Na^+ channels in planar lipid bilayers. *Am. J. Physiol.* **265,** C85–C91.

Paller, M. S. (1994). Lateral mobility of Na,K-ATPase and membrane lipids in renal cells. Importance of cytoskeletal integrity. *J. Membrane Biol.* **142,** 127–135.

Palmer, L. G. (1995). Epithelial Na channels and their kin. *NIPS* **10,** 61–67.

Paulmichl, M., Li, Y., Wickman, K., Ackerman, M., Peralta, E., and Clapham, D. E. (1992). New mammalian chloride channel identified by expression cloning. *Nature* (*London*) **356,** 238–241.

Prat, A. G., Ausiello, D. A., and Cantiello, H. F. (1993a). Vasopressin and protein kinase A activate G protein-sensitive epithelial Na^+ channels. *Am. J. Physiol.* **265,** C218–C223.

Prat, A. G., Bertorello, A. M., Ausiello, D. A., and Cantiello, H. F. (1993b). Activation of epithelial Na^+ channels by protein kinase A requires actin filaments. *Am. J. Physiol.* **265,** C224–C233.

Prat, A. G., Xiao, Y.-F., Ausiello, D. A., and Cantiello, H. F. (1995). cAMP-independent regulation of CFTR by the actin cytoskeleton. *Am. J. Physiol.* **268,** C1552–C1561.

Ren, R., Mayer, B., Cicchetti, P., and Baltimore, D. (1993). Identification of a ten amino acid proline rich SH3 binding site. *Science* **259,** 1157–1161.

Renard, S., Voilley, N., Bassilana, F., Lazdunski, M., and Barbry, P. (1995). Localization and regulation by steroids of the α, β and γ subunits of the amiloride-sensitive Na^+ channel in colon, lung, and kidney. *Pflügers Arch.* **430,** 299–307.

Riordan, J. R. (1993). The cystic fibrosis transmembrane conductance regulator. *Annu. Rev. Physiol.* **55,** 609–630.

Rotin, D., Bar-Sagi, D., O'Brodovich, H., Merilainen, J., Lehto, V. P., Canessa, C. M., Rossier, B. C., and Downey, G. P. (1994). An SH3 binding region in the epithelial Na^+ channel (αrENaC) mediates its localization at the apical membrane. *EMBO J.* **13,** 4440–4450.

Sariban-Sohraby, S., and Fisher, R. S. (1995). Guanine-nucleotide-dependent carboxymethylation: A pathway for aldosterone modulation of apical Na^+ permeability in epithelia. *Kidney. Int.* **48,** 965–969.

Sariban-Sohraby, S., Burg, M., Wiesmann, W. P., Chiang, P. K., and Johnson, J. P. (1984). Methylation increases sodium transport into A6 apical membrane vesicles: Possible mode of aldosterone action. *Science* **225,** 745–746.

Sariban-Sohraby, S., Abramow, M., and Fisher, R. S. (1992). Single channel behavior of a purified epithelial Na^+ channel subunit that binds amiloride. *Am. J. Physiol.* **263,** C1111–C1117.

Sariban-Sohraby, S., Fisher, R. S., and Abramow, M. (1993). Aldosterone-induced and GTP-stimulated methylation of a 90-kDa polypeptide in the apical membrane of A6 epithelia. *J. Biol. Chem.* **268,** 26613–26617.

Shimkets, R. S., Warnock, D. G., Bositis, C. M., Nelson-Williams, C., Hansson, J. H., Schambelan, M., Gill, J. R. Jr., Ulick, S., Milora, R. V., Findling, J. W., Canessa, C. M., Rossier, B. C., and Lifton, R. P. (1994). Liddle's syndrome: Heritable human hypertension caused by mutations in the β subunit of the epithelial Na^+ channel. *Cell* **79,** 6407–6414.

Simon, H., Gao, Y., Franki, N., and Hays, R. M. (1993). Vasopressin depolymerizes apical F-actin in rat inner medullary collecting duct. *Am. J. Physiol.* **265,** C757–C762.

Smith, P. R., Saccomani, G., Joe, E.-H., Angelides, K. J., and Benos, D. J. (1991). Amiloride-sensitive sodium channel is linked to the cytoskeleton in renal epithelial cells. *Proc. Natl. Acad. Sci. U.S.A.* **88,** 6971–6975.

Smith, P. R., Stoner, L. C., Viggiano, S. C., Angelides, K. J., and Benos, D. J. (1995). Effects of vasopressin and aldosterone on the lateral mobility of epithelial Na^+ channels in A6 renal epithelial cells. *J. Membrane Biol.* **147,** 195–205.

Sorscher, E. J., Accavitti, M. A., Keeton, D., Steadman, E., Frizzell, R. A., and Benos, D. J. (1988). Antibodies against purified epithelial sodium channel protein from bovine renal papilla. *Am. J. Physiol.* **254,** C835–C843.

Staub, O., Verrey, F., Kleyman, T. R., Benos, D. J., Rossier, B. C., and Krahenbuhl, J.-P. (1992). Primary structure of an apical protein from *Xenopus laevis* that participates in amiloride-sensitive sodium channel activity. *J. Cell Biol.* **119,** 1497–1506.

Stoner, L. C., Engbretson, B. G., Viggiano, S. C., Benos, D. J., and Smith, P. R. (1995). Amiloride-sensitive apical membrane sodium channels of everted *Ambystoma* collecting tubule. *J. Membrane Biol.* **144,** 147–156.

Stutts, M. J., Canessa, C. M., Olsen, J. C., Hamrick, M., Cohn, J. A., Rossier, B. C., and Boucher, R. C. (1995a) CFTR as a cAMP-dependent regulator of sodium channels. *Science* **269,** 847–850.

Stutts, M. J., Lazarowskl, E. R., Paradiso, A. M., and Bouche, R. C. (1995b). Activation of CFTR Cl^- conductance in polarized T84 cells by luminal extracellular ATP. *Am. J. Physiol.* **268,** C425–C433.

Suzuki, M., Miyazaki, K., Ikeda, M., Kawaguchi, Y., and Sakai, O. (1993). F-actin network may regulate a Cl^- channel in proximal tubule cells. *J. Membrane Biol.* **134,** 31–39.

Tousson, A., Fuller, C. M., and Benos, D. J. (1996). Apical recruitment of CFTR in T84 cells is dependent on cAMP and microtubules but not Ca^{2+} or microfilaments. *J. Cell Sci.* In press.

Van Renterghem, C., and Lazdunski, M. (1991). A new non voltage-dependent epithelial-like Na^+ channel in vascular smooth muscle cells. *Pflügers Arch.* **419,** 401–408.

Verrey, F., Groscurth, P., and Bolliger, U. (1995). Cytoskeletal disruption in A6 kidney cells: Impact on endo/exocytosis and NaCl transport regulation by antidiuretic hormone. *J. Membrane Biol.* **145,** 193–204.

Vigne, B., Champigny, G., Marsault, R., Barbry, P., Frelin, C., and Lazdunski, M. (1989). A new type of amiloride-sensitive cationic channel in endothelial cells of brain microvessels. *J. Biol. Chem.* **264,** 7663–7668.

Waldman, R., Champigny, G., Bassilana, F., Voilley, M., and Lazdunski, M. (1995). Molecular cloning and functional expression of a novel amiloride-sensitive Na^+ channel. *J. Biol. Chem.* **270,** 27411–27414.

Wang, W. (1995). View of K^+ secretion through the apical K channel of the cortical collecting duct. *Kidney Int.* **48,** 1024–1030.

Wang, W.-H., Cassola, A. C., and Giebisch, G. (1994). Involvement of actin cytoskeleton in modulation of apical K channel activity in rat collecting duct. *Am. J. Physiol.* **267,** F592–F598.

Weeds, A., and Maciver, S. (1993). F-actin capping proteins. *Curr. Opin. Cell Biol.* **5,** 63–69.

Weismann, W. P., Johnson, J. P., Miura, G. A., and Chiang, P. K. (1985). Aldosterone-stimulated transmethylations are linked to sodium transport. *Am. J. Physiol.* **248,** F43–F47.

Wes, P. D., Chevesich, J., Jeromin, A., Rosenberg, C., Stetten, G., and Montell, C. (1995). TRPC1, a human homolog of Drosophila store-operated channel. *Proc. Natl. Acad. Sci. U.S.A.* **92,** 9652–9656.

Wills, N. K., Purcell, R. K., Clausen, C., and Millinoff, L. P. (1993). Effects of aldosterone on the impedance properties of cultured renal amphibian epithelia. *J. Membrane Biol.* **133,** 17–27.

Yin, H. L., Zaner, K. S., and Stossel, T. P. (1980). Ca^{2+} control of actin gelation: Interaction of gelsolin with actin filaments and regulation of actin gelation. *J. Biol. Chem.* **255,** 9494–9500.

Zhang, F., Lee, G. M., and Jacobson, K. (1993). Protein lateral mobility as a reflection of membrane microstructure. *Bioessays* **15,** 579–588.

CHAPTER 17

Role of Actin Filament Organization in Ion Channel Activity and Cell Volume Regulation

Horacio F. Cantiello and Adriana G. Prat
Renal Unit, Massachusetts General Hospital East, Charlestown, Massachusetts 02129 and Department of Medicine, Harvard Medical School, Boston, Massachusetts 02115

I. INTRODUCTION

The actin-based cytoskeleton, consisting of actin filaments and associated proteins, is assembled in a network localized in the cortical cytoplasm lying beneath the plasma membrane of cells. This dynamic structure plays an essential role in the regulation of cellular events, including the stability of cell shape and the onset of cell motility (Stossel, 1984), the distribution of integral membrane proteins (Bennett and Lambert, 1991; Nelson and Veshnock, 1987), and the control of hormone action (Hall, 1984). A current view of the structural interaction between actin networks and the cellular plasma membrane involves either the direct binding of actin to the plasma membrane (Schwartz and Luna, 1988) or the anchoring of proteins, including actin-binding protein (ABP or filamin) (Weihing, 1985), spectrin (fodrin, the nonerythroid form of spectrin), and ankyrin (Bennett and Lambert, 1991; Morrow, 1989). These ABPs couple the actin cytoskeleton to a variety of transmembrane proteins, including ion transport molecules (Nelson and Veshnock, 1987; Srimivasan *et al.,* 1988), such as the band 3 anion exchanger (Drenckhahn *et al.,* 1985), epithelial Na^+,K^+-ATPase (Morrow *et al.,* 1989), and rat brain voltage-sensitive Na^+ channels (Edelstein *et al.,* 1988). Other ion transport proteins, including the Na^+,K^+Cl^-–co-transporter (Jorgensen *et al.,* 1984), the Na^+–H^+ exchanger (Watson *et al.,* 1992), and the Na^+ channel complex of epithelial A6 cells (Smith *et al.,* 1991), are also linked to cytoskeletal components. It thus appears probable that actin filament networks have a functional modulatory role in the control of membrane function. Functional interactions between the actin cytoskeleton and ion channels have been reported for Ca^{2+} channels from neurons (Johnson and Byerly, 1993; Rosenmund and Westbrook, 1993) and Na^+ channel activity in epithelial cells (Cantiello *et al.,* 1991b).

At the center of this interface there is actin, an intracellular protein accounting for more than 20% of the total cell protein (Pollard and Cooper, 1986). Under physiological conditions, including high ionic strength and such substrates as ATP, actin readily polymerizes into long helical (rigid) filaments of several micrometers in length. Once polymerized, filamentous (F) actin remains in dynamic equilibrium with monomeric (G) actin. Although only polymerized actin is known to have a biological role, controversy still exists as to what is the average length of physiologically relevant actin filaments (Glenney and Weber, 1981). Actin can be found in several different states within the intracellular compartment. Dynamic transitions entailing changes from one conformation to another may be central to such mechanically related events as cell locomotion and rounding, cytokinesis and spreading, and intracellular transport to the cell surface (Cramer *et al.,* 1994). Actin filaments may be also linked to osmotically induced phase transitions (Ito *et al.,* 1987), which in turn may represent novel electro-osmotic signaling events (Cantiello *et al.,* 1991a). ABPs are essential in this regard as they may also help modify the 3-dimensional conformation of actin filaments by cross-linking F-actin into either networks of gel conformation or tight bundles such as those found in microvilli (Matsudaira, 1991; Weeds, 1982).

A large number of proteins that modulate actin filament organization have been described (Weeds, 1982). These proteins, generally known as ABPs, bind to and interact with the various forms of actin in a highly specific manner. ABPs can bind to either G- or F-actin, thus regulating actin polymerization, or may also help link actin filaments to other relevant intracellular structures, including microtubules (Pollard *et al.,* 1984), and to the plasma membrane (Geiger, 1985; Hitt and Luna, 1994). ABPs, which bind G-actin and function by stabilizing the pool of monomeric actin, include profilin and DNAse I. These proteins therefore help maintain a balance between G- and F-actin concentrations. Another important family of ABPs, however, actually binds already formed actin filaments (F-actin). Most F-actin–binding proteins are known to have at least two actin-binding domains, thus allowing coupling to more than one actin filament. Depending on the distance between the actin-binding domains, these proteins may enable either the bundling of actin (tightly packed arrays of filaments), as observed in the complexes of F-actin found in microvilli, or the cross-linking of actin into 3-dimensional actin filament networks with gel behavior (Matsudaira, 1991; Weihing, 1985). In contrast to actin bundling, the actin cross-linking property of these F-actin–binding proteins elicits the networking of actin filaments in loose, open structures in which actin filaments are orthogonally oriented (Matsudaira, 1991). This arrangement of the actin filaments resists local deformation and is the basis for the gellike properties of the cytoplasm (Luby-Phelps, 1994). One of the most important proteins in this group is filamin (Weihing, 1985) and its homologue, ABP (Hartwig and Kwiatkowski, 1991; Shizuta *et al.,* 1976). Filamin and ABP are members of a group of homodimeric proteins (~500 kDa) that are known to cross-link actin filaments *in vitro* (Hartwig, 1995). Studies have determined that filamin also helps actin filaments interact with specific membrane glycoproteins, thus enabling a functional interaction with the plasma membrane distinct from its actin filament cross-linking capability (Ohta *et al.,* 1991).

II. ROLE OF ACTIN IN Na^+ CHANNEL REGULATION

A functional interaction between actin filament organization and stretch-activated channels has been previously observed (Guharay and Sachs, 1984; Sackin, 1989; Ubl *et al.,* 1988). It was originally concluded that stretch activation of ion channels implicated changes in the elastic activity of the cortical cytoskeleton (Guharay and Sachs, 1984). This interaction, however, may be linked to a direct coupling of cytoskeletal components to transport proteins or may be the result of sequential interaction of actin filaments with membrane regulatory pathways. The hypothesis that the actin cytoskeleton was directly involved in the regulation of epithelial Na^+ channels was first supported by immuno–co-localization studies showing that Na^+ channels always appeared on the cell surface in close proximity to actin filaments (Cantiello *et al.,* 1991b; Smith *et al.,* 1991). This finding raised the possibility that a potential Na^+ channel–actin filament interaction might be an early feature of epithelial cell development (Cantiello *et al.,* 1991b). While this interaction may serve as a means to control the spatial distribution of Na^+ channels in the apical membrane of epithelial cells, which is a requirement in polarized epithelia, the possibility also existed that the co-localization of actin filaments with the apical Na^+ channels would be of a functional nature.

To assess the functional role of endogenous cortical actin networks on apical Na^+ channel activity in A6 cells, actin filament organization was modified with cytochalasin D (CD), a fungal toxin known to depolymerize actin filaments (Goddette and Frieden, 1986; Schliwa, 1982). Acute addition of CD induced Na^+ channel activity (Cantiello *et al.,* 1991b). Because DNase I inhibited the CD-induced Na^+ channel activity, the data allowed us to speculate that short actin filaments might be involved in effecting ion channel activation. This was further supported by the fact that the CD effect was transient and was only observed in cells treated with the drug for short periods of time. Longer exposures to CD, however, resulted in dramatic changes in cell shape and the loss of Na^+ channel activation (Prat *et al.,* 1993). Addition of purified "short" actin filaments to the cytoplasmic side of excised inside-out patches also induced and/or increased Na^+ channel activity (Fig. 1) (Cantiello *et al.,* 1991b). Whenever actin was added after

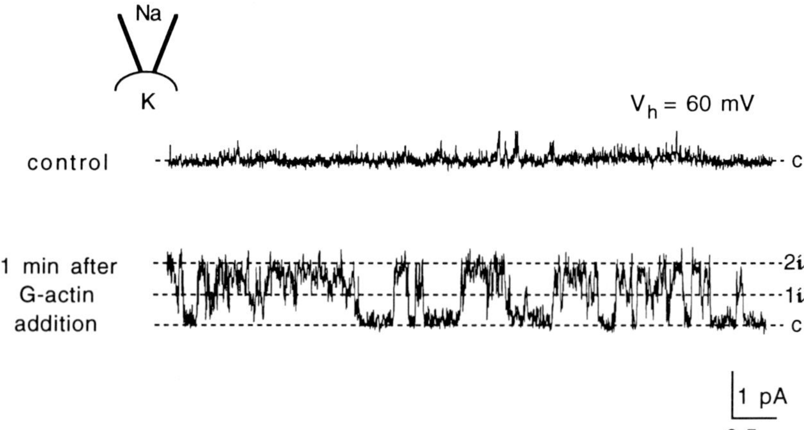

FIGURE 1 Effect of actin and ATP on Na^+ channel activity in epithelial A6 cells. The functional role of actin on Na^+ channel regulation was assessed by addition of actin (1 mg/ml) to the cytoplasmic side of quiescent excised inside-out patches from A6 epithelial cells (top tracing), which induced Na^+ channel activity within 5 min (bottom tracing). (From Cantiello *et al.,* 1991b).

being polymerized to achieve predominantly long filaments, however, no Na^+ channel activation was observed. Thus, the length of the actin filament appears to be the relevant parameter for Na^+ channel activation. Actin filaments of various lengths were therefore tested on the Na^+ channel activity by the direct use of ABPs with known specific interactions with the actin molecule. Addition of DNAse I, which inhibits actin polymerization, to the cytoplasmic surface of apical membrane patches prior to the addition of actin prevented the stimulatory effect of exogenous actin on Na^+ channels. In addition, DNAse I–actin complexes were also completely ineffective in activating Na^+ channel activity. Thus, the contention that short actin filaments were the active species responsible for Na^+ channel regulation was further strengthened. This hypothesis was confirmed by the addition of actin polymerized in the presence of different concentrations of gelsolin, a Ca^{2+}-sensitive protein that shortens actin filaments depending on the actin–gelsolin molar ratio (Yin *et al.,* 1980, 1981). Actin–gelsolin complexes at a ratio of 4:1, consistent with short oligomers (trimers, tetramers) induced Na^+ channel activity in excised patches with no spontaneous activity (Cantiello *et al.,* 1991b). The critical ratio of actin to gelsolin to elicit Na^+ channel activity was $<8:1$.

These findings provided the first evidence for a functional interface entailing "short actin" filaments as relevant ligands that contribute to the onset and regulation of a specific membrane function. Furthermore, the data also raised the possibility that ABPs can be used as tools to investigate whether specific interactions between the plasma membrane, ion channels, and the various components of the actin cytoskeleton do indeed exist. Exploration of the regulatory role of actin filaments on ion transport mechanisms of known sequences, including Na^+,K^+-ATPase and the anion channel cystic fibrosis transmembrane conductance regulator (CFTR), has provided more recent evidence for a direct interaction between actin and these ion transport molecules (Cantiello, 1995; Prat *et al.,* 1995).

III. Regulatory Role of Actin in Epithelial Na^+,K^+-ATPase Activity

Studies have established that actin filaments also stimulate Na^+,K^+-ATPase activity by a mechanism that appears to entail a direct association of actin to the enzyme (Cantiello, 1995). This interaction suggests that a conformational change of the enzyme is elicited by actin, which is reflected in an increased affinity for its intracellular substrate, Na^+. The molecular nature of the stimulatory effect of actin on Na^+,K^+-ATPase, however, has unmasked further details. In agreement with our previous findings on the regulatory role of "short" actin filaments on Na^+ channel activity (Cantiello *et al.*, 1991b), neither monomeric nor prepolymerized actin was responsible for the stimulatory effect on Na^+,K^+-ATPase. The stoichiometry of interaction between actin and Na^+,K^+-ATPase provided a calculated 4.47 Hill coefficient, thus indicating an interaction for activation that required between four and five actin monomers per enzyme. Monomeric actin, however, also binds to Na^+,K^+-ATPase, although without eliciting a stimulatory response. Thus stimulation is effected after elongation of the actin polymer to a size considered to be "short." Stimulation of Na^+,K^+-ATPase by actin would likely increase until a length of the polymer is reached that no longer supports any further stimulation. Sequence homology studies between the actin-binding domain of several ABPs and Na^+,K^+-ATPase, and experimental evidence from immunoblotting analysis, strengthen the contention that actin directly interacts with the enzyme (Cantiello, 1995). This has also been demonstrated for the regulation of the anion channel CFTR by actin (Prat *et al.*, 1995).

IV. ROLE OF ACTIN IN CFTR FUNCTION

The anion-selective channel CFTR, whose dysfunction is responsible for the onset of cystic fibrosis, is regulated by cAMP through the activation of cAMP-dependent protein kinase (PKA). The nature of CFTR activation by PKA is unknown. We recently evaluated the effect of actin and associated proteins on CFTR activation and compared these results with the disruption of the endogenous actin cytoskeleton (Prat *et al.*, 1995). The severing of endogenous actin filaments with short incubations with CD, or direct addition of exogenous actin, induced CFTR activation (Prat *et al.*, 1995). Thus, any process entailing an increase in the availability of "short" actin filaments by either disruption of preexisting filaments or the formation of new actin filaments activated CFTR. A decrease in the number of "short" actin filaments by preventing actin polymerization with DNAse I or bundling filaments with filamin, in contrast, inhibited CFTR function. Thus neither long linear nor monomeric actin is responsible for CFTR activation, similar to findings reported in our previous studies with epithelial Na^+ channels (Prat *et al.*, 1993) and Na^+,K^+-ATPase (Cantiello, 1995). A novel finding of this study, however, was the demonstration that complete cytoskeletal disruption blunted the activation of CFTR by PKA (Fig. 2). This suggested that a mechanism linking these two regulatory pathways was present. For example, actin itself is a suitable substrate for specific phosphorylation mediated by PKA, and phosphorylated actin is a poor substrate for sustaining actin polymerization (Prat *et al.*, 1993). A mixed pool of phosphorylated

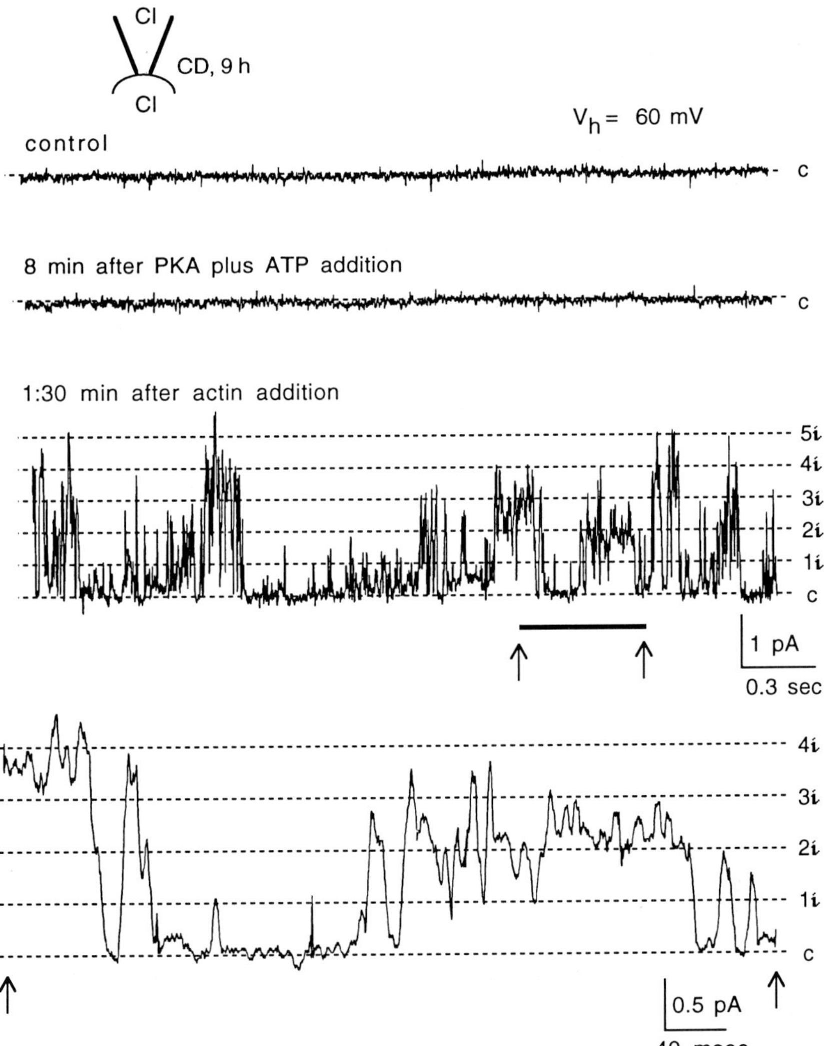

FIGURE 2 Effect of actin cytoskeleton disruption on the PKA activation of CFTR-expressing cells. To assess the role of actin filament organization on the response of CFTR to cAMP activation, CFTR-expressing cells were first treated with CD (10 μg/ml) for 9 h (Prat *et al.*, 1995). Addition of PKA (10 μg/ml) in the presence of ATP (1 mM) to excised inside-out patches from CD-treated CFTR cells (top tracing) had no effect on the ion channel activity after 8 min (second tracing). Subsequent addition of actin (1 mg/ml; bottom tracing), in contrast, stimulated Cl^- channel activity within 2 min. The bottom tracing shows an expansion of the segment marked between arrows of the third tracing.

and native actin monomers, therefore, will have a decreased ability to polymerize, resulting in a longer lived pool of "short" actin filaments, the active substrate for inducing ion channel activation (Prat *et al.,* 1993).

The comparison between the amino acid sequences of CFTR and various ABPs (Prat *et al.,* 1995) indicated that distinct domains of CFTR may represent novel actin-binding domains, thus raising the possibility that this channel protein may be directly regulated by the binding of actin. Both nucleotide-binding domains of CFTR were found to share domains with two families of ABPs, including severin and filamin (Prat *et al.,* 1995). The data indicate that CFTR is regulated by actin filaments whose organization, in turn, is associated with the cAMP activation of this transport protein. Phosphorylation of actin, in addition to phosphorylation of the channel, may play a significant role in the modulation of channel activity.

V. ROLE OF ABP IN CELL VOLUME REGULATION OF HUMAN MELANOMA CELLS

A. *Role of the Cytoskeleton*

The data summarized earlier support the concept that ion transport proteins may be directly linked to the actin cytoskeleton and thus effect a functional interface whereby changes in actin filament organization may be responsible for changes in membrane function. By extension, the plasma membrane–cortical cytoskeleton interface may also sense dynamic changes based in the bulk of the cytosolic actin cytoskeleton. The onset of cell activation, for example, usually involves changes in both membrane function and intracellular structures, including the actin cytoskeleton. From cell crawling to neutrophil activation, muscle contraction, and epithelial cell secretion, dramatic shifts in the actin cytoskeleton gel–sol transformations are triggered by cell activation (Hall, 1984; Mercier *et al.,* 1989; Stossel, 1993). These, in turn, may act as effectors, leading to a concerted effort by the cell to elicit a specific function. Interactions between the actin cytoskeleton, actin cross-linking proteins, and the various ion transport molecules are likely to take place as a result of environmental stress whereby cell shape or motility could be challenged. A particularly relevant example is the volume-regulatory response of cells to hypotonic stress (regulatory volume decrease (RVD)), as it involves the concerted exit of ions, frequently mediated by an initial ion channel activation (Christensen, 1987; Melmed *et al.,* 1981; Reuss, 1988; Ubl *et al.,* 1988; Welling *et al.,* 1985), with resultant water loss from the intracellular compartment (for reviews, see Lewis and Donaldson, 1990; Schultz, 1989; Strange, 1994). The volume-regulatory response also involves dynamic changes in actin filament organization (Sackin, 1989; Ubl *et al.,* 1988). This is supported by reversible changes in actin filament organization that are associated with hypotonicity (Ziyadeh *et al.,* 1992), and the effect of the actin filament disrupters cytochalasins B and D, which strongly blunt the cell volume recovery after osmotic swelling (Foskett and Spring, 1985; Linshaw *et al.,* 1991).

Cunningham *et al.* described several human melanoma cell lines completely lacking expression of ABP (ABP-280, nonmuscle filamin), a homodimer composed of 280-kDa subunits that, as noted earlier, promotes the cross-linking of actin filaments at wide angles and that also helps attach actin filaments to membrane glycoproteins (Cunningham *et al.,* 1992; Fox, 1985; Ohta *et al.,* 1991). ABP-deficient ($ABP^{(-)}$) cells display an impaired cell motility, a dysfunctional actin organization, and a pattern of continuous cell blebbing, further suggesting an impaired functional interaction between the plasma membrane and the cortical cytoskeleton. Genetic rescue of $ABP^{(-)}$ cells by transfection of full-length ABP cDNA resulted in the isolation of clonal sublines expressing ABP. These cells had almost complete cessation of cell blebbing and a large increase in cell motility (Cunningham *et al.,* 1992). Because cell blebbing may imply an unregulated bulk fluid movement across the plasma membrane, the $ABP^{(-)}$ melanoma cells have provided us with a useful example of a cell model in which an abnormal volume-regulatory response may be associated with dysfunctional actin filament–membrane protein interactions—in particular ion channel regulation and deranged cytoskeletal dynamics, both of relevance to a proper RVD response. Summarized next is our effort to assess the molecular physiology of the interaction between ABP, actin filaments, and the osmotically induced regulation of ion channel activity of $ABP^{(-)}$ and $ABP^{(+)}$ human melanoma cells.

The cell volume of both $ABP^{(-)}$ and $ABP^{(+)}$ cells, determined by Coulter counter cell sizing of freshly trypsinized cells in an isosmotic solution, and at room temperature, was stable for more than 40 min. The cell volume of $ABP^{(-)}$ cells was highly similar to that of $ABP^{(+)}$ cells, therefore, that the swelling equilibrium of the intracellular cytoskeleton of either cell type remains similar. Hypotonic stimuli were then administered in order to determine both the cell volume expansion and the RVD response of either cell type. In the presence of a hypotonic stimulus (50% dilution), $ABP^{(-)}$ cells were able to elicit an RVD response, although significantly less than that of the $ABP^{(+)}$ cells (Fig. 3A). The cell volume of $ABP^{(-)}$ and $ABP^{(+)}$ cells following the hypotonic stimulus reached its peak in approximately 1–2 min (Table I) and decreased to a plateau volume only 9% lower (Table II) after approximately 10 min (Fig. 3B). Thus, $ABP^{(-)}$ or $ABP^{(+)}$ cells were able to elicit an RVD response. An increase in the hypotonic stimulus

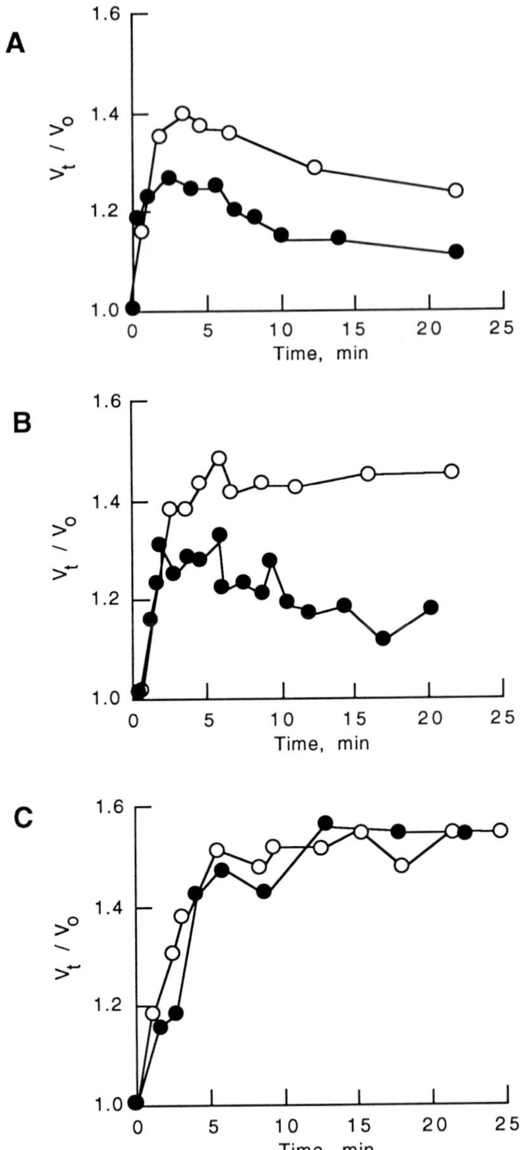

FIGURE 3 Regulatory cell volume response of ABP$^{(-)}$ and ABP$^{(+)}$ melanoma cells. The cell volume-regulatory response of ABP$^{(+)}$ and ABP$^{(-)}$ human melanoma cells was assessed following various hypotonic stimuli. The bathing solution was diluted 50 (A), 66 (B), and 75% (C). V_t/V_0 is the ratio between cell volume at various times after hypotonic shock (V_t) and the control volume under isosmotic conditions (V_0). Open and filled circles indicate experimental results obtained for either ABP$^{(-)}$ or ABP$^{(+)}$ cells, respectively. The melanoma cell lines, originally derived from the transfection of a parent ABP$^{(-)}$ cell line, M2, with a

TABLE I
Peak Volume of ABP$^{(-)}$ and ABP$^{(+)}$ Human Melanoma Cells after Hypotonic Shock

Condition	Peak volume (μm^3)		p
	ABP$^{(-)}$	ABP$^{(+)}$	
Control	2,450 ± 21 (5)	2,567 ± 22 (5)	0.005
50%	3,381 ± 37 (3)	3,196 ± 38 (3)	0.05
66%	3,528 ± 36 (4)	3,337 ± 30 (3)	0.01
75%	3,798 ± 24 (2)	3,675 ± 49 (2)	NS

The various hypotonic shock conditions are indicated as percent dilution of bathing solution.

(66% dilution), however, disabled ABP$^{(-)}$ cells, which were no longer able to elicit an RVD upon hypotonic osmotic swelling, in contrast to ABP$^{(+)}$ cells, which recovered to 15% of control volume (Fig. 3B; Table II). The cell volume of ABP$^{(-)}$ cells following the hypotonic shock reached its peak, 44% higher than control (Table I), and remained constant for 20 min (Fig. 3B; Table II). In contrast, ABP$^{(+)}$ cells reached a 30% higher peak (Table I) but then decreased to 16% (volume plateau) of the original value (Table II). Further hypotonic stress (75% dilution), however, also eliminated the RVD response of the ABP$^{(+)}$ cells (Fig. 3C). Under these conditions, both ABP$^{(-)}$ and ABP$^{(+)}$ cells reached a peak in approximately 10 min (55 and 43%, respectively; not significant) (Table I) and remained constant for another 20 min (62.4 and 49%, respectively; not significant) (Table II). Thus, both basal cell volume and maximal peak expansion (75%) were identical in ABP$^{(-)}$ and ABP$^{(+)}$ cells. Cell volume regulation, however, was observed at lower hypotonic stimuli, and in particular at 66% dilution, in which only ABP$^{(+)}$ cells were able to volume regulate.

The data suggest, therefore, that the elastic response of the phenotypically normal (ABP$^{(+)}$) human melanoma cells was highly dependent on the elastic compliance of the cross-linked, intracellular actin networks as a result of

---

mammalian expression vector (LK444) that either did or did not contain the cDNA for full-length ABP, were grown as previously described (Cunningham *et al.*, 1992). Cell volume was measured as previously described (Cantiello *et al.*, 1993). The isosmotic bathing solution contained, in millimoles per liter,: 135 NaCl, 5 KCl, 0.8 $MgSO_4$, 1.2 $CaCl_2$, and 10 Hepes (pH 7.4). The various hypotonic solutions were prepared by diluting the isosmotic bathing solution in distilled water.

TABLE II

Plateau Volume of ABP$^{(-)}$ and ABP$^{(+)}$ Human Melanoma Cells after Hypotonic Shock

Condition	Plateau volume (μm^3) ABP$^{(-)}$	Plateau volume (μm^3) ABP$^{(+)}$	p
Control	2,450 ± 21 (5)	2,567 ± 22 (5)	0.005
50%	3,087 ± 49 (3)	2,888 ± 38 (3)	0.05
66%	3,528 ± 35 (3)	2,978 ± 33 (4)	0.001
75%	3,979 ± 26 (2)	3,825 ± 50 (2)	NS

The various hypotonic shock conditions are indicated as percent dilution of bathing solution.

ABP activity. The RVD response of either cell line was largely dependent on the magnitude of the osmotic swelling, as both cell types were able to elicit an RVD response depending on the strength of the osmotic stress. These data strongly suggested that the selective RVD response of ABP$^{(+)}$ cells was likely associated with an increase in the dynamic response to the hypotonic shock, which, in turn, had to be linked to the presence of ABP. ABP-cross-linked actin gels should therefore be more responsive to sudden changes in environmental conditions, by helping return the swelling equilibrium of the intracellular cytoskeleton. To assess this hypothesis, experimental values in Fig. 3 were fitted to a theoretical equation (Table III) that closely followed volume expansion as well as the RVD response of either cell type. As indicated in Table III, the recovery time ($1/\lambda$, in minutes) of ABP$^{(+)}$ cells was between 46 and 52% faster than that of ABP$^{(-)}$ cells for identical stress conditions (50 and 66% dilution, respectively). However, the highest osmotic shock (75% dilution) disabled both cell lines from volume regulation (Fig. 3C; Table II), which is in agreement with identical decay constants (Table III). The data are consistent with an elastic response of the intracellular actin gel, which is reflected in the speed by which the cell recovers from a hypotonic stimulus. This, in turn, must be somehow related to membrane-associated events, as the RVD response is only possible after bulk solution is extruded from the cell. The possibility that the actin cytoskeleton interacted with the plasma membrane by eliciting RVD-associated ion channel regulation was further studied by assessing directly the effect of hypo-osmotic stress on ion channel regulation of either ABP$^{(+)}$ or ABP$^{(-)}$ cells.

TABLE III

Decay Constant of Cell Volume-Regulatory Response (RVD) as a Function of Hypotonic Stimulus[a]

	Dilution (%)		
Cell line	50	66	75
$ABP^{(-)}$	6.6	10.1	18.7
$ABP^{(+)}$	3.6	4.9	17.0

[a] Data correspond to best-fitting values of $1/\lambda$ (expressed in minutes), obtained by least squares nonlinear fitting of the equation

$$V_t = V_0 + A\ (t)\ e^{-\lambda\ (t)}$$

(where V_t is expressed as a ratio control volume ($V_0 = 1$)), which closely followed the cell volume changes and the RVD response of the experimental points in Fig. 3.

B. *Effect of Hypotonicity on Whole-Cell Currents of $ABP^{(-)}$ and $ABP^{(+)}$ Melanoma Cells*

Based on the fact that $ABP^{(+)}$ and $ABP^{(-)}$ cells had a similar basal cell volume, the swelling equilibrium of the intracellular gel had to be similar. Thus, only dynamic changes in such equilibrium would unmask the cell's inability to respond to net movements in both intracellular ions and water. The RVD response of human melanoma cells, therefore, should strongly depend on the ability of the intracellular gel to respond to environmental changes that are reflected in the gel's swelling equilibrium, a concerted effort by which a new swelling equilibrium of the intracellular gel effects a bulk fluid redistribution that is linked to the cell's ability to move ions (and water) out of the cell. Therefore, cross-talk between the actin cytoskeleton and plasma membrane ion transport mechanisms is mandatory to preserve intracellular hydroelectrolytic homeostasis, as well as the consequent cell volume regulation. This was reflected in different responses by $ABP^{(-)}$ and $ABP^{(+)}$ cells to transient modifications of cell volume. The role of osmotically regulated ion channels in both spontaneous and osmotically induced whole-cell currents was sought in the $ABP^{(-)}$ and $ABP^{(+)}$ cells with the whole-cell voltage clamp technique (Cantiello *et al.,* 1993).

Whole-cell currents were initially obtained under isosmotic conditions (Fig 4A, B). Contrary to the similar resting volumes of either cell type, the basal whole-cell conductance of $ABP^{(+)}$ cells was approximately 52%

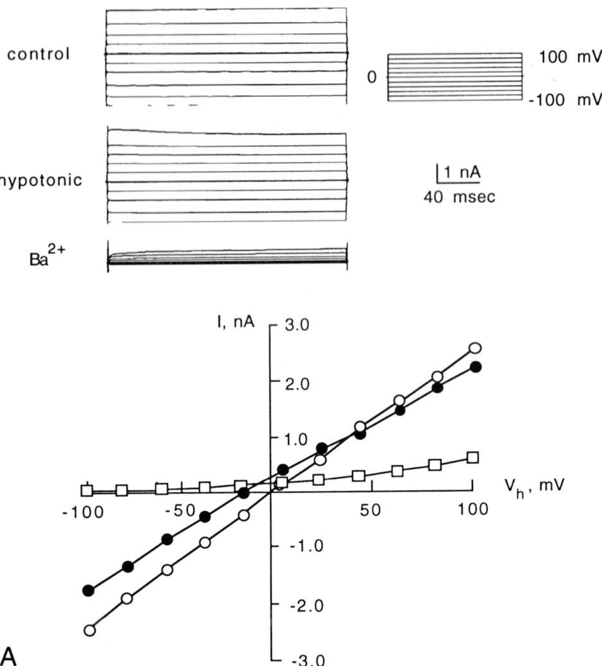

FIGURE 4 Effect of hypotonicity on whole-cell currents of ABP$^{(-)}$ and ABP$^{(+)}$ melanoma cells. Whole-cell currents were measured under isosmotic (open circles) or 1 min after hypotonic (filled circles) conditions. Whole-cell currents were inhibited by addition of 1-m*M* $BaCl_2$ (open squares). Data are the average of 6 independent measurements for the ABP$^{(-)}$ cells (A) or 10 experiments for the ABP$^{(+)}$ cells (B). (C) The effect of time after hypotonic stimulus on the whole-cell conductance of melanoma cells was also assessed. The whole-cell conductance of ABP$^{(-)}$ cells (open bars) or ABP$^{(+)}$ cells (dashed bars) was calculated from the slopes obtained from the whole-cell currents as shown in A and B. Addition of $BaCl_2$ (1 m*M*) reduced the whole-cell conductance of both cell lines to <15% of the isotonic value. Values were expressed as the mean ± SEM, where *n* represented the number of experiments conducted. Whole-cell experiments were obtained as previously described (Cantiello *et al.*, 1993). The patch pipette contained, in millimoles per liter, 60 K_2SO_4, 40 KCl, 5 NaCl, 1.0 $CaCl_2$, 1.0 $MgCl_2$, and 10 Hepes (pH 7.4). The isosmotic and hypotonic bathing solutions used in both the volume-regulatory response experiments and whole-cell clamps were a 66% dilution of this solution. The whole-cell, current experiments were conducted in a randomized, double-blind protocol prior to matching with the experiments conducted on the cell volume response shown in Fig. 3B.

lower than that of ABP$^{(-)}$ cells (Fig. 4), thus suggesting a higher and likely unregulated K^+ conductance in ABP$^{(-)}$ cells. Because actin filaments should be readily cross-linked in the presence of ABP (Matsudaira, 1991; Weeds, 1982), the lower whole-cell conductance of ABP$^{(+)}$ cells had to be linked to a regulatory role of a functional ABP. Filamin, for example, inhibited

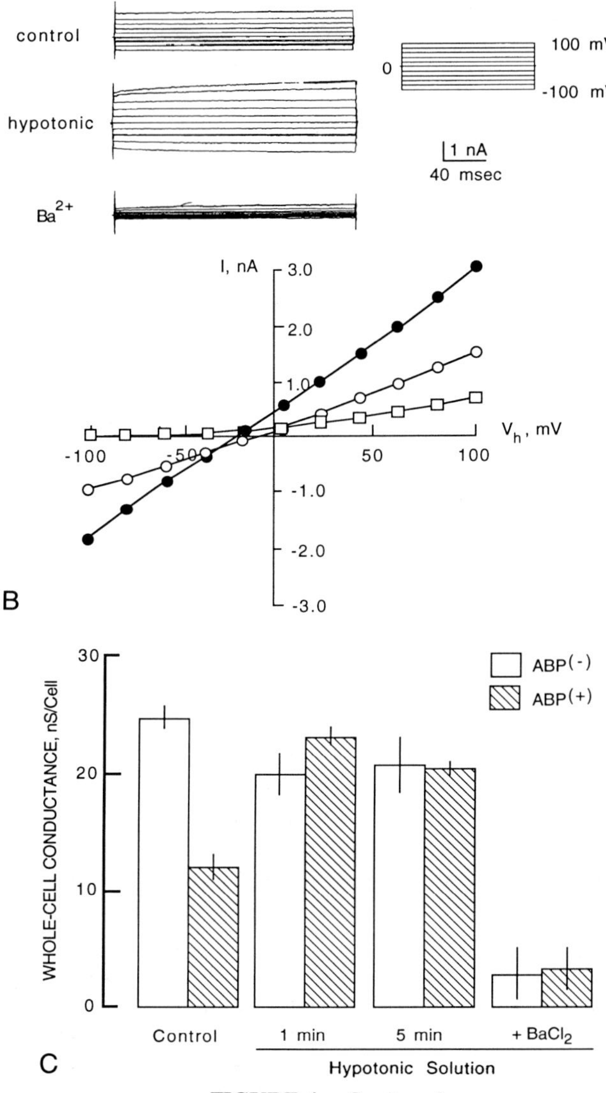

FIGURE 4. *Continued*

both spontaneous and actin-induced Na^+ channel activity in epithelial cells (Cantiello *et al.*, 1991b). Filamin also inhibits the cytoskeletal activation of CFTR (Prat *et al.*, 1995); therefore, a stabilized actin filament organization may contribute to maintain ion channels in the closed state. The finding that the basal whole-cell conductance of $ABP^{(-)}$ cells doubled that of phenotypically rescued ($ABP^{(+)}$) cells indicated that the inability of $ABP^{(-)}$ cells to maintain a normal cross-linked gel is also associated with a high basal K^+ permeability.

The change in whole-cell conductance was then assessed following hypotonic shock conditions (Fig. 4) identical to those under which $ABP^{(+)}$ but not $ABP^{(-)}$ cells elicited an RVD response (Fig. 3B). The already higher whole-cell conductance of $ABP^{(-)}$ cells did not increase after the hypotonic stimulus (it actually decreased by approximately 20%) after 1 min of osmotic shock and was indistinguishable from control conditions (isosmotic) after 5 min of exposure to the hypotonic stimulus (Fig. 4C). In contrast, the $ABP^{(+)}$ cells responded to the hypotonic shock by doubling the whole-cell conductance, which was sustained for up to 5 min of stress. Therefore, the difference in the osmotically induced changes in whole-cell conductance of either cell type was associated with the activation of previously silent ion channels only in $ABP^{(+)}$ cells. In the presence of Ba^{2+}, a known K^+ channel blocker, however, the whole-cell conductance decreased to similar basal values in either $ABP^{(-)}$ or $ABP^{(+)}$ cells (Fig. 4), further suggesting that $ABP^{(-)}$ cells had an increased K^+ permeability prior to stimulation by hypotonic shock.

C. Effect of Hypotonicity on K^+ Channel Activity in $ABP^{(-)}$ and $ABP^{(+)}$ Melanoma Cells

The cell differences in volume data, and the basal and osmotically regulated whole-cell conductance of $ABP^{(-)}$ and $ABP^{(+)}$ cells, were consistent with a dysfunctionally increased ion channel activity in the $ABP^{(-)}$ cells. Thus, patch clamp studies were also conducted under cell-attached conditions to assess resting as well as hypotonically stimulated ion channel activity. Under control conditions, both cell types expressed spontaneously active cation-selective channels, which were further characterized as K^+ selective (Fig. 5). There was no difference in the current–voltage relationship for these K^+ channels in the $ABP^{(-)}$ and $ABP^{(+)}$ cells, an indication that ABP expression did not modify the functional properties of the ion channels. Ion channel activity under basal conditions, however, was consistently higher in $ABP^{(-)}$ cells. An average of two channels per patch was observed in $ABP^{(-)}$ cells, while $ABP^{(+)}$ cells only had 0.40 ($p < .01$). Exposure of $ABP^{(-)}$ cells to a hypotonic stimulus identical to that used for the whole-cell measurements did not induce ion channel activation as predicted from the whole-cell data in Fig. 4. The genetically rescued $ABP^{(+)}$ cells exposed to the same hypotonic stress, however, displayed K^+ channel activation (Fig. 6) as manifested by an increase in the apparent channel number (2.75 ± 0.48, $n = 10$, vs 0.50 ± 0.28 for isosmotic conditions,

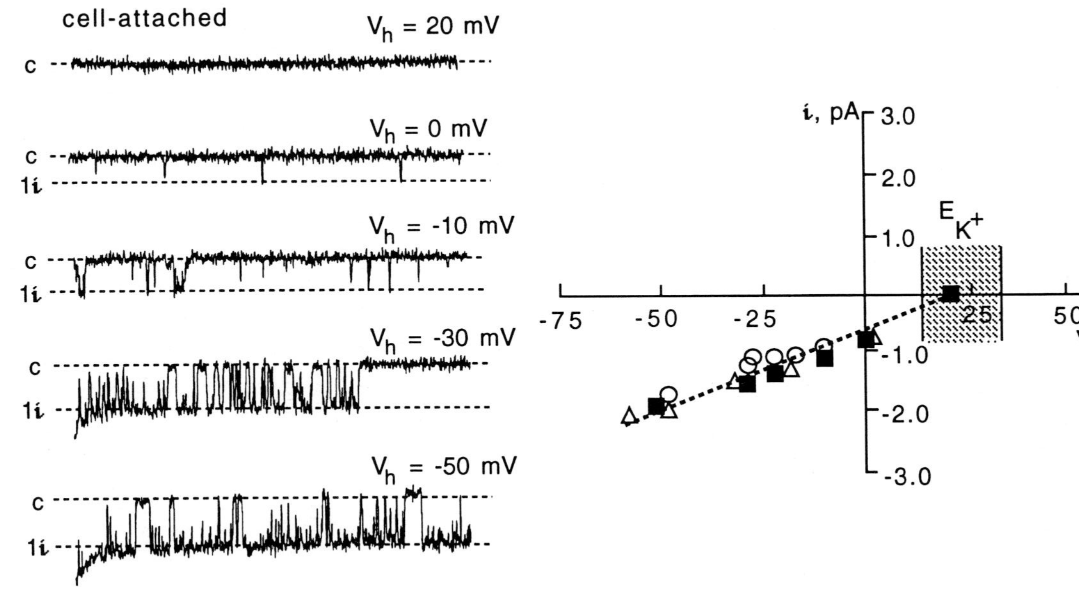

FIGURE 5 Spontaneous K^+ channels in $ABP^{(-)}$ and $ABP^{(+)}$ cells. K^+ channels were observed under basal conditions for $ABP^{(-)}$ (triangles) and $ABP^{(+)}$ (circles and squares) (n = 3). From single-channel traces as shown for $ABP^{(+)}$ cells (left), the current–voltage relationship was obtained (right), which was identical for $ABP^{(+)}$ and $ABP^{(-)}$ cells. The reversal potential is in close agreement with the K^+ equilibrium potential calculated assuming −50 to −70 mV of membrane potential, 5-mM extracellular K^+, and 140-mM intracellular K^+, as indicated. Open probability of single channels dramatically increased with cell depolarization. Closed state of the channels is indicated by c; each single channel opening is indicated by ***i***. Actual currents and command voltages were obtained as previously reported (Cantiello *et al.*, 1990, 1991b).

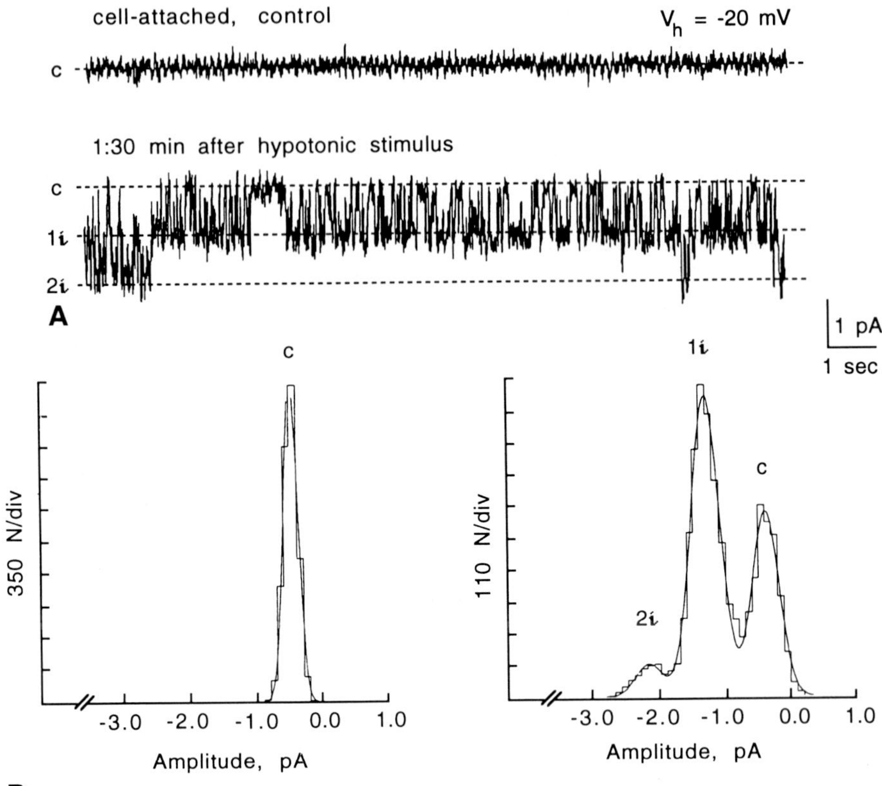

FIGURE 6 Effect of a hypotonic stimulus on single K^+ channel activation in $ABP^{(+)}$ cells. (A) Patch clamp technique in the cell-attached configuration was conducted as described in Fig. 5. Single K^+ channel currents were observed under cell-attached conditions after a hypotonic shock as described in Fig. 3B in the perfusion bath. The open probability of the channel increased by 400%. (B) Amplitude histograms of the total current are shown for both traces in A (left, control; right, hypotonic stimulus).

$n = 4$; $p < .01$). These findings are consistent with a change in the ionic conductance of $ABP^{(+)}$ cells following a hypotonic shock, which is elicited by activation of transient K^+-selective currents. The data indicate that the maximal whole-cell conductance and single-channel properties were both identical, further suggesting that ABP transfection only modified the regulation and basal ability of the K^+ channel to remain open. The data, however, support the contention that the basal ionic permeability does not modify cell volume, only stresses the adaptive response to sudden changes.

VI. PERSPECTIVE: MOLECULAR PHYSIOLOGY OF THE ACTIN–ION TRANSPORT INTERFACE

Studies from our laboratory have begun to address the possibility that high-affinity interactions, including direct binding of actin to transport proteins, may be a novel signaling mechanism linking the cytosol with plasma membrane function. Actin binds to the α-subunit of Na^+,K^+-ATPase, a process that may be an essential first step in its regulatory role in the enzyme. Two putative actin-binding sites are found in the α-subunit of Na^+,K^+-ATPase. The sequence in myosin located between amino acids 698 and 705 (Leu-Glu-Gly-Ile-Arg-Ile-Cys-Arg) represents a high-affinity region for actin binding that is responsible for the activation of the myosin-mediated ATPase activity (Eto *et al.,* 1991). This sequence, located in the S region of the myosin heavy chain, has a strong similarity with a sequence found in the α-subunit of epithelial Na^+,K^+-ATPase at amino acids 883 to 888 (Leu-Glu-Gly-Ile-Arg-Gly) in a region believed to be intracellular. There is also a putative binding site in the α-subunit of Na^+,K^+-ATPase with strong similarity to actin-binding domains of gelsolin, severin, and villin. This site is located in the main cytoplasmic loop of the α-subunit, a region sharing residues that correspond to the actin-binding domain of ankyrin (Morrow, 1989).

A comparison between the amino acid sequences of CFTR and various actin-binding proteins also suggests that distinct domains of CFTR may represent relevant actin-binding domains and raises the possibility that this channel protein may also be considered a novel ABP (Prat *et al.,* 1995). Both nucleotide-binding domains of CFTR were found to share distinct actin-binding domains with two families of ABPs, including severin and filamin. Because changes in the actin cytoskeleton may be either dependent or independent of cAMP activation, an alternative pathway for CFTR activation and regulation may occur through cAMP-independent regulation of actin filaments (Prat *et al.,* 1995). This is consistent with a sequence homology between CFTR and actin-binding domains that correspond to proteins with different functional effects on actin filament organization. Therefore, more than one regulatory effect may take place at such interactive domains.

The possibility of binding and elongation of actin to ion transport proteins provides a new means for regulation, as actin filaments can take various conformations. As indicated earlier, monomeric actin binds to Na^+,K^+-ATPase; however, this is not functionally active. The possibility of elongation is therefore more appealing; as the actin filament is formed, activation and regulation will ensue until a cytoskeletal structure is formed that no longer regulates, and even inhibits, transport function. The data to date indicate that only short actin filaments are capable of supporting ion transport activation; however, activation of Na^+,K^+-ATPase by actin was optimal at ratios between Na^+,K^+-ATPase and actin higher than 1:500. Considering that Na^+,K^+-ATPase is an integral membrane protein complex while actin is a cytosolic protein, their functional interaction will depend on the volume of the actin-containing cortical cytoskeleton. Thus functional ratios between the membrane proteins and actin will change exponentially by taking into account the cell surface–cell volume ratio. This leads to the possible relevance of changes in the bulk of the cytosol as potential effectors of signals directed to the functional ion channel–cortical cytoskeleton interface.

The ability of cells to change shape, to crawl, or to otherwise exercise motility, for example, is thought to largely depend on the cell's ability to undergo localized gel–sol transformations associated with dynamic changes in the actin cytoskeleton. Similar changes are also implicated in cell activation, whenever the onset of cell function is linked to a modified cell shape. Changes in actin–cross-linked network organization, for example, may take place whenever a normal motile cell response is expected. The consequent collapse of actin gel networks must be coordinated with cytosolic bulk fluid redistribution, which in turn must be linked to selective ion transport regulation. The mechanisms associated with the actin network–membrane function cross-talk are largely unknown. Despite relevant cytoskeletal changes that are associated with cell activation, most regulatory cell events, especially those taking place and concerning intracellular hydroelectrolytic homeostasis, are thought to involve and/or be controlled by the plasma membrane. Thus, hypo- as well as hypertonic changes are thought to be regulated by specific changes in plasma membrane function. This membrane response, however, can only take place in synchrony with controlled changes in the cytoskeletal gel conformations underlying most of the structural aspects of the cell's cytoskeleton. Each component has its own physicochemical as well as biochemical properties that must be taken into consideration for a normal cell response to ensue.

Normally, gels, especially those composed of highly charged, rigid, linear polymers such as actin filaments, maintain a 3-dimensional structure that depends on the regulatory proteins and associated mechanisms. It is expected that, as changes in cell volume occur, changes in the swelling equilibrium of the intracellular gel will follow. Studies on polyacrylamide and other highly charged synthetic polymers, for example, have established that changes in the osmotic properties of the medium and/or pH will modify, and therefore induce dramatic changes in, the physicochemical properties of sugh gels (Tanaka, 1987; Tanaka *et al.,* 1980; Tanaka and Ohmine, 1982). Thus, the intracellular gel can be envisioned as an elastic structure that swells or shrinks with the concomitant intragel water movement as the gel collapses. The reflection coefficient of an actin gel in the absence of actin cross-linking proteins, for example, will follow a linear (ideal) correlation with osmotic pressure. This behavior is dramatically modified by either the presence of ABPs and/or the ability of actin networks to break (mechanically) into smaller fragments (Ito *et al.,* 1987). Therefore, as the actin gel conformation changes, this change will modify the ability of bulk fluid to move throughout the gel, which, in turn, will be reflected in membrane-specific events entailing changes in membrane function associated with both movement of intracellular electrolytes and the consequent movement of water. We have attempted to address some of these issues by taking advantage of human melanoma cells naturally devoid of normal intracellular actin networks. The experimental evidence points to a relevant role of modifying the swelling equilibrium of the intracellular actin gels that is reflected in the activation and/or regulation of plasma membrane ion transport. In combination with our recent studies indicating a direct interaction between actin filaments and various ion transport mechanisms, a relevant functional interface can begin to take shape in which intracellular structures, such as the actin cytoskeleton, may now be envisioned as effectors of signaling events that target and control plasma membrane function.

Acknowledgments

The authors acknowledge the *Journal of Biological Chemistry* and the American Physiological Society for allowing the reproduction of published material. The authors gratefully acknowledge Dr. Dennis A. Ausiello for critical reading of the manuscript and helpful suggestions, and Mr. George R. Jackson, Jr., for superb technical help.

References

Bennett, V., and Lambert, S. (1991). The spectrin skeleton: From red cells to brain. *J. Clin. Invest.* **87,** 1483–1489.

Cantiello, H. F. (1995). Actin filaments stimulate Na^+,K^+ ATPase. *Am. J. Physiol.* **269,** F637–F643.

Cantiello, H. F., Patenaude, C. R., Codina, J., Birnbaumer, L., and Ausiello, D. A. (1990). $G_{\alpha i3}$ regulates epithelial Na^+ channels by activation of phospholipase A_2 and lipoxygenase pathways. *J. Biol. Chem.* **265,** 21624–21628.

Cantiello, H. F., Patenaude, C. R., and Zaner, K. S. (1991a). Osmotically induced electrical signals from actin filaments. *Biophys. J.* **59,** 1284–1289.

Cantiello, H. F., Stow, J., Prat, A. G., and Ausiello, D. A. (1991b). Actin filaments control epithelial Na^+ channel activity. *Am. J. Physiol.* **261,** C882–C888.

Cantiello, H. F., Prat, A. G., Bonventre, J. V., Cunningham, C. C., Hartwig, J., and Ausiello, D. A. (1993). Actin-binding protein contributes to cell volume regulatory ion channel activation in melanoma cells. *J. Biol. Chem.* **268,** 4596–4599.

Christensen, O. (1987). Mediation of cell volume regulation by Ca^{2+} influx through stretch-activated chanels. *Nature* (*London*) **330,** 66–68.

Cramer, L. P., Mitchinson, T. J., and Theriot, J. A. (1994). Actin-dependent motile forces and cell motility. *Curr. Opin. Cell Biol.* **6,** 82–86.

Cunningham, C. C., Gorlin, J. B., Kwiatkowski, D. J., Hartwig, J. H., Janmey, P. A., Byers, H. R., and Stossel, T. P. (1992). Actin-binding protein requirement for cortical stability and efficient locomotion. *Science* **255,** 325–327.

Drenckhahn, D., Schluter, K., Allen, D. P., and Bennett, V. (1985). Colocalization of Band 3 with ankyrin and spectrin at the basal membrane of intercalated cells in the rat kidney. *Science* **230,** 1287–1290.

Edelstein, N. G., Catteral, W. A., and Moon, R. T. (1988). Identification of a 33 kD cytoskeletal protein with high affinity for the sodium channel. *Biochemistry* **27,** 1818–1822.

Eto, M., Morita, F., Nishi, N., Tokura, S., Ito, T., and Takahashi, K. (1991). Actin polymerization promoted by a heptapeptide, an analog of the actin-binding S site on myosin head. *J. Biol. Chem.* **266,** 18233–18236.

Foskett, J. K., and Spring, K. R. (1985). Involvement of calcium and cytoskeleton in gallbladder epithelial cell volume regulation. *Am. J. Physiol.* **248,** C27–C36.

Fox, J. E. (1985). Identification of actin-binding protein as the protein linking the membrane skeleton to glycoproteins on platelet plasma membranes. *J. Biol. Chem.* **260,** 11970–11977.

Geiger, B. (1985). Microfilament-membrane interactions. *Trends Biochem. Sci.* **10,** 456–461.

Glenney, J. R. J., and Weber, K. (1981). Calcium control of microfilaments: Uncoupling of the F-actin-severing and bundling activity of villin by limited proteolysis *in vitro. Proc. Natl. Acad. Sci. U.S.A.* **78,** 2810–2814.

Goddette, D. W., and Frieden, C. (1986). Actin polymerization: The mechanism of action of cytochalasin D. *J. Biol. Chem.* **261,** 15974–15980.

Guharay, F., and Sachs, F. (1984). Stretch-activated single ion channel currents in tissue-cultured embryonic chick skeletal muscle. *J. Physiol.* **352,** 685–701.

Hall, P. F. (1984). The role of the cytoskeleton in hormone action. *Can. J. Biochem. Cell Biol.* **62,** 653–665.

Hartwig, J. (1995). Actin-binding proteins 1: Spectrin superfamily. *In* "Protein Profile," (P. Sheterline, ed.), pp. 711–778. Academic Press, London.

Hartwig, J., and Kwiatkowski, D. (1991). Actin-binding proteins. *Curr. Opin. Cell Biol.* **3,** 87–97.

Hitt, A. L., and Luna, E. J. (1994). Membrane interactions with the actin cytoskeleton. *Curr. Opin. Cell Biol.* **6,** 120–130.

Ito, T., Zaner, K. S., and Stossel, T. P. (1987). Nonideality of volume flows and phase transitions of F-actin solutions in response to osmotic stress. *Biophys. J.* **51,** 745–753.

Johnson, B. D., and Byerly, L. (1993). A cytoskeletal mechanism for Ca^{2+} channel metabolic dependence and inactivation by intracellular Ca^{2+}. *Neuron* **10,** 797–804.

Jorgensen, P. L., Petersen, J., and Rees, W. D. (1984). Identification of a Na^+, K^+, Cl^-, -cotransport protein of M_r 34,000 from kidney by photolabeling with [^{3}H] bumethanide: The protein is associated with cytoskeleton components. *Biochim. Biophys. Acta* **775,** 105–110.

Lewis, S. A., and Donaldson, P. (1990). Ion channels and cell volume regulation: Chaos in an organized system. *News Physiol. Sci.* **5,** 112–119.

Linshaw, M. A., Macalister, T. J., Welling, L. W., Bauman, C. A., Herbert, G. Z., Downey, G. P., Koo, E. W. Y., and Gotlieb, A. I. (1991). Role of cytoskeleton in isotonic cell volume control of rabbit proximal tubules. *Am. J. Physiol.* **261,** F60–F69.

Luby-Phelps, K. (1994). Physical properties of cytoplasm. *Curr. Opin. Cell Biol.* **6,** 3–9.

Matsudaira, P. (1991). Modular organization of actin crosslinking proteins. *Trends Biol. Sci.* **16,** 87–92.

Melmed, R. N., Karanian, P. J., and Berlin, R. D. (1981). Control of cell volume in the J774 macrophage by microtubule disassembly and cyclic AMP. *J. Cell Biol.* **90,** 761–768.

Mercier, R., Reggio, H., Devilliers, G., Bataille, D., and Mangeat, P. (1989). Membrane-cytoskeleton dynamics in rat parietal cells: Mobilization of actin and spectrin upon gastric acid secretion. *J. Cell Biol.* **108,** 441–453.

Morrow, J. S. (1989). The spectrin membrane skeleton: Emerging concepts. *Curr. Opin. Cell Biol.* **1,** 23–29.

Morrow, J. S., Cianci, C. D., Aidito, T., Mann, A. S., and Kashgaridan, M. (1989). Ankyrin links fodrin to the α subunit of the Na^+,K^+-ATPase in MDCK kidney cells and in intact renal tubule cells. *J. Cell Biol.* **108,** 455–465.

Nelson, W. J., and Veshnock, P. L. (1987). Ankyrin binding to (Na^+,K^+) ATPase and implications for the organization of membrane domains in polarized cells. *Nature* (*London*) **328,** 533–536.

Ohta, Y., Stossel, T. P., and Hartwig, J. H. (1991). Ligand-sensitive binding of actin-binding protein to immunoglobulin G Fc receptor I (FcγRI). *Cell* **67,** 1–20.

Pollard, T. D., and Cooper, J. A. (1986). Actin and actin-binding proteins: A critical evaluation of mechanisms and functions. *Annu. Rev. Biochem.* **55,** 987–1035.

Pollard, T. D., Selden, S. C., and Maupin, P. (1984). Interaction of actin filaments with microtubules. *J. Cell Biol.* **99,** 33s–37s.

Prat, A. G., Bertorello, A. M., Ausiello, D. A., and Cantiello, H. F. (1993). Activation of epithelial Na^+ channels by protein kinase A requires actin filaments. *Am. J. Physiol.* **265,** C224–C233.

Prat, A. G., Xiao, Y.-F., Ausiello, D. A., and Cantiello, H. F. (1995). cAMP-independent regulation of CFTR by the actin cytoskeleton. *Am. J. Physiol.* **268,** C1552–C1561.

Reuss, L. (1988). Volume regulation in non-epithelial cells. *Renal Physiol. Biochem.* **3-5,** 187–201.

Rosenmund, C., and Westbrook, G. L. (1993). Calcium-induced actin depolymerization reduces NMDA channel activity. *Neuron* **10,** 805–814.

Sackin, H. (1989). A stretch-activated K^+ channel sensitive to cell volume. *Proc. Natl. Acad. Sci. U.S.A.* **86,** 1731–1735.

Schliwa, M. (1982). Action of cytochalasin D on cytoskeletal networks. *J. Cell Biol.* **92,** 79–91.

Schultz, S. G. (1989). Volume preservation: Then and now. *News Physiol. Sci.* **4,** 169–172.

Schwartz, M. A., and Luna, E. J. (1988). How actin binds and assembles onto plasma membranes from *Dictyostelium discoideum. J. Cell Biol.* **107,** 201–209.

Shizuta, Y., Shizuta, H., Gallo, M., Davies, P., and Pastan, I. (1976). Purification and properties of filamin, an actin binding protein from chicken gizzard. *J. Biol. Chem.* **251,** 6562–6567.

Smith, P. R., Saccomani, G., Joe E., Angelides, K. J., and Benos, D. J. (1991). Amiloride-sensitive sodium channel is linked to the cytoskeleton in renal epithelial cells. *Proc. Natl. Acad. Sci. U.S.A.* **88,** 6971–6975.

Srimivasan, Y., Elmer, L., Davis, J., Bennett, V., and Angelides, K. (1988). Ankyrin and spectrin associate with voltage-dependent Na^+ channels in brain. *Nature* (*London*) **333,** 177–180.

Stossel, T. P. (1984). Contribution of actin to the structure of cytomatrix. *J. Cell. Biol.* **99**(Suppl.), 15s–21s.

Stossel, T. P. (1993). On the crawling of animal cells. *Science* **260,** 1086–1094.

Strange, K. (1994). "Cellular and Molecular Physiology of Cell Volume Regulation." CRC Press, Boca Raton, FL.

Tanaka, T. (1987). Collapse of gels and the critical endpoint. *Phys. Rev. Lett.* **40,** 820–823.

Tanaka, T., and Ohmine, I. (1982). Salt effects on the phase transition of ionic gels. *J. Chem. Phys.* **77,** 5725–5729.

Tanaka, T., Filmore, D., Sun, S. T., Nishio, I., Swislow, G., and Shah, A. (1980). Phase transitions in ionic gels. *Phys. Rev. Lett.* **45,** 1636–1639.

Ubl, J., Murer, H., and Kolb, H. A. (1988). Ion channels activated by osmotic and mechanical stress in membranes of opossum kidney cells. *J. Membrane Biol.* **104,** 223–232.

Watson, A. J. M., Levine, S., Donowitz, M., and Montrose, M. H. (1992). Serum regulates Na^+/H^+ exchange in Caco-2 cells by a mechanism which is dependent on F-actin. *J. Biol. Chem.* **267,** 956–962.

Weeds, A. (1982). Actin-binding proteins—regulators of cell architecture and motility. *Nature (London)* **296,** 811–816.

Weihing, R. R. (1985). The filamins: Properties and functions. *Can. J. Biochem. Cell Biol.* **63,** 397–413.

Welling, P. A., Linshaw, M. A., and Sullivan, L. P. (1985). Effect of barium on cell volume regulation in rabbit proximal straight tubules. *Am. J. Physiol.* **249,** F20–F27.

Yin, H. L., Zaner, K. S., and Stossel, T. P. (1980). Ca^{2+} control of actin gelation: Interaction of gelsolin with actin filaments and regulation of actin gelation. *J. Biol. Chem.* **255,** 9494–9500.

Yin, H. L., Albretch, J. H., and Fattoum, A. (1981). Identification of gelsolin, a Ca^{2+}-dependent regulatory protein of actin gel-sol transformation, and its intracellular distribution in a variety of cells and tissues. *J. Cell Biol.* **91,** 901–908.

Ziyadeh, F. N., Mills, J. W., and Kleinzeller, A. (1992). Hypotonicity and cell volume regulation in shark rectal gland: Role of organic osmolytes and F-actin. *Am. J. Physiol.* **262,** F468–F479.

CHAPTER 18

Role of the Cytoskeleton in Membrane Alterations in Ischemic or Anoxic Renal Epithelia

R. Brian Doctor, Robert Bacallao*, and Lazaro J. Mandel

Department of Cell Biology, Division of Physiology and Cellular Biophysics, Duke University Medical Center, Durham, North Carolina 27710 and *Department of Medicine, Indiana University School of Medicine, Indianapolis, Indiana 46202

I. INTRODUCTION

Previous chapters in this volume document and describe the critical role that is performed by the cytoskeleton in sorting, distributing, retaining, and regulating integral membrane proteins. For individual epithelial cells, these capabilities enable asymmetrical organization. For groups of epithelial cells, these capabilities enable the epithelia to specifically adhere to neighboring cells and regulate the vectorial movement of water, ions, and solutes across the epithelial sheet. In the kidney, these coordinated intracellular and intercellular properties concurrently allow the essentially complete reabsorption of filtered water, glucose, and amino acids; elimination of filtered urea; and the complete single-pass excretion of organic acids from the plasma. Renal ischemia, the restriction of blood flow to the kidneys, results in the loss of renal function. Increasingly, many of these functional alterations have been linked to changes in the membrane–cytoskeleton. This chapter first describes the models primarily employed to study the effects of renal cell energy depletion and reviews intracellular ramifications of renal ischemia or anoxia. The focus is on the alterations in signal transduction primarily involved in the regulation of protein phosphorylation and dephosphorylation. Next, the ramifications of these alterations upon the regulation and structure of the cytoskeleton and how the cytoskeletal changes result in a loss of cellular and tissue function are discussed. Interestingly, these studies have not only begun to unravel the cellular events that underlie the ischemia-induced pathological processes but have also dissected and defined several physiological characteristics of the membrane–cytoskeleton.

II. MODELS OF ISCHEMIA

The histological progression of the injury to the renal epithelia that occurs during renal ischemia was first described in the late 1970s (Glaumann *et al.*, 1975; Glaumann and Trump, 1975; Venkatachalam *et al.*, 1978). The proximal tubule (PT), the nephron segment that first receives the glomerular filtrate and that is responsible for the reabsorption of two-thirds of the filtered water and solutes, was shown to be the first nephron segment to be affected by the cessation of renal blood flow. Since that time, extensive efforts have been made to determine the molecular and cellular alterations that evoke the histological changes.

A. *In Vivo* Ischemia

Three models have been used extensively to investigate these alterations. The first model is energy deprivation of the *in vivo* kidney, accomplished through either renal artery occlusion (ischemia) or pharmacological interference with cellular energy metabolism (Kellerman, 1993). In this model, all the renal components, including the glomeruli, the entire nephron, and the vasculature, remain intact. This model permits the study of the interactions among these renal components and most closely reflects the progression of injury and the resultant consequences that occur clinically. These complex interactions are especially important during the recovery phase following renal ischemia. However, the asymmetrical distribution and regulation of the vasculature, the multitude of different epithelial cell types, and the complex interactions between the vasculature and the epithelial segments render the use of this model disadvantageous for many mechanistic studies. Nevertheless, with the use of novel biochemical and immunological techniques, this model has provided qualitative and quantitative information regarding the cellular content and location of specific proteins and lipids during energy deprivation.

B. Isolated Proximal Tubule Suspensions

The second predominant model employs suspensions of purified PT fragments. The PT has been studied extensively because it is the first cell type in the ischemic kidney to display histological evidence of impairment. Reversible histological changes occur within 15 min after the onset of ischemia while irreversible injury occurs after 60 min of ischemia (Kellerman and Bogusky, 1992; Glaumann and Trump, 1975; Glaumann *et al.,* 1975; Venkatachalam *et al.,* 1978). Primarily isolated from either the rat or the rabbit, these suspensions contain purified PT fragments, with glomeruli, endothelial cells, and other nephron segments largely eliminated by preparative density centrifugation. Sacrificing the role of the vasculature and hemodynamic effects, this model focuses on the effects of energy depletion on the PT cells. Although mitochondrial inhibitors are available, the elimination of additional confounding factors, the dependence of PT solely on oxidative metabolism, and the ease of application has made "true" anoxia, the gassing with a nitrogen atmosphere, the preferred method of metabolic inhibition in this model (Weinberg, 1985; Takano *et al.,* 1985; Doctor and Mandel, 1991). This model allows a wide range of morphological, biochemical, and physiological measurements to be performed on the same tissue. A primary limitation of this model is the relatively short period of viability of the proximal tubular suspensions (<12 h). This precludes their use in long-term recovery studies.

C. Renal Cell Cultures

Immortalized renal cell cultures, the third model, have been employed for mechanistic studies of the epithelial response to energy deprivation because of their homogeneity, reproducibility, ease of growth and maintenance, and simple geometries. Furthermore, their immortality allows long-term recovery studies to be performed, and the receptiveness of these cell lines to stable transfection will allow studies to incorporate molecular manipulation as a tool to assess causality (Ragno *et al.,* 1992; Friederich *et al.,* 1992; Costa de Beauregard *et al.,* 1995). The two most commonly employed renal cell lines are Madin–Darby canine kidney (MDCK) cells, stemming from canine kidneys and displaying primarily distal tubular properties, and LLC-PK1 cells, stemming from porcine kidneys and displaying primarily proximal tubular properties. Because of technical constraints, energy deprivation in cell cultures is primarily induced by chemical anoxia (Venkatachalam *et al.,* 1988; Canfield *et al.,* 1991; Doctor *et al.,* 1994). Since cell cultures display simultaneous oxidative and glycolytic metabolism, both pathways must be inhibited to rapidly and extensively reduce the cellular ATP content. A negative aspect of the results obtained from renal cell cultures is that many of the cellular properties of the immortalized cell lines have dedifferentiated and may no longer reflect the properties of the *in vivo* tissue.

In general, there is an inverse correlation between the malleability and the applicability of the model. Events observed in the isolated PT model must be verified *in vivo* to determine if they occur in the environment of the intact organ. Events observed in the cell culture models must be interpreted cautiously and verified in the intact tissue before the observed cellular events are conferred to the *in vivo* tissue.

III. DEPLETION OF CELLULAR ATP

A. Cytosolic ATP Concentration During Energy Deprivation

Measurements of average ATP concentration in the *in vivo* kidney subjected to ischemia have been made with use of nuclear magnetic resonance. Control cortical levels are approximately 2 mmol/liter, and this value is inhibited by 85% after 45 min of energy deprivation (Siegel *et al.,* 1983). A similar result is obtained in the anoxic PT suspension, starting at about 3 mmol/liter and being inhibited by about 90% after 40 min of anoxia (Mandel *et al.,* 1988). If the remaining 10% of ATP were averaged over the cytoplasm, it would have a concentration of about 0.3 mmol/liter, which is relatively high because it is approximately the half-maximal concentration ($K_{1/2}$) of Na,K-ATPase for ATP (Soltoff and Mandel, 1984), and much higher than the $K_{1/2}$ for many ATP-requiring enzymes, especially kinases, which have $K_{1/2}$ values in the 1- to 10-μM range (Takai *et al.*, 1977).

Rather than a homogeneous distribution, the measured cellular ATP value likely represents an average of compartmentalized ATP. Measurements from functional assays clearly indicate that, during energy deprivation, ATP becomes rate limiting for a number of cellular functions, suggesting that much of the ATP is compartmentalized within organelles and is essentially depleted from the cytosol. One organelle that has been linked to ATP and ADP compartmentation is the mitochondria (Jones, 1986). Experiments with liver cells indicate that during energy deprivation the mitochondria seem to seal themselves off and maintain their ATP levels even after 1 h of anoxia or ischemia (Jones, 1986). Furthermore, cultured renal cells, whose metabolism is predominantly glycolytic, have a much lower mitochondrial density than *in vivo* or freshly isolated PT cells and have a larger drop in the average ATP concentration during anoxia (Venkatachalam *et al.*, 1988; Doctor *et al.*, 1994). Therefore, it is possible that much of the measured ATP in these renal cells is in the mitochondria during these short periods of energy depletion.

Energy compartmentation has also been suggested by the ATP dependence of Na,K-ATPase activity in intact PT cells, obtained by graded energy depletion (Fig. 1; Soltoff and Mandel, 1984). A linear relation was found

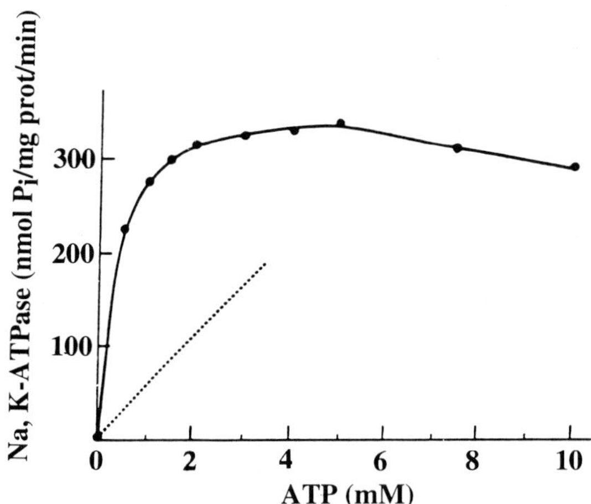

FIGURE 1 Dependence of Na,K-ATPase activity (solid line) of PT membranes on the ATP concentration. Na^+ pump activity of intact PTs, measured as initial K^+ uptake (broken line), as a function of their measured ATP content, calculated as concentration using an intracellular volume of 2.4 μl/mg protein. (From Soltoff and Mandel, 1984.)

between active K^+ uptake (a measure of the Na,K-ATPase activity) and ATP content of the cells. This relationship contrasts sharply with the saturation and $K_{1/2} = 0.3\text{-m}M$ value obtained in the same renal tubule suspension after permeabilization of the plasma membrane and exogenous addition of ATP. The latter dependence essentially reproduced the $K_{1/2}$ value obtained by other investigators *in vitro*. The major difference between these two conditions was that, in the intact cells, the ATP content was measured and averaged over the intracellular water space, whereas in permeabilized cells the ATP concentration available to Na,K-ATPase was set by the supplemented extracellular ATP concentration. These experiments suggested that ATP was rate limiting for Na,K-ATPase even during normoxia, possibly because the ATP concentration in the vicinity of the ATPase was much lower than the average ATP content measurement would indicate. Although other possibilities were also considered by these investigators, the most plausible scenario involved energy compartmentation. This rate limitation for ATP would be expected to become exacerbated during partial energy depletion and especially during the essentially complete energy depletion that is obtained during anoxia. This was indeed found to be the case, since anoxia caused a rapid release of intracellular K^+, comparable in rate to that elicited by ATPase inhibition through ouabain addition (Harris *et al.*, 1981). Assuming that another variable did not become rate limiting, Na,K-ATPase may be used as a bioassay for ATP concentration *in situ* during energy depletion. The rapid inhibition of Na,K-ATPase during anoxia suggests that the cytosolic ATP concentration was a magnitude or more below the $K_{1/2}$ value of 0.3 mmol/liter.

Measurements of the overall PT kinase activity suggest an even lower estimate of the cytosolic ATP concentration (Kobryn and Mandel, 1994). During anoxia, the overall PT kinase activity is inhibited by 92%. This inhibition is most likely due to the decrease in ATP. Assuming that the $K_{1/2}$ value for ATP measured for kinases *in vitro* applies within the tubular cells, this result may be used as another bioassay for the intracellular ATP concentration during anoxia. Since the $K_{1/2}$ values for kinases are in the 1- to $10\text{-}\mu M$ range (Takai *et al.*, 1977), such a profound inhibition would require a decrease in cytosolic ATP concentration below that range, possibly below 1 μmol/liter.

B. Consequences of ATPase Inhibition

As discussed earlier (see Section III,A), inhibition of Na,K-ATPase causes a rapid loss of cellular K^+, which is virtually complete after 5 min, by equilibrating with the extracellular value (Soltoff and Mandel, 1984;

Spencer *et al.*, 1991). This result suggests that the normally large K^+ conductance across the basolateral membrane of PTs remains large during anoxia. The K^+ conductance during hypoxia was measured to increase via the opening of a K^+-selective channel (Reeves and Shah, 1994). The inhibition of Na,K-ATPase also caused an increase in the intracellular Na^+ and Cl^- concentrations and equilibrated with the extracellular media (Spencer *et al.*, 1991). For Cl^-, this equilibration may be accelerated by the opening of a Cl^--selective channel during anoxia (Miller and Schnellmann, 1993). The equilibration of Na^+ across the plasma membrane eliminates the Na^+ gradient necessary for the normal function of the Na^+-linked transporters. Inhibition of H^+-ATPase and H^+ extrusion through Na^+–H^+ exchange is thought to produce an intracellular acidification; however, measurements suggest that the intracellular pH of PTs does not change during chemical anoxia (Weinberg, 1985). This result is explained, in part, from the relatively low glycolytic capacity of the PT which would limit proton production (Gullans and Mandel, 1992). Inhibition of Ca^{2+}-ATPase and Ca^{2+} extrusion by Na^+–Ca^{2+} exchange likewise may be expected to cause an increase in cytosolic free calcium (Ca_f) during energy deprivation. The extent and magnitude of this increase has been the object of conflicting results. Ca_f changes measured during energy deprivation range from no measurable change (Jacobs *et al.*, 1991) to a doubling of the resting value to a value of about 200 nmol/liter (Goligorski and Hruska, 1988), to five times the resting value, reaching about 500 nmol/liter (Borle *et al.*, 1986; Garza-Quintero *et al.*, 1993). The investigators who found an increase observed a delay of 15–30 min before this increase occurred. These differences may be due to the variability of conditions leading to the activation of a cytoskeleton-associated Ca^{2+} channel (Mills and Mandel, 1994). Nevertheless, it is unclear whether this increase in Ca_f is of sufficient magnitude and duration to activate Ca^{2+}-dependent enzymes in the PTs. This point is discussed later in this chapter, in the context of possible individual actions of these enzymes (see Sections IV,A and IV,C).

C. Consequences of Kinase Inhibition

Anoxia or ischemia causes virtually complete inhibition of all ATPases and kinases in isolated PTs (see Section III,A). These inhibitions have a profound impact on the analysis of the molecular mechanisms of anoxic cell injury in terms of the consequent signal transduction alterations.

1. Resultant Alterations in Signal Transduction

Figure 2 demonstrates the multitude of signal transduction pathways involving kinases (Cohen, 1992). Each of these pathways is presumably

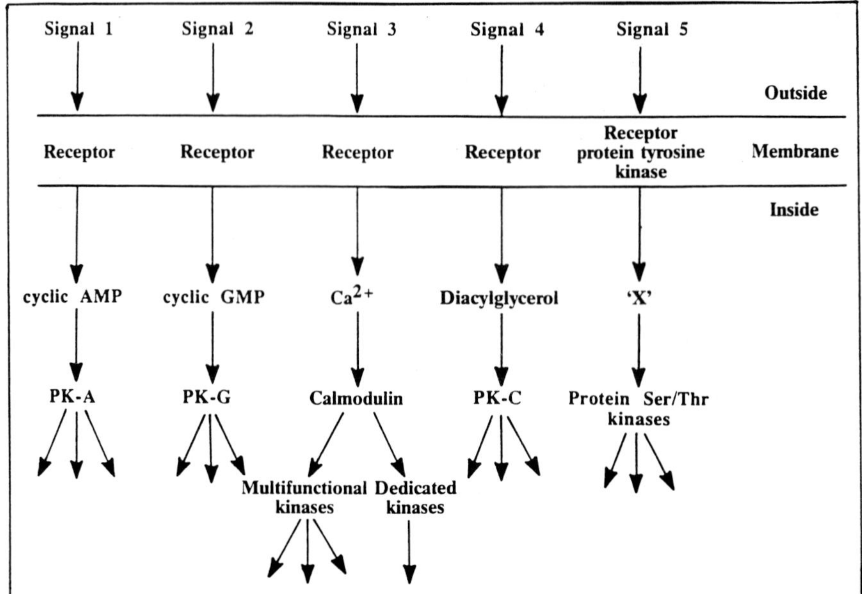

FIGURE 2 Five of the principal signaling systems that operate in eukaryotic cells. Abbreviations: PK-A, cAMP-dependent protein kinase; PK-G, GMP-dependent protein kinase; PK-C, protein kinase C; 'X', hypothetical second messenger. (From Cohen, 1992.)

inhibited during energy deprivation and highlights the magnitude of the effects of energy deprivation on the signal transduction events that are affected. For example, the activation of proteolytic enzymes would not occur through enhanced kinase activity, including cell adhesion molecule kinase. Similar considerations would apply to any processes that were activated through phosphorylation by kinases, whether cAMP-dependent protein kinase, protein kinase C, or tyrosine kinases.

In contrast, although individual phosphatases may be modulated differentially, the overall phosphatase activity is unaffected by anoxia in proximal renal tubules (Kobryn and Mandel, 1994). There are numerous examples of signal transduction pathways that involve complex interactions between kinases and phosphatases (Cohen, 1992). Therefore, abolition of kinase activity during energy deprivation would create an imbalance with profound implications for signal transduction events. These results suggest that a major change expected during energy deprivation would be a decrease in cellular phosphorylation.

There is ample evidence that phosphorylation plays a key role in numerous cellular processes, ranging from the modulation of transport activity

(Jennings and Schulz, 1991) to the determination of cellular morphology (Fernandez *et al.,* 1990) and attachment to the substratum through focal contacts, which are enriched in vinculin, talin, paxillin, and tyrosine kinases (Burridge *et al.,* 1992). Furthermore, numerous phosphoproteins, such as ZO-1, E-cadherin, and the catenins, are key components of intercellular junctions, leading various investigators to suggest modulation of their assembly and disassembly and/or modulation of junctional properties through phosphorylation–dephosphorylation (Citi, 1992; Nigam *et al.,* 1991; Stevenson *et al.,* 1989; Mandel *et al.,* 1993). Furthermore, increased activity of type-1 phosphatase has been demonstrated to bring about the disassembly of the actin network and the dephosphorylation of myosin light chain (Fernandez *et al.,* 1990). A clear correlation has been found between the phosphorylation of ezrin, a putative linker protein between the F-actin core and the plasma membrane in renal PT microvilli, and its association with the cytoskeleton (Chen *et al.,* 1994, 1995). Anoxia causes ezrin dephosphorylation, dissociation of ezrin from the cytoskeleton, and microvillar breakdown. All of these effects are mimicked by kinase inhibition, suggesting that this was the main molecular mechanism leading to microvillar anoxic dysfunction. These considerations suggest that the effects on the cellular kinases and phosphatases and the resultant protein dephosphorylation events are the principle modulators of the alterations in signal transduction during energy deprivation.

IV. CYTOSKELETAL ALTERATIONS DURING ATP DEPLETION

A. Apical Membrane

The cytoskeletal protein content, configuration, and response to ATP depletion within the apical, lateral, and basal membrane regions of renal epithelia are strikingly different. Histologically, the first observed cellular alteration in the ischemic kidney is the blebbing, clubbing, sloughing, and retraction of the apically localized microvilli that make up the normally lush brush border of renal PT cells (Venkatachalam *et al.,* 1978; Kellerman and Bogusky, 1992). These alterations have been attributed to the dissociation of the plasma membrane from the microvillar cytoskeleton (Chen *et al.,* 1994). In the PT, the microvillar core consists of actin filaments that are organized and tethered to the membrane by several actin-associated proteins, including villin, fimbrin, and ezrin (Rodman *et al.,* 1986; Coudrier *et al.,* 1988). The microvilli of intestinal epithelial cells contain a 110-kDa myosin-I–like protein that apparently links the membrane to the microvillar core (Bretscher, 1991). A 105-kDa homologue of this protein has been found in the kidney but has yet to be localized to the renal microvilli (Coluccio, 1991). Largely uncharacterized, the ends of the microvillar actin filaments terminate in the electron-dense microvillar tips and the terminal web.

1. Villin

Villin, the best biochemically characterized and most prevalent of the microvillar actin-associated proteins, is essential for the formation of the microvilli (Bretscher and Weber, 1979). The transfection of villin into villin-deficient fibroblasts resulted in the formation of organized microvilli (Friederich *et al.,* 1989). Conversely, when Caco-2 cells, a villin-containing intestinal cell line with prodigious microvilli, were transfected with antisense villin DNA, the expression of villin protein was completely inhibited and the phenotypic expression of microvilli was absent (Costa de Beauregard *et al.,* 1995). Related to gelsolin, villin is a calcium-regulated protein that caps, nucleates, and severs actin (Yin and Stossel, 1979; Glenney and Weber, 1981, Matsudaira *et al.,* 1985). Unlike gelsolin, villin can also bundle actin filaments (Bretscher and Weber, 1980). Biochemical analysis of the Ca^{2+} dependence of the capping and severing properties of villin indicated that bundling and capping occurred with 10- to 30-n*M* intracellular calcium ion ($Ca^{2+}{}_i$) while the cutting rate was half-maximal at 200 μmol/liter (Northrup *et al.,* 1986). As described earlier, the free $Ca^{2+}{}_i$ level increased during ATP depletion from 100 to 500 nmol/liter, well below the level required for villin-severing activity. A caveat to this conclusion is the potential for regional differences in $Ca^{2+}{}_i$ within the PT. Electron probe analysis of total calcium in the various cellular regions of PT cells demonstrated that the total calcium concentration in the microvilli was 10 mmol/kg under control conditions (Spencer *et al.,* 1991). This was about twice the level of other regions of the cell. After 20 min of anoxia, the total calcium was unchanged in all regions except the microvilli. The microvillar total calcium increased to 17 mmol/kg. It is unclear how this change in calcium relates to the calcium-buffering capacity and the Ca_f concentration in the microvilli. Although the activity status of villin is unclear, the association of villin with the microvillar actin persists during ATP depletion. In LLC-PK1 cells subjected to 30 and 60 min of anoxia, the Triton X-100 solubility of villin, a measure of the villin that is dissociated from the cytoskeleton, did not increase. This cytoskeletal association and the co-localization of villin with F-actin persisted during the anoxic period despite the partial retraction of the microvilli (Golenhofen *et al.,* 1995). This suggests that actin and villin continue to interact after ATP has been depleted and after the cellular environment has been modified.

2. Ezrin

The most defined example of the pathway between the depletion of ATP and the disruption of a cytoskeletal complex has been shown for ezrin. Ezrin, a phosphoprotein in the ezrin–moesin–radixin family, is a putative membrane-linking protein with both membrane- and actin-binding sites (Bretscher, 1983; Sato *et al.,* 1992; Algrain *et al.,* 1993). Predominantly localized to the microvillar region in PT cells, ezrin is largely associated with the cytoskeleton, with 91% found in the Triton-insoluble fraction (Chen *et al.,* 1995). Furthermore, 73% of the cytoskeleton-associated ezrin was phosphorylated while only 14% of the dissociated ezrin was phosphorylated. After 60 min of ATP depletion, total ezrin phosphorylation was reduced from 72% to 21% and, concurrently, the cytoskeletal association decreased from 91% to 58%. This shift in both the dephosphorylation and cytoskeletal dissociation of ezrin was inhibited by calyculin A, a serine–threonine phosphatase inhibitor. In summary, these results suggest that the >90% anoxia-induced drop in the cellular ATP levels completely inhibited the kinase activity. This loss of activity resulted in the net dephosphorylation of ezrin. In a variety of model systems, the increased phosphorylation status of ezrin is correlated with microvillus formation (Turunen *et al.,* 1994; Bretscher, 1989; Hanzel *et al.,* 1991; Pakkanen *et al.,* 1988; Urushidani *et al.,* 1989). The dephosphorylation of ezrin, which is correlated with its dissociation from the cytoskeleton, may be responsible for a reduction in the strength of membrane retention by the microvillar cytoskeleton and may initiate the histological alterations in the microvilli (Chen *et al.,* 1994, 1995).

B. Tight Junction

Defining the barrier between the apical and basolateral membranes, the tight junction has a dual function within renal epithelial cells. The first function is to act as an intracellular "gate" to modulate the paracellular permeability of water and ions and block the paracellular passage of larger molecules. The second function is that of a molecular "fence" that maintains the protein and lipid asymmetry between the apical and basolateral membranes. The sensitivity of the "gate" function to ATP depletion is observed *in vivo* with increased paracellular permeability after as little as 5 min of ischemia and *in vitro* with a decrease in the transepithelial resistance (TER) in anoxic MDCK and LLC-PK1 cells (Molitoris *et al.,* 1989; Canfield *et al.,* 1991; Bacallao *et al.,* 1994, Golenhofen *et al.,* 1995). ATP depletion studies in MDCK cells, however, demonstrated that the molecular fence and gate functions of the tight junction are distinct and respond to ATP depletion independently (Mandel *et al.,* 1993). Within 10 min of ATP depletion the TER was immeasurable. Despite this dramatic loss of the gate function, freeze fracture double replicates showed that the strands of the tight junction remained continuous and uninterrupted. After 50 min of ATP depletion, the molecular fence function of the tight junction, as determined by the inability of apically labeled lipids to cross the tight junction into the basolateral domain, remained intact. It was not until 60 min of ATP depletion that the molecular fence function of the tight junction was compromised (Mandel *et al.,* 1993). Although a causal relationship has not been established, the persistence of the molecular fence function of the tight junction correlates with the intracellular dispersion of ZO-1, a peripheral tight junctional protein, and disruption of the actin ring that is present at the level of the tight junction (Bacallao *et al.,* 1994).

The degree and time course of an uncoupling of the gate and fence functions of tight junction has not been determined *in vivo.* A loss of lipid and protein polarity between the apical and basal domains, however, has been demonstrated in the ischemic kidney (Molitoris *et al.,* 1988,1991). Since it is unclear if and when the tight junction fence function is lost in the *in vivo* kidney, it is unclear whether the loss of polarity seen *in vivo* is the result of a transjunctional diffusion of lipids and proteins in the plane of the membrane or the internalization and redistribution of proteins. Studies within anoxic MDCK cells and ischemic human kidney samples have demonstrated the internalization of integral proteins and the concurrent redistribution of actin from its normal submembranous localization to a cytoplasmic or perinuclear localization (Bacallao *et al.,* 1994; Mandel *et al.,* 1994). The internalized proteins include E-cadherin, a cell adhesion protein distributed on the lateral membrane of MDCK cells, and Na,K-ATPase, a basolaterally distributed protein (Mandel *et al.,* 1993). Likewise, the normally basolateral distribution of Na,K-ATPase, ankyrin, and fodrin is redistributed into the cytoplasm of PT cells in human kidneys awaiting transplantation (Alejandro *et al.,* 1995). Although the mechanism of this internalization is yet to be determined, it is temporally correlated with the redistribution of F-actin away from its predominantly cortical localization (Bacallao *et al.,* 1994). The diminished integral membrane protein concentration would both change the physical characteristics of the membrane and reduce the capacity of the cell to execute the functions that are performed by these internalized proteins (Sheetz, 1993). Diminished functions would likely include the modulation of the intracellular ionic environment, detection and adherence to adjacent cells or substratum, and the vectorial transport of water, ions, and solutes across the epithelium.

C. Spectrin Cytoskeleton

This redistribution of Na,K-ATPase is of further interest since Na,K-ATPase is normally tethered and retained in the basolateral membrane of renal epithelia by the spectrin–ankyrin cytoskeleton (Bennett, 1985; Dreckhahn and Bennett, 1987; Koob *et al.,* 1987; Nelson and Hammerton, 1989). The events that transpire to allow Na,K-ATPase to detach from the spectrin–ankyrin lattice and become internalized are unclear. Ank1, the red blood cell isoform of ankyrin, localized to the distal elements of the nephron, was completely degraded in the thick ascending limb cells during renal ischemia (Doctor *et al.,* 1993). In contrast, Na,K-ATPase and spectrin levels were unchanged. Although the fate of ankyrin in PT cells is unknown, these results suggest that, similar to ezrin, the linkage between the integral membrane proteins and the cytoskeleton is a vulnerable point in the membrane–cytoskeleton complex. The recent isolation, sequencing, and localization of the Ank3 isoform of ankyrin to the mouse PT may soon allow the direct investigation of the fate and role of ankyrin in the ATP-depleted PT (Kordeli *et al.,* 1995; Piepenhagen *et al.,* 1995). Both ankyrin and spectrin are calpain sensitive (Croall and DeMartino, 1991). In the ischemic rat kidney, no calpain fragments are observed for Ank1. Classical calpain fragments for spectrin are observed in the PT but appear only after 60 min of ischemia. This suggests that calpain is activated only after the PT become irreversibly damaged and that, in contrast to *in vitro* studies, calpain is not instrumental in the induction of cell injury (Edelstein *et al.,* 1995).

D. Microtubule Network

Similar to anoxic MDCK cells, the chelation of calcium results in an internalization of E-cadherin (Mandel *et al.,* 1994; Citi, 1992). The mechanism of the chelation-mediated internalization is dependent upon microtubules. Conversely, in MDCK and JTC cells, the localization and function of microtubules are apparently unaffected by ATP depletion (Mandel *et al.,* 1994). Similarly, the microtubule network in fibroblasts is stabilized under ATP-depleted conditions (Bershadsky and Gefland, 1981). Furthermore, microtubule depolymerization by nocodazole prior to ATP depletion does not affect the rate of membrane internalization or actin depolymerization (Mandel *et al.,* 1994). This resistance to ATP depletion may be due to utilization of GTP, rather than ATP, as an energy source for tubulin polymerization and the comparative conservation of GTP versus ATP during anoxia (Doctor *et al.,* 1994). These data suggest that more than one mechanism is capable of the internalizing integral proteins and that one of the mechanisms for internalization is dependent upon protein phosphorylation.

E. Intermediate Filaments

In contrast to the other cytoskeletal systems of the renal epithelia, little is known about the fate and role of the intermediate filaments in ischemic cell dysfunction and injury.

F. Basal Membrane Cytoskeleton

The basal membranes of most cultured cells are highlighted by the presence of actin stress fibers that terminate into focal adhesion sites (Burridge *et al.,* 1988). These focal contacts are composed of a complex of proteins that mechanically attach the cells to the extracellular matrix or substratum. Included in these proteins are the integrins, integral membrane proteins that bridge the membrane to attach the extracellular matrix to the cytoskeleton, talin, α-actinin, and vinculin. This complex performs a dual role in mechanically linking the cells with the substratum as well as transducing signals from the extracellular matrix (Wang *et al.,* 1993; Lin and Bissell, 1993). Although the fate of these specific proteins is unclear, *in vivo* ischemia induces the detachment of viable cells from their substratum, allowing them to enter the lumen of the nephron and block the flow of filtrate (Oliver *et al., 1951; Racusen et al.,* 1991). Pretreatment with Arg-Gly-Asp (RGD) peptides reduces the nephron blockade, decreases the pressure in the lumen of the PT, and improves renal function following the ischemic period (Goligorsky and DiBona, 1993; Noiri *et al.,* 1994). In cultured primate renal cells, cell injury results in the redistribution of integrin from the basal region of the cell into the apical and cytoplasmic domains and an increase in cell detachment (Gailit *et al.,* 1993).

G. Actin Cytoskeleton during ATP Depletion

Within each region of the renal epithelium described earlier, alterations in function or distribution of various proteins are accompanied by changes in F-actin at that site. Although all regional pools of F-actin appear sensitive to the depletion of ATP, the onset of change varies dramatically (Bacallao *et al.,* 1994; Golenhofen *et al.,* 1995) (Fig. 3). In the ischemic kidney, the reduction of F-actin in the apical, brush border region of PT cells is observed after as little as 5 min of ischemia (Kellerman and Bogusky, 1992). After 15 min of ischemia, the microvilli begin displaying histological changes, including membrane blebbing, clubbing, soughing, and retraction of the microvilli. This reduction of microvillar actin coincides with the appearance of cytoplasmic F-actin. In LLC-PK1 cells, which also contain apical microvilli, alterations in the microvillar F-actin coincide with the retraction of the microvillar structures into a supranuclear position (Golenhofen *et al.,* 1995). Along the lateral membranes of MDCK and LLC-PK1 cells, F-actin is diminished after only 20–30 min of anoxia and coincides with the internalization of laterally localized integral proteins (Mandel *et al.,* 1994). Although a direct linkage remains to be shown, the loss of the basal stress fibers in MDCK and LLC-PK1 cells is paired with the detachment of PT cells observed in the ischemic kidney and redistribution of the integrin proteins along the entire membrane surface of the detached cells (Bacallao *et al.,* 1994; Doctor *et al.,* 1994; Golenhofen *et al.,* 1995; Goligorski and DiBona, 1993; Gailit *et al.,* 1993; Niori *et al.,* 1994, Goligorski *et al.,* 1993). In MDCK cells, F-actin of the apical ring is diminished only after 60 min of ATP depletion. Coincidently, the tight junction strand structure and linear localization of ZO-1, a peripheral tight junctional protein, is disrupted.

An apparent ramification in all of these regions of the cell is the loss of the mechanical functions of the actin cytoskeleton. Using micropipette techniques, the strength of retention of the plasma membrane by the cytoskeleton was determined (Doctor *et al.,* 1996). In isolated PT fragments, the depletion of ATP resulted in a significant loss in the strength of membrane retention after only 10 min, and membrane retention was less than 10% of the initial strength after 30 min. Disruption of F-actin with cytochalasin D also weakened the membrane retention, suggesting that the effect of ATP depletion was mediated primarily through its effects on the actin cytoskeleton. This correlation was also observed in suspended MDCK cells in which ATP depletion and cytochalasin D addition caused similar changes in strength of membrane retention and a similar disruption of the cortically localized F-actin.

V. RECOVERY FROM ISCHEMIC INJURY OR ATP DEPLETION

A. Models of Cellular Recovery

Prolonged ischemia or ATP depletion results in the disruption of membrane integrity, loss of cytoplasmic components, and demise of the cell.

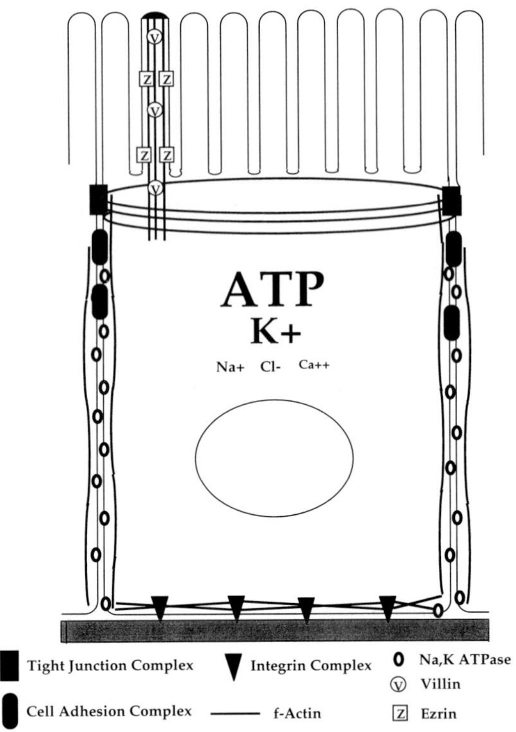

FIGURE 3 Differential effects of energy depletion on the actin-associated cytoskeletal elements. These diagrams are a compilation of cellular and actin cytoskeletal alterations observed in different renal epithelial models during energy depletion. Many of the observations require verification in the intact kidney. (A) Under normoxic conditions, the intracellular concentrations of ATP, K^+, Na^+, Cl^- and Ca^{2+} are approximately 10, 140, 5, 135, and 100 nmol/liter, respectively. Within the apical microvilli, villin and ezrin are tightly associated with the core filaments. The tight junction configuration results in a relatively high transepithelial resistance and paracellular permeability. Integral proteins such as E-cadherin and Na,K-ATPase are restricted to the lateral and basolateral membrane domains, respectively. Actin filaments run parallel to the lateral membrane axis. The integrin complexes are restricted to the basal membrane and link the extracellular matrix to the actin cytoskeleton.

(B) Within minutes after the onset of energy depletion, the ATP concentration is reduced by 90% or more. K^+, Na^+ and Cl^- levels are essentially equilibrated with those of the extracellular environment. Ca^{2+} concentration increases to between 150 and 500 nmol/liter. After a modest period of energy depletion (15–30 min), the apical membrane becomes blebbed, clubbed, and less organized. Basolaterally localized integral proteins are detected within the apical membrane. Although retracted, the microvillar core filaments are still organized and associated with villin. In contrast, ezrin has dephosphorylated and dissociated from the microvillar core. Although the gate function of the tight junctions has become more permeable to paracellular movement, the structure and fence function of the tight junction is resistant to modest periods of energy depletion. The cortical actin along the lateral membranes is very sensitive to energy depletion. These actin fibers diminish and, concurrently, F-actin appears within the cytoplasm. At the same time, Na,K-ATPase and E-cadherin are internalized into the cytoplasm from the lateral membrane. Basal stress fibers are relatively resistant to modest periods of energy depletion. Extended periods result in the loss of stress fibers, redistribution of integrin proteins, and cell detachment.

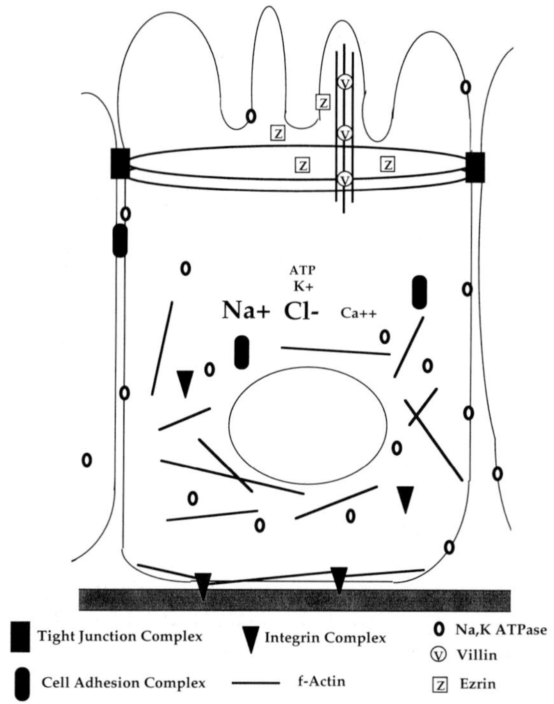

However, if blood flow or oxygen is restored after moderate periods of ischemia or anoxia, the structure and function of the cells will recover. *In vivo,* reversible ischemic injury is reproduced by transiently clamping the renal artery for sublethal periods (<45 min). Following the ischemic period, the clamp is released, reperfusion of the kidneys is permitted for variable periods, and then the kidneys are excised and processed for evaluation (Donohoe *et al.,* 1978; Venkatachalam *et al.,* 1978; Molitoris *et al.,* 1989; Doctor *et al.,* 1993). Anoxic PT suspensions have been readily recovered by simply reoxygenating the suspension (Doctor and Mandel, 1991; Chen *et al.,* 1994, 1995).

To investigate the mechanisms that underlie the loss and recovery of morphology and function, renal cell cultures have been developed. Energy depletion in cell cultures was predominantly induced using mitochondrial and glycolytic inhibitors. One model employed antimycin A, a site II mitochondrial inhibitor, and deoxyglucose, a competitive inhibitor of glycolysis (Venkatachalam *et al.,* 1988; Canfield *et al.,* 1991). During the recovery phase, the inhibitor concentration was reduced through washing. After 4 h, these cells recovered 50% of their intracellular ATP stores (Venkatachalam *et al.*, 1988; Canfield *et al.*, 1991). A second model employed rotenone, a site I mitochondrial inhibitor, and deoxyglucose (Doctor *et al.*, 1994). During the recovery period, the inhibition at site I of the respiratory chain allowed reducing equivalents to bypass the blockade and enter the respiratory chain at site II. This model recovered both oxygen consumption rate and over 50% of the ATP after 15 min of recovery, and complete ATP recovery was achieved after 2 h in the repletion media (Doctor *et al.*, 1994).

B. Cytoskeletal Recovery in Renal Epithelia

1. Microvilli

In PT or LLC-PK1 cells, short periods of ischemia (15–30 min) or ATP depletion (30 min) are associated with the clubbing, sloughing, blebbing, and retraction of the apical microvilli. After 30–60 min of recovery, the microvillar actin relocalizs from the apical cell body into microvilli and the microvillar morphology is reassembled (Glaumann *et al.*, 1975; Glaumann and Trump, 1975; Venkatachalam *et al.*, 1988; Golenhofen *et al.*, 1995). Villin, which remains co-localized and associated with the retracted microvillar actin during anoxia, continues its actin association and relocalizes back into the microvilli (Golenhofen *et al.*, 1995). Ezrin is dephosphorylated and dissociated from the cytoskeleton during the anoxic period (Chen *et al.*, 1994). Despite this dissociation, after 30 min of anoxia, ezrin remains localized in the apical region of the PT. Although a causal relationship has not been definitively established for the function of ezrin in the microvilli, ATP recovery after this period results in the rephosphorylation of ezrin, the reassociation with the cytoskeleton, and the recovery of microvillar morphology (Chen *et al.*, 1994, 1995).

2. Actin Cytoskeleton Recovery

As described earlier, the duration of ATP depletion in LLC-PK1 cells has differential effects on the distinct actin cytoskeletal domains. In agreement with the hypothesis of Bacallao and Fine (1989), regional differences in the reorganization of the actin cytoskeleton are observed. After 30 min of anoxia and 60 min of recovery, the actin structures of the microvilli, apical ring, and basal stress fibers are completely recovered (Golenhofen *et al.*, 1995). The cortical actin along the lateral walls is also recovered but appears somewhat thinner. After 60 min of anoxia and 60 min of recovery, the actin response is heterogeneous. Remarkably, the microvilli and apical ring show complete recovery. The cortical actin staining is relatively weak, with some clumpiness. Most notably, the basal stress fibers are dramatically reduced or absent. These abnormalities in the actin cytoskeleton are likely to translate into function deficits in the respective regions of the cells.

3. Functional Recovery

Histologically, the sublethally injured PT cells recover within hours (Glaumann *et al.,* 1975; Glaumann and Trump, 1975). After modest periods of anoxia, the actin cytoskeletal domains recover within hours after the repletion of ATP (Golenhofen *et al.,* 1995). Despite the morphological recovery, there can be a delay of up to 7 days in the recovery of renal function (Spiegel *et al.,* 1989). During the energy depletion phase, there is a loss in the gate function of the tight junction. Recovery studies *in vivo* and in LLC-PK1 cells suggest that the gate function is reestablished within hours after reperfusion or reoxygenation (Canfield *et al.,* 1991; Golenhofen *et al.,* 1995). Given that the paracellular pathway is functional, the loss of renal function is likely a cellular phenomena. Unlike the cytoskeletal proteins, which are rapidly relocalized, the distribution and function of the integral proteins are apparently delayed. The best characterized of these proteins is Na,K-ATPase. During energy depletion, Na,K-ATPase is depolarized and internalized, being found in the apical membrane and in the cytoplasm (Spiegel *et al.,* 1989; Molitoris *et al.,* 1991, 1992; Mandel *et al.,* 1994; Alejandro *et al.,* 1995). Following ischemia, the reestablishment of renal function coincides with repolarization of Na,K-ATPase into the basolateral membrane (Speigel *et al.,* 1989). Much remains to be determined regarding the functional repolarization of Na,K-ATPase. It is unclear if the depolarized and internalized Na,K-ATPase is redistributed to the basolateral membrane or if new protein synthesis is required. Furthermore, within renal epithelia, Na,K-ATPase is normally retained within the basolateral membrane by an ankyrin-dependent linkage to the fodrin cytoskeleton (Bennett, 1985; Drenckhahn and Bennett, 1987; Nelson and Hammerton, 1989). It will be important to determined how the localization and function of the fodrin cytoskeleton responds to ischemia and reperfusion or anoxia and reoxygenation.

C. Another Paradigm for Ischemic Cell Recovery

Bacallao and Fine (1989) proposed that renal epithelial cells lose their differentiated features during ischemic injury and would recapitulate the normal developmental pathway during recovery from ischemic injury. Many aspects of this hypothesis have been experimentally verified for the effects of injury on the differentiated phenotype of epithelial cells; however, the mechanisms of recovery have not been elucidated. Some experimental work examining epithelial cells recovery from *in vivo* injury has been published, and the development of reversible models of ATP depletion *in vitro* offers researchers the opportunity to characterize and isolate the proteins that mediate cell injury (Spiegel *et al.*, 1989, Canfield *et al.*, 1991; Chen *et al.*, 1994; Golenhofen *et al.*, 1995).

VI. CONCLUSIONS

The characterization of the *in vivo* model of reversible ischemic injury has greatly increased our understanding of the molecular events associated with ischemia. The development of a reversible model of ATP depletion promises the opportunity to isolate and biochemically characterize the protein(s) responsible for the cellular changes associated with ATP depletion. Important questions remain; for example, it is not clear why the actin cytoskeletal network depolymerizes, especially since ATP is not essential for actin polymerization *in vitro*. The question as to the effect of ATP depletion on the protein-sorting mechanisms remains open. Additionally, further work is necessary to determine the effect of ischemia–reperfusion on the microtubule and intermediate filament network. As our understanding of ischemic injury improves, it is hoped that therapeutic interventions that speed the repair processes or impede the processes that mediate cellular injury will be developed.

References

Alejandro, V. S. J., Nelson, W. J., Huie, P., Sibley, R. K., Dafoe, D., Kuo, P., Scandling, J. D. Jr., and Myers, B. D. (1995). Postischemic injury, delayed function and Na+/K+ ATPase distribution in the transplanted kidney. *Kidney Int.* **48,** 1308–1315.

Algrain, M., Turunen, O., Vaheri, A., Louvard, D., and Arpin, M. (1993). Ezrin contains cytoskeleton and membrane binding domains accounting for its proposed role as a membrane-cytoskeleton linker. *J. Cell Biol.* **120,** 129–139.

Bacallao, R., and Fine, L. G. (1989). Molecular events in the organization of renal tubular epithelium: from nephrogenesis to regeneration. *Am. J. Physiol.* **257,** F913–F924.

Bacallao, R., Garfinkel, A., Monke, S., Zampighi, G., and Mandel, L. J. (1994). ATP depletion: a novel method to study junctional properties in epithelial tissues. *J. Cell Sci.* **107,** 3301–3313.

Bennett, V. (1985). The membrane skeleton of human erythrocytes and its implications for more complex cells. *Ann. Rev. Biochem.* **54,** 273–304.

Bershadsky, A. D., and Gefland, V. I. (1981). ATP-dependent regulation of cytoplasmic microtubule disassembly. *Proc. Natl. Acad. Sci. U.S.A.* **78,** 3610–3613.

Borle, A. B., Freudenrich, C. C., and Snowdowne, K. W. (1986). A simple method for incorporating aequorin into mammalian cells. *Am. J. Physiol.* **251,** C323–C326.

Bretscher, A., and Weber, K. (1979). Villin: the major microfilament-associated protein of the intestinal microvillus. *Proc. Natl. Acad. Sci. U.S.A.* **76,** 2321–2325.

Bretscher, A., and Weber, K. (1980). Villin is a major protein of the microvillus cytoskeleton which binds both G- and F- actin in a calcium dependent manner. *Cell* **20,** 839–847.

Bretscher, A. (1983). Purification of an 80, 000 dalton protein that is a component of the isolated microvillus and its localization in non-muscle cells. *J. Cell Biol.* **97,** 425–432.

Bretscher, A. (1989). Rapid phosphorylation and reorganization of ezrin and spectrin accompany morphological changes in A431 cells induced by EGF. *J. Cell Biol.* **108,** 921–930.

Bretscher, A. (1991). Microfilament structure and function in the cortical cytoskeleton. *Annu. Rev. Cell Biol.* **7,** 337–374.

Burridge, K., Fath, K., Kelly, T., Nuckolls, G., and Turner, C. (1988). Focal adhesions: transmembrane junctions between the extracellular matrix and the cytoskeleton. *Annu. Rev. Cell Biol.* **4,** 487–525.

Burridge, K., Turner, C. E., and Romer, L. H. (1992). Tyrosine phosphorylation of paxillin and pp120FAK accompanies cell adhesion to extracellular matrix. *J. Cell. Biol.* **119,** 893–903.

Canfield, P. E., Geerdes, A. M., and Molitoris, B. A. (1991). Effect of reversible ATP depletion on tight-junction integrity in LLC-PK1 cells. *Am. J. Physiol.* **261,** F1038–F1045.

Chen, J., Doctor, R. B., and Mandel, L. J. (1994). Cytoskeletal dissociation of ezrin during renal anoxia: role in microvillar injury. *Am. J. Physiol.* **267,** C784–C795.

Chen, J., Cohn, J. A., and Mandel, L. J. (1995). Dephosphorylation of ezrin as an early event in renal microvillar breakdown and anoxic injury. *Proc. Nat. Acad. Sci. U.S.A.* **92,** 7495–7499.

Citi, S. (1992). Protein kinase inhibitors prevent junction dissociation induced by low extracellular calcium in MDCK epithelial cells. *J. Cell Biol.* **117,** 168–178.

Cohen, P. (1992). Signal integration at the level of protein kinases, protein phosphatases, and their substrates. *Trends Biochem. Sci.* **17,** 408–413.

Coluccio, L. (1991). Identification of the microvillar 110-kDa calmodulin complex (myosin-1) in kidney. *Eur. J. Cell Biol.* **56,** 286–294.

Costa de Beauregard, M. A., Pringault, E., Robine, S., and Louvard, D. (1995). Suppression of villin expression by anti-sense RNA impairs brush border assembly in polarized epithelial intestinal cells. *EMBO J.* **14,** 409–412.

Coudrier, E., Kerjaschki, D., and Louvard, D. (1988). Cytoskeleton organization and submembranous interactions in intestinal and renal brush borders. *Kidney Int.* **34,** 309–320.

Croall, D. E., and DeMartino, G. N. (1991). Calcium-activated neutral protease (calpain) system: structure, function, and regulation. *Physiol. Rev.* **71,** 813–847.

Doctor, R. B., and Mandel, L. J. (1991). Minimal role of xanthine oxidase in rat renal tubular reoxygenation injury. *J. Am. Soc. Nephrol.* **1,** 959–969.

Doctor, R. B., Bennett, V., and Mandel, L. J. (1993). Degradation of spectrin and ankyrin in the ischemic rat kidney. *Am. J. Physiol.* **264,** C1003–C1013.

Doctor, R. B., Bacallao, R., and Mandel, L. J. (1994). Method for recovering ATP content and mitochondrial function after chemical anoxia in renal cultured cells. *Am. J. Physiol.* **266,** C1803–C1811.

Doctor, R. B., Zhelev, D. V. and Mandel, L. J. (1996). Loss of plasma membrane structural support in ATP depleted renal epithelial cells. Submitted for publication.

Donohoe, J. F., Venkatachalam, M. A., Bernard, D. B., and Levinsky, N. G. (1978). Tubular leakage and obstruction after renal ischemia: structural-functional correlations. *Kidney Internat.* 13, 208–222.

Drenckhahn, D., and Bennett, V. (1987). Polarized distribution of Mr 210,000 and 190,000 analogs of erythrocyte ankyrin along the plasma membrane of transporting epithelia, neurons and photoreceptors. *Eur. J. Cell Biol.* **43,** 479–486.

Edelstein, A., Wieder, E. D., Yaqoob, M. M., Gengaro, D. E., Burke, T. J., Nemenoff, R. A., and Schrier, R. W. (1995). The role of cysteine proteases in hypoxia-induced rat renal proximal tubular injury. *Proc. Natl. Acad. Sci. U.S.A.* **92,** 7662–7666.

Fernandez, A., Brautigan, D. L., Mumby, M., and Lamb, N. J. C. (1990). Protein phosphatase type-I, not type 2A, modulates actin microfilament integrity and myosin light chain phosphorylation in living non-muscle cells. *J. Cell Biol.* **111,** 103–112.

Friederich, E., Huet, C., Arpin, M., and Louvard, D. (1989). Villin induces growth and actin redistribution in transfected fibroblasts. *Cell* **59,** 461–475.

Friederich, E., Vancompernolle, K., Huet, C., Goethals, M., Finidori, J., Vanderkerckhove, J., and Louvard, D. (1992). An actin-binding site containing a conserved motif of charged amino acid residues is essential for the morphogenic effect of villin. *Cell* **70,** 81–92.

Gailit, J., Colflesh, D., Rabiner, I., Simone, J., and Goligorsky, M. S. (1993). Redistribution and dysfunction of integrins in cultured renal epithelial cells exposed to oxidative stress. *Am. J. Physiol.* **264,** F149–F157.

Garza-Quintero, R., Weinberg, J. M., Ortega-Lopez, J., Davis J. A., and Venkatachalam, M. A. (1993). Conservation of structure in ATP-depleted proximal tubules: role of calcium, polyphosphoinositides, and glycine. *Am. J. Physiol.* **265,** F605–F623.

Glaumann, B., and Trump, B. F. (1975). Studies on the pathogenesis of ischemic cell injury. III. Morphological changes of the proximal pars recta tubules (S3) of the rat kidney made ischemic *in vivo. Virchows Arch. B Cell Pathol.* **19,** 303–323.

Glaumann, B., Glaumann, H., Berezesky, I. K., and Trump, B. F. (1975). Studies on the pathogenesis of ischemic cell injury. II. Morphological changes of the pars convoluta (S1 and S2) of the proximal tubule of the rat kidney made ischemic *in vivo. Virchows Arch. B. Cell Pathol.* **19,** 281–302.

Glenney, J. R., Jr., and Weber, K. (1981). Calcium control of microfilaments: uncoupling of the F-actin severing and bundling activity of villin by limited proteolysis *in vitro. Proc. Natl. Acad. Sci. U.S.A.* **78,** 2810–2814.

Golenhofen, N., Doctor, R. B., Bacallao, R., and Mandel, L. J. (1995). Actin and villin compartmentation during ATP depletion and repletion in renal cultured cells. *Kidney Int.* **48,** 1837–1845.

Goligorsky, M. S., and DiBona, G. F. (1993). Pathogenetic role of Arg-Gly-Asp recognizing integrins in acute renal failure. *Proc. Natl. Acad. Sci. U.S.A.* **90,** 5700–5704.

Goligorsky, M. S., and Hruska, K. A. (1988). Hormonal modulation of cytoplasmic calcium concentration in renal tubular epithelium. *Miner. Electrolyte Metab.* **14,** 58–70.

Goligorsky, M. S., Lieberthal, W., Racusen, L., and Simon, E. E. (1993). Integrin receptors in renal tubular epithelium: new insights into pathophysiology of acute renal failure. *Am. J. Physiol.* **264**, F1–F8.

Gullans, S. R., and Mandel, L. J. (1992). Coupling of energy to transport in proximal and distal nephron. *In* "The Kidney: Physiology and Pathophysiology", Second Edition. (D. Seldin and G. Giebisch, eds.) pp. 1291–1337. Raven Press, New York.

Hanzel, D., Reggio, H., Bretscher, A., Forte, J. G., and Mangeat, P. (1991). The secretion stimulated 80K phosphoprotein of parietal cell is ezrin, and has properties of a membrane cytoskeletal linker in the induced apical microvilli. *EMBO J.* **10,** 2363–2373.

Harris, S. I., Balaban, R. S., Barrett, L., and Mandel, L. J. (1981). Mitochondrial respiratory capacity and Na+- and K+-dependent adenosine triphosphatase-mediated ion transport in the intact renal cell. *J. Biol. Chem.* **256,** 10319–10328.

Jacobs, W. R., Sgambati, M., Gomez, G., Vilaro, P., Higdon, M., Bell, P. D., and Mandel, L. J. (1991). Role of cytosolic Ca in renal tubule damage induced by anoxia. *Am. J. Physiol.* **260,** C545–C554.

Jennings, M. L., and Schulz, R. K. (1991). Okadaic acid inhibition of KCl cotransport. Evidence that protein dephosphorylation is necessary for activation of transport by either cell swelling or N-ethylmaleimide. *J. Gen. Physiol.* 97, 799–817.

Jones, D. P. (1986). Renal metabolism during normoxia, hypoxia, and ischemic injury. *Ann. Rev. Physiol.* **48,** 33–50.

Kellerman, P. S. (1993). Exogenous ATP preserves proximal tubule microfilament structure and function *in vivo* in a maleic acid model of ATP depletion. *J. Clin. Invest.* **92,** 1940–1949.

Kellerman, P. S., and Bogusky, R. T. (1992). Microfilament disruption occurs very early in ischemic proximal tubule cell injury. *Kidney Int.* **42,** 896–902.

Kobryn, C., and Mandel, L. J. (1994). Decreased protein phosphorylation induced by anoxia in proximal tubules. *Am. J. Physiol.* **267,** C1073–C1079.

Koob, R., Zimmermann, M., Schoner, W., and Drenckhahn, D. (1987). Colocalization and coprecipitation of ankyrin and Na+,K+ ATPase in kidney epithelial cells. *Eur. J. Cell Biol.* **45,** 230–237.

Kordeli, E., Lambert, S., and Bennett, V. (1995). Ankyrin-G. A new ankyrin gene with neural-specific isoforms localized at the axonal initial segment and node of Ranvier. *J. Biol. Chem.* **270,** 2352–2359.

Lin, C. Q., and Bissell, M. J. (1993). Multi-faceted regulation of cell differentiation by extracellular matrix. *FASEB J.* **7,** 737–743.

Mandel, L. J., Takano, T. Soltoff, S. P. and Murdaugh, S. (1988). Mechanisms whereby exogenous adenine nucleotides improve rabbit renal proximal function during and after anoxia. *J. Clin. Invest.* **81,** 1255–1264.

Mandel, L. J., Bacallao, R., and Zampighi, G. (1993). Uncoupling of the molecular 'fence' and paracellular 'gate' functions in epithelial tight junctions. *Nature* **361,** 552–555.

Mandel, L. J., Doctor, R. B., and Bacallao, R. (1994). ATP depletion: a novel method to study junctional properties in epithelial tissues. II. Internalization of Na,K ATPase and E-cadherin. *J. Cell Sci.* **107,** 3315–3324.

Matsudaira, P., Jakes, R., and Walker, J. E. (1985). A gelsolin-like Ca2+-dependent actin-binding domain in villin. *Nature* **315,** 248–250.

Miller, G. W., and Schnellmann, R. (1993). Cytoprotection by inhibition of chloride channels: the mechanism of action of glycine and strychnine. *Life Sci.* **53,** 1211–1215.

Mills, J. W., and Mandel, L. J. (1994). Cytoskeletal regulation of membrane transport events. *FASEB J.* **8,** 1161–1165.

Molitoris, B. A., Hoilien, C. A., Dahl, R., Ahnen, D. J., Wilson, P. D., and Kim, J. (1988). Characterization of ischemia-induced loss of epithelial polarity. *J. Membrane Biol.* **106,** 233–242.

Molitoris, B. A., Falk, S. A., and Dahl, R. H. (1989). Ischemia induced loss of epithelial polarity. Role of the tight junction. *J. Clin. Invest.* **84,** 1334–1339.

Molitoris, B. A., Geerdes, A., and McIntosh, J. R. (1991). Dissociation and redistribution of Na,K ATPase from its surface membrane actin cytoskeletal complex during cellular ATP depletion. *J. Clin. Invest.* **88,** 462–469.

Molitoris, B. A., Dahl, R., and Geerdes, A. (1992). Cytoskeleton disruption and apical redistribution of proximal tubule Na+,K+ ATPase during ischemia. *Am. J. Physiol.* **263,** F488–F495.

Nelson, W. J., and Hammerton, R. W. (1989). A membrane-cytoskeletal complex containing Na+,K+ ATPasem ankyrin, and fodrin in MDCK cells: Implications for the biogenesis of epithelial cell polarity. *J. Cell Biol.* **108,** 893–902.

Nigam, S. K., Denisenko, N., Rodriguez-Boulan, E., and Citi, S. (1991). The role of phosphorylation in development of tight junctions in cultured renal epithelial (MDCK) cells. *Biochem. Biophys. Res. Commun.* **181,** 548–553.

Noiri, E., Gailit, J., Sheth, D., Magazine, H., Gurrath, M., Miller, G., Kessler, H., and Goligorsky, M. S. (1994). Cyclic RGD peptides ameliorate ischemic acute renal failure in rats. *Kidney Int.* **46,** 1050–1058.

Northrop, J., Weber, A., Mooseker, M. S., Franzini-Armstrong, C., Bishop, M. F., Dubyak, G. R., Tucker, M., and Walsh, T. P. (1986). Different calcium dependence of the capping and cutting activities of villin. *J. Biol. Chem.* **261**, 9274–9281.

Pakkanen, R., von Bonsdorff, C. H., Turenen, O., Wahlstrom, T., and Vaheri, A. (1988). Redistribution of Mr 75, 000 plasma membrane protein, cytovillin, into newly formed microvilli in herpes simplex and Semliki Forest virus infected human embryonal fibroblasts. *Eur. J. Cell Biol.* **46,** 435–443.

Piepenhagen, P. A., Peters, L. L., Lux, S. E., and Nelson, W. J. (1995). Differential expression of Na+,K+ ATPase, ankyrin, fodrin, and E-cadherin along the kidney nephron. *Am. J. Physiol.* **269,** C1417–C1432.

Racusen, L. C., Fivush, B., Li, Y.-L., Slatnik, I., and Solez, K. (1991). Dissociation of tubular cell detachment and tubular cell death in clinical and experimental "acute tubular necrosis". *Lab. Invest.* **64,** 546–556.

Ragno, P., Estreicher, A., Gos, A., Wohlwend, A., Belin, D., and Vassalli, J-D. (1992). Polarized secretion of urokinase-type plasminogen activator by epithelial cells. *Exp. Cell Res.* **203,** 236–243.

Reeves, W. B., and Shah, S. V. (1994). Activation of potassium channels contributes to hypoxic injury in proximal tubules. *J. Clin. Invest.* **94,** 2289–2294.

Rodman, J., Mooseker, M., and Farquhar, M. G. (1986). Cytoskeletal proteins of the rat kidney proximal tubule brush border. *Eur. J. Cell Biol.* **42,** 319–327.

Sato, N., Funayama, N., Nagafuchi, A., Yonemura, S., Tsukita, S., and Tsukita, S. (1992). A gene family consisting of ezrin, radixin, and moesin. Its specific localization at actin filament/plasma membrane association sites. *J. Cell Sci.* **103,** 131–143.

Sheetz, M. P., (1993). Glycoprotein motility and dynamic domains in fluid plasma membranes. *Annu. Rev. Biophys. Biomol. Struct.* **22,** 417–431.

Siegel, N. J., Avison, M. J., Reilly, H. F., Alger, J. R., and Shulman, R. G. (1983). Enhanced recovery of renal ATP with postischemic infusion of ATP-$MgCl_2$ determined by ^{31}P-NMR. *Am. J. Physiol.* **14,** F530–F535.

Soltoff, S. P., and Mandel, L. J. (1984). Active ion transport in the renal proximal tubule. III. The ATP dependence of the sodium pump. *J. Gen. Physiol.* **84,** 643–662.

Spencer, A. J., LeFurgey, A., Ingram, P., and Mandel, L. J. (1991). Elemental analysis of organelles in proximal tubules. II. Effects of oxygen deprivation. *J. Am. Soc. Nephrol.* **1,** 1321–1333.

Spiegel, D. M., Wilson, P. D., and Molitoris, B. A. (1989). Epithelial polarity following ischemia: a requirement for normal cell function. *Am. J. Physiol.* **256,** F430–F436.

Stevenson, B. R., Anderson, J. M., Braun, I. D., and Mooseker, M. S. (1989). Phosphorylation of the tight junction protein ZO-1 in two strains of Madin-Darby canine kidney cells which differ in transepithelial resistance. *Biochem. J.* **263,** 597–599.

Takai, Y., Kishimoto, A., Inoue, M., and Nishizuka, Y. (1977). Studies on a cyclic nucleotide-independent protein kinase and its proenzyme in mammalian tissues. *J. Biol. Chem.* **252,** 7603–7609.

Takano, T., Soltoff, S. P., Murdaugh, S., and Mandel, L. J. (1985). Intracellular respiratory dysfunction and cell injury in short term anoxia of rabbit renal proximal tubules. *J. Clin. Invest.* **76,** 2377–2384.

Turunen, O., Wahlstrom, T., and Vaheri, A. (1994). Ezrin has a COOH-terminal actin-binding site that is conserved in the ezrin protein family. *J. Cell Biol.* **126,** 1445–1453.

Urushidani, T., Hanzel, D. K., and Forte, J. G. (1989). Characterization of an 80-kDa phosphoprotein involved in parietal cell stimulation. *Am. J. Physiol.* **256,** G1070–G1081.

Venkatachalam, M. A., Bernard, D. B., Donohoe, J. F., and Levinsky, N. G. (1978). Ischemic damage and repair in the rat proximal tubule: differences among the S1, S2, and S3 segments. *Kidney Int.* **14,** 31–49.

Venkatachalam, M. A., Patel, Y. J., Kreisberg, J. I., and Weinberg, J. M. (1988). Energy threshold that determine membrane integrity and injury in a renal epithelial cell line (LLC-PK1). *J. Clin. Invest.* **81,** 745–758.

Wang, N., Butler, J. P., and Ingber, D. E. (1993). Mechanotransduction across the cell surface and through the cytoskeleton. *Science.* **260,** 1124–1127.

Weinberg, J. (1985). Oxygen deprivation-induced injury to isolated rabbit kidney tubules. *J. Clin. Invest.* **76,** 1193–1208.

Yin, H. L., and Stossel, T. P. (1979). Control of cytoplasmic actin gel-sol transformation by gelsolin, a calcium-dependent regulatory protein. *Nature* **281,** 583–586.

Index

T

V

ISBN 0-12-153343-3

90018

9 780121 533434